PHYLOGENETIC INFERENCE, SELECTION THEORY, AND HISTORY OF SCIENCE

Selected Papers of A.W.F. Edwards with Commentaries

A. W. F. Edwards is one of the most influential mathematical geneticists in the history of the discipline. One of the last students of R. A. Fisher, Edwards pioneered the statistical analysis of phylogeny in collaboration with L. L. Cavalli-Sforza, and helped establish Fisher's concept of likelihood as a standard of statistical and scientific inference. In this book, edited by philosopher of science Rasmus Grønfeldt Winther, Edwards's key papers are assembled alongside commentaries by leading scientists, discussing Edwards's influence on their own research and on thinking in their field overall. In an extensive set of interviews with Winther, Edwards offers his thoughts on his contributions, their legacy, and the context in which they emerged. This book is a resource both for anyone interested in the history and philosophy of genetics, statistics, and science, and for scientists seeking to develop new algorithmic and statistical methods for understanding the genetic relationships between and among species both extant and extinct.

A. W. F. EDWARDS, F.R.S., is a Fellow of Gonville and Caius College, Cambridge. In addition to some 250 scientific papers on a wide range of topics, he has published four books including *Likelihood* and *Foundations of Mathematical Genetics*, both with Cambridge University Press. Edwards has been interested in the history and philosophy of science for as long as he has been a contributor to science.

RASMUS GRØNFELDT WINTHER is a philosopher of science, researcher, writer, educator, diver, seeker, and explorer. www.rgwinther.com

PHYLOGENETIC INFERENCE, SELECTION THEORY, AND HISTORY OF SCIENCE

Selected Papers of A.W.F. Edwards with Commentaries

Edited by

RASMUS GRØNFELDT WINTHER

CAMBRIDGE
UNIVERSITY PRESS

CAMBRIDGE
UNIVERSITY PRESS

University Printing House, Cambridge CB2 8BS, United Kingdom

One Liberty Plaza, 20th Floor, New York, NY 10006, USA

477 Williamstown Road, Port Melbourne, VIC 3207, Australia

314–321, 3rd Floor, Plot 3, Splendor Forum, Jasola District Centre, New Delhi – 110025, India

79 Anson Road, #06–04/06, Singapore 079906

Cambridge University Press is part of the University of Cambridge.

It furthers the University's mission by disseminating knowledge in the pursuit of education, learning, and research at the highest international levels of excellence.

www.cambridge.org
Information on this title: www.cambridge.org/9781107111721
DOI: 10.1017/9781316276259

First published 2018

Printed in the United Kingdom by TJ International. Padstow Cornwall

A catalogue record for this publication is available from the British Library.

Library of Congress Cataloging-in-Publication Data
Names: Edwards, A. W. F. (Anthony William Fairbank), 1935– author. | Winther, Rasmus, editor.
Title: Phylogenetic inference, selection theory, and history of science : selected papers of A.W.F. Edwards, with commentaries / edited by Rasmus Grønfeldt Winther
Description: Cambridge : Cambridge University Press, 2018. | Includes index.
Identifiers: LCCN 2017055938 | ISBN 9781107111721
Subjects: LCSH: Edwards, A. W. F. (Anthony William Fairbank), 1935– | Population genetics – Mathematical models. | Genetics – Mathematical models. | Cladistic analysis.
Classification: LCC QH438.4.M3 E376 2018 | DDC 577.8/8–dc23
LC record available at https://lccn.loc.gov/2017055938

ISBN 978-1-107-11172-1 Hardback

P. ii Photo of A. W. F. Edwards © Joe Felsenstein.

Contents

Selected Featured Papers of A. W. F. Edwards

Note that the paper number refers to the complete publication list starting on page 465.

[25] (1963) The reconstruction of evolution. (Abstract) *Heredity* 18, 553, and *Annals of Human Genetics* 27, 104–105. (With L. L. Cavalli-Sforza)

[27] (1964) Reconstruction of evolutionary trees. In *Phenetic and Phylogenetic Classification*, eds. V. H. Heywood and J. McNeill, The Systematics Association Publication No. 6, London: The Systematics Association, 67–76. (With L. L. Cavalli-Sforza) Reprinted in *Cladistic Theory and Methodology*, eds. T. Duncan and T. F. Stuessy, New York: Van Nostrand Reinhold (1985).

[28] (1964) Analysis of human evolution under random genetic drift. *Cold Spring Harbor Symposia on Quantitative Biology* 29, 9–20. (With L. L. Cavalli-Sforza and I. Barrai)

[29] (1965) Analysis of human evolution. Proceedings of the 11th International Congress of Genetics, The Hague 1963. *Genetics Today* 3, 923–933. (With L. L. Cavalli-Sforza)

[30] (1965) A method for cluster analysis. *Biometrics* 21, 362–375. (With L. L. Cavalli-Sforza)

[32] (1966) Studying human evolution by computer. *New Scientist* (19th May) 30, 438–440. Reprinted in *Evolution*, ed. M. Ridley, Oxford: Oxford University Press (1997).

[35] (1966) Estimation procedures for evolutionary branching processes. *Bulletin of the International Statistical Institute* (Proceedings of the 35th session) 41, 803–808. (With L. L. Cavalli-Sforza)

[36] (1967) Fundamental Theorem of Natural Selection. *Nature* 215, 537–538.

[37] (1967) Phylogenetic analysis: models and estimation procedures. *Evolution* 21, 550–570. (With L. L. Cavalli-Sforza) Also in *American Journal of Human Genetics,* Supplement 19, 233–257.

[42] (1969) Statistical methods in scientific inference. *Nature* 222, 1233–1237.

[46] (1970) Estimation of the branch points of a branching diffusion process. *Journal of the Royal Statistical Society B* 32, 155–174.

[49] (1971) Review of *Evolution and the Genetics of Populations*, Volume 2, by S. Wright. *Heredity* 26, 332–338.

[53] (1972) Affinity as revealed by differences in gene frequencies. In *The Assessment of Population Affinities in Man*, eds. J. S. Weiner and J. Huizinga, Oxford: Clarendon Press, 37–47. (With L. L. Cavalli-Sforza)

[56] (1972) Likelihood. *Bulletin of the Institute of Mathematics and its Applications* 8, 329–331.

[60] (1974) The history of likelihood. *International Statistical Review* 42, 9–15. Reprinted in *Likelihood* (expanded edition), Baltimore, MD: Johns Hopkins University Press (1992).

[96] (1986) Are Mendel's results really too close? *Biological Reviews* 61, 295–312.

[117] (1989) Probability and likelihood in genetic counselling. *Clinical Genetics* 36, 209–216.

[140] (1994) The Fundamental Theorem of Natural Selection. *Biological Reviews* 69, 443–474.

[141] (1995) Assessing molecular phylogenies. *Science* 267, 253.

[146] (1996) The origin and early development of the method of minimum evolution for the reconstruction of phylogenetic trees. *Systematic Biology* 45, 79–91.

[173] (2001) Darwin and Mendel united: the contributions of Fisher, Haldane and Wright up to 1932. In *Encyclopedia of Genetics*, ed. E. C. R. Reeve, London: Fitzroy Dearborn, 77–83.

[192] (2003) Human genetic diversity: Lewontin's fallacy. *BioEssays* 25, 798–801.

[197] (2004) Parsimony and computers. In *Milestones in Systematics*, eds. D. M. Williams and P. L. Forey, The Systematics Association Special Volume, Series 67, Boca Raton, FL: CRC Press, 181–190.

[225] (2009) Statistical methods for evolutionary trees. *Genetics* 183, 5–12.

[227] (2011) Mathematizing Darwin. *Behavioral Ecology and Sociobiology* 65, 421–430.

[235] (2013) Robert Heath Lock and his textbook of genetics, 1906. *Genetics* 194, 529–537.

[238] (2014) R. A. Fisher's gene-centred view of evolution and the Fundamental Theorem of Natural Selection. *Biological Reviews* 89, 135–147.

[248] (2016) Analysing nature's experiment: Fisher's inductive theorem of natural selection. *Theoretical Population Biology* 109, 1–5.

Contributors

Walter Bodmer, Weatherall Institute of Molecular Medicine, University of Oxford, UK

Maurizio Esposito, Department of Philosophy, University of Santiago, Chile

Warren J. Ewens, Departments of Biology and Statistics, University of Pennsylvania, PA, USA

Joseph Felsenstein, Departments of Genome Sciences and Biology, University of Washington, WA, USA

Jean-Baptiste Grodwohl, Department of History and Philosophy of Science, University of Cambridge, UK

Rasmus Nielsen, Departments of Integrative Biology and Statistics, University of California, Berkeley, CA, USA

Samir Okasha, Department of Philosophy, University of Bristol, UK

Mark Pagel, School of Biological Sciences, University of Reading, UK

Noah A. Rosenberg, Department of Biology, Stanford University, CA, USA

Elizabeth Thompson, Department of Statistics, University of Washington, WA, USA

Ziheng Yang, Department of Genetics, Evolution and Environment, University College London, UK

Dedicated to

Luigi Luca Cavalli-Sforza
Pioneer of phylogenetic inference

***

We are grateful to Ian Hacking for his wonderfully thoughtful contributions on the history and philosophy of statistics and science.

Rasmus Grønfeldt Winther and A. W. F. Edwards

Introduction

RASMUS GRØNFELDT WINTHER

The 1960s and 1970s were a heady time for theoretical biology. Theoretical ecology and population genetics were undergoing fundamental conceptual and methodological changes. Sophisticated mathematical techniques were introduced, computers increased in power and became readily available, and the 'molecularisation' of biology was underway. A. W. F. Edwards at the University of Cambridge was an influential researcher in this early interweaving of mathematics, computation, and molecular data. Edwards contributed to three key areas of inquiry. First, trained under noted statistician and biologist R. A. Fisher, and working alongside L. L. Cavalli-Sforza, another close associate of Fisher's, Edwards developed key methods and models for the reconstruction of the evolutionary history of organisms. This was indeed the birth of modern statistical *phylogenetic inference*. Second, Edwards contributed to theory on sex-ratio evolution and the Fundamental Theorem of Natural Selection. This helped advance our understanding of the operation of natural selection, i.e., *selection theory*. Third, Edwards authored numerous papers on many important researchers in the *history of science*, particularly in the histories of mathematics and genetics. Edwards's subjects included Pierre de Fermat, Blaise Pascal, Thomas Bayes, Francis Galton, John Venn, and, especially, R. A. Fisher.[1] In all of this, Edwards also contributed, indirectly at least, to probability theory and statistics. For instance, he articulated key insights on maximum likelihood and clustering, and helped clarify the Fisher–Neyman controversy.

Given the importance of Edwards's work, it is unfortunate that much of it is difficult to locate, requiring hours of effort from dedicated researchers and expert librarians. The purpose of this book is to make a representative subset of Edwards's papers available and to place Edwards's contribution into its philosophical, historical, and sociological context. To provide this context, commentaries by distinguished

[1] For a small sample of Edwards's contributions to the history of science, see [96] and [235].

experts are provided, as is an in-depth set of interviews of Edwards carried out by the editor, Rasmus Grønfeldt Winther, over three days at the University of Cambridge. Given that Edwards has published over 200 papers, only the most interesting and influential papers could be selected for this volume. Many of the selected papers are on phylogenetic inference, with the remainder distributed among selection theory, the history of science, and probability theory and statistics.

This introduction situates Edwards's impact with respect to three key themes: (1) statistics as a methodology for phylogenetic inference, (2) Fisherian population genetics as a 'gene's-eye view' of selection theory, and (3) human genetic diversity and population structure. It then provides descriptions of each commentary, and concludes by situating and explaining the editor's role in this book, as a philosopher of science, and by briefly discussing future research possibilities.

As in many other episodes of scientific discovery, historical details matter. In the interviews, Edwards recounts some of the variegated early history of statistical phylogenetic inference. Both Edwards and Cavalli-Sforza were inspired by Fisher's suggestion to represent gene frequency data (abstracted from many blood-group polymorphisms) in Cartesian spaces, or state spaces, in which populations occupy distinct but potentially overlapping regions.[2] Reconstructing historical connections could then be seen as a statistical inference problem, which results in producing branching trees via averaging or minimising pairwise distances between population clouds in these state spaces ([25], [27], [28], [29], [53]; Figure 1 of [28], p. 29, this volume; Fig. 1 of [29], p. 43, this volume; Day 2 of the interviews.[3] Cavalli-Sforza had the original idea of trying to 'extract a tree structure from the [blood group] information and see how it related to the geography of the continents.'[4] But Edwards added much theoretical nuance, including the principle of minimum evolution. Edwards and Cavalli-Sforza's subsequent innovation of building an evolutionary model with a Brownian-motion approximation of random genetic drift, upon which maximum likelihood could then be applied as the estimation procedure for inferring the most likely evolutionary trees, can be partly attributed to Motoo Kimura as well as to Fisher's earlier influence ([27], [28], [37], [46], [141]; cf. [35], [56], [60], [225]; Interviews, pp. 432–433, 443, this volume). Edwards also implemented statistical phylogenetic models in computational code, e.g. his program EVOTREE ([32], [197]; Interviews, pp. 435–439, this volume).

In the chapter 'A digression on history and philosophy', Felsenstein comments on 'the remarkably creative work' of Edwards and Cavalli-Sforza, noting that

[2] Interviews, pp. 429–431, this volume.

[3] Numbers in brackets refer to the numerical system for Edwards's articles that is used in this volume, indicating the articles reprinted, as listed in the table of contents; see A. W. F. Edwards: Scientific Publications (pp. 465–482, this volume) for a full list of Edwards's work.

[4] Interviews, p. 429, this volume.

'Edwards and Cavalli-Sforza's paper of 1964 ([27]) is remarkable in that it introduces the parsimony method, the likelihood method, and the statistical inference approach to inferring phylogenies, all in one paper.'[5] Interestingly, [27] contains another first: hypothesised genealogical relationships among modern human populations were computed from blood group allelic gene frequency data and subsequently superimposed on a world map (Figure 1, p. 26, this volume; see volume back cover).[6] The authors' intent was to cartographically represent the history and geography of modern humans (p. 319 (Bodmer); p. 344 (Pagel); interviews, pp. 441–442, 450–451, this volume). In their commentaries, Felsenstein, Pagel, Thompson, and Yang place these advances in phylogenetics in the context of the logic of statistical inference ([42], [146], [227]).[7] Nielsen writes that 'cladistic ideology' has been replaced by more 'pragmatic' statistical methods of phylogenetic tree inference (p. 340; cf. [146], pp. 236–238, [197], p. 259, this volume). It would make for fascinating history of science to track and explore the rise of statistical phylogenetics in the context of the fall – or, at minimum, the retransformation – of cladistics.[8] Which causal connections, if any, exist between the rise of one and the fall of the other? Is the emergence of statistical theory the single most significant historical factor here, or might the flood of genetic data and the shifting patterns of funding at universities, museums, field stations, and other research institutions play a role? Finally, is a pluralistic future possible, one in which we take seriously morphological characters and the concept of homology, as well as certain taxonomic, nomenclature, and documenting practices from cladistic systematics?[9] In cases such as this, historians, philosophers, and sociologists of

[5] Felsenstein, J. 2004, *Phylogenetic Inference*. Sunderland, MA: Sinauer Associates, pp. 125, 128.

[6] This map also appears in [28], together with the actual computed tree of descent (Figures 2 and 3, p. 31, this volume). A topologically distinct tree was published a year later ([29], Figure 5, p. 48, this volume). For further discussion, see Bodmer and Pagel commentaries; Sommer, M. 2015. Population-genetic trees, maps, and narratives of the great human diasporas. *History of the Human Sciences* 28, 108–145; Sommer, M. 2016. *History Within: The Science, Culture, and Politics of Bones, Organisms, and Molecules*. Chicago, IL: University of Chicago Press.

[7] For a synoptic, longer historical view of statistics and probability theory, see Hacking, I. 1990. *The Taming of Chance*. Cambridge: Cambridge University Press.

[8] Elliott Sober covers many of the philosophical points: e.g. Sober, E. 1988. *Reconstructing the Past: Parsimony, Evolution, and Inference*. Cambridge, MA: MIT Press; 2008. *Evidence and Evolution: The Logic Behind the Science*. Cambridge: Cambridge University Press. Sober 1988 explicitly addresses Edwards's phylogenetic work, both alone and with Cavalli-Sforza. On the history of cladistics, see Hull, D. 1988. *Science as a Process: An Evolutionary Account of the Social and Conceptual Development of Science*. Chicago, IL: University of Chicago Press; Williams, D. M. and Forey, P. L. (eds.). 2004. *Milestones in Systematics*. Boca Raton, FL: CRC Press. Note that Williams and Forey contains [197].

[9] Some philosophers of science provide conceptual frameworks for analysing the assumptions and worldviews of a plurality or diversity of research programs (e.g. Longino, H. 2002. *The Fate of Knowledge*. Princeton, NJ: Princeton University Press; Hacking, I. 2004. *Historical Ontology*. Cambridge, MA: Harvard University Press; and Winther, R. G., The structure of scientific theories, In *The Stanford Encyclopedia of Philosophy* (Winter 2016 edition), ed. E. N. Zalta (https://plato.stanford.edu/archives/win2016/entries/structure-scientific-theories/); Winther, R. G. (in press). *When Maps Become the World*. Chicago, IL: University of Chicago Press.

science could engage in recovery, critical awareness, and repurposing of old and even currently marginalised scientific theories and methods.[10]

Turning to mathematical population genetics, according to Edwards, the 'gene's-eye' or 'gene-centred' view 'is implicit in any genetical model of the analytical kind favoured by Haldane and Wright', although 'in Fisher's ["inductive and statistical"] work it was explicit' ([238], p. 297, this volume). Edwards has helped explore and clarify Fisher's Fundamental Theorem of Natural Selection ([36], [140], [238], [248]). Edwards favours a Fisherian gene's-eye selection theory. Fisher stressed adaptive evolutionary change as driven by mutation and natural selection among individuals in large panmictic (random-mating) populations. Gene frequencies are here understood as the one single inductively measurable feature 'capable of rigorous analysis' ([238], p. 297, this volume). Edwards warns, however, that we must be wary of attributing an 'oversimplified "gene-centred" view' to Fisher ([227], p. 281, this volume; Esposito commentary). Despite relying on the power of averaging, Fisher acknowledged gene interactions and genotype-by-environment interactions. Thus, Edwards argues, we should not read Fisher as either a genetic determinist or an arch-reductionist.

As explored in the Ewens, Okasha, and Grodwohl commentaries, the gene's-eye view connects intimately to the assumptions and mathematics of the fundamental theorem, especially in light of George Price's reinterpretation.[11] Historically, the 'shifting balance theory' of Sewall Wright contrasted with Fisher's explanations and predictions of the evolutionary process. Wright's model was more complex, with three phases: (I) random genetic drift within local populations (demes), moving them across rugged adaptive landscapes; (II) mass Fisherian selection within demes, moving each population up to the closest adaptive peak; and (III) interdemic (group) selection, wherein demes at higher peaks send out relatively more migrants than demes at lower peaks, thereby shifting other populations to the gene frequencies and gene combinations of the higher peaks ([49], [173]; Grodwohl commentary).[12] Furthermore, ongoing debates surrounding inclusive fitness theory, levels of selection, group selection, epistasis, indirect genetic effects, and other phenomena nuance or challenge the gene-centred view.[13] Whatever our

[10] Hasok Chang's notion of *complementary science* is useful here (Chapter 6 of Chang, H. 2004. *Inventing Temperature: Measurement and Scientific Progress*. Oxford: Oxford University Press, 235–250).

[11] Price, G. R. 1972. Fisher's 'fundamental theorem' made clear. *Annals of Human Genetics* 36, 129–40.

[12] Provine, W. 1986. *Sewall Wright and Evolutionary Biology*. Chicago, IL: University of Chicago Press; Hodge, M. J. S. 1992. Biology and philosophy (including ideology): a study of Fisher and Wright. In *The Founders of Evolutionary Genetics*. ed. S. Sarkar, Kluwer, Dordrecht, 231–293; Winther, R. G. 2006. Fisherian and Wrightian perspectives in evolutionary genetics and model-mediated imposition of theoretical assumptions. *Journal of Theoretical Biology* 240, 218–232; Winther, R. G., Wade, M. J. and Dimond, C. C. 2013. Pluralism in evolutionary controversies: styles and averaging strategies in hierarchical selection theories. *Biology and Philosophy* 28, 957–979.

[13] Because inclusive fitness is often seen as a part of a gene's-eye perspective, theoretical controversies vis-à-vis inclusive fitness also involve disagreements about the centrality and scope of application of the gene's-eye

theoretical and philosophical perspectives, let us revisit Edwards's work for a learned and influential theoretical characterisation of the gene-centred view.

Inferences about the history and population structure of our species are associated with rampant conceptual confusions. With 'Human genetic diversity: Lewontin's fallacy' ([192]), Edwards helped eliminate confusions regarding population clustering and individual classification. Consider a thought experiment. Imagine pulling miscellaneous metallic objects out of two immense and otherwise indistinguishable bags labelled 'Bag A' and 'Bag B'. Every object you pull out has physical properties such as weight, sound when struck, melting point, and colour. These features can be measured for each object, and you know the averages and variances for each property, for each bag. For instance, you know that the average weight of objects in the two bags is almost identical, as are the weight distributions. You also know that sounds emitted by Bag A objects when hit are almost always very tinny, though a few objects emit the same deep rumbling sound as *all* the objects of Bag B. Moreover, you are told that these properties vary independently of one another.[14] This might be because each feature is controlled by a different set of chemical elements or distinct object substructures, for example, and hence the different characteristics are not causally related, even within the same bag. Now, a friend hands you an object she has drawn from one of the bags, and she asks you to guess which one. She also hands you a scale and a metallic rod. Do you think you would be able to guess correctly using just the scale? No. What if you hit the object with the rod? Better, but this is still a very error-prone classification strategy. What if you simultaneously evaluated the mysterious object's weight, sound when struck, melting point, and colour? The basic argument of [192] is that by assessing increasingly many genetic loci of a human (or, analogously, by examining sufficiently many distinct and independent physical features of an object drawn from one of our bags), we can, with decreasing error, correctly classify from which original biological population(s) or bag a human or metallic object was sampled. As [192] concludes, citing Rosenberg *et al.* 2002: '[I]t was only in the accumulation of small allele-frequency differences across many loci that population structure was identified' ([192], p. 253, this volume).[15]

view. Consider debates about models for the evolution of eusociality (Nowak, M. A., Tarnita, C. E., Wilson, E. O. 2010. The evolution of eusociality. *Nature* 466, 1057–1062; Abbot, P., Abe, J., Alcock, J., *et al.* 2011. Inclusive fitness theory and eusociality. *Nature* 471, doi: 10.1038/nature09831) or for the evolution of reduced virulence (Wild, G., Gardner, A., West, S. A. 2009. Adaptation and the evolution of parasite virulence in a connected world. *Nature* 459, 983–986; Wade, M. J., Wilson, D. S., Goodnight, C., *et al.* 2010. Multilevel and kin selection in a connected world. *Nature* 463, doi: 10.1038/nature08809). For the possibly most influential public statement of inclusive fitness as an essential feature of the gene-centred view, see Dawkins, R. 1976. *The Selfish Gene*. Oxford: Oxford University Press. Contrasting perspectives, such as multilevel selection, look beyond single additive and averaged genes (e.g. Winther, Wade, and Dimond 2013).

14 Strictly speaking, the features are independent, *conditional on the bag*.

15 This is the penultimate sentence (p. 2384) of Rosenberg, N. A., Pritchard, J. K, Weber, J. L., *et al.* 2002. Genetic structure of human populations. *Science* 298, 2381–2385.

As Lewontin pointed out, there is indeed very little classificatory information in a locus with similar allelic frequencies across populations: 'based on randomly chosen genetic differences, human races and populations are remarkably similar to each other' (Lewontin 1972, p. 397).[16] But as we increase the number of loci measured for allele frequencies across different populations, our probability of misclassifying individuals diminishes ([192], Figure 1, p. 251, this volume; Rosenberg commentary; Appendix 1).[17] Edwards's earlier work on clustering analysis and classification resonates with his more recent arguments ([30]).[18] In short, the robust result that (approximately, and averaging the variances across loci) 85% of all genetic variance is found within human subpopulations (e.g. Han Chinese or Sami), 10% across subpopulations within a continental region, and only 5% across the three Old World continents[19] is *not* inconsistent with the increase of classificatory accuracy as we examine gene frequencies of more and more (roughly) independent loci. Some commentators point out that Lewontin and Edwards were asking different questions; others observe that their approaches correspond to distinct research programs or paradigms.[20] Despite the fraught ethical concerns involved in this

[16] Lewontin, R. C. (1972) The apportionment of human diversity. *Evolutionary Biology* 6, 381–398.

[17] See also Fig. 1, p. 34 of Edge, M. E. and Rosenberg, N. A. 2015. Implications of the apportionment of human genetic diversity for the apportionment of human phenotypic diversity. *Studies in History and Philosophy of Science Part C: Studies in History and Philosophy of Biological and Biomedical Sciences* 52, 32–45.

[18] The two are distinguished because inferring or postulating the populations from which multiple samples are drawn (i.e. clustering) is logically distinct from having a sample and inferring the population(s) from which that sample was drawn, as in our bag example (i.e. classification). Computer programs such as *structure* are intended to infer population structure, though they can also be used for classification. Lewontin 1972 used populations postulated by linguistic and anthropological data, and he investigated the degree of overlap among these populations with respect to their genetic structure. Lewontin relied on previously specified clusters and was not interested in the statistical procedures of either clustering or classification. See Rosenberg commentary; Winther, R. G. 2011. ¿La cosificación genética de la 'raza? Un análisis crítico. In *Genes (&) mestizos: Genómica y raza en la biomedicina Mexicana*, ed. C. López Beltrán, UNAM: Mexico City (in Spanish), 237–258; Winther, R. G. 2014. The genetic reification of "race"? A story of two mathematical methods. *Critical Philosophy of Race* 2, 204–223, reprinted in Appendix 2; Kaplan, J. M. and Winther, R. G. 2013. Prisoners of abstraction? The theory and measure of genetic variation, and the very concept of "race," *Biological Theory* 7, 401–412.

[19] As pointed out in Bodmer's commentary, almost certainly the earliest published data for this statistical result is found in Table 12 'World Variation of Gene Frequencies' of Cavalli-Sforza 1966, p. 367 (Cavalli-Sforza, L. L. 1966. Population structure and human evolution. *Proceedings of the Royal Society of London B* 164, 362–379). Averaging the standard variances (Table 12, penultimate column) for the 15 loci investigated gives the ~15% across-population variance Lewontin also found. For studies verifying the robustness of these statistics of human genetic diversity, which also include populations from Oceania and the Americas, see also: Barbujani, G. A., Magagni, A., Minch, E. and Cavalli-Sforza, L. L. 1997. An apportionment of human DNA diversity. *Proceedings of the National Academy of Sciences of the USA* 94, 4516–4519; Rosenberg, N. A. 2011. A population-genetic perspective on the similarities and differences among worldwide human populations. *Human Biology* 83, 659–684.

[20] Regarding the first: Feldman, M. W. and Lewontin, R. C. 2008. Race, ancestry, and medicine. In *Revisiting Race in a Genomic Age*, eds. B. A. Koenig, S. S.-J. Lee, and S. S. Richardson, New Brunswick, NJ: Rutgers University Press, 89–101; regarding the second: Winther 2011, 2014; Winther, R. G. 2018. Race and biology. In *The Routledge Companion to the Philosophy of Race*, eds. P. C. Taylor, L. Martin Alcoff, and L. Anderson, Abingdon: Taylor and Francis, 305–320. For practical appraisals of basic patterns, and methods of measuring, human genetic diversity, see, e.g., Barbujani, G., Ghirotto, S., and Tassi, F. 2013. Nine things to remember about human genome diversity. *Tissue Antigens* 82: 155–164; Bodmer, W. F. 2015. Genetic characterization of human populations: from ABO to a genetic map of the British people. *Genetics* 199, 267–279; Cavalli-Sforza, L. L. and Feldman, M. 2003. The application of molecular genetic approaches to the study of human evolution. *Nature Genetics* 33, 266–275.

topic, Edwards's controversially titled paper, [192], acted as a wake-up call to embrace the logical and statistical consistency of variance-partitioning and clustering-classification approaches to human genetic diversity.[21] Regardless of our political and moral stance on human populations, clusters, or races, we scientists and philosophers must directly address all known evolutionary and genetic information. Our arguments will be stronger for it.

Let us now turn to a description of each commentary, in the following rough order: general background, phylogenetic inference, selection theory, history of science, and human genetic diversity.

Walter Bodmer provides a bird's-eye perspective of Edwards's career and contributions. Bodmer takes Edwards's major scientific contribution to be the construction of phylogenetic trees of human populations using gene-frequency data. After discussing 'Evolutionary trees from human blood 17 group frequency data', he analyses Edwards's diverse body of work in the following three sections: 'Likelihood', 'Historical papers', and 'Human genetic diversity: Lewontin's fallacy – comments based on Bodmer (2015)'. Throughout his commentary, Bodmer supplies relevant details about his interactions with Edwards, Fisher, and Cavalli-Sforza.

Joe Felsenstein describes how he came to work on the statistical approach to inferring phylogenies after meeting Edwards and Cavalli-Sforza, both of whom had wanted to use trees to describe the differentiation of gene frequencies among local populations. Edwards invented a parsimony method and Cavalli-Sforza a distance matrix method; they then sought to discover whether maximum likelihood recommended either approach. Using a Brownian-motion approximation, they encountered technical difficulties, but could see that it was a third distinct method. Felsenstein argues that with Edwards's 1970 paper ([46]) discussing Bayesian inference approaches, the four major approaches to inferring phylogenies had effectively been invented in one body of work.

Rasmus Nielsen traces how insights from population genetics informed methods for estimating trees in the early seminal work of Edwards and Cavalli-Sforza. Nielsen argues that partly as a casualty of the dominance of cladistic ideology, phylogenetics and population genetics subsequently diverged and became separate fields of scientific inquiry. The advent of genomic data from multiple loci has

[21] See the penultimate paragraph of [192], including this sentence: '[I]t is a dangerous mistake to premise the moral equality of human beings on biological similarity because dissimilarity, once revealed, then becomes an argument for moral inequality'; see also [117]. On normative concerns, including Lewontin's tireless critique of 'hereditarianism' and its linking of race and IQ, see Kaplan and Winther 2014, pp. 1043–1046. Kaplan, J. M. and Winther, R. G. 2014. Realism, antirealism, and conventionalism about race. *Philosophy of Science* 81, 1039–1052; and Winther, R. G. and Kaplan, J. M. 2013. Ontologies and politics of bio-genomic 'race'. *Theoria. A Journal of Social and Political Theory (South Africa)* 60, 54–80. It might be worth noting that while Edwards's work on clustering and classification is only a small part of his total opus, his contribution has been particularly striking to philosophers of science.

renewed the interest within the field of phylogenetics for understanding population-level processes, and some state-of-the-art modern methods for estimating the historical relationship between evolutionary units are essentially simple extensions of the methods developed by Edwards and Cavalli-Sforza half a century ago. Nielsen's commentary discusses this historical development and the relationship between phylogenetics and population genetics in general.

Mark Pagel describes how the foundations Edwards and Cavalli-Sforza had laid down influenced Paul Harvey and him to write their book *The Comparative Method in Evolutionary Biology*.[22] Modern comparative statistical methods for analysing evolutionary trends among species owe an intellectual debt to the methods of maximum likelihood that Edwards and Cavalli-Sforza were pioneering in the mid 1960s to infer phylogenetic trees. The connection was inevitable: where phylogenetic inference makes use of a suite of characters to infer patterns of relatedness among species, comparative methods typically study just one or two traits to develop inferences about how those traits evolved or co-evolved along the branches of a phylogeny.

Elizabeth Thompson describes her experience from 1970 to 1974 as a graduate student under Edwards's mentorship, and the deep influence he had on her scientific and statistical thinking. In terms of population genetics, Edwards's earlier work with Cavalli-Sforza and his broad and deep knowledge of topics such as descent at the individual level led to Thompson's abiding fascination with genetic and genealogical structure among individuals and between populations. In statistical thinking, Edwards's clear views on model-based inference and more specifically on likelihood have influenced not only Thompson's own work but also broad areas of statistical genetics.

Ziheng Yang recaps [46] for the biologist reader. He traces the efforts made by Edwards and Cavalli-Sforza in the 1960s to apply Fisher's maximum likelihood method to estimate the relationships among human populations. When the confusion between parameters and random variables is cleared up, the procedure described in [46] is recognised as a Bayesian approach to phylogenetics, using the Yule branching process to specify the prior for trees. Yang discusses the impact of Edwards's paper on the introduction of the Bayesian approach to molecular phylogenetic reconstruction in the 1990s.

Warren Ewens's commentary takes the form of a conversation in which he encourages Edwards to write a population genetics-based biography of Fisher. Fisher is highly admired by both Ewens and Edwards, and Ewens considers the first four chapters of Fisher's classic *The Genetical Theory of Natural Selection*,

[22] Harvey, P. H. and Pagel, M. D. 1991. *The Comparative Method in Evolutionary Biology*. Oxford: Oxford University Press.

discussing how, in such a biography, Edwards would change and improve the chapter contents. Ewens formulates some criticisms of the book. His commentary describes aspects of Fisher's work (in particular his Fundamental Theorem of Natural Selection) that appear to be opaque or even wrong. He also discusses how Fisher's treatment of the stochastic theory can be simplified and brought up to date.

Samir Okasha examines one of Edwards's lifelong interests, Fisher's Fundamental Theorem of Natural Selection. Owing to the opacity of Fisher's presentation, there has been considerable confusion over the years about the true meaning, scope, and validity of the fundamental theorem. Happily, this confusion has been largely resolved, thanks in no small part to Edwards's work. Okasha traces the development of Edwards's ideas on this topic from the 1960s to the present day.

Maurizio Esposito focuses on two of Edwards's hermeneutical suggestions related to Fisher's 1930 magnum opus: first, that Fisher's gene-centric view was more sophisticated than many interpreters have maintained; and second, that Fisher's eugenics should not be taken as the driving motive of Fisher's science. In taking seriously these two suggestions, the chapter complements Edwards's interpretive effort with an analysis of Fisher's use of the human sciences. In so doing, it shows how Fisher's ideas on human evolution were unexpectedly anti-reductionist and his epistemic approach interestingly pluralistic.

Jean-Baptiste Grodwohl revisits Edwards's contributions to the history of population genetics, by relating them to his lifelong reading of Fisher. Grodwohl's chapter reads as an intellectual biography of Edwards, from his first steps in Fisher's Department of Genetics up to his most recent historical work on the Fundamental Theorem of Natural Selection.

Noah Rosenberg discusses Edwards's article on multivariate classification and human genetic diversity ([192]), distinguishing Edwards's classification question from the variance-partitioning problem posed by Lewontin. Rosenberg comments on earlier iterations from the 1970s and 1980s of the conversation between Lewontin and others with views similar to Edwards's. In doing so, he confirms the importance of both perspectives, remarking on contexts in which one or the other appears to be more salient. The chapter ends with a description of the influence of Edwards's model of multivariate classification on a new extension in which classification proceeds from phenotypes rather than genotypes.

On being provided with an opportunity to respond to the commentaries, Professor Edwards preferred to add a previously unpublished autobiographical talk, which can be found in Appendix 3.

As for the editor, how did this book about Edwards come to be edited by a philosopher of science rather than a bona fide biologist? Ian Hacking (Stanford; University of Toronto; Collège de France), one of my mentors and a distinguished philosopher of science, author of the inspiring book *Representing and Intervening*, brought [192] to my attention, and we discussed it at length.[23] Hacking, furthermore, was once Edwards's peer at the University of Cambridge, which provided me easy introduction and a strong recommendation to Edwards. I quickly became fascinated with Edwards and Cavalli-Sforza's early technical work in statistical phylogenetics. The probability theorist and data analyst Amir Najmi added immeasurably to my understanding of the quantitative methodologies of this book (his geometric interpretation of Edwards's 'Lewontin's fallacy' argument is included in Appendix 1), as did biologist Michael 'Doc' Edge (Stanford University; University of California, Davis). Historian and philosopher of science Carlos López Beltrán (Universidad Nacional Autónoma de Mexico, Mexico City) helped me to frame relevant ethical and political concerns. Conversations with Richard Lewontin (Harvard) and Noah Rosenberg helped clarify further subtleties, technical and conceptual alike. In a number of publications, some of them co-authored with philosopher Jonathan Kaplan (Oregon State University), I have explored the philosophical implications of the simultaneous truth of variance-partitioning (Lewontin) and clustering and classification (Edwards) approaches to human genetic diversity. (One of these is reprinted in Appendix 2.) Another of my immediate concerns was how various social and common conceptions of race in modern humans seemed to turn on details of the theoretical and experimental methodologies explored in this volume.

It has been an honour and a significant education to work with and learn so much from Professor Edwards. Corresponding with Edwards, reading his work, and interviewing him proved to be deeply enlightening. Katrina Halliday's expert judgment at Cambridge University Press also added immensely to this book, and Natasha Whelan, Zoë Lewin, Claire Heidi Hafner, Lucas McGranahan, and Oliver Baker have provided significant editorial assistance of various kinds.

In addition to the education I received during the editing of this book, I already had considerable interest and foundation in the pertinent science. Furthermore, as *Phylogenetic Inference, Selection Theory, and History of Science* analyses controversies and intellectual lineages, it seems important to declare by whom and in what I received my scientific tutelage. I was taught evolutionary biology in the American tradition of Sewall Wright. My teachers included Marcus Feldman, Michael J. Wade, Peter Godfrey-Smith, and Elisabeth A. Lloyd, at Stanford

[23] Hacking, I. 1983. *Representing and Intervening: Introductory Topics in the Philosophy of Natural Science*. Cambridge: Cambridge University Press.

University and Indiana University, and, later, Richard Lewontin at Harvard University. Works by Stephen J. Gould and by Richard Levins and Lewontin were early inspirations. A minor in mathematics strengthened my understanding of probability theory and statistics. I learned cladistic phylogenetics and taxonomy, mostly according to principles associated with Willi Hennig, under the tutelage of Michael Ghiselin, David Hull, and Olivier Rieppel, as well as through research on Pacific and Caribbean sponge-inhabiting barnacles at the California Academy of Sciences (San Francisco) under Robert van Syoc.[24]

Thus, the approaches I was taught differ from those of Edwards and others who advocate an abstract and additive gene-centred view and who favour statistical phylogenetics over cladistics. However, being of a philosophical sensibility, initiation did not a partisan make. I have always believed in the importance of keeping an open mind, and in identifying both the reasons for conflict among competing paradigms, and the areas where they could potentially overlap and be integrated. Intellectual life is an ongoing exploration, a process with no clear path and no end in sight. Whatever your affiliation and interest in evolution and genetics, the influence of Edwards's work and the perspective he defends are a wonderful case study in the practices and progress of science.

As Edwards concludes on Day 3, interviews, 'I am very impressed with philosophers of science, how carefully they read the literature' (p. 459, this volume). Let us all aspire to live up to this description – as generous yet critical readers of the tradition – whatever our disciplinary background.

Finally, how might this window on the history of twentieth century quantitative methods in genetics, phylogenetics, and evolution illuminate the future? Edwards's work and influence enable us to study complex relations among multiple sciences: genetics, data science, computing, statistics, evolutionary theory, and philosophy. Such relations promise to become only more important in the twenty-first century, with the ubiquity of technology for communicating and collaborating in real time, and with the exponentially increasing volumes and diversity of data, demanding constant innovation to be exploited. Edwards's contributions are foundational, and much can still be learned from them. For instance, readers of *Phylogenetic Inference, Selection Theory, and History of Science* might wish to follow Edwards's use of geometric visualisations, metaphors, and conceptual analysis in teaching likelihood and other basic statistical tools. Or one might desire to study his discussions of key concepts, such as population and sample; parameter and statistic; and estimator,

[24] See, e.g., van Syoc, R. J., Winther, R. G. 1999. Sponge-inhabiting barnacles of the Americas: a new species of *Acasta* (Cirripedia, Archaeobalanidae), first record from the eastern Pacific, including discussion of the evolution of cirral morphology. *Crustaceana* 72, 467–486; Winther, R. G. 2009. Character analysis in cladistics: abstraction, reification, and the search for objectivity *Acta Biotheoretica* 57, 129–162.

estimand, and estimate. In inspiring young scientists to obtain a foundation in statistics, few have written as evocatively as Edwards, who cautions, in *Likelihood* (1972, pp. xi–xii), against building a biological, scientific house on the sands of complex reality without at least consulting a clear set of statistical 'maps'. Perhaps this window on one revolution will inspire or provide guideposts for another.

Part 1

Selected Papers of A. W. F. Edwards

Paper 25 (1963)

The reconstruction of evolution
(Abstract) *Heredity* 18, 553, and *Annals of Human Genetics* 27, 104–105.
(With L. L. Cavalli-Sforza)

Note from A. W. F. Edwards: This joint paper was read by AWFE at the summer meeting of the Genetical Society of Great Britain in London in July 1963. It presents the algorithmic method of 'minimum evolution' for reconstructing an evolutionary tree, later to be called 'parsimony'. The typescript has not survived. Further background information may be found for all the 'evolutionary tree' papers in [225].

HEREDITY VOLUME 18 PART 4 NOVEMBER 1963

THE RECONSTRUCTION OF EVOLUTION

A. W. F. EDWARDS and L. L. CAVALLI-SFORZA

International Laboratory of Genetics and Biophysics, Naples, Pavia Section: Instituto di Genetica, Universita di Pavia

Darwin's theory of evolution invites us to believe that closely similar species are closely related. In cases in which there is no fossil record, or other means of following evolutionary history, this concept provides the only means of drawing inferences about the course of evolution, and it is important to see whether it can be defined sufficiently precisely to be used for estimating the most likely form of an evolutionary tree solely from data on the similarities and differences between species. A detailed consideration of this problem leads to the conclusion that the following principle may be employed :

THE PRINCIPLE OF MINIMUM EVOLUTION

The most plausible estimate of the evolutionary tree is that which invokes the minimum nett amount of evolution.

This principle will be discussed, the system of estimation to which it leads expounded, and some examples of its application given.

Paper 27 (1964)

Reconstruction of evolutionary trees
In *Phenetic and Phylogenetic Classification*, eds. V. H. Heywood and J. McNeill, The Systematics Association Publication No. 6, London: The Systematics Association, 67–76.
(With L. L. Cavalli-Sforza)
Reprinted in *Cladistic Theory and Methodology*, eds. T. Duncan and T. F. Stuessy, New York: Van Nostrand Reinhold (1985).

Note from A. W. F. Edwards: This joint paper was read by AWFE at the meeting of the Systematics Association in Liverpool in April 1964. It introduces the statistical model of a branching random walk to represent an evolutionary tree together with the proposal to estimate the tree by the method of maximum likelihood, but since the associated computer programs were not yet working the program for finding the minimum-evolution tree is used in the example. In a moment of youthful exuberance, the tree for 15 human populations characterised by their blood-group gene frequencies was displayed at the meeting on a map of the world, and found its way into the printed record.

I am at a loss to explain why this tree does not in fact correspond to the one presented earlier in [29]. In particular, the original form has the Maoris separating first from the three American populations.

PHENETIC AND PHYLOGENETIC CLASSIFICATION, pp. 67-76

Issued DECEMBER 1964

Reconstruction of Evolutionary Trees

A. W. F. EDWARDS & L. L. CAVALLI-SFORZA

(*International Laboratory of Genetics and Biophysics, Naples* (*Pavia Section*), *Istituto di Genetica, Università di Pavia, Italy*)

CONTENTS

1. INTRODUCTION

THE extent to which the classification of a group of organisms may be used to make inferences about phylogenetic relationships has long been a controversial point in biology. Today, particularly in view of the development of 'numerical taxonomy' it is important to consider, more deeply than hitherto, the whole question of evolutionary inferences and their relation to taxonomy. This paper contains some suggestions for studying the problem, which lead to practical methods of estimation. First, however, it is convenient to consider the logical nature of taxonomic studies in general.

2. TAXONOMY

We suspect that the present confusion of ideas about the logical basis of taxonomy is partly due to a failure to distinguish between different types of taxonomic enquiry, and that this failure may have been induced by the fact that all taxonomic studies use much the same type of data. Every kind of taxonomic enquiry must have its own particular purpose and logical basis; indeed, the logical structure must arise naturally from the particular purpose, which therefore needs to be specified in detail. A classification created with no particular purpose in mind is likely to be logically indefensible and partially useless; for when it is put to a specific* use, for which it was not, of course, designed, it will

* We use *specific* and its derivatives in their everyday senses.

most probably be inefficient. In the same way, a motor vehicle built without a specification is unlikely to turn out to be a fast sports car: it is more likely to be a heterogeneous collection of ill-suited mechanical bits and pieces.

Though the need to specify the purpose of a taxonomic study has, from time to time, been admitted, some confusion has been generated by the concept of the 'general purpose' classification. To say that the purpose of a classification is "general" is, in our view, too vague to be of use in its construction. If we may return to the motor vehicle, a 'general purpose' car can be specified only because the particular purposes to which it may be put, and their relative importance, can themselves be specified rather accurately. Thus the designer may anticipate that the proposed vehicle will spend 50% of its time carrying only the driver, 20% carrying two people, and so forth, and that it will be required to carry so much in the way of goods so often, and go up hills of such-and-such a gradient once a year on average. On the basis of such information an acceptable compromise design will be achieved. There are three important things to note in this example. First, that 'general purpose' is only meaningful if it is regarded as a collection of weighted specific purposes: secondly, that the outcome of a general purpose specification is unlikely to fulfil any of the specific purposes as well as a specially designed vehicle could; and, thirdly, that the justification of the general purpose vehicle is an economic one, because for optimum service all the time one would need as many vehicles as there are uses, but this would be very expensive: only rich farmers can afford a Jaguar *and* a Land-Rover.

In taxonomy the situation is much the same; the only justification for a general purpose classification is that we cannot afford the time or the money to create the separate specific classifications that we really need. It seems to us to have no purely scientific justification, and it is surely an illusion to imagine that a 'general purpose' classification approaches an unattainable 'ideal' one in the same way that a circle drawn on a piece of paper approaches a perfect circle. We thus hold that the most essential prerequisite of a classification is its purpose, clearly defined, and that this is no less true in the case of a 'general purpose' classification, which should only seek to be a compromise solution to the problem of satisfying specific purposes economically.

Given the purpose to which a classification is to be put, its design, the choice and weighting of the characters to be used, and indeed its logical structure, follow, just as the design of a motor car follows from its specification. Naturally, there is likely to be room for subjective judgements in the choice and weighting of characters, and we must not be surprised to find two taxonomists disagreeing even when the specification of the classification seems quite rigid, just as there is room for motor car designers to interpret a specification subjectively. But the scope for subjective decisions is directly related to the comprehensiveness of the specification, which should be complete if objectivity is very important. Conversely, when there is virtually no specification at all, as in the 'general purpose' case as used hitherto, there is immense scope for subjective decisions, and attempts to find a logical justification where none exists are doomed to death by drowning in a sea of indefensible statements.

Rather as Fowler apologizes for taking all his examples of bad English from *The Times*, and insists that this is not because *The Times* contains more bad English than most newspapers, but because it is the newspaper he prefers, so we apologize for taking some statements about general purpose classification from the admirable summary of Numerical Taxonomy by Sneath and Sokal (1962) and quoting them:

"Such [natural] classifications may be termed 'general' and are for general purposes, in distinction to classifications made with a special purpose in mind, which may be called 'special' classifications. This implies among other things that in constructing general classifications characters have equal weight...".

"The taxonomic equivalence of all characters can be seen most clearly if we attempt to construct an objective criterion for weighting characters. If we cannot decide how to weight the features we must give them equal weight, unless we propose to allocate weight on irrational grounds".

"We hold the view that a 'natural' or orthodox taxonomy is a general arrangement intended for general use by all kinds of scientists; it cannot therefore give greater weight to features of one sort, or it ceases to be a general arrangement".

The logical frailty of these attempts to justify a certain procedure when the purpose of the intended classification has not been carefully specified should be a warning to us all.

There seem to us to be three major types of taxonomic study, each of entirely different logical structure, and it is of the utmost importance to differentiate them. The first, which we may call Pure Taxonomy, is solely concerned with the processing of information. It is descriptive. Each organism can only be described in terms of the characters it exhibits, but since some of these characters will be common to two or more of the organisms, it is clearly unnecessary to describe each organism separately, and some form of classification will present all the information in a simpler form. This approach has been developed by Maccacaro (1958) and Rescigno and Maccacaro (1961), and has not, in our view, received the attention it deserves. It should be noted that the problem of choosing and weighting characters is irrelevant in *pure* taxonomy: *given* the characters, the problem is to classify the organisms according to them in such a way that the maximum amount of information is retained. Thus, given a set of snooker balls, it is possible to classify the balls by colour without loss of any information about this character.

The second type of taxonomy may be called "applied", for it seeks to classify the organisms in such a way that the resulting classification shall be the best obtainable for the specified purpose at hand. Thus tyre manufacturers are primarily interested in classifying motor vehicles by the size and number of their wheels, ferry operators by their lengths, and customs officials by their contents. Once the purpose of the classification has been clearly specified, the taxonomist can set about choosing and weighting the characters he considers appropriate, though, as we mentioned above, if the specification is not sufficiently comprehensive, different taxonomists may have different views about the characters.

The third type of taxonomy involves what Sneath and Sokal (1962) call "phylogenetic speculation". We may refer to it as phylogenetic taxonomy, and note that it does not primarily involve classification, since its avowed purpose is the estimation of the evolutionary tree which supposedly unites the organisms being studied. Characters must be chosen and weighted according to their evolutionary significance, and judgement of this is likely to be subjective. However, the more that is known about the genetics of the organisms, the easier it will be to agree about the relevance of characters. In Pavia over the past two years we have studied in considerable detail the problem of the statistical estimation of the form and dimensions of evolutionary trees, and we describe our current approach below. The logical basis of such studies seems to be rather simple, probably because their purpose can be so simply and clearly defined, but it is important to realise that the procedure is not primarily classificatory, although, once an estimate of the form of an evolutionary tree has been produced, organisms can of course be grouped according to relative evolutionary divergence.

Our three-taxon classification of taxonomy, and our belief that it holds the key to a satisfactory logical structure of taxonomic studies, will, no doubt, be criticized. It may well be asked, for example, whether the classification of sheep as sheep and goats as goats * is pure, applied, or phylogenetic in type, and, if applied, what is the specific purpose of the classification? Is it not really a general purpose classification? Our answer to this is that the classification *was* indeed applied, but that the original division into sheep and goats was for a purpose which is lost in antiquity. However, it was found that the division was useful in other ways, or, to look from a different point of view, classification of the animals for other purposes frequently led to the same division. The division was evidently rather stable, so much so that it was appropriate to name the animals in the two classes as sheep and goats. Such a classification is acceptable to us for two main reasons: first, because sheep and goats are really rather different, and intermediate forms do not exist, so that there is never any doubt as to whether an animal is a sheep or a goat. The subject matter of the classification is 'robust' with respect to the different purposes to which the classification is to be put. Only robust classes of objects acquire names. But secondly, and more important, the classification is acceptable because we understand its limitations. We know that classification into sheep and goats is not much use if we are interested in the proportion of the combined flock that is male, or the proportion that is under a certain age. In fact

* The terms "sheep" and "goat" refer here to the West European domesticated races of *Ovis aries* L. and *Capra hircus* L., respectively.

our classification is not really general purpose at all; though it arose out of an applied taxonomy, because of the stability of the classes, it has become a pure taxonomy, descriptive in intent. When we say "sheep" we mean, according to the Oxford Dictionary, a "timid gregarious beast kept in flocks for the mutton or lamb and wool and leather it yields". A goat, on the other hand, is "a lively wanton strong-smelling usually horned and bearded ruminant quadruped".

Our present problem, however, is not that of considering how classifications arose historically, because we are trying to set up a classification which is logically acceptable, and does not rely on age-old decisions. It is as though we were confronted with a field full of sheep and goats for the first time, and were asked to classify the animals. Our reaction must be to ask to what purpose the classification will be put, or, alternatively, to ask what are the characters that we should consider. Thus we can embark on an applied or a pure taxonomy, whichever is appropriate. But if we are asked to classify the animals "for general purposes" we must insist on knowing *what* purposes, though, of course, there is nothing wrong in admitting that the animals seem at first sight to be divisible into two homogeneous classes, and that this classification will no doubt serve the majority of purposes included under the title "general". *But* this fact derives from the particular data we are using, and cannot be used as a justification for a taxonomic procedure whose logical structure must be independent of the organisms being considered. The fact that different characters and different taxonomic methods lead to the same segregation of the animals into sheep and goats tells us something about sheep and goats, but nothing about the logical basis of taxonomy.

3. Reconstruction of Evolutionary Trees

Having detached phylogenetic taxonomy from purely classificatory studies, we may now proceed to examine it more closely. The characters which are fundamental to differentiation are the base-sequences of DNA. These are at present unavailable directly, although methods have recently been suggested to measure the extent to which two DNA's match each other for base sequence. In the absence of such data, use must be made of genotypic, or even phenotypic, differences. At the genotypic level the relative importance of various gene substitutions in evolution is a matter which only detailed study can reveal; it will depend upon the past and present selective forces, mutation rates and population structures and sizes. At the phenotypic level such information is even less readily available.

As has been found with classification, it is convenient to represent the variation amongst the organisms geometrically, the chosen characters being represented by axes in a multidimensional space, and the organisms, specified by these characters, being represented as points in this hyperspace. If a time dimension, everywhere normal to the character space, is added, the course of evolution — were it but known — could be seen as a 'tree', sometimes branching as groups of organisms diverged, sometimes growing together as hybridization took place, and with many branches ending in extinction before the remainder intercepted the 'now' character space to indicate the current disposition of the organisms. This approach presupposes that the character-space is continuous, which will normally be an acceptable approximation. In cases in which discreteness is apparent the space will take the form of a lattice of points.

The proper basis for a study of evolutionary divergence will be provided by means of that transformation of the space-time which, as far as is known, makes a unit vector, in whatsoever direction (normal to time) and in whatsoever part of the space-time, correspond to a fixed amount of evolutionary divergence — divergence being properly defined for the problem under consideration. Such a transformed space-time would thus be everywhere isotropic with respect to evolutionary progress. Transformation is, in this context, equivalent to character-weighting.

Put thus formally, it will be obvious that the knowledge necessary for the reconstruction of evolution is rarely available, but, as in other problems of statistical estimation, deductions can always be made provided the assumptions on which they rest are remembered.

The present problem is to estimate the form of the evolutionary tree given only the information contained in the 'now' character-space. Fossils, which would appear at other time levels in space-time, will not be treated at present, although we would like to stress particularly that it is clear from this approach that there is no substantial logical difference about estimating the course of evolution with or without fossil evidence.

In our isotropic space we may think of evolution as a branching random walk, with a constant probability of branching and a constant rate o walking: that is, after the elapse of a certain time interval, the probability distribution in space of the position of a population will be normal, with mean at its original position and variance proportional to the time elapsed. The constant of proportionality will depend on many parameters, such as the population size and the type and intensity of selection, and will usually be unknown. Since an evolutionary tree uniting n points (without loops) is bound to contain $n-1$ branching points, the probability of branching is not a parameter which has to be estimated, although it is worth noting that the theory of birth and death processes may have some interesting things to say about the expected form of an evolutionary tree.

Now the probability density at a distance d spatially and t temporally from a point in the space-time of p spatial dimensions is given by

$$\sigma^{-p}\ (2\pi t)^{-\frac{1}{2}p} \exp\{-d^2/(2t\sigma^2)\}$$

where σ^2 is the constant of proportionality mentioned above. The log-likelihood is therefore given by

$$-\{d^2/(2t\sigma^2) + \tfrac{1}{2}p \log(2t\sigma^2) + \tfrac{1}{2}p \log\pi\}$$

Writing T for $2t\sigma^2$, omitting the constant, and changing the sign, the expression becomes

$$d^2/T + \tfrac{1}{2}p \log T$$

Each arm of a postulated evolutionary tree will have an expression of this type associated with it, d being its spatial length and T proportional to its temporal length. Thus the likelihood of the tree will be maximized if the sum of these expressions, over all the arms, is minimized (since we have changed the sign, and the branching points, being constant in number, are irrelevant).

We have found that the mathematics of the estimation procedure can be made quite compact using a matrix notation, and we do not anticipate any major difficulties in the corresponding computer programmes, although these are not yet working. However, as will be mentioned below in connexion with our example, other methods have been used which may be expected to give similar results.

Thus maximum-likelihood estimation is not difficult once the topological form of the tree has been specified, but unfortunately there will generally be too many forms for the maximum likelihood of each to be evaluated: with n points to unite there are $(2n-5)!/(n-3)!\ 2^{n-3}$ trees, or more than two million with ten points. We therefore have to resort to some prior method of clustering. Initially we developed our own method, based on the analysis of variance (Edwards and Cavalli-Sforza, 1963b), but we have finally adopted a method which is particularly simple and rapid. Prim (1957) has shown that the network of minimum length uniting n points, but with the segments constrained to meet only at these given points, can be found by listing all the distances between points in increasing order, and successively allocating segments to these distances, omitting any segment which completes a loop. The resulting net gives some indication of the types of tree it will be worthwhile to use in the maximum-likelihood estimation procedure. The relationship is somewhat intuitive at the moment, but once one is "on the track" of a good topology, experience shows what sort of changes will be likely to increase the overall likelihood. In particular, if the length of an arm is estimated as zero, a change in the topology of that part of the tree is indicated.

We are now left with the problem of finding the correct transformation of the character space. The transformation is, of course, a reflection of the genetic assumptions which are being made, and these will be peculiar to each case. For the moment it will be sufficient to indicate how we have proceeded in a particular example.

4. The Recent Evolution of Man

(a) *Materials*

As our example, we decided to study the phylogeny of fifteen samples of human populations, using as characters the frequencies of various blood-group alleles. This choice was made because of the availability of the data, the intrinsic interest of the problem, and the amount of knowledge about the genetic situation. Data on the blood-group systems A_1A_2BO, Rh (four sera), MNSs, Fy and Di were obtained for fifteen populations, three from each continent, with the invaluable help of Dr. Mourant and Mrs. Sobczak, of the Medical Research Council Blood Group Reference Library, London. In all, twenty 'genes' were distinguished. Nearly every sample consisted of more than 70 individuals, and was tested for each of the five groups; although there is a certain amount of heterogeneity with respect to the sera used, and the methods of calculating the gene frequencies, these data were considered adequate for a trial of the methods.

The five blood groups were assumed to be independent, because of the absence of evidence for linkage between them. Each allele within a system, and each of the systems, was given equal weight. In view of the apparent absence or weakness of selective forces in these blood groups, and, in cases where differential mortality has been demonstrated, of the impossibility of setting up selective models of sufficient validity, it has been assumed for our preliminary work that selective forces are absent, and that the observed divergences are, or can be treated as being, due to drift. Boyd's comment (Boyd, 1963), that "unless the blood groups are adaptive, they are not going to be very useful in racial classification", is not acceptable: any variable, even a random one, which shows a persistently high correlation between successive generations, may be useful. If differential selective forces are operating, the simultaneous use of several independent loci should render their effects on the analysis minimal. All the populations have been treated as though they were, and always have been, of the same size and structure.

(b) *Methods*

It is now necessary to set up a character-space on the basis of these assumptions, following the methods outlined above. Since the normal scale of gene frequencies is anisotropic in the sense that the effects of drift are differently measured in different parts of the scale, a transformation is necessary. The 'square-root' transformation given by

$$\sin^2\theta = p, \qquad \cos^2\theta = q,$$

is appropriate for two alleles at a locus (Fisher, 1954), since it stabilizes the binomial variance throughout the range, and Fisher (personal communication) has pointed out that it may be extended to the case of many alleles. With n alleles, populations will be represented as points on $(1/2^n)$th of the surface of the unit hypersphere in n dimensions, the angular distance between populations with gene frequencies (p_1, q_1, r_1,...) and (p_2, q_2, r_2,...) being given by

$$\text{Cos}\,a = \sqrt{p_1p_2} + \sqrt{q_1q_2} + \sqrt{r_1r_2} + \ldots$$

Since the character-space is curved, it will be necessary to transform it into a Euclidean space before the available maximum-likelihood treatment can be applied. As in map projections, some desirable qualities of the attribute-space will be lost in the transformation. The most appropriate transformation is worth a separate study, but the simplest approach is to use, as the measure of distance between two points, the chord — the Euclidean straight line — rather than the arc of the great circle on the surface of the hypersphere. This is equal to $\sqrt{(2 - 2\cos\alpha)}$. The maximum possible error thus incurred is given by the ratio of the length of the arc of a quadrant of a circle to that of the corresponding chord, or $\sqrt{2} : \frac{1}{2}\pi$, amounting to a maximum under-estimate of distance of 10%.

Given the Euclidean distance between two populations for each locus, the total distance over all loci is found by taking the square root of the sum of the squared distances for individual loci, by Pythagoras' theorem, since the character-spaces for the loci are mutually orthogonal. The resulting array of pairwise distances between populations is ready for finding the Prim network. In order to use the maximum-likelihood computer program it is necessary to set up a system of Cartesian co-ordinates for the points representing the populations, and this may be done on a computer from the pairwise distances. The space thus created is the final version of the character-space, but with an arbitrary frame of reference. The unit distance, however, is not arbitrary, and corresponds to an amount of evolution equal to that incurred in a single gene substitution at a single locus.

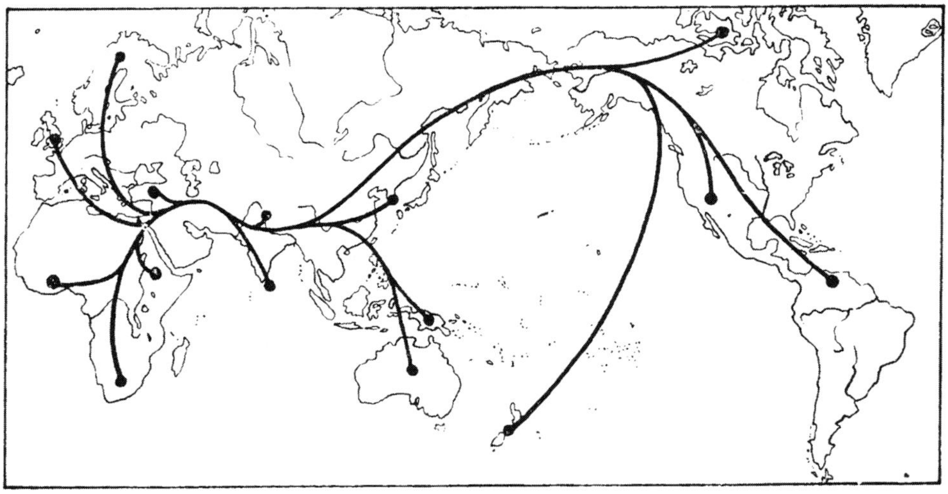

FIG. 1. Topology of the minimum-evolution tree uniting fifteen human populations; constructed on the basis of the frequency of blood-group alleles.

Unfortunately, as has been mentioned, our maximum-likelihood programmes are not yet working, but, using this character-space, our earlier "method of minimum evolution" (Edwards and Cavalli-Sorza, 1963a) gave a 'best' tree of the topological form shown in the figure (Fig. 1). It is probable that this method gives a tree which is approximately the same as the projection of the maximum-likelihood tree onto the 'now' character space. Whilst we would be the first to admit that we have been fortunate to get such a reasonable result from preliminary data, we feel that it is at least encouraging, and justifies our intention to develop our methods further, and apply them to more extensive bodies of information.

5. Acknowledgements

This work has been supported by a grant from the U. S. Atomic Energy Commission and by Euratom-CNR-CNEN contract no. 012-61-12 BIAI.

REFERENCES

Boyd, W. C., 1963. Genetics and the human race. *Science,* **140**: 1057-1064.

Edwards, A. W. F., and Cavalli-Sforza, L. L., 1963a. The reconstruction of evolution. Unpublished paper read at 142nd meeting of the *Genetical Society of Great Britain,* London, July 1963. (Abstract in *Ann. hum. Genet.,* **27**: 104-105, and in *Heredity, Lond.,* **18**: 553.)

—— —— 1963b. A method for cluster analysis. Unpublished paper read at *5th Int. Biometric Conf.,* Cambridge, 1963.

Fisher, R. A., 1954. *Statistical Methods for Research Workers,* ed. 12. Oliver & Boyd, Edinburgh. 356 pp.

Maccacaro, G. A., 1958. La misura della informazione contenuta nei criteri di classificazione. *Annali Microbiol.,* **8**: 231-239.

Prim, R. C., 1957. Shortest connection networks and some generalizations. *Bell Syst. tech. J.,* **36**: 1389-1401.

Rescigno, A., and Maccacaro, G. A., 1961. The information content of biological classifications. In: C. Cherry (ed.), *Information Theory,* 437-446. Butterworth and Company, London.

Sneath, P. H. A., and Sokal, R. R., 1962. Numerical taxonomy. *Nature, Lond.,* **193**: 855-860.

Paper 28 (1964)

Analysis of human evolution under random genetic drift

Cold Spring Harbor Symposia on Quantitative Biology 29, 9–20.

(With L. L. Cavalli-Sforza and I. Barrai)

Note from A. W. F. Edwards: This joint paper was read by LLCS at the 29th Cold Spring Harbor Symposium on Quantitative Biology in June 1964. The figure of the tree displayed on a world map from [27] is reproduced. Alongside it is a figure of the corresponding minimum-evolution tree, which differs from the original in Paper 29.

Analysis of Human Evolution Under Random Genetic Drift

L. L. Cavalli-Sforza, I. Barrai and A. W. F. Edwards
International Laboratory of Genetics and Biophysics, Pavia Section, Istituto di Genetica, Università di Pavia, Italy

INTRODUCTION

At a time which anthropologists cannot yet state exactly, but which may lie between twenty thousand and five hundred thousand years ago, the human species must have started to undergo the splitting process which gave rise to the human races as we observe them today.

Our ignorance of the times and places at which the splits leading to the major racial groups occurred will perhaps be reduced by future paleontological discoveries. It is, however, theoretically possible to reconstruct the probable mode of origin of a group of populations, even in the absence of a fossil record, provided we have a satisfactory model of evolution, a sufficient body of data on hereditary characters of the populations, and a method to fit the model to the data (Cavalli-Sforza and Edwards, 1963, 1964, in preparation, Edwards and Cavalli-Sforza, 1963, 1964). We have endeavored to produce a model and a method to enable us to reconstruct the most likely sequence of splits which led to the observed populations, and have applied this method to man, the organism for which the greatest body of data on gene frequencies is available for a variety of populations. We would like to give a short summary of the results to date.

It may be anticipated that the analysis of the data suggests, or at least is not incompatible with the idea that random genetic drift has more importance than would have been expected. Drift is the random variation of gene frequencies due to the sampling of genes which accompanies reproduction. In this way, variation arises and accumulates over time between the populations into which a species is subdivided for geographical or other reasons. The rate of increase of variation with time is greater with decreasing size of the populations as well as with decreasing cross migration between them. Given an adequate demographic description of a population, the extent of variation attributable to drift can be predicted. If drift is not masked by evolutionary events of a different nature, it should be possible to predict the amount of it when examining the variation in gene frequencies from place to place. Man is the organism for which demographic data of the type necessary for the evaluation of the expected amount of drift are most easily available; even in man, however, there are considerable difficulties in defining and measuring those parameters which are essential for predicting drift. We will concentrate, in the rest of the paper, on our results with this other problem.

MODEL AND METHOD FOR THE ANALYSIS OF EVOLUTION

The problem we have in mind is the reconstruction of the most likely sequence of $(n - 1)$ splits, which we shall call a "tree", leading to a group of n populations differentiated by an evolutionary process. The data on which to base our inference can be, for instance, gene frequencies. In theory there may be information on only one gene, but clearly the precision of the answer will increase with the number of genes. We have not yet considered complications arising from "loops" in the evolutionary process, for example hybridization between nonadjacent branches, but they could be added to the picture.

Figure 1 represents the process of evolution leading to four different populations in space and

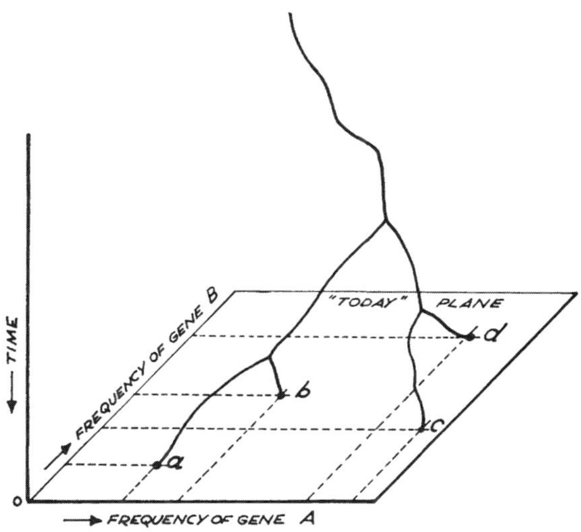

FIGURE 1. A schematic representation of the evolutionary process leading to four populations living today from one original population living some time ago.

time. Only two dimensions are available in the figure to plot gene frequencies; in real cases many more than two genes would be used. The third dimension is used for time.

If we are ignorant of the true sequence of splits and want to infer it from the data, there are twelve different trees which could have led to the four populations, but on the basis of the resemblances between populations many of these trees are not acceptable. Thus, with the data given in the figure, a tree in which the first split separates a, d from b, c is intuitively unlikely. For a quantification of this process, to enable us to manage the mathematical likelihood of each possible tree and choose the most likely one, an evolutionary model is necessary. We have chosen a model based on the hypothesis that changes in gene frequencies take place by processes which can be considered statistically independent for different populations.

"Random genetic drift", as defined before, is an evolutionary process fulfilling these conditions, but is not the only one. We are adding the word "random" at the suggestion of Kimura (1964), as the word drift alone is somewhat misleading. It might even be further defined as "sampling" drift, as there is another type of genetic drift which has elements of randomness but is of another nature, being "selective" in origin. This is the case of *variable* selective coefficients, with a certain probability distribution; under such selective forces, as Kimura (1954) has shown, a variation of gene frequencies sets in which increases with time, and the effect is formally very similar to that of sampling drift. Clearly, sampling drift and selective drift can both determine differences between populations which are acceptable in the model defined above.

In addition to these mechanisms, populations can develop differences when selection coefficients are stably different in the environments they occupy, or in the conditions in which they live. This mechanism causes no disturbance to our aims, on the contrary it can add very useful information, except for specific situations leading to evolutionary convergence, which we shall not discuss further here.

The specific mathematical model we are using is as follows:

(1) We assume a distribution for the gene frequency of the next generation, whose mean will not differ systematically from the gene frequency of the previous generation if there is no "directional" selection, while the variance will depend on the amount of drift present. If there is directional selection, it will determine a shift in the mean of the gene frequency distribution in successive generations, and our present model assumes that there is no variability in directional selection between populations.

(2) The variance of the distribution after t generations will be the sum of the variances in each generation, and the distribution will tend to normality because of the central limit theorem. If the variance in each generation is constant, or varies in a random way, the variance after t generations will be proportional to t.

There is some limitation to this last statement due to the fact that the variance of a proportion p is, as is well known, a function of p. To make it independent, and simplify our problems, we use, instead of the proportion of an allele p, its angular transform ϕ, given by $p = \cos^2 \phi$. Even so, there will be some difficulties for gene frequencies very near to zero or one, or due to the fact that proportionality of the variance with time will not be valid for very long times, when the variance of gene frequencies (or their transforms) is approaching its asymptote. In practice these limitations are not serious, however, and they could be met by other transformations, but the numerical work would be much heavier.

Given the above assumptions, one can then estimate the most likely tree and the parameters belonging to it; the time intervals separating every two successive splits in the tree (in relative measure), the variance of gene frequency accumulating over time for each gene, and the gene frequencies of the presumed "ancestors" at each split. Such "ancestral" gene frequencies, however, cannot take account of changes due to directional selection common to all the populations being studied.

Estimation by maximum likelihood is still causing some difficulties, because of the very high number of parameters to be estimated and some peculiarities of the system. However approximate methods have been used with success and we anticipate that the results thus obtained are very close to those obtainable by maximum likelihood. Such approximate methods correspond to models slightly different from the one employed in the maximum likelihood procedure, and we can only offer an intuitive justification for them. One such model uses the "method of minimum evolution" (Edwards and Cavalli-Sforza, 1963), in which the gene frequencies of ancestral types are computed so that evolutionary divergence invokes the minimum total amount of evolution.

ANALYSIS OF HUMAN DATA

Although the data on human gene frequencies in existence could fill several books of tables, the extraction of a number of populations known for

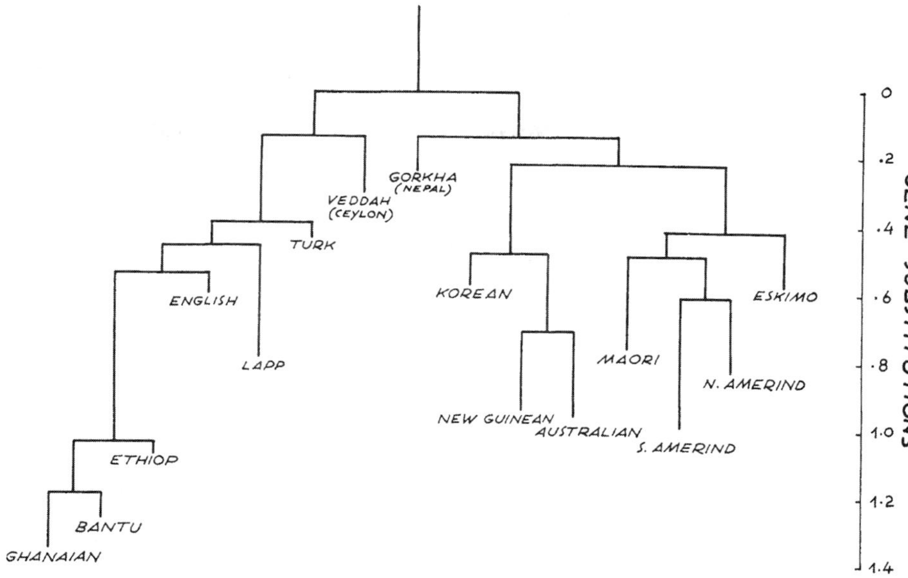

FIGURE 2. Most likely tree of descent for fifteen human populations selected to represent the world population. Method of minimum path.

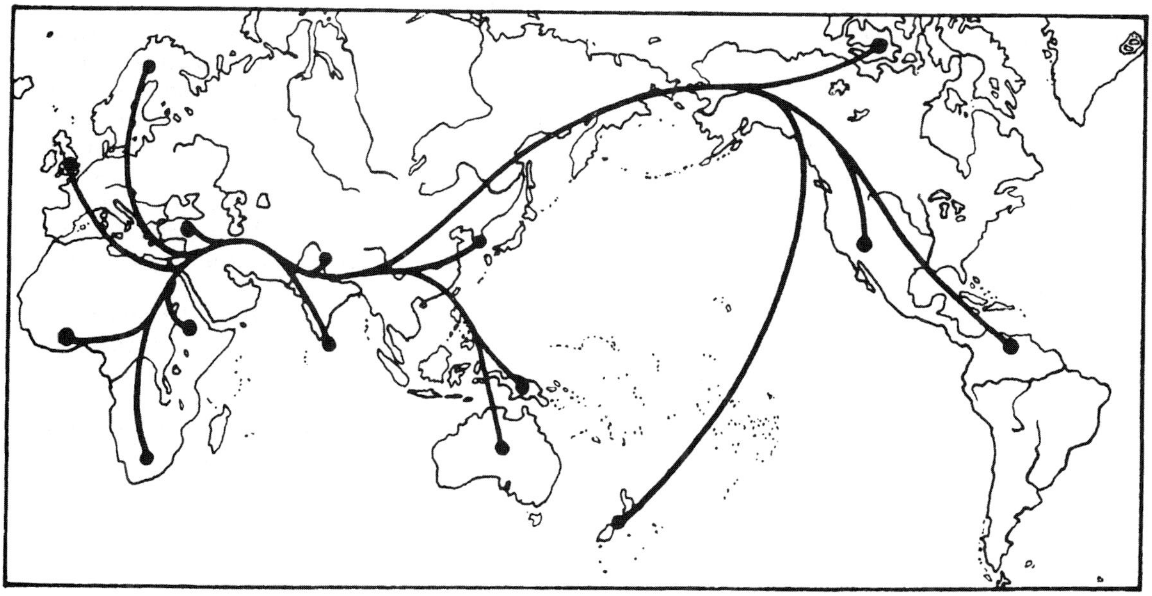

FIGURE 3. Topology of the minimum-evolution tree uniting fifteen populations. The same tree as in Fig. 2 projected on the world map to show its correspondence with the probable manner of diffusion of man.

many genes is a rather frustrating task, because few populations have been analyzed for the same genes or alleles. As the number of possible trees increases very fast with n, the number of populations and computing time per tree increases fairly rapidly with n, we employed a relatively small group of populations, namely fifteen, which we chose as a representative sample of the world population: three widely separated ethnic groups from the aboriginal populations of each of the five continents.

The populations were chosen so as to give information on all the following blood group systems: A_1A_2BO, MNS, Rh with 8 "chromosomes", Fy and Di: a total of twenty alleles at five loci. All the populations had a sample size of at least 70, to keep sampling errors small with respect to the variation in which we were interested.

Today there are enough genes known in man almost to double this number of alleles, but it would have been difficult to find populations meeting the above requirements for more genes. The pooling of data from several related populations would have allowed us to obtain information on more and more genes, but this procedure is time-consuming and not very satisfactory. We have preferred, at this stage, to concentrate on improving the method of estimation, which has been an unusually challenging task.

A superficial analysis of other characters has, however, been done (see Cavalli-Sforza and Edwards, 1963).

The most likely tree obtained with the five blood group systems listed above is given in Fig. 2, with distances between splits measured as the sum of gene substitution equivalents. The unit is a single gene substitution, namely the complete substitution of one allele by another.

The same tree is given in Fig. 3 on a world map, ignoring the lengths of the branches, merely to show how the most likely tree seems to describe quite accurately the probable ways of diffusion of the populations considered. It should be added that the estimation of a tree is sufficiently time-consuming, computer-wise, that only a very small fraction of all possible trees can at present be investigated, and we use approximate methods for the detection of the most likely tree. But we know that even if a slightly more likely tree than our "most likely" one can be found, the difference is bound to be a small one, and insignificant if compared with sampling errors.

POSSIBLE CONTRIBUTION OF DRIFT

In the evolutionary model used for maximum likelihood estimation, the estimates of the time intervals are based on the assumption that the rates of evolutionary change due to drift and random changes in selective values are substantially constant in the various ethnic groups. This may be a rough approximation, but the fact that we are averaging over several genes should to some extent decrease the error. Time is measured on a relative scale; in order to convert it to number of generations we would have to know the amount of change per generation, about which we have no exact information. There are, however, some elements that may be used to gauge the orders of magnitude of the desired quantities, and also to test the possible importance of various evolutionary factors.

The forces which determine differences between populations have been, in part, mentioned before. It may be useful to list them here:

(1) Random ("sampling") drift will determine a variation of gene frequencies which is the same for every gene, provided the variance of gene frequencies is suitably standardized (either by dividing σ_p^2 by $\bar{p}(1 - \bar{p})$, i.e. giving it as a fraction of the maximum, or by using the angular transformation of gene frequencies). Such a standardized variance, being a sampling variance, will depend only on the *size* of the population. It is known that population size has the special meaning in this context (see later) of "effective breeding size."

(2) The random variation of selective coefficients with time and/or space can simulate to a large extent the effects of drift, but will not be necessarily the same for all genes; it may therefore introduce differences in the behavior of different genes, adding its variance to that caused by drift.

(3) If some genes are subject to constant selection of the "stabilizing" kind that is leading to a stable polymorphism (for example because of heterozygous advantage), their variances will be reduced by the stabilizing effect of selection.

(4) If some genes are subject to constant selection of the "directional" kind this will not contribute to the variance of gene frequencies, at least not in an important way, except for differences in selection coefficients of different populations, which will then add to the variance of only these particular genes.

The main point is that selection, whether of the types given at points 2, 3, or 4, will contribute to a decrease or increase of the variance *only* of the particular gene subject to selection.

Variances of the frequencies of different genes were computed for the world population from the

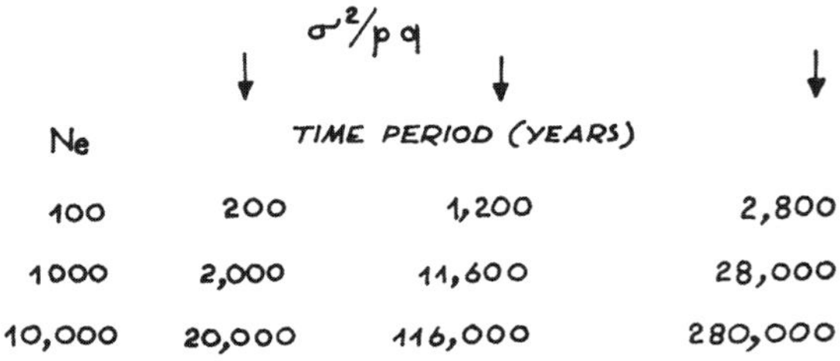

FIGURE 4. Upper part of the graph: values of $F = \sigma_p^2/p(1 - p)$ for the alleles used in the calculations on which Figs. 2 and 3 are based. Lower part: given the three effective population sizes $N = 100$, 1,000, 10,000, times (in years) necessary to obtain, by random sampling drift, the F values of 0.05, 0.25, 0.50.

fifteen populations given before (see Fig. 4). Alleles of the same multiallelic system were considered as independent genes, in spite of the usually small negative correlation existing between them because of their belonging to the same locus.

One can indicate the range in which these variances should be found if "sampling" drift alone determined the variance of gene frequencies. To this effect we need an estimate of the number of intervening generations, and of the average "effective" population size. If one assumes that the earliest split leading to the differentiation of the present human groups took place at a date ranging from twenty thousand to five hundred thousand years ago, taking twenty years for the average generation time, the number of generations ranges from 1,000 to 25,000. Some anthropologists may want a longer time; probably none would accept a shorter time than the minimum given here. A correction has to be made for the fact that not all populations split at the same time, but that most splits occurred later; it can be shown that the correction is not large and will be ignored at this level of approximation.

The average population size that may have been effective in the period just mentioned is even less well known, and we may have to accept that it falls in an interval ranging from $N = 100$ to $N = 10{,}000$. It should be remembered that effective

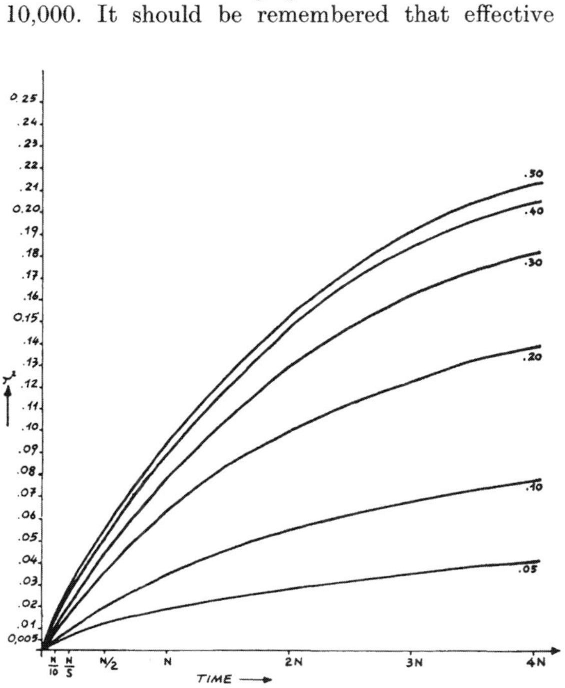

FIGURE 5. Variance of gene frequencies σ_p^2, expected as a function of time, under random sampling drift, with different initial gene frequencies p. Time is given as t/N, where t is in generations and N is effective population size. Computed from Kimura's (1955) distribution functions.

population size is not the direct population size of the isolate, but a fraction of it (see e.g. Kimura and Crow, 1963) which, for human populations, may be in the neighborhood of 60–150% of the census figure. Moreover, it can be decreased by the fact that averaging over different populations (in time and space) we have to take the harmonic mean of the single population sizes, and it may be increased because of the effect of migration, which obviously reduces the consequences of isolation.

The increase in the variance of gene frequencies with time, under random genetic drift alone, can be obtained from the equations published by Kimura (1955) for the distribution of gene frequencies. A computation of the variances from these distributions permitted us to derive Fig. 5 in which the variances are represented as a function of time, for various initial values of p. It will be noted that the equations can be given as a function of t/N, where t is the time in generations and N the population size, so that in the abscissa of Fig. 5 time is given as a function of N, for greater generality.

The theoretical curves expressing the increase of the variance of gene frequencies with time can be given, to a first approximation, by the formula:

$$\sigma^2 = \bar{p}(1 - \bar{p})(1 - e^{-t/2N}$$

where $\bar{p}$ is the mean gene frequency. The approximation is valid when t is of the order N or larger.

It will be noted from Fig. 4 that the variances of gene frequencies observable for the majority of genes are compatible with the variation due to random drift, even if effective population size were as large as 10,000, for time periods in the range that would be accepted by anthropologists.

A DIRECT INVESTIGATION OF GENETIC DRIFT

Perhaps many would agree on the following conclusion from the above analysis. If, in recent human evolution, drift has played a role unmitigated by stabilizing selection, then effective breeding size must have been large, say of the order of 10,000. If, on the other hand, effective size was small, say of the order of 100, then at least some of the polymorphisms we have investigated are buffered by some type of stabilizing selection, and therefore their variance is lower than would be expected under the effect of drift alone.

If we look for direct proofs of stabilizing selection in the common polymorphisms, we find some in the MN and ABO systems, (Morton and Chung, 1959; Chung and Morton, 1961). Of course it is possible that in modern populations, or at least in populations living with modern standards, selective pressures of the kind leading to the stabilization of polymorphisms have changed or even have ceased to exist. Also, measurements of selective values of the various genotypes that are in existence today are still rather imprecise. They are all of retrospective nature, and the ideal method of obtaining them, which is a prospective one, has not yet been used, to our knowledge.

On the other hand, estimates of effective breeding sizes are almost completely lacking, and methods of estimation are still in their infancy. The gaps between genetics and demography have not been bridged yet to any large extent, but it is clear that some demographic estimates should provide information of importance on this topic, and it is also clear that such estimates are more easily available for man than for any other organism.

One attempt was started, some ten years ago, on an analysis of a human population in which demographic knowledge was exceptionally good, or rather, could be made very good by putting in a certain amount of effort. In Roman Catholic countries, in the middle of the XVI century, parish priests were given a mandate, and a set of rules, for keeping the local demography. Most of the parish books are still in existence; and they usually prove reasonably reliable.

An area chosen for the investigation on the use of such information for genetic purposes was the valley of the Parma river, from the town of Parma to the mountains. It supplied a gradient of conditions, especially of population density. There is little immigration and there has been little change in the last three centuries especially in the less dense, higher parts of the valley (in terms of population and village sizes). Almost all villages and their churches were already in existence in the eleventh century, and the latest foreign settlement, of which a trace remains, must have taken place some four centuries before that time. The majority of the settlers may have been there for several millenia before.

Demographic analysis has been started but is not yet complete. A genetic survey of common markers was completed and gave a rather clear picture (Cavalli-Sforza & Conterio 1959, Cavalli-Sforza 1958, 1963a) which can be summarized as follows: there is no variation between villages for blood groups in the more dense part of the valley. In the less dense part, corresponding to the mountains, the variation between villages is significant for all blood groups tested (ABO, MN, Rh systems). There seems to be a tendency of the variation to increase with decreasing density of population.

It is clear that the data deserve a more refined treatment, if possible, especially with a view to testing the validity of available models.

There are three major types of model of what is called "population structure."

(1) The island model, in which populations of equal size (N) exchange genes at an equal rate, m, which is therefore independent of the distance between islands.

(2) The continuous model, in which population density is constant on a line, or on a surface, and diffusion of individuals occurs according to some law, usually the normal distribution. Migration is measured by the distribution of the distance between birth places of parent and offspring, an interesting demographic quantity which no demographer had used before. Both these models were put forward by S. Wright (1943, 1946).

(3) The "stepping stone model" was recently developed by Kimura and Weiss (1964); it has the advantage of assuming a discontinuous distribution of population as in the island model (an assumption much nearer to reality than the continuous model) and of taking account of the effect of isolation by distance as in the continuous model. In the stepping stone model, exchange between colonies occurs between adjacent ones. Colonies can be spread on a line, or in two dimensions; a third has been also introduced to include social conditions.

Because of some difficulties connected with the use of our migration data to this aim, we cannot yet easily use the stepping stone model; we therefore chose the next best, namely the continuous model.

The analysis was limited to the area showing a significant genetic variation. This area includes the communities of Corniglio, Monchio, Palanzano, and Tizzano, and is made up of about 66 villages ("frazioni") for a total of some 15,000 individuals. Some of the villages are extremely small; the unit we have considered as village is a little larger than the smallest cluster of houses: it is the parish, and of this there are 37 active today in the fraction of the valley which we have considered. A representation of most of the parish villages in which blood was sampled, and of the roads between them (Fig. 6) shows clearly how the geographic distribution is somewhat intermediate between the one- and the two-dimensional. Most of the villages are on roads at the bottom of the valleys; where the valley branches the road usually branches, but there are only few such branchings. An analysis considering the mode of increase with distance of the number of inhabitants has shown that the area behaves in a way which is exactly intermediate between a one-dimensional distribution, in which the number of inhabitants living at distance r is independent of r, with that of a two-dimensional distribution, in which it grows proportionately to the distance (Cavalli-Sforza and Conterio, 1959).

The variation in the gene frequencies of the parish villages in which blood was sampled is given

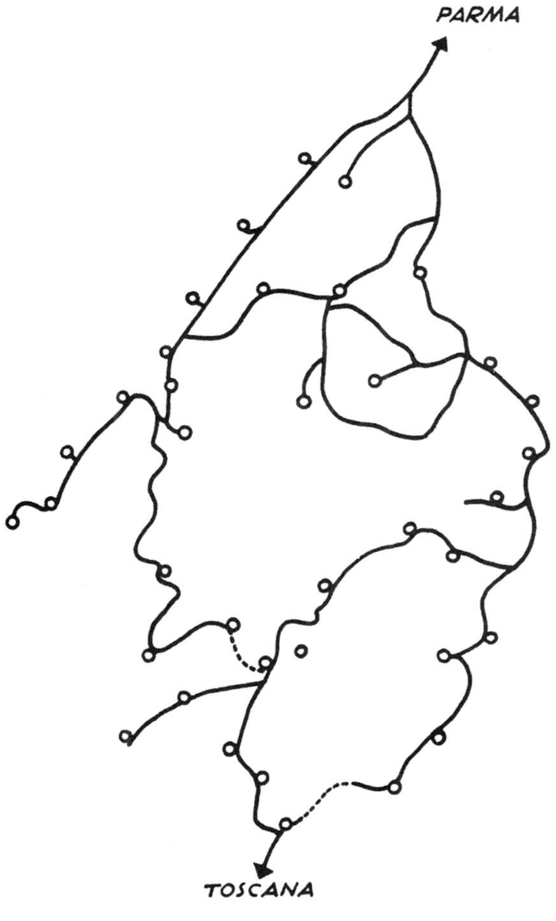

FIGURE 6. The network of roads in the upper Parma valley connecting the main parish villages from which individuals were sampled for the blood group gene survey. The geographical distribution is neither one-dimensonal nor two-dimensional.

by the chi-square analysis of Table 1. There is significant heterogeneity for 3 alleles out of 4 in the ABO system; the MN system, in which only the M, N alleles were tested, is also significantly heterogeneous, and all the Rh chromosomes (only four considered in the analysis) show significant variation.

In the multiallelic systems, the negative correlations between alleles may create some uncertainty in the interpretation and there may be difficulty in estimating the significance of variation for the rarest alleles. Limiting conclusions to the most frequent alleles, namely four only for the Rh system (R_1, R_2, r and the rest), one can compute a variance of the gene frequencies between villages from which the random sampling variance has been extracted, and divide it by $\bar{p}(1-\bar{p})$ to express it, as usual, as a fraction of the maximum it can reach. Such quantities are indicated in Table 1, for each allele that shows a variation higher than that due to random sampling. Averaging this variance over all alleles one obtains the quantity 0.036.

On the basis of this variance between 34 villages, the continuous model, assuming permanent pairs and exclusion of brother-sister mating, would assign a "neighborhood size" of 140 if the distribution of the population is linear, and of 55 for a distribution in two dimensions. We are grateful to Dr. F. Scudo who carried out the necessary computations using S. Wright's model (1946).

One should not, however, compare these neighborhood sizes directly with villages. In fact, in the continuous model, in which population density is constant, there cannot exist a "village", by definition. As a substitute, a "neighborhood" is taken as the number of individuals living in a strip (linear model) or a circle (area model) within a distance of 2σ from the center, σ being the standard deviation of the parent-offspring migration distribution. This corresponds, in the linear model, to the strip from which, with probability 92.4%, the parent of an individual born at the center comes, and in the area model to the area from which, with probability 86.5%, a parent of the individual at the center comes.

A neighborhood might be just a little larger than a village in our case. In fact, 73% of its individuals have their mother, and 82% have their father, born in the same village. On average, 77.5% of the parents are born in the same village. From these migration data, it appears that a "neighborhood" should be just a little larger than our village. In fact, extrapolating on the basis of a normal diffusion curve, this population of parents would correspond to a strip of 1.21 σ for a linear

TABLE 1. HETEROGENEITY BETWEEN GENE FREQUENCIES OF VILLAGES IN THE PARMA VALLEY

Blood group alleles	χ^2	df	Significance	$\frac{\sigma^2}{pq}$
MN	102.76	36	yes	0.048
A_1	98.74	33	yes	0.052
A_2	40.23	33	no	—
B	76.08	33	yes	0.034
O	110.25	33	yes	0.061
$r^{(+)}$	81.77	33	yes	0.038
R_1	75.46	33	yes	0.033
R_2	54.54	33	yes	0.017
other Rh	80.01	33	yes	0.037
			mean:	0.0356 ± 0.006

$^{(+)}$Rh frequencies computed with inefficient method.

model and of 1.73 σ for an area model. A neighborhood would then correspond to 1.65 (2/1.21) villages for the linear case and 1.33 for the area case. Our typical village has some 400 individuals, considering a parish as a village (respectively 230 for a "frazioni"); correcting it on the basis of migration data as mentioned above, it would give rise to a neighborhood of 660 (respectively 380) for the linear case, and 530 (306) for the area case.

A further correction is necessary to take account of the fact that the neighborhood refers to reproductively mature individuals. Using here the correction suggested by Kimura and Crow (1963), this would decrease the neighborhood computed on the basis of village size and migration data at most to 266 for the linear and 214 for the area case. These values are always higher than the values computed from the variance for blood group genes, but the difference is not large.

The difference could be accounted for, qualitatively, by one or more of these explanations:

(1) There are social limitations to intermarriage, thus reducing further the "isolate size";

(2) There has not been time for evening out differences that originated at the beginning, when villages may have had a smaller initial size;

(3) Selective differences might exist between the villages. Of course, it is strange that they should exist for *all* the genes tested.

On the other hand, stabilizing selection would act in the opposite direction, namely decreasing the variance between villages. These data do not, therefore, suggest that it is present and important; on the contrary, they suggest that the effect of drift may be larger than it would be inferred to be from unsophisticated demographic data.

It may be worth adding that the theory of continuous population structure provides a direct estimate of neighborhood size, which should not be applied in this case. Such estimates use population density and the standard deviation of migration distributions (σ). In the linear and area cases respectively, they are: $N = 3.5\ d\sigma$; $N = 12.6\ d\sigma^2$. Such estimates would be very severely affected by the fact that the distribution of migration is not normal, but is far from it, having an almost infinite variance. Distributions which can fit the observed data (see Fig. 7), both on theoretical and empirical grounds, have been discussed elsewhere (Cavalli-Sforza, 1963b). We have found it useful, in the present circumstances, to assume that the distributions are sums of normal distributions. A minimum of three normal distributions is necessary (Fig. 8) to represent the observed data satisfactorily. This may seem an exercise in the acrobatics of curve fitting; but it has the advantage, in fact, that it simplifies considerably the mathematics of theories connected with gene diffusion over several generations (see later). It also takes account of the fact that there is heterogeneity of means of transportation, which can in principle be represented in the same way.

When the observed distributions are fitted in these ways using the sum of two normals, or the sum of three normals, or more complex but related frequency curves, the first component (short range migration) has overwhelming importance. It absorbs about 80% of the migrants, averaging between the father-offspring and the mother-offspring curves. Its standard deviation, 0.65,

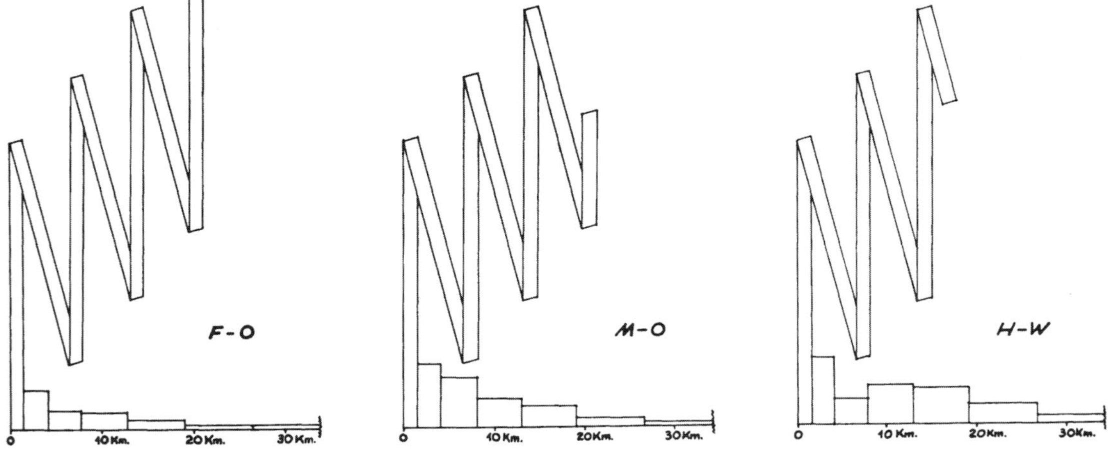

FIGURE 7. Distribution of migration distances in the Parma valley (data in Cavalli-Sforza, 1963b, sample of living population). F–O: distance between birthplaces of father and offspring; M–O, same for mother and offspring; H–W same for husband-wife.

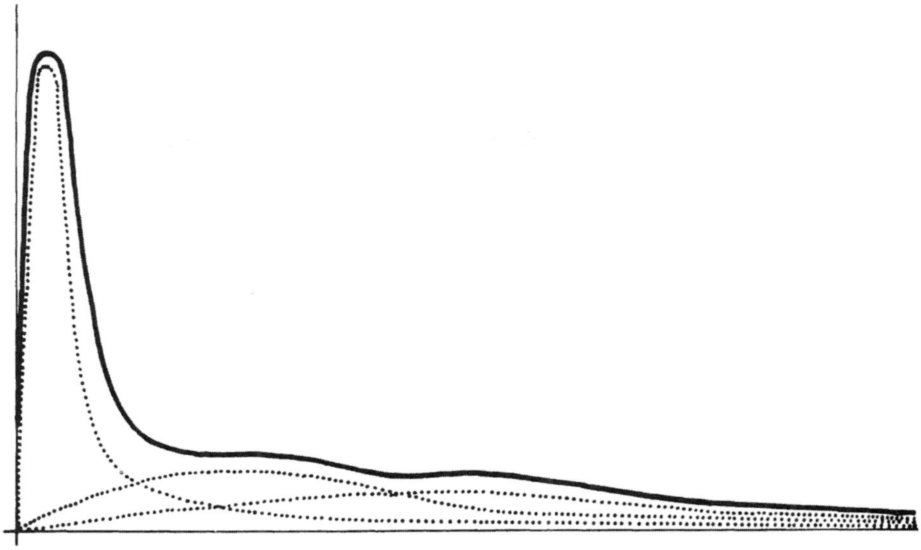

FIGURE 8. A way of reproducing the distribution of migration distances given in Fig. 7 as the sum of three "normal" distributions. By "normal" is meant here distribution of distances from the center of a two-dimensional normal uncorrelated surface.

is at least a couple of orders of magnitude lower than that of the medium-range component; it corresponds fairly closely to the value that would be calculated, given the neighborhood size obtained above from the variance in blood group gene frequencies, the population density and the formulas mentioned before.

It is tempting to assume that the short range migration is the one that matters most when the distribution of migration distances is not a simple normal, but can be analyzed into few normal components.

As a conclusion, the size of the elementary population, or population effective size, is an important quantity, which is difficult to estimate. In the case considered it is almost satisfactory to identify it with the "village", namely the parish. Here, the parish village happens to be fairly homogeneous socially, at least as long as it is small; people marry usually in the village and children are normally born in the same village as the parents. But some of the villages, the smallest ones, are clearly too small to be considered as independent units, while some are too large to be considered as homogeneous units. When villages are larger than those considered here (which usually is the case) homogeneity is unacceptable, if only from a social point of view. No standard demographic quantity could therefore be used to estimate effective population sizes.

It is, therefore, important to consider another completely independent approach to the problem of estimating population effective sizes, which was suggested some time ago by Dahlberg (1929).

FREQUENCY OF CONSANGUINEOUS MATINGS AS A WAY TO ESTIMATE "POPULATION EFFECTIVE SIZE"

"Isolate sizes" were estimated by Dahlberg, using the frequency of consanguineous marriages. The rationale of the method is that if there is no tendency to choose or avoid consanguineous individuals as mates, the probability of a consanguineous marriage is the ratio of the number of consanguineous individuals to the total number of individuals (N) among which a spouse is chosen. This definition of "isolate size" is therefore very near to that of population size, although a finer comparison may be necessary.

An uncritical use of Dahlberg's formulae could lead to completely erroneous results, as shown by Morton (1955). It is, however, possible to predict to a considerable degree of precision the frequencies of various consanguineous marriages, once the factors affecting them have been identified (Barrai et al. 1962). The following three factors are of importance: age at marriage, migration, and some psychological and social factors which differ from place to place. The first two groups of factors can be studied independently using demographic sources, and the conclusion obtained can help the prediction of the expected frequencies of a great number of consanguineous marriages.

Consanguineous degrees may require a dispensation from the religious authority. In particular, for Roman Catholics, marriages of uncle-niece, aunt-nephew, first cousins, first cousins once removed, and second cousins today require a dispensation (see Moroni, 1960a, b). Dispensation records usually include genealogies from which a great number of different trees can be distinguished. In particular, for our purposes, we distinguished four types of first cousins and four of uncle-niece, aunt-nephew; twelve of 1½ cousins, nine of second cousins.

Theories have been developed (Hajnal, 1963; and independently by our group in collaboration with Kimura) to predict the frequencies of these 29 types on the basis of the following quantities: mean and variance of generation times of males and females (τ_m, τ_f, σ_m^2, σ_f^2); variance of age difference between sibs, and between husband and wife (the sum of these will be called σ_T^2); finally, as a minimum picture of migration, the proportion of fathers p_m (and of mothers p_f) whose offspring is born in the same village; as well as the proportion of marriages in which husband and wife are both born in the same village. Alternatively, one may want to describe more fully the migration distribution.

On the basis of these demographic quantities, which can be estimated from independent data, and also directly from the body of consanguineous marriages, one can test if the frequencies of consanguineous matings are consistent with this model, thus determining or ruling out the importance of social or psychological factors favoring, or barring, consanguineous marriages of some specific type. One can also compare the estimates of the demographic parameters mentioned above and obtained from demographic sources with those obtained by fitting the theory to the observed marriage frequencies.

We have proceeded to an analysis of consanguineity dispensations collected over 100 years in the upper Parma valley, in the same area over which our blood samples were collected. About 2300 marriages were thus obtained. We limit here the analysis to marriages as close or closer than second cousins, in order to avoid some difficulties connected with 2½ and 3rd cousin data; also to marriages with only one type of consanguineity. The observed frequencies are given in Table 2, along with expected frequencies computed from the demographic parameters (which were actually estimated from a larger sample) given in Table 3.

The agreement between observed and expected frequencies (Table 2) is not perfect because of one source of bias, the existence of which is known to us from a more complete analysis of these and other

TABLE 2. FREQUENCIES OF CONSANGUINEOUS MARRIAGES, UPPER PARMA VALLEY 1850–1950

	No. of male ancestors in branch leading to		Absolute frequencies	
	husband	wife	observed	expected
uncle-niece	–	0	5	6.47
	–	1	4	2.68
aunt-nephew	0	–	1	0.23
	1	–	0	0.08
first cousins	0	0	109	110.41
	1	0	99	99.55
	0	1	157	142.12
	1	1	96	153.37
1½ cousins husband in shorter branch	0	0	21	11.74
	1	0	30	28.67
	0	1	20	14.32
	1	1	55	39.78
	0	2	5	4.11
	1	2	22	12.87
1½ cousins wife in shorter branch	0	0	3	2.01
	0	1	14	7.16
	1	0	2	2.06
	1	1	20	8.23
	2	0	2	0.50
	2	1	4	2.25
2nd cousins	0	0	41	43.64
	1	0	63	87.57
	2	0	47	40.04
	0	1	93	109.83
	1	1	227	247.03
	2	1	129	125.91
	0	2	60	61.76
	1	2	171	154.94
	2	2	107	87.67
$\chi^2_{[20]} = 88.26$			1607	

TABLE 3. PARAMETERS FOR THE THEORY ON FREQUENCY OF CONSANGUINEOUS MARRIAGES

			estimated from consanguineous marriages Parma Valley 1850–1950	estimated from demographic data obtained from a sample of present day population living in same area
Generation time, males	mean	τ_m	36.10	33.24 ± 0.19
	variance	σ_m^2	53.32	46.60
Generation time, females	mean	τ_f	30.95	28.73 ± 0.16
	variance	σ_f^2	42.59	33.78
Variance of age difference between sibs			53.08	25.78 } 59.09
Variance of age difference between mates				33.31 }
Probability of non-migration per generation				
	males	p_m	0.845	0.821 ± 0.019
	females	p_f	0.696	0.727 ± 0.070

data (Barrai et al. 1962) and which tends to alter the relative ratios, especially among the four types of first cousins, which are also the most frequent ones.

Even though the fit is not perfect, it can be considered satisfactory at this stage. It is also interesting to compare estimates of the parameters obtained from a sample of the present day population living in the area of the Parma valley, with those obtained by fitting the formula given before to a sample of consanguineous marriages which took place from 1850 to 1950 in the same area, and of which those given in Table 2 are a fraction.

It will be noted that there is some divergence between the two sets of values. Differences are in the expected direction. Thus, the generation times computed from consanguineous marriage data refer to the ancestors of the consanguineous mates and, therefore, to a period of about two hundred years. It is not surprising, therefore, to find a longer generation time and a larger variance than in the living population, in which there has been a considerable reduction of birth rates and, with it, of age at childbirth.

It is then possible to compute N, a measure of the effective breeding size, from the relative frequency of consanguineous marriages. The estimate thus obtained varies from 120 to 240, depending on assumptions made. It is thus still within the range obtained from the evaluation, with the continuous model, of the variance of blood group genes (60–130). This agreement makes it reasonable to think that, if one can have an understanding of the factors determining the pattern of consanguineous marriages, their frequencies can be used to estimate population effective sizes, avoiding difficulties created by social stratification and by geographic distribution.

CONCLUSIONS AND SUMMARY

If, as a working hypothesis, it is assumed that genetic diversity among present-day populations has been largely caused by random processes (both random genetic drift and the random variation of selective values), and if adequate data are available, it is possible to estimate the path of evolution. A preliminary analysis of human data, based on this admitted over-simplification, has given a picture of human evolution which is probably acceptable to many anthropologists.

Kinetic considerations have been advanced to show that drift might have determined the present diversity for polymorphic loci even in the presence of large effective population sizes. Furthermore, a search for drift in a modern Italian population shows the presence of variation which could be ascribed to drift. The fact that it sometimes exceeds what would be expected from drift alone maybe due to demographic data not considering the estimation of the decrease in effective population size owing to social stratification, and other reasons.

An analysis of the frequencies of consanguineous marriages in the same area, after taking into account age and migration effects, gave estimates of the effective population sizes of comparable magnitude to those obtained from the variances of the blood-group genes.

ACKNOWLEDGMENTS

We wish to take this opportunity to thank Drs. F. Conterio, A. Moroni, G. Zei and L. Zonta for their invaluable help in various aspects of this work and Dr. A. E. Mourant for help in securing the data which were used for Figs. 2, 3, and 4.

This work has been supported in part by a grant from the Rockefeller Foundation and for another part by grants from the U.S. Atomic Energy Commission and by EURATOM-CNR-CNEN Contract no. 012-61-12 BIAI.

REFERENCES

BARRAI, I., L. L. CAVALLI-SFORZA, and A. MORONI. 1962. Frequencies of pedigrees of consanguineous marriages and mating structure of the population. Ann. Hum. Genet. *25:* 347–376.

Cavalli-Sforza, L. L. 1958. Some data on the genetic structure of human populations. X Intern. Congr. Genet. *1:* 388–407.

——. 1963a. Genetic drift for blood groups. *In* E. Goldschmidt (ed.) The genetics of migrant and isolate populations. The Williams and Wilkins Company, Baltimore.

——. 1963b. The distribution of migration distances, models, and applications to genetics. p. 139–198. Entretiens de Monaco en Sciences Humaines, Première Session, 24–29 Mai, 1962, Human Displacements.

Cavalli-Sforza, L. L., and F. Conterio. 1959. Analisi della fluttuazione di frequenze geniche nella popolazione della Val Parma. Atti X Ass. Genet. Italy *V:* 1–11.

Cavalli-Sforza, L. L., and A. W. F. Edwards. 1963. Analysis of human evolution. Paper presented at the XI Intern. Congr. Genetics, The Hague.

Chung, C. S. and N. E. Morton. 1961. Selection at the ABO locus. Am. J. Hum. Genet. *13:* 9–27.

Dahlberg, G. 1929. Inbreeding in man. Genetics *14:* 421–454.

Edwards, A. W. F., and L. L. Cavalli-Sforza, 1963. The reconstruction of evolution (abstr.). Heredity *18:* 553.

——, ——. 1964. A method for cluster analysis. Biometrics, in press.

Hajnal, J. 1963. Concepts of random mating and the frequency of consanguineous marriages. Proc. Royal Soc. B *159:* 125–177.

Kimura, M. 1954. Process leading to quasi-fixation of genes in natural populations due to random fluctuation of selection intensities. Genetics *39:* 280–295.

——. 1955. Random genetic drift in multi-allelic locus. Evolution *9:* 419–435.

——. 1964. Diffusion models in population genetics. J. Appl. Prob., in press.

Kimura, M., and J. F. Crow. 1963. The measurement of effective population number. Evolution *17:* 279–288.

Kimura, M. and G. H. Weiss. 1964. The stepping stone model of population structure and the decrease of genetic correlation with distance. Genetics *49:* 561–576.

Moroni, A. 1960a, Analisi metodologica del rilievo della consanguineità. Parte prima—Legislazione e materiale religioso e civile italiano per il rilievo della consanguineità. Folia Hered. et Pathol. *9:* 199–247.

——. 1960b. Sources, reliability and usefulness of consanguineity data with special reference to Catholic records. Intern. Symp. on Demography and Genet., Geneva.

Morton, N. E. 1955. Non-randomness in consanguineous marriages. Ann. Hum. Genet. *20:* 116–124.

Morton, N. E. and C. S. Chung, 1959. Are the MN blood groups maintained by selection? Am. J. Human Genet. *11:* 237–251.

Wright, S. 1943. Isolation by distance. Genetics *28:* 114–138.

——. 1946. Isolation by distance under diverse systems of mating. Genetics *31:* 39–59.

Paper 29 (1965)

Analysis of human evolution
Proceedings of the 11th International Congress of Genetics, The Hague, 1963. *Genetics Today* 3, 923–933. (With L. L. Cavalli-Sforza)

Note from A. W. F. Edwards: This joint paper was read by LLCS at the XI International Congress of Genetics at The Hague in the Netherlands in September 1963. It applies the method of minimum evolution to data on five blood-group systems from 15 human populations. The resulting tree is displayed. It also mentions the 'additive tree' method, which had been the first suggestion of Cavalli-Sforza when he initiated the research.

ANALYSIS OF HUMAN EVOLUTION

L. L. CAVALLI-SFORZA

and

A. W. F. EDWARDS

International Laboratory of Genetics and Biophysics (Pavia Section)
Istituto di Genetica, Università di Pavia, Italy

In the early days of biology, when the concept of evolution was young, the tracing of relationships and ancestries of living organisms became a common exercise. A forest of trees was thus produced, which purported to represent the lines of descent of families, genera and species.

But this work could hardly be satisfactory to the critical worker. No single character, or simple combination of characters, would usually be adequate for the purpose: no precise guiding principles or methods were available, and in any case the synthesis of a large number of observations and characters proves difficult for the human mind and is likely to be largely subjective.

In recent times, the development of research on what is often called "numerical taxonomy" has shown how to deal with a large number of characters observed in a large number of species. Various methods have been devised to measure overall similarities or differences between objects (be they living organisms or not), and to group them in hierarchical categories which resemble, and are probably close to, the real trees of descent. The primary aim of a classification method is, however, different from that of tracing evolutionary histories.

We have asked ourselves whether the evolution of a group of populations of living organisms can be handled by exact methods, so as to be able to measure the amount of evolution that has separated any two populations, and to reconstruct the path which has led to the differentiation into several populations—be these separated by specific or subspecific differences.

Clearly, one must ask guidance, in this aim, from genetic principles; one must define a scale and a unit of measurement and search for suitable methods. While developing these methods, we also found that human data could be a useful testing material, in addition to having that special attraction which inevitably attaches to the history of our own species. Few, if any, other organisms show, in fact, a similar wealth of genetic observations on natural populations. We attempt to present here the results of our work on this line, even if they are very preliminary and caution in their interpretation is still essential.

First we want to define our scale and unit of measurement. We wish to measure the amount of divergence that has formed in the evolution of two populations. The simplest unit that presents itself for consideration is the "*gene substitution*", taking the word gene for "unit of inheritance" so as to avoid being restricted to gene mutations. Many gene substitutions may have taken place in the process that has led to the formation of two different populations from the original one: but many may also be incomplete at the time

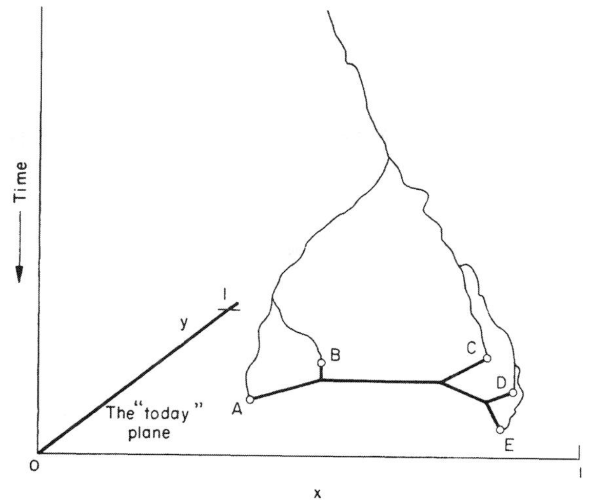

Fig. 1. An evolutionary tree

of analysis, and differences in gene frequencies may count as a fraction of gene substitutions, to be added to those substitutions that are already complete. Many blood group genes in Man offer examples of such incomplete gene substitutions, which are perhaps on the way to completion.

It is clearly difficult to estimate the *total* number of gene substitutions which differentiate two populations. But one can use a suitable *sample* of genes to measure the relative divergence between populations, if such a sample is unbiased, and sufficiently large, it should supply an index proportional to the total amount of evolution that has taken place in the separation of two populations. Such an index of the amount of evolution should therefore be the sum of gene substitutions, or fractions thereof, by which two populations differ for the genes considered.

It will be useful, for understanding the process of reconstructing an evolutionary tree, to make recourse to geometric representation. The process of evolution and differentiation into physically or otherwise separated populations can be represented in a space using as many dimensions as are the genes involved, plus one dimension for time. If we want a graphical aid we must consider only two genes, the frequencies of which will be the x- and y-axes while the third axis will be used for time. Thus, Fig. 1 shows the possible differentiation of five populations from an original one. The lines representing the process are actually discontinuous, but as soon as the number of individuals in the populations is not too small they become almost continuous.

Actual data will be points in such a space, and the problem of reconstruction might seem just that of interpolating a figure, similar to an inverted tree, through these points. But, actually the problem is more complicated and is best split into two: the cases in which gene frequencies are available also for the past and those in which they are available only for present-day populations. The second case is clearly the one for which material is available and at present we will restrict our consideration to it.

Data will then be given only for the plane corresponding to time zero, or the "today" plane, and on this plane it will be possible to represent the tree of descent as a net, which will terminate in the n points corresponding to the n species and will be made of $2n$-3 "edges". The shape of the net on the "today" plane will be, in a way, a "projection" of the net in the space–time representation. From such a projection one can hope to get a correct representation of the splits between populations that have taken place in the real space-time net.

How can one reconstruct such a net on the "today" plane from the actual data? It is worth noting that, given a set of n points, the number of nets of topologically different shapes that can connect them is very high as soon as the points are not too few With ten populations the number of possible nets is over two million. We therefore need a criterion to choose the best between them, as well as a practical way to find it.

We have used two different, but related, criteria. One of them is easier to understand intuitively, and can be called the "principle of minimum evolution". the best net is that which involves the least sum of edges to connect all the points. Its physical counterpart would be the net formed by the minimum amount of string necessary to connect all the points in an n-dimensional space.

To apply this principle the work of the computer can be considerably helped by first reconstructing the approximate shape of the tree. A method of cluster analysis * was devised which separates the group of populations into smaller and smaller clusters by a succession of binary splits. The reconstruction of the net in the multidimensional space and the computation of the coordinates of the nodes follows.

The other criterion for finding the best net on the "today" plane relies formally on the assumption that evolution is a random process, namely that in separate populations gene frequencies evolve independently. Under these conditions, the total difference existing between two populations is the sum of the differences accumulated in evolving from a common ancestral population. The pathway connecting two points in the net is then equated to the actual difference between them. A system of a large number of equations (as many as the number of pairs of populations being examined) is then created, where the unknowns are the edges of the net. For more than three species, the system has more equations than unknowns and can be solved by methods such as least squares or maximum likelihood, which minimize discrepancies between observed and expected differences in population pairs. We will call this criterion that of the "additive tree".

Both criteria should, in general, lead to similar results.

They both fail to reproduce the correct shape of the space–time tree when (1) hybridization between subspecies takes place, and (2) convergent or parallel evolution occurs. The first possibility is amenable to quantitative analysis, but we will not consider it here. The second is more elusive and more dangerous. its actual importance is, however, a matter of conjecture.

We have no time to expand on the logic of the theory, or to give here the details of the methods used for reconstructing evolutionary trees, nor to mention how some of the practical difficulties met when handling multiallelic loci were solved, by using a method suggested to us by R. A. Fisher. The sequence of operations is summarized in Fig. 2.

* See A. W F. Edwards, L. L. Cavalli-Sforza, Preprints of the Vt Int. Biometric Conference, Cambridge, 1963.

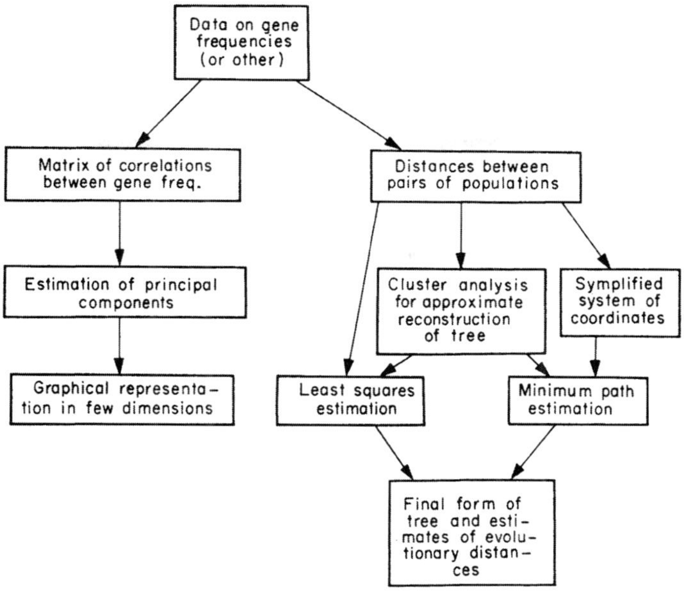

FIG. 2. Reconstruction of phylogenetic trees: flow chart of the analysis.

We will rather concentrate on some results obtained to date in the analysis of observations in Man. For a survey of the world we had to restrict our choice to a limited number of populations. The amount of numerical work, even if it is carried out by computer, is heavy and increases very fast with the number of populations.

We hope however to increase substantially in the near future the number of populations with which we can cope.

We decided to select fifteen populations, three for each continent and as wide apart as possible. All populations had to meet the criterion, that a sample of reasonable size had been tested for all the five most useful blood group systems. These were ABO (with four alleles); Rh (five antisera, eight chromosomes of which only six informative); MNS (four chromosomes), Diego and Duffy with one antiserum each. A total of eighteen alleles was thus available. Other blood group systems are either insufficiently tested, or are fairly homogeneous in their world distribution. For the choice of the data and their extraction from the literature we must thank Dr. Mourant and Mrs. Sobczak for invaluable help.

Distances between these populations, taken in pairs, were then computed. They summarize the evolutionary divergence over these five blood group systems.

It would be out of place to show here the table of such distances, as it would be exceedingly difficult to extract from it any synthesis of the data. Rather, it may help to give an overall view of the 15 populations if we use a graphical aid to portray synthetically the blood group data referring to them.

The method we shall use to this aim has been known in the statistical literature for a while, but, in the absence computers it was used only scantily, so that it is not widely known.

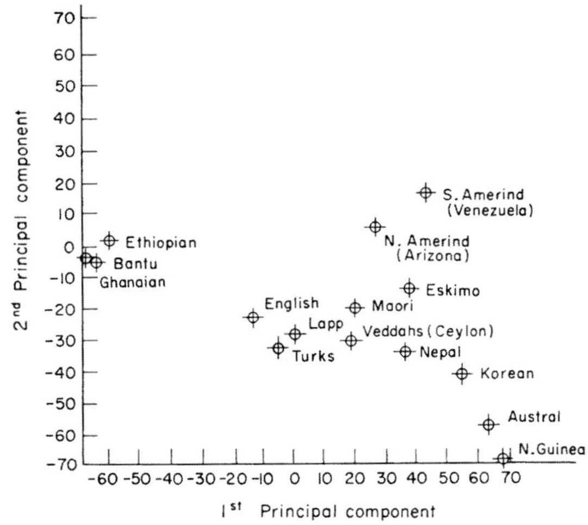

FIG. 3. Chart of first two principal components synthesizing frequencies of five blood group systems (A_1A_2BO, Rh, MNS, Fy, Di).

It is called the method of principal components and permits us to reduce the number of dimensions used to represent points from a multidimensional space to as few as required, minimizing and at the same time measuring the loss of information thus incurred.

If we reduce our original number of dimensions to two, we suffer a loss of half of the total information but can represent our fifteen populations in a graph (Fig. 3).

In this graph it is seen that most populations from the same continent tend to cluster together: as the three African groups, the three European ones, the three American ones and the Asiatics, while of the three groups from Oceania the Maoris, as might be expected, are separated from the Australians and Melanesians.

If we try to recover some of the lost information we can introduce a third principal component and thus have a three dimensional model in which 66 per cent, or about 2/3, of the total variation is recovered (Fig. 4). The clustering of some populations is now more visible. It is in some such space that we must draw our net connecting the points for reconstructing the evolutionary pathway, but we will use for it all the dimensions.

The reconstruction of the phylogenetic tree was carried out by least squares and eventually by minimum path. The segments are proportional to the quantity of evolution; they give the number of gene substitutions for five selected loci. The results are largely similar to those visible in the graph by principal components, although the methods are independent, and are given in Fig. 5.

The greatest difference is the one that separates Australians and Africans. It should be noted that the origin of the tree, which we have set near to the centre of gravity of the system, is arbitrary, but the splits and the computed segment lengths would be unchanged wherever the origin.

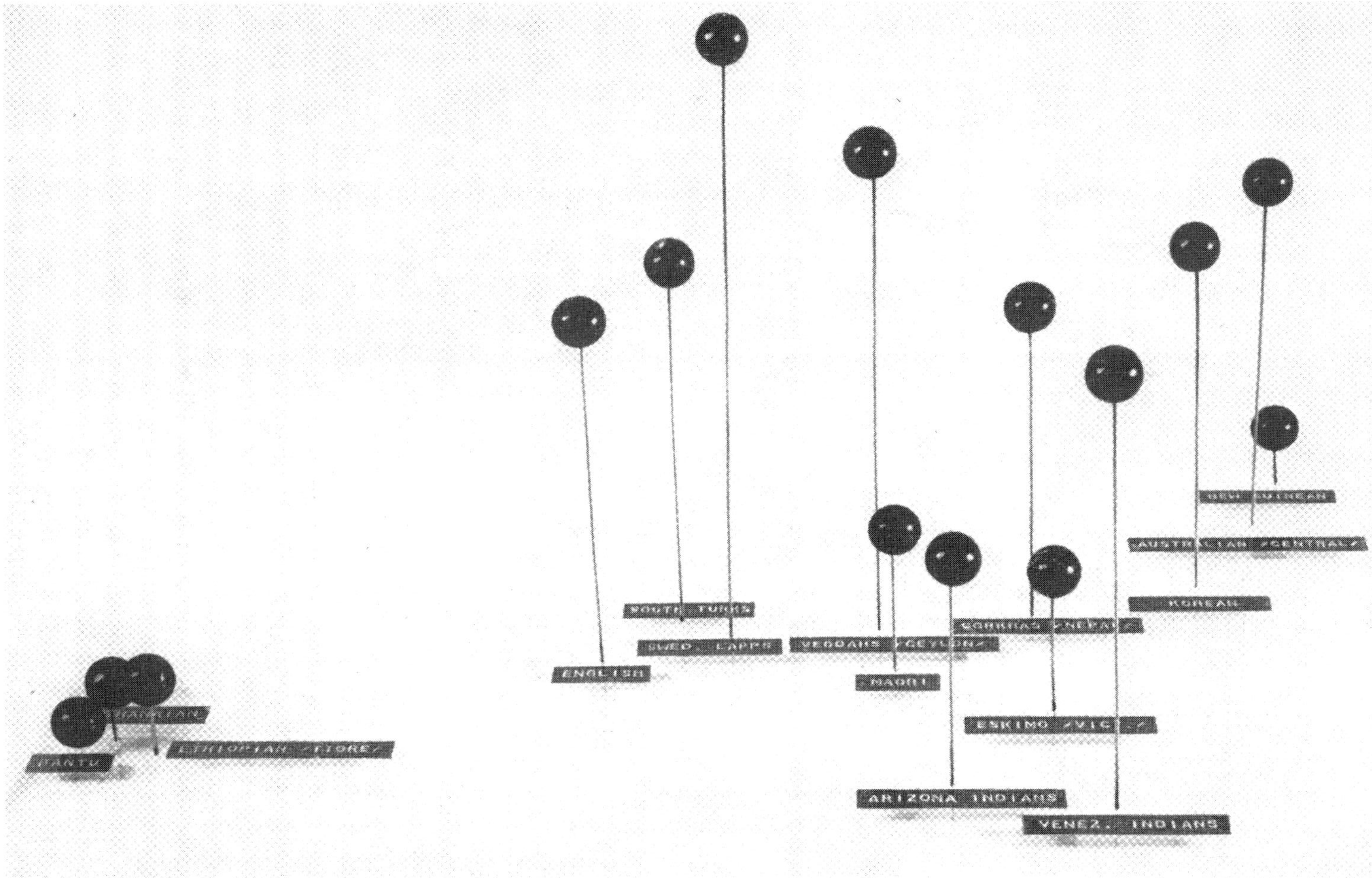

BANTU
ENGLISH
SWED. LAPPS
VEDDAHS /CEYLON/
MAORI
ARIZONA INDIANS
VENEZ. INDIANS
KOREAN

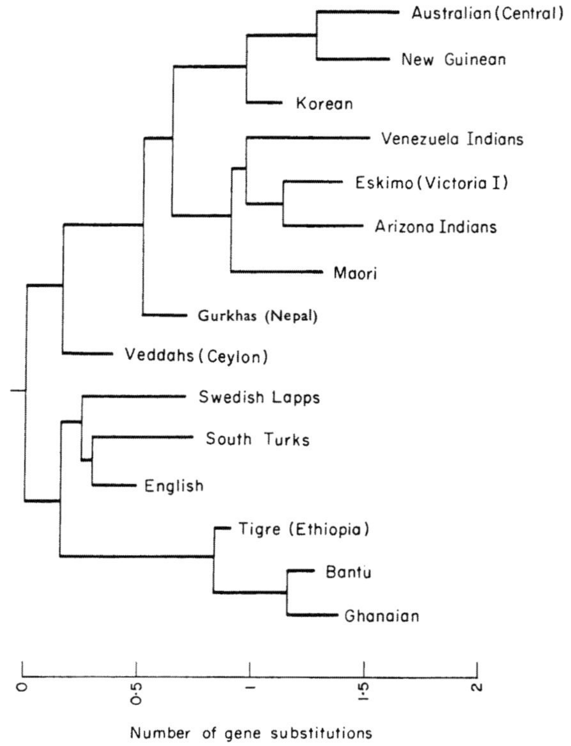

FIG. 5. Tree computed from blood group gene frequencies.

The clustering of the populations from the five continents is the same in the tree as in the analysis by principal components. Note in particular the wide separation between North and South American Indians, the position of the Maoris, fairly close to these and to representatives of non-Mongolian Asiatics. Quite remarkable is the vicinity of Mongolians and Australians: it will be interesting to see if this will remain after introduction of further Mongolian groups into the analysis. The common origin of the three European groups, even if they have been widely separated between themselves, is also noteworthy. All in all, the tree shows three main branches. Europeans, Africans and Asiatics, the latter including Americans and splitting into a variety of subgroups.

The addition of new populations is not likely to alter drastically the shape of the tree, as we have seen by adding to the original 15, twenty other populations, mostly American aborigines, Melanesians and Polynesians. The clustering of populations belonging to the same continent in the map obtained with first two principal components is quite satisfactory, except for Polynesians which are scattered, exactly in the way one might expect to find a group of small populations, largely exposed to drift as well as to hybridization with a variety of neighbours. There is a remarkable similarity between the blood group map

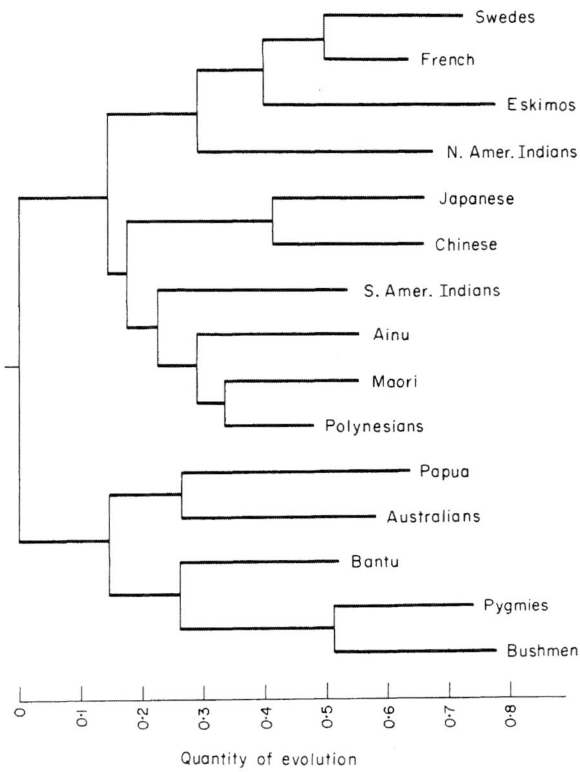

FIG. 6. Tree computed from anthropometric characters.

thus obtained, and the geographical map; this fact is not too surprising, as physical proximity implies, on average, higher rates of gene diffusion. But the acid test of the tree will come from the throwing of new genes into the analysis. Some data on other polymorphisms that we have so far neglected will be considered in the near future, but a substantial increase of the number of genes must await the detection of new polymorphisms and the augmentation of data for already known ones.

Meanwhile, it is worth considering an independent source of data: those obtained from physical anthropology. These are, after all, the basis from which current knowledge on human races has been built, and in principle, they can be subjected to the same kind of analysis as the blood groups and, in general, data on gene polymorphisms.

There are some difficulties, however, due to the variety of techniques used, of characters measured and their variations, the correlations between characters which greatly reduce their utility, and the lack of a critical collection of the data available, which are scattered in an immense literature.

Another difficulty is more fundamental in nature. It refers to the weighting of characters. In the case of gene frequencies, we have not hesitated to give equal weight to each gene (or, perhaps more exactly, to each allele). We are interested, in the end, in knowing the sum of gene substitutions. When evaluating differences in anthropometric traits between two populations, we would therefore like to give each character a weight proportional to the number of gene substitutions. In fact, we would like to transform a difference in means into a number of gene substitutions. This can be done, and only very roughly, for very few traits. In the absence of such information the weighting of characters is largely left to either subjective or arbitrary appraisal.

In spite of these difficulties, we have given the problem a preliminary analysis by taking 15 populations, as similar as possible to those analyzed before. Their means for 26 anthropometric characters (listed in Table 1) were extracted from Martin's *Handbook af Physical Anthropology* by Dr. F. Conterio.

TABLE 1. *List of 26 Anthropometric Characters used for Mapping 15 Human Populations*

Stature	Skull: height
Weight	vertical longitudinal index
Relative sitting height	fronto-parietal transversal index
Relative shoulder width	Facial index
Relative pelvic width	Prognathism angle
Chest girth	Nose height
Relative arm length	Nose width
Humerus-radius ratio	Nasal index
Hand relative length	Orbital index
Hand relative width	Skin colour
Femur-tibia ratio	Hair thickness
Scapular index	Mouth width
Skull. transversal longitudinal index	Ear length

The analysis of anthropometric data shows both similarities and discrepancies with the blood groups data. In general, populations from the same continent tend to cluster together, though less sharply The major discrepancy is that the divergence between Australians and Africans is now much less pronounced, while that between Africans and Europeans is emphasized. The discrepancy between South and North American Indians is accentuated, and so is the similarity between these two groups and Mongoloid populations. The similarity between North American Indians and Europeans is certainly unexpected. In general, there is a fair amount of agreement between blood group data and anthropometrics, but there are also marked discrepancies (see tree in Fig. 6).

The gaps between these two approaches indicate the way to future research. A larger body of data has to be scrutinized. The experience of numerical taxonomists shows that, when a sufficient number of characters is considered, independent sets of characters tend to give the same classification. So it should be here. The blood group data, or the anthropometric data, or possibly both, will have to be enlarged. The former can be increased but slightly on the basis of present knowledge: the analysis can easily be extended, however, at least to a number of other populations. With anthropometric data, a more elaborate survey of the literature than that done so far may increase the accuracy of the results by increasing the number of characters and using within-groups correlations. Should the two approaches—using genetic markers and anthropometric data—give, eventually, a more complete agreement, considerable support could be gained for our conclusions. It should also be remembered that one particular section of the anthropometric data, that relating to bone measurements, forms our biggest hope of extending the method to reconstructing the space-time tree, in view of the availability of the fossil data.

We have given some of the reasons for our belief that the history of descent of living beings can be reconstructed by exact methods.

It is possible that only a fraction of the conclusions reached here on the evolutionary history of Man will stand the test of future research. On the other hand, it is clear that the longest segments of our trees are those which should be considered as best established, and the shortest ones those that are more likely to be altered by the addition of new data. We hope it will be possible to reinforce considerably our conclusions in the near future.

DISCUSSION

N. E. Morton *(U.S.A.)*: Can you make allowance for the occurrence of hybridization?

L. L. Cavalli-Sforza: Hybridization will be represented by a loop in the evolutionary tree, and this will upset the additivity in that section of the tree: sectional non-additivity may, therefore, reflect hybridization. We have not yet proceded to cover this aspect of the problem, but intend to do so soon.

N. E. Morton *(U.S.A.)*: Can you assign a standard error to the edge lengths?

L. L. Cavalli-Sforza: Yes, for a given topological form of tree these are given by the least-squares estimation procedure (subject to certain independence conditions which may only be approximately satisfied). The estimation of the topology, however, is also open to error, and this must not be overlooked.

J. B. S. Haldane *(India)* Since the amount of selection needed to change a gene frequency from 0.0001 to 0.1 is considerably more than that needed to change it from 0.1 to 0.9, would not the logarithm of the ratio of gene frequencies, in some cases, give a better measure of the "position" of a population than the gene frequency? Such a method is, of course, inapplicable where genes are apparently completely fixed, or for characters whose genetics has not been analysed.

L. L. Cavalli-Sforza: We have considered this measure, but apart from the difficulties at extreme gene frequencies, it appears hard to extend it to multiple alleles. Perhaps Dr. Edwards would like to add a word on this.

A. W F. Edwards: Ideally one wants a transformation of the gene frequency such that the transformed variate changes at a constant rate, and although the logarithm of the gene ratio may be appropriate in some instances, the optimum transformation clearly depends on the system of selection operating. In the case of the blood groups it seems best to assume the absence of selective forces, in which case one wants to standardize for the effect of drift over the whole range of gene frequency. The transformation we have used, which is the multidimensional analogue of the square-root chart, does this.

M. Siniscalco *(Italy)*: I would like to know how you would correct for differential fitness in your method. For example, in Sardinia the use of the genes for thalassemia or G-6-Pd-deficiency would certainly affect the estimate of "genetical distance"

L. L. Cavalli-Sforza: Loci for which parallel selection is to be clearly expected, as thalassemia, might perhaps be left out if the sample of genes used in the analysis is small. With a larger sample there might be enough such genes to balance each other and decrease the bias. However, as Dr. Edwards has pointed out, the problem will be best dealt with by choosing the appropriate transformation.

C. Barigozzi *(Italy)*: I would like to know something more about the samples used. For instance, France includes the Basques, who are fairly different from the other ethnic groups in France.

L. L. Cavalli-Sforza: The blood group data used were from particular populations tested for all, or nearly all, the genes we have used. They were not selected as being specially representative of the larger areas in which they lay, nor have we averaged gene frequencies over these larger areas. In the case of the anthropometric data, the origins of the samples are less well defined, and reference should be made to Martin's Handbook.

L. D. Sanghvi *(Switzerland)* I should like to mention my work on the comparison of genetical and morphological differences (*Am. J. Phys. Anthrop.* 1953) and the subsequent corroboration of this work by W Pollitzer, (*Am. J. Phys. Anthrop.* 1957). Now the work of Cavalli-Sforza and Edwards confirms this general comparability by a more powerful technique and the use of a computer.

H. Levene *(U.S.A.)*: Dr. Pollitzer found that anthropometric measurements made by different investigators were not comparable. This may help to explain why your tree based on anthropometric measurements was less satisfactory than that based on blood groups.

J. V Neel *(U.S.A.)*: How much can one blood-group system dominate the analysis? For example, the divergence of the East Indian and American Indian populations must be largely due to the ABO system.

L. L. Cavalli-Sforza: The weights given to each blood group in the analysis by principal components are computed taking account of all populations. The blood group alleles given a relatively higher weight are therefore those that are more useful for the differentiation of all the populations considered. In the analysis of the evolutionary tree, however, all alleles are given equal weight.

J. V Neel *(U.S.A.)* To what extent do you think the rather striking similarities between the East Indians and the American Indians are due to parallel selection?

L. L. Cavalli-Sforza. It is difficult to answer the specific question posed, but I should like to emphasize that the approach suggested in this paper leads to estimates biased by an amount determined by the extent to which parallel evolution occurs (at least with respect to the space-time tree). With respect to blood group genes, it might seem that the selective effect of incompatibility would determine parallel selection. But, from the little we know about selective forces active in blood groups, it would seem that incompatibility itself is relatively unimportant with respect to other factors, which tend to affect the selective advantages of the various alleles in different directions for different evironments. If that is so, selective forces for blood group genes would act in a way formally reminiscent of drift, and the situation is thus nearer to that required for the application of the theory.

L. G. Silvestri *(Italy)*: I would call your attention to the philosophical assumptions of this kind of work, in particular to the so-called *Principle of Minimum Evolution*. Personally, I have some doubts that it is possible, from the study of the final state of a biological system, to reconstruct its history. Principal component analysis is only an elegant mathematical device to diminish the number of variables necessary to specify a system. This condensation of information is only relative to the "today" situation, and how far are we allowed to extrapolate into past history? The derived tree is only the most likely one on the basis of specific assumptions. My impression is that the *Principle of Minimum Evolution* has no general validity, but that it can sufficiently approximate the situation for short periods of evolution, and in some particular cases.

L. L. Cavalli-Sforza: We have only used Principle of Component Analysis to give a graphical representation of the situation, and the reconstruction of the tree is not dependent on it. The reconstruction process is, of course, but an example of statistical estimation. The likelihood of any tree can be estimated, and the dependence of this estimate on the assumptions made is by no means unique to this kind of inference.

Some principle clearly is needed if the course of evolution is to be traced at all, and we believe The Principle of Minimum Evolution to be quite satisfactory.

Perhaps Dr. Edwards would like to add something on the logic of this type of inference.

A. W F. Edwards: Unfortunately there is no time to enlarge on the problems of inference: we dealt with these in more detail in London at the last meeting of the Genetical Society and will do so in print at the earliest convenience. It seems to us, however, that our principle has been implicitly used in all evolutionary studies, including cases in which fossil evidence has been available, and that all we have done is to admit it explicitly and develop it as the basis for statistical estimation.

Paper 30 (1965)

A method for cluster analysis
Biometrics 21, 362–375.
(With L. L. Cavalli-Sforza)

Note from A. W. F. Edwards: In the early application of computers to multivariate data many methods for clustering were suggested. This one was presented by AWFE at the 5th International Biometric Conference in Cambridge, September 1963, at the same time as a similar suggestion was made by J. H. Ward (Hierarchical grouping to optimize an objective function, *Journal of the American Statististical Association* 58, 236–244, 1963). Ward's method involved building up clusters but in this paper successive splits into smaller clusters were sought, the natural approach to adopt when modelling a tree-like structure.

A METHOD FOR CLUSTER ANALYSIS

A. W. F. EDWARDS[1] AND L. L. CAVALLI-SFORZA

International Laboratory of Genetics and Biophysics (Pavia Section), Istituto di Genetica, University of Pavia, Italy.

SUMMARY

A method for investigating the relationships of points in multi-dimensional space is described. Using an analysis of variance technique, the points are divided into the two most-compact clusters, and the process repeated sequentially so that a 'tree' diagram is formed. The application of the method to problems of classification is particularly stressed, and numerical examples are given.

1. INTRODUCTION

A number of the problems with which we have been concerned recently require the consideration of the relationships between points in space. For example, the dispersion of human chromosomes as they appear on a photograph of mitosis involves a two-dimensional array of points; the dispersion of stars involves a three-dimensional array (when information on distance is available); in taxonomy, n species which have each been scored for m characters may be represented by n points in a space of m dimensions, and information on the natural groupings of the species may be obtained from the groupings of these points; and so with patients who exhibit the presence or absence of a number of symptoms.

Indeed, any investigation into the classification of objects, for which taxonomy may be taken as the type example, or any investigation into the dispersion of things, requires the elucidation of spatial groupings. Sometimes, as in taxonomy, the groups and sub-groups must be found, and at other times, as with chromosomes, the problem is more purely statistical—do certain chromosomes tend to cluster together? In this paper we concentrate on the former (and simpler) aspect of the problem, except for a slight digression in Section 8 on the dispersion of chromosomes, a matter of current interest. The latter, or probabilistic, aspect, with its vast literature, is not of immediate concern to us here.

We are aware of several investigations bordering on this subject, especially in the field of numerical taxonomy, where the papers have been reviewed by Sneath and Sokal [1962]. Rao [1952] has described an

[1]Present address: Department of Mathematics, Stanford University, Stanford, California, U.S.A.

intuitive approach to Cluster Analysis (using the generalised distance of Mahalanobis) which is particularly relevant, and will be mentioned below. However, our specific approach does not seem to have been considered before, so that a comprehensive review of previous work would be out of place.

Both Hotelling's [1933] Principal Component Analysis and Fisher's [1936] Discriminant Analysis are cognate with Cluster Analysis. Principal Component Analysis is, in effect, a method of projecting points in multi-dimensional space into a space of fewer dimensions so that the maximum amount of information is retained. The first component can be described as the orthogonal projection into one dimension that maximises the variance of the projected points; removing this dimension and its associated variance, the second component is the projection into one dimension that maximises the variance of the projected new array of points, and so on. Discriminant Analysis, on the other hand, seeks that linear combination of characters (to use taxonomy as an example) which best enables a species to be classified. It provides a sorting procedure into previously-chosen classes, and thus differs from Cluster Analysis, in which no such classes are admitted. However, a close affinity between the two analyses will become evident later.

2. GENERAL CONSIDERATIONS

Cluster Analysis, in the first instance, seeks to identify clusters of points in space. A prerequisite of a successful analysis is, naturally, a method of presenting the results, and we have found a 'tree' diagram useful. In this, as will be seen, each cluster of points is grouped on an individual branch. The peculiar advantage of this representation is that a multi-dimensional tree, reaching out to every point in the space, can be mapped, with all its ramifications, on paper in two dimensions. The analogy with a taxonomic tree will be evident. In particular instances it may be meaningful to assign lengths to the branches, and this will be mentioned below.

Starting at the base of the tree, the first bifurcation will represent the splitting of the points into two clusters; each branch so formed will then split again as the two clusters are themselves each resolved into two more clusters, and the process continues until the individual points are reached. Thus the main problem to be solved is how to divide a cluster of points into two in the most acceptable manner, for the repetition of this process will give the complete analysis.

3. THE DIVISION OF A CLUSTER

Clearly, the best division of a cluster for our purposes will be that in which the two resulting clusters are as dense as possible, according to some criterion. We have found that the analysis of variance provides an excellent criterion, which is meaningful in operation, simple in calculation, and useful for testing significance. From the analysis of variance it is known that the sum of the squared distances of points on a line from their mean can be partitioned, when the points are classified into two groups, into two within-group sums of squares and a between-groups sum of squares (see Fisher [1954]). Further, since all the quantities involved are squared distances, it is evident that this is also true for points in any number of dimensions, because these squared distances can all be partitioned into squared distances along the Cartesian axes, so that if the partition is possible along each axis it is possible amongst the points as a whole. Thus we see that when points are divided into two clusters the sum of the squared distances from their mean can be partitioned into the sum of the squared distances of the points of one cluster from *their* mean, the similar sum for the other cluster, and the between-clusters sum of squares; this is but a single-classification analysis of variance conducted in many dimensions. The natural criterion for division is clearly the between-clusters sum of squares, and we will regard the best split as that for which this sum is a maximum (and the within-clusters sum of squares consequently a minimum). Continued splitting according to this criterion will lead to a tree diagram, and with each branching will be associated a between-clusters sum of squares, which will be a measure of the 'importance' of the split. Further, since at the end of this process each cluster contains only one point, there is no within-clusters sum of squares left, and the total of the sums of squares associated with each branching must exactly equal the original sum of squares: all the original variation is accounted for. Table 2, which will be referred to below, exemplifies this type of hierarchical breakdown of the total variation. The natural measure of cluster density is clearly the variance, or the within-cluster sum of squares divided by the number of points (we divide by n rather than $n - 1$ because we are not primarily concerned with estimation).

This criterion of maximising the between-clusters sum of squares is essentially that used in Discriminant Analysis (Fisher [1936]), but whereas we use it to determine the natural clustering of objects from raw observations, Discriminant Analysis uses it to determine what linear function of the observations best discriminates between two *a priori* clusters.

4. THE CALCULATION OF THE SUMS OF SQUARES

This technique is as simple in practice as it is in theory. Since n points can be divided into two clusters in $2^{n-1} - 1$ ways, and the sum of squares must be calculated in each case (at the present state of our knowledge), it is essential to have a quick method of performing the calculation. In fact it is simpler to minimise the within-clusters sum of squares than to maximise the between-clusters sum of squares. Our method depends upon the fact that the variance of n points on a line is equal to the sum of all the squared distances between points taken two at a time (but each distance being used once only), divided by n^2. Thus the sum of squares about the mean is equal to the sum of the squared distances between points, divided by n. As before, since we are dealing exclusively with squared distances, the result must hold for points in any number of dimensions. Thus all the necessary information for *all* the splits is contained in the half-matrix of the squared pair-distances. This is reasonable, for the half-matrix clearly contains all the information necessary to reconstruct the cluster as far as the relative positions of the points are concerned; what it does not contain is information about the axes—these are, of course, no longer relevant, and have been dropped.

Such a matrix is given in Table 1, the numbers (which are in reality

TABLE 1

THE HALF-MATRIX OF SQUARED DISTANCES (DERIVED FROM THE DATA OF TABLE 3)

A	*B*	*C*	*D*	*E*	*F*	*Points*
	5	11	11	14	14	*A*
		10	6	13	15	*B*
			6	17	21	*C*
				13	15	*D*
					6	*E*
						F

the squares of distances) being taken as integers for the sake of the example: thus the distance between A and B is $\sqrt{5}$ units. This matrix may be taken as an arbitrary numerical example for the moment; in fact it is derived from real data which will be mentioned below. The sum of the numbers is 177 so that the sum of squares is 177/6, or $29\frac{1}{2}$. Investigating the split $ABC : DEF$, the sum of squares for the first cluster is $(5 + 11 + 10)/3$, or $8\frac{2}{3}$, and for the second cluster is $(13 + 15 + 6)/3$, or $11\frac{1}{3}$. The total within-clusters sum of squares is therefore 20. All the other thirty possible splits can be similarly treated, and the best one chosen, the between-clusters sum of squares measuring the 'importance' of the split. In this case the best split is $ABCD : EF$ with a within-clusters sum of squares of $15\frac{1}{4}$, leaving a between-clusters sum of squares of $14\frac{1}{4}$. Splitting the clusters so found, the final tree is formed; this is given in Table 2, together with the between-clusters sum of squares for each split. The total of these values is, of course, $29\frac{1}{2}$.

TABLE 2

AN EXEMPLARY TREE

- $14\frac{1}{4}$
  - $6\frac{3}{4}$
    - $2\frac{1}{2}$
      - A
      - B
    - 3
      - C
      - D
  - 3
    - E
    - F

This technique is clearly well-suited to electronic computing (which is necessary owing to the number of trials involved), especially as the $2^{n-1} - 1$ splits of n points into two clusters may be characterised by the binary numbers from 1 to $2^{n-1}-1$. In our experience about $(n-1)^2 2^{n-11}$ seconds of computing time are needed to split n points when the squared distances run to two significant figures, so that up to 16 points can be treated in a reasonable time. Thereafter a sequential method is necessary to avoid testing all the splits, and this has also been developed. It is interesting to note that we are limited by computer speed rather than computer capacity. (The above computing time was obtained on an Olivetti Elea 6001, which we are using, and which has an Access Time of 5μ-seconds). A computer is also useful, of course, for calculating the $\frac{1}{2}n(n - 1)$ squared distances between points from the coordinates, except in cases of direct measurement on small number of points. We may add that an algorism for finding the best split directly would be very welcome.

It is encouraging to note that when all the points are equidistant one from another, no split can be preferred by our method: if the distance between any two points is x, then splitting n points into clusters of r and $n - r$ will give a within-clusters sum of squares of

$$\tfrac{1}{2}r(r - 1)x^2/r + \tfrac{1}{2}(n - r)(n - r - 1)x^2/(n - r) = \tfrac{1}{2}(n - 2)x^2,$$

which is independent of r. Indeed, for *any* array of points this formula must give the mean within-clusters sum of squares for all the possible splits into clusters of r and $n - r$, if x^2 is taken to be the mean of the squared distances. For, considering all the possibilities, all the squared distances will occur within a cluster with equal frequency. Further, since the total sum of squares is equal to $\frac{1}{2}(n - 1)$ times the mean square distance x^2, the mean within-cluster sum of squares must be $(n-2)/(n-1)$ times the total sum of squares, giving a mean between-clusters sum of squares of $1/(n - 1)$ of the total. Since this is independent of r, it must also be the general mean over all possible splits. It can provide a datum from which to gauge the 'importance' of the best split.

5. 'TAXONOMIC' CONSIDERATIONS

In 'taxonomic' situations, in which a number of objects have been scored for the presence or absence of various characters (each accorded equal weight), all the coordinates of the points are 0 or 1, and the calculations become even simpler: since these numbers are unchanged by squaring them, the squared distance between two points is equal to the number of characters in respect of which the two objects differ, In taxonomy itself, this measure of 'distance' is already in use, so that our methods are particularly appropriate; of course, they may also be used with metric characters without difficulty.

In fact the example given above was taken from the bacteriological data of Lysenko and Sneath [1959]. The six organisms were *Escherichia Coli, Salmonella, Klebsiella, Hafnia, Proteus vulgaris* and *Morganella*, in that order, and we have omitted the results of the tests 'gas from glucose', 'sucrose', 'dulcitol', 'inositol', 'salicin', 'raffinose' and 'trehalose', which gave variable results in many cases. Information from such tests can in fact be used without difficulty, but we wish here to preserve the simplicity of the example. The consequent table of scores, from which the half-matrix of squared distances was constructed, is given in Table 3.

TABLE 3

Matrix of Scores for Six Species of Bacteria

A	111011111000110010001111
B	101011111010010110001110
C	011111111000001111011000
D	101001111010001111011010
E	100001010111110111100001
F	100000000011110001100011

In some instances it may be informative to classify the characters according to the objects, as well as the objects according to the characters: this will give an idea of the independence of the characters, a subject which will be taken up in Section 7. Finally, having identified several 'taxa' amongst the objects, Discriminant Analysis can be used to determine the most informative characters for the classification of future objects into these 'taxa'. But here a word of warning is necessary, for Discriminant Analysis provides a 'closed system' to which further objects do not contribute any information; it is a synthesis of past experience which does not 'learn' from further cases.

6. SCALE AND DISTANCE

Cluster Analysis is not invariant with respect to changes of scale along the axes, and the user must satisfy himself that his scales are appropriate. Conversely, in 'taxonomic' uses, a character may easily be weighted by changing the scale along the corresponding axis. In these uses the independence of the classifying characters can also be considered, and will be treated in the next section.

From the between-clusters sum of squares (SS) it is easy to calculate the distances from the means of the two clusters to that of the combined cluster, and these distances may be associated with the corresponding segments of the tree. They are given by

$$d_m^2 = \frac{nSS}{m(m+n)} \quad \text{and} \quad d_n^2 = \frac{mSS}{n(m+n)},$$

where the clusters contain m and n points. In this way a metric tree can be found, if it is meaningful for the problem in hand. For fuller discussion of metric trees, especially in evolution, see Edwards and Cavalli-Sforza [1964] and Cavalli-Sforza and Edwards [1964].

It is evident that our technique of Cluster Analysis by sequential splitting is more appropriate to some types of problem than to others, and the same may be said of the resultant tree diagram. The method is well-suited to cases in which the array of points might indeed have arisen by a branching process, such as in evolutionary studies or in some studies on the spatial distribution of organisms; in other cases it is merely a way of reducing the data and detecting clusters. In the former cases the assignment of some point, almost mid-way between two clusters, to a particular cluster indicates the most probable origin of the point on the available information, but in the latter cases there can be no such interpretation, and a tree diagram is correspondingly less informative. Nevertheless, Cluster Analysis is the best method we know for treating such data, and no difficulties should arise as long as its limitations are clearly understood. In particular, in cases in which a prior branching process cannot be postulated, the interpretation of the tree needs special care: two points, one from each of two clusters, which lie at the positions of closest approach of the clusters, may be widely separated on the tree. It hardly needs adding that the linear arrangement of the points on the tree is largely arbitrary, although this can be rectified to some extent, for example by arranging them according to the first component of a Principal Component Analysis (see e.g. Rao [1952]).

7. INDEPENDENCE, AND A FURTHER EXAMPLE

In many situations to which Cluster Analysis may be applied, each point will represent not one observation, but the mean of many. Thus, in the following example from Rao [1952], each point represents a particular Indian caste or tribe, and was derived from the measurement of many individuals. This will not be so in pure taxonomy, where each organism is scored for the presence or absence of each of many characters, and similar organisms will be scored identically, unless metrical characters are used.

Thus examples may be classified into those with a 'within-points' dispersion matrix, and those without. In most cases with a 'within-points' matrix it will be desirable to eliminate the correlations between classifying characters by a transformation to uncorrelated standardised variables. However, no general rules can be laid down, and each case must be treated on its merits. Indeed, there is no hard and fast distinction between the two types: we may look upon data having a 'within-points' matrix as a series of individual points on which *a priori* clustering has been carried out, and similarly, after clusters of acceptable size have been isolated from data having no 'within-points' matrix, the within-clusters dispersion matrix becomes the 'within-points' matrix of the points representing the cluster means. To make Cluster Analysis fully comparable to Discriminant Analysis it would be necessary, for every possible split, first to transform according to the within-clusters dispersion matrix. The 'most-discriminating' split would then be chosen. We have not pursued this approach.

In simple taxonomic situations, characters will usually be correlated 'between-points'. Such correlations can be eliminated by an orthogonal transformation which rotates the reference-frame of axes bodily, but this leaves the relative positions of the points unchanged, and is therefore irrelevant to Cluster Analysis. However, it may in certain circumstances be thought desirable to standardize the resultant variables for variance: this is tantamount to changing the scales along the new axes so that all the components of the Principal Component Analysis (which gives these axes) have the same variance, and may be accomplished by calculating the values of D^2 (see below) using the 'between-points' dispersion matrix.

The technique of transforming to a set of uncorrelated standardized variables by using the within-groups dispersion matrix is well-known (for example, see Rao [1952]), and there is no need to describe it here. As has been mentioned above, such a transformation consists of an orthogonal rotation of the axes, followed by changes of scale along the new axes: the old axes thus become oblique to one another, and all the distances between points are affected, in general. Cluster Analysis must be carried out on the new distances. The use of such a transformed distance was first proposed by Mahalanobis, and it is usually referred to as 'Mahalanobis' generalized distance', or 'D'. Rao gives a full numerical example, calculating the values of D^2 between twelve castes and tribes of the United Provinces in India, using nine anthropometric characters. Since these values of D^2 are squared distances in the transformed space they can be used, without modification, in Cluster Analysis. They are given in Table 4.

Rao clusters the castes and tribes using largely intuitive methods, and follows this with a Principal Component Analysis. He comments that 'No formal rules can be laid down for finding the clusters because a cluster is not a well-defined term'. It is thus interesting to compare his results with those obtained by the 'rules' of Cluster Analysis. Of course, with only twelve castes an intuitive solution is not difficult. The clusters that Rao finds are given in Table 5, the figures being the *within-cluster* sums of squares expressed as percentages of the total

TABLE 4

VALUES OF D^2 FOR TWELVE INDIAN CASTES AND TRIBES (AFTER RAO)

Note: These values have been derived from Rao's by multiplying his by a constant so that the largest value becomes 99; in this way the greatest possible accuracy using two figures (as required by our computer program) is attained.

B_1	B_2	C_1	C_2	D	Bh	Ch	M	A_1	A_2	A_3	A_4		
	5	65	42	54	84	57	54	22	28	40	62	B_1	Brahmin, Basti
		68	31	53	72	54	49	15	19	28	51	B_2	Other Brahmin
			25	85	96	99	84	50	56	63	79	C_1	Bhatu
				40	46	88	70	24	29	31	54	C_2	Habru
					22	72	46	55	45	43	50	D	Dom
						94	59	48	42	33	42	Bh	Bhil
							8	64	40	51	42	Ch	Chattri
								46	25	27	17	M	Muslim
									6	9	29	A_1	Ahir
										2	11	A_2	Kurmi
											8	A_3	Other Artisan
												A_4	Kahar

TABLE 5

CLUSTERING OF TWELVE INDIAN CASTES AND TRIBES, AS FOUND BY RAO

Caste or tribe	Within-cluster SS (% of total SS)
Ahir, Kurmi, Other Artisan, Kahar	6.48
Chattri, Muslim	1.60
Dom, Bhil	4.39
Brahmin, Basti, Other Brahmin	1.00
Bhatu, Habru	4.99.
	18.46

TABLE 6

'TREE' OF TWELVE INDIAN CASTES AND TRIBES

27.92	21.58	14.04	2.39	0.40	Kurmi
					other Artisan
				Kahar	
			1.60	Chattri	
				Muslim	
		4.39	Dom		
			Bhil		
	17.11	4.59	1.00	Brahmin, Basti	
				Other Brahmin	
			Ahir		
		4.99	Bhatu		
			Habru		

TABLE 7

CLUSTERING OF TWELVE INDIAN CASTES AND TRIBES, AS FOUND BY CLUSTER ANALYSIS

Caste or tribe	Within-cluster SS (% of total SS)
Kurmi Other Artisan Kahar	2.79
Chattri Muslim	1.60
Dom Bhil	4.39
Brahmin, Basti Other Brahmin Ahir	5.59
Bhatu Habru	4.99
	19.36

sum of squares. The sum of these values is 18.46%, so that 81.54% of the variation has been accounted for by this arrangement, which is evidently very successful. Cluster Analysis, on the other hand, leads to the tree given in Table 6, the between-clusters sums of squares being expressed as percentages of the total. It will be seen that the four major splits account for 80.65% of the total variation; the five clusters so derived are given in Table 7.

As opposed to Rao, we have clustered the Ahir with the Brahmin, but by so doing the percentage of variation accounted for drops from 81.54 to 80.65. But here is no paradox, for there is nothing in our method which ensures that, after several splits, the between-clusters sum of squares shall be an absolute maximum. It is so after one split, but to find the absolute maximum after several splits it would, in general, be necessary to consider all the possible arrangements. Now the number of ways of arranging n different objects into exactly r groups is $n!/r!$ times the coefficient of x^n in the expansion of $(e^x - 1)^r$ (see Whitworth [1901]). When $n = 12$ and $r = 5$, as in the present case, there are thus 1,379,400 ways, or far too many to be tried in a reasonable time. With $r = 3$ groups there are $\frac{1}{2}(3^{n-1} - 2^n + 1)$ ways, or already 28,501 for eleven objects, compared with 1023 splits into two groups. A systematic investigation of multiple splits is thus seen to be impossible except with very small numbers of objects, and our method avoids it.

Some further information about the split into two clusters can be obtained by considering the ten best splits, rather than just the single best one. A computer takes little additional time to record these, and it is interesting to compare the difference in the within-clusters sum of squares between the best split and the second best, and the mean difference between the second and third, third and fourth, ... ninth and tenth splits. If the former difference exceeds the latter by an appreciable factor, some confidence may be had in the best split. In the present example the difference between the first two splits is 2.48 (in arbitrary units), and the mean difference amongst the others 0.88. The first three splits all group the Ahir with the Brahmin, Bhatu and Habru, the second adding the Kurmi as well, and the third the other Artisans. The difficulty of knowing to which cluster the Ahir should be ascribed is well-illustrated in Rao's Principal Component chart (his Figure 2 [1952] p. 370). Had there been little to choose between the first few splits, the consequences of each could have been followed through without difficulty, for subsequent splits take a negligible time compared with the first.

It seems to us that both the power and some of the limitations of Cluster Analysis are evident in this example.

In passing, it should be noted that Rao uses 'average within-cluster D^2' as a measure of cluster density: this is $2n/(n-1)$ times our variance; and that he uses the square root of the average between-clusters D^2 as a measure of distance, whereas we prefer the 'real' distance between the means of the clusters, defined by

$$d^2 = \frac{m+n}{mn} \text{ (between-clusters } SS\text{)},$$

which is independent of the variances of the two clusters.

8. TESTING A NULL HYPOTHESIS

As was mentioned in the introduction, in some problems we wish to test for the tendency of points, chosen *a priori*, to lie close together. For example, with human chromosomes we might wish to test whether homologous chromosomes tend to lie closer together (on a photograph) than chance would lead us to expect. Statistical procedures have been suggested by Barton and David [1962] and Schneiderman and Smith [1962], the former authors actually using the ratio of twice the within-homologues sum of squares to the total, the moments of whose sampling

TABLE 8

ANALYSIS OF VARIANCE OF CHROMOSOME DISPERSION

	df	*SS*	Variance
Between homologues	22	322.04	14.64
Within homologues	23	184.04	8.00
Total	45	506.08	

$F = 1.83$, $10\% > P > 5\%$.

distribution they investigate. But it seems to us that a straightforward analysis of variance (as Barton and David indeed suggest in their final paragraph, but do not use) is appropriate. Using their example, we obtain the analysis given in Table 8. It will be noted that significance at the 5% level is not quite attained, although it is in Barton and David's test. We might add that, when considering the dispersion of real objects, the fact that two or more objects cannot occupy precisely the same point must introduce a bias into the test.

9. EPILOGUE

Apart from its use in investigating the spatial relationships of real objects, Cluster Analysis lends itself to any problem involving classification. It searches for hidden similarities, and sorts items into abstract groups. Such problems occur in nearly all scientific disciplines, and we feel that our approach may provide a fruitful point of view.

ACKNOWLEDGMENT

The work on which this paper is based was supported by a grant from the U. S. Atomic Energy Commission.

REFERENCES

Barton, D. E. and David, F. N. [1962]. The analysis of chromosome patterns in the normal cell. *Ann. Hum. Genet., Lond.* *25*, 323–29.

Cavalli-Sforza, L. L. and Edwards, A. W. F. [1964]. Analysis of human evolution. *Proc. XI Int. Congr. Genetics*, The Hague, 1963.

Edwards, A. W. F. and Cavalli-Sforza, L. L. [1964]. Reconstruction of evolutionary trees. In *Systematics Association Publ. No. 6: Phenetic and Phylogenetic Classification.*

Fisher, R. A. [1936]. The use of multiple measurements in taxonomic problems. *Ann. Eugen., Lond.* *7*, 179–88.

Fisher, R. A. [1954]. *Statistical Methods for Research Workers.* 12th edition. Edinburgh: Oliver and Boyd.

Hotelling, H. [1933]. Analysis of a complex of statistical variables into principal components. *J. Educ. Psych.* *24*, 417. (Quoted by Kendall [1955])

Kendall, M. G. [1955]. *The Advanced Theory of Statistics.* 2nd ed. London: Griffin and Company.

Lysenko, O. and Sneath, P. H. A. [1959]. The use of models in bacterial classification. *J. gen. Microbiol. 20,* 284–90.

Rao, C. R. [1952]. *Advanced Statistical Methods in Biometric Research.* New York: John Wiley and Sons.

Schneiderman, L. J. and Smith, C. A. B. [1962]. Non-random distribution of certain homologous pairs of normal human chromosomes in metaphase. *Nature, Lond. 195,* 1229–30.

Sneath, P. H. A. and Sokal, R. R. [1962]. Numerical taxonomy. *Nature, Lond. 193,* 855–60.

Whitworth, W. A. [1901]. *Choice and Chance.* 5th edition. Reprinted 1951: New York: Hafner Publishing Company.

Paper 32 (1966)

Studying human evolution by computer
New Scientist (19th May) 30, 438–440.
Reprinted in *Evolution*, ed. M. Ridley, Oxford: Oxford University Press (1997).

Note from A. W. F. Edwards: A popular article containing a simple demonstration that the accumulation of random changes can provide the information for constructing a tree, contrary to the opinion sometimes expressed that only selective changes could be informative. It also takes aim at the suggestion that the method of minimum evolution or parsimony can be justified by assuming that nature proceeds according to some minimum principle.

STUDYING HUMAN EVOLUTION BY COMPUTER

by Dr A. W. F. Edwards
Department of Statistics, University of Aberdeen

Numerical methods can release information about the course of human evolution hidden in contemporary genetic and other data. The most probable relationships between different human populations can be traced by seeking the "tree" which maximizes the likelihood of arriving at the present distribution

"In the future", wrote Darwin in *The Origin of Species*, "I see open fields for far more important researches. Much light will be thrown on the origin of man and his history." True to this prediction, we have, in the intervening years, learnt a great deal about the inter-relationships of the races of man at present inhabiting the Earth. Much of this knowledge has come from cultural and geographical evidence, archaeological studies providing the former and our ideas of the likely migration routes the latter. But an increasing proportion of what we know is being contributed by comparative studies on living populations, in which physical and cultural differences, including linguistic ones, are examined to see what light they can throw on the evolutionary paths which led to the observed diversity of man.

The discovery of the ABO blood groups and the realization, in 1919, that the frequencies of the genes controlling this blood-group system were characteristically different in different populations, added another field for study, that of *genetic* differences between races. The importance of this field has been recognized only slowly by anthropologists. But since "evolutionary trees" purport to show biological relationships it is obvious that information about genes, which are passed from generation to generation according to certain rules, is potentially of greater importance than information provided by other studies.

In making use of facts relating to the present races of man, and trying to deduce an evolutionary tree from them, we are invoking the conservative nature of human biological and cultural evolution. We admit that the characteristics of a race change from generation to generation, but that they do so slowly, so that each generation provides some, but not all, of the information about what the previous generation was like.

We can best see how such contemporary information can be used by constructing an artificial example. Suppose

Figure 1 *Darwin's first diagram of an evolutionary tree. The ancestral species is shown at 1, and living descendants are represented by barred ends to branches, the remaining branches leading to extinction. From "First Notebook on Transmutation of Species", 1837*

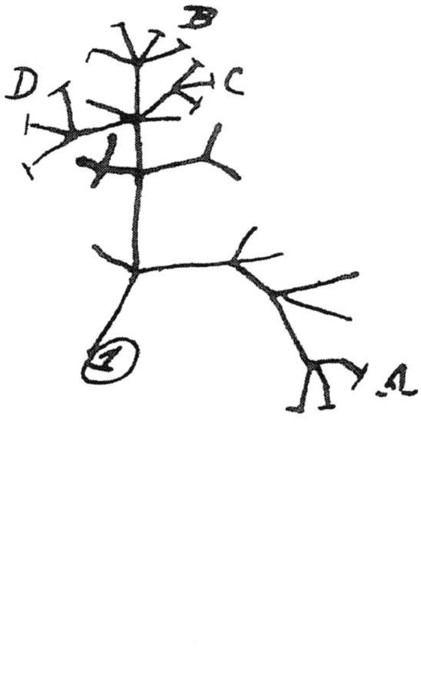

Figure 2 *A model for random evolution, showing how chance differences accumulate in such a way that closely-related populations (A and B) tend to be more alike than distantly-related ones (A or B, and C)*

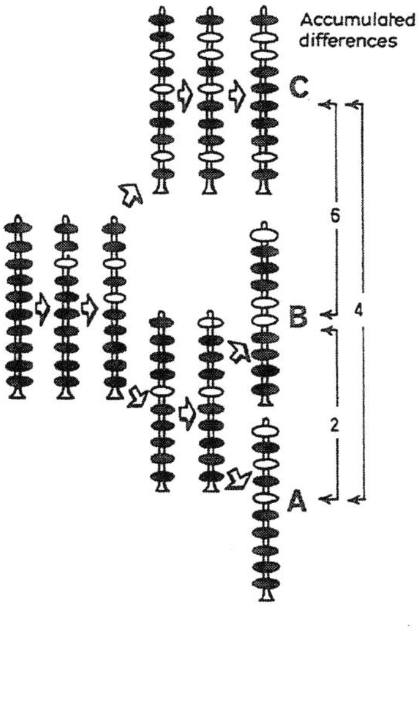

Figure 3 *The "method of minimum evolution" suggests the evolutionary tree of shortest total length linking all the populations. It indicates the positions of all the splits except the first, which may be taken to be at the point X which divides the tree so that the mean distance to the populations on one side equals the mean distance to those on the other*

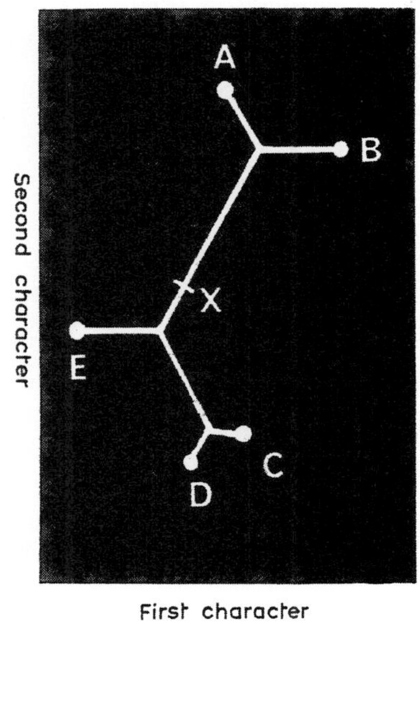

that the original population from whom the present races of man derived had decided on the following custom: a pole was set up on which were threaded 100 black discs with holes in them. Every year at the appointed time a ceremony was held in which a number between one and 100 was chosen at random and the corresponding disc was changed for one of the opposite colour, white if it was black and black if it was white (as it might have been after a number of years). The only other rule attaching to this custom was that when a group migrated away from the parent population they took with them a replica of the pole complete with coloured discs arranged according to the year in question, and maintained the custom, as did any group resulting from further splits.

Such a system conserves information, and yet admits a slow change. We can see that any characteristic—cultural, physical, or genetical—must have this property of changing slowly before it can be considered as a source of evolutionary information. If it changes too fast, information about the past will be lost too quickly, but if it changes too slowly it will not be useful because there will be little or no diversity amongst the descendant populations.

Let us suppose that after many years had passed we examined the poles and discs in a number of distinct populations known to have descended from a common ancestral population by successive splitting (with no subsequent rejoining or hybridization), and let us also suppose that we knew the rules governing the custom. We should then be able to draw up a table showing the number of discs of different colour in comparable positions for each pair of populations (Figure 2), and we could then start constructing an evolutionary tree from this information. We know that populations that have recently separated would show a small number of differences, and by inverting the inference (which may be done with proper statistical methods) we conclude that populations little-different are likely to be closely related. Indeed, for each tree we care to postulate, we can evaluate the corresponding mathematical "likelihood", which provides a measure for choosing between different trees.

This reasoning can be made watertight provided we know the rules. It is the type of reasoning that has been applied, on intuitive grounds, to all characteristics that have an inherited component (either genetically or socially), but of course its validity in many of these cases may be called into question precisely because we do *not* have a perfect knowledge of the rules. Thus the argument has been applied to physical characters with conflicting results, which is not surprising seeing that little is known about the inheritance pattern of most physical traits.

The poles-and-discs custom represents the ideal system from the point of view of the student of evolution, and it is thus appropriate to seek its closest natural analogue if we are to learn about the past from present data. From amongst the biological material we must choose that about whose inheritance most is known, and it is here that genetical differences come into their own, and where physical differences of unknown inheritance are a poor substitute. For nature, in endowing us with genes arranged on chromosomes rather as the discs are arranged on the pole, has unwittingly presented us with an excellent tool for reconstructing the past—excellent provided we know the rules of inheritance.

Given information on gene frequencies in various populations—and in man this information is "exploding" at present owing to the great variety of biochemical traits, including blood groups, which are being discovered—we can estimate the evolutionary tree from the interpopulation differences just as we proposed to do in the case of the discs. But it will only be as good as our knowledge of the rules for the inheritance of the particular genes used. However, it is not sufficient to know that the genes are inherited in Mendelian fashion, because we must also know whether there has been environmental selection for or against the gene in question and, if so, whether it was of the same magnitude in each population.

In our artificial example, it would not matter very much if the table of random numbers, or the roulette wheel, used to determine which disc to change in any year, were biased towards a particular set of numbers, provided we knew which set, and provided we knew that each population used a set of random numbers with the same bias. We can seldom be confident that these conditions are satisfied in reality, because we know so little about selection in the past. For this reason we are on the safest ground if we make use of those genes which seem to have little or no value in relation to environmental pressures. Now one of the mysteries of the genetics of man is why there are so

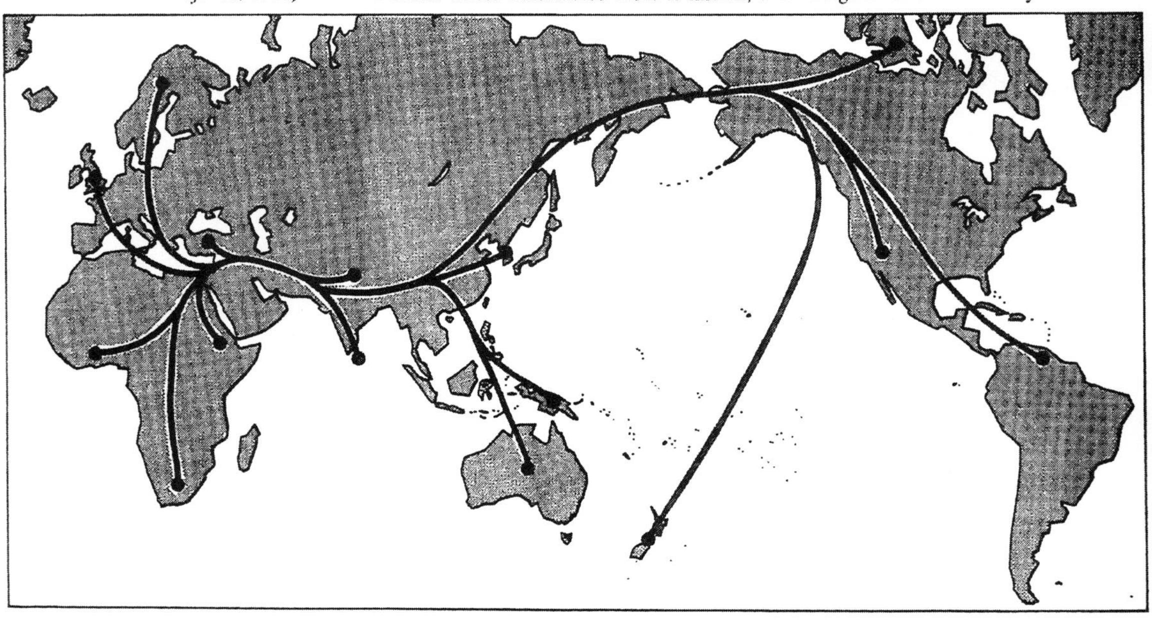

Figure 4 *The "minimum-evolution" tree for fifteen human populations, based on blood-group data. In order to draw this tree on a map of the world the* lengths *of the branches have been distorted, but not the ways in which they connect. Metrically, the longest branch is that leading to the three African populations. (Adapted from Edwards & Cavalli-Sforza, 1964)*

many genes in this category, so it is not difficult to find some suitable ones. For example, the great majority of genes controlling the blood groups and other biochemical factors seem, at the present state of our knowledge, to be selectively "neutral".

When this line of work was started, no one foresaw how important it was to have a specific model for evolution on which to base inferences about evolutionary trees. The trend of thinking was but an extension of the intuitive methods of assessing "closeness of relationship" and "degree of resemblance" current in evolutionary studies since Darwin's day. One attempt at specifying these ideas with sufficient precision to enable automatic computing techniques to be used was the "method of minimum evolution" put forward by me and Professor Cavalli-Sforza, of the University of Pavia.

We suggested that "a plausible estimate of the evolutionary tree is that tree which unites the populations and invokes the minimum net quantity of evolution". For discontinuous characters, such as the discs on the pole, the method selects the tree with the minimum total number of changes, whilst with continuous characters, such as height or the frequencies of particular genes, it suggests the tree the sum of whose branch lengths is a minimum. When many characters are considered simultaneously, this tree will occur in a many-dimensioned space, and the problem becomes that of finding the "string network" that joins all the points (representing populations) using the minimum amount of string (Figure 3). Unfortunately this is well known as an unsolved mathematical problem, but we can proceed some way towards the solution with the help of computers.

This minimum principle has been used to investigate the evolutionary tree uniting 15 human populations, the data being blood-group gene frequencies (Figure 4). It has also been used to trace the evolution of the protein haemoglobin (in which one seeks to minimize the total number of actual chemical changes in the genetic message coding for this molecule) and to study the evolution of the horse. In retrospect, however, the method has little to support it except Occam's razor, unless it can be shown to give much the same result as a proper estimation procedure based on a specific model for evolution, as is probably the case with the blood-group data. Some writers have even tried to justify it on the assumption that nature is simply parsimonious—which, if it were true, would imply no evolution at all!

The use of computers in "numerical taxonomy" has provoked a useful discussion on the aims of the classification of species (see "Classifying species by computer", by T. L. Jahn, *New Scientist* Vol. 29, p. 151). In a similar way, early attempts at reconstructing evolutionary trees using computers are leading to a clarification of our basic ideas as to how it should be done. It has become particularly clear that any attempt at producing an evolutionary tree must be based on a specific model, for only then can proper statistical procedures be adopted, and only then are the assumptions implicit in the method clear for all to see.

Cavalli-Sforza and I have set up such a model for data on gene frequencies, which is applicable to genes with small or random selective effects. The first results using this model are to be published shortly, and it seems that the rigour gained will have been worth the effort that has gone into overcoming the unexpectedly heavy statistical difficulties we encountered.

As the methods and computer programmes develop from the prototypes of today, it will become possible to handle the vast amount of information contained in the published data on blood-group gene frequencies and other genetic characteristics in human populations, by automatic methods. The resultant evolutionary trees will certainly not provide the last word on human evolution. Besides being based on too simple a model, they will not include all the important information on cultural and geographic affinities. Nevertheless, they will provide the genetic information in an understandable form, so that those concerned with the synthesis of the available knowledge on the recent evolution of man will have the genetic contribution at their fingertips.

Paper 35 (1966)

Estimation procedures for evolutionary branching processes

Bulletin of the International Statistical Institute (Proceedings of the 35th session) 41, 803–808.

(With L. L. Cavalli-Sforza)

Note from A. W. F. Edwards: Presented by LLCS at the 35th session of the International Statistical Institute in Belgrade in 1965, this joint paper gives the first mathematical description of the statistical model for an evolutionary tree and the attempt to estimate its parameters by the method of maximum likelihood.

After three years working with Cavalli-Sforza in Pavia, Italy, I moved to Stanford University, California, for the academic year 1964–1965. Our work had outgrown the computing facilities available in Pavia and at Stanford I was able to upgrade my four-population test program to the full 15 populations of our data. When this program repeatedly failed to converge on reasonable solutions I wasted much time seeking a programming error, but there was none. This paper identifies the cause of the difficulty, the existence of singularities in the multidimensional function whose maximum was being sought.

It is observed that given the time coordinates of the nodes of the tree their spatial coordinates are readily found by maximum likelihood, so that perhaps the node times should be estimated by a different method. One is suggested but 'has not yet been investigated in detail' (p. 76, this volume).

The suggestion in the Discussion by Yu. V. Linnik gives a foretaste of the solution that was eventually to emerge through the work of Joseph Felsenstein.

ESTIMATION PROCEDURES FOR EVOLUTIONARY BRANCHING PROCESSES

L. L. CAVALLI-SFORZA and A. W. F. EDWARDS
International Laboratory of Genetics and Biophysics, University of Pavia, Italy)*

INTRODUCTION AND STATEMENT OF THE PROBLEM

In an attempt to rationalize the means by which anthropologists and others have drawn inferences about evolutionary trees from comparative data on living populations, we have been led to introduce a model for evolutionary divergence which, from a purely mathematical point of view, is equivalent to a branching Brownian-motion or Weiner process occurring in a multidimensional Euclidean space, the points at which branches split (always into two) being random in time[1, 2]. The fundamental problem, which we wish to discuss today, is: given the disposition of n points in a p-dimensioned space, and given the rate of the Brownian-motion process, what estimate of the single-origin evolutionary tree which gave rise to these n points can be made?

We may assume, without loss of generality, that the rate of the process is unity. It is immediately clear that the estimation problem involves both the choice of the form of the tree and the estimation of its branch lengths; indeed, the latter problem cannot be treated unless the choice of the form has been made, for we have been unable to devise any unified estimation procedure which would handle both aspects of the problem simultaneously. Even so, it is possible to use the same *criterion* of excellence in both cases, and likelihood naturally presents itself. We will, therefore, first describe our attempts to find the tree for which the likelihood is a maximum.

APPLICATION OF THE METHOD OF MAXIMUM-LIKELIHOOD

We may note that there are a great many different tree forms for even moderate n: consider a tree with (n-1) end-points, which will have (2n-3) branches (including the trunk). It is clear that the nth end-point and its associated branch can be added so as to bisect any of the existing branches, and that each of the (2n-3) possibilities thus accounted for will be dissimilar. Hence the total number of possibilities is $3.5.7.... (2n\text{-}3) = (2n\text{-}3)! / [2^{n-2} (n\text{-}2)!]$. When $n = 10$ this number is 34,459,425. Since for n points every tree which contains only splits into exactly two branches has (n-1) such splits, the contribution of the form of the tree to the likelihood will be the same for each tree, and may therefore be ignored for present purposes. Our first problem is now to maximise the likelihood given the tree form. The probability density of a distance d from an origin in a space of p dimensions after a time t has elapsed is $(1/\sqrt{2\pi t})^p \, exp\,(\text{-}d^2/2t)$, owing to the Gaussian nature of the Brownian-motion process. The log-likelihood of a branch of length d and interval t is therefore $-(d^2/2t + \frac{1}{2}p \log 2t + \frac{1}{2}p \log \pi)$. Omitting the constant term and changing the sign, this becomes $d^2/2t + \frac{1}{2}\, p \log 2t$.

*) This work has been supported by Euratom-Cnr-Cnen Contract No. 012-61-12 BIAI and from the U.S. Atomic Energy Commission.

Let X_{ir} be the rth spatial coordinate of the ith node (or end-point) and t_i its time coordinate, measured backwards from the time of the observations. For the branch joining nodes i and j $(t_j > t_i)$ the above expression then becomes

$$\tfrac{1}{2} \sum_r (x_{jr} - x_{ir})^2 \,/\, (t_j - t_i) + \tfrac{1}{2} p.\ \log 2\,(t_j - t_i)$$

Summing over all the branches (denoted by $[i, j,]$ the quantity to be minimized is

$$L = \sum_{[i,j]} [\sum_r (x_{jr} - x_{ir})^2 / (t_j - t_i) + \tfrac{1}{2} p \sum_{[i,j]} \log 2\,(t_j - t_i). \qquad (1)$$

If j, k, and l are the three nodes (or, in the case of k and l, possibly end-points) connected by single branches to node i $(t_j\ t_i\ t_k,\ t_l)$, then

$$\frac{\partial L}{\partial x_{ir}} = (x_{ir} - x_{jr})/(t_j - t_i) + (x_{ir} - x_{kr})/(t_i - t_k) + (x_{ir} - x_{lr})/(t_i - t_l) = 0.$$

Each node yields a similar equation linear in the $x_{.r}$, the net result being a system of linear equations which enables the rth spatial coordinates of the nodes to be expressed in terms of the rth spatial coordinates of the observed end-points and the time coordinates of the nodes, by inverting a matrix. Thus by assuming values for the time coordinates, the numerical values found can be inserted in the first and second partial derivatives of L with respect to the t's (these will not be quoted explicitly, but may be found from (1)) giving the scores and the information matrix. Normally, iteration could then take place on the t's alone, this substitution being freshly made each cycle. By means of this reduction of the likelihood equations the order of the information matrix shrinks from $(n-1)\,(p+1)$ to $(n-1)$. However, it transpires that iterating on the time coordinates of the nodes almost invariably leads to trivial solutions in which many of the nodes coincide, for reasons which we must now examine in detail.

THE FUNDAMENTAL PROBLEM REQUIRING SOLUTION

It is clear that, since the log-likelihood of the tree is found by summing the log-likelihoods of the individual branches, each node should, in the final solution, be in the maximum-likelihood position with respect to the three adjacent nodes if these are regarded as being fixed. In particular, given the positions and times of the second and third splits, those of the first should be determinable. Let us therefore consider the simple case of the determination of the origin of a Brownian-motion process when we have observed just two different end-points: at time t_1 particle 1 was looked for and found to be at x_1, and at time t_2 particle 2 was looked for and found to be at x_2. Let the origin of the process, which we are required to estimate, be at (X, T). The negative log-likelihood, less a constant, is readily seen to be

$$L = (X - x_1)^2 / 2(T - t_1) + (X - x_2)^2 / 2(T - t_2) + \log 2(T - t_1) + \log 2(T - t_2).$$

This surface can be plotted for varying X and T, but it is already obvious that it contains an unfortunate characteristic: suppose $t_1 > t_2$ (so that it is earlier in time, which, it may be remembered, is being measured backwards), and suppose that we trace the path $X - x_1 = T - t_1$ along the surface towards (x_1, t_1): then the first term in L will tend to zero, the second will tend to the constant

$x_1 - x_2)^2/2\ (t_1 - t_2)$, but the third will become indefinitely large and negative as T approaches t_1, although the fourth will tend to $\log 2\ (t_1 - t_2)$. Thus, provided the two observations were not made simultaneously, in which case $t_1 = t_2$, the negative log-likelihood can be made as large and negative as we please, and hence the log-likelihood as large as we please, simply by letting the origin approach the first observation along a certain path; the surface contains in fact a singularity at the point (x_1, t_1), and any iterative procedure, unless it is started near a well-behaved peak elsewhere in the surface, will lead to the coincidence of the origin and the first observation. It is convergence to such singularities which has led to the trivial solutions so frequently obtained when studying complete trees. Needless to say, such solutions are intuitively unacceptable. We do not, and should not, believe that the antecedent process leading to an array of points known to have been generated by a branching Brownian-motion process was one in which the splits followed each other so rapidly as to appear simultaneous, like in the "big bang" theory of creation. Nor should we believe, as the simpler example would have us, that all the information about the origin of the process when two end-points are observed is contained in the first end-point, for this conclusion would have to be maintained however many more points were observed, which is absurd.

The natural tendency at this stage would be to brand the method of maximum-likelihood as inappropriate for the problem, but this would do it less than justice. If our criterion of the acceptability of any method is whether we like the answer it gives, then the method is, at least partly, redundant. Maximum-likelihood provides point estimates which are meaningful if, and only if, the likelihood surface in the region of the maximum is regular; for the variance of a maximum-likelihood estimate is calculated from the curvature of the log-likelihood surface, and to quote a point estimate and its variance is only entirely meaningful when the likelihood surface is multivariate normal, and thus single-peaked. In the present problem this is far from being the case: the likelihood surface is so irregular that it is meaningless to ask for a point estimate, and the only satisfactory solution would be the exhibiting of the likelihood surface itself. But this is quite impossible owing to the large number of parameters involved. If a solution is to be achieved, it will thus have to be a compromise between what is desirable and what is practicable, odious as this may be for the pure statistician.

SUGGESTIONS FOR A SOLUTION

Given that a perfect solution to our problem is impossible, the most important attribute of any proposed solution is that the way by which it was derived is clear for all to see. It has already been noted that, given the time coordinates of the nodes, the spatial coordinates may be derived by the method of maximum likelihood without iteration, and this applies to both the complete problem and its smaller counterpart. It is therefore natural to seek estimates of the time coordinates other than by maximum likelihood.

In the simple problem with only two end-points we need two independent equations of estimation. As has been indicated above, maximizing the likelihood with respect to the spatial coordinate of the origin x, whilst holding the time coordinate T fixed, provides one satisfactory equation:

$$\frac{\partial L}{\partial X} = (X - x_1)/(T - t_1) + (X - x_2)/(T - t_2) = 0\,,$$

A second equation may be found by equating the expected value of some function of the distance between the two points, $(x_1 - x_2)$ to its observed value. The choice of function is arbitrary, though, of course, it influences the result: the square of the distance is the most tractable, in which case we find

$$E(x_1 - x_2)^2 = E(x_1 - X + X - x_2)^2 = E(x_1 - X)^2 + 2\,E(x_1 - X)(X - x_2) + \\ + E(X - x_2)^2 = (T - t_1) + (T - t_2),$$

since the central term is zero with independent movements of the particles, and since the rate of the process has been assumed to be unity. Equating observed and expected values, the second equation is thus

$$(x_1 - x_2)^2 = 2T - t_1 - t_2, \text{ or } T = \tfrac{1}{2}[t_1 + t_2 + (x_1 - x_2)^2].$$

This is not entirely satisfactory because it allows T to be less than t_1, which is impossible. It is therefore necessary to stipulate that our estimate of T shall be t_1 or $\frac{1}{2}[t_1 + t_2 + (x_1 - x_2)^2]$, whichever is the greater. Thus $T = t_1$ and hence $X = x_1$ if $(x_1 - x_2)^2 < (t_1 - t_2)$, but otherwise $T = \frac{1}{2}[t_1 + t_2 + (x_1 - x_2)^2]$ and hence $X = \frac{1}{2}(x_1 + x_2) + \frac{1}{2}\dfrac{t_1 - t_2}{x_1 - x_2}$, a result which is intuitively rather reasonable.

The application of this, and similar, solutions to the main problem has not yet been investigated in detail, but we anticipate the estimates of the node times will be obtained which will command a fair degree of acceptability, and from then it is but an easy step, already tried, to obtaining the conditional maximum-likelihood estimators of the spatial coordinates of the nodes, and the conditional maximum value of the likelihood for the tree form under consideration.

THE CHOICE OF TREE FORM

Naturally, we wish to choose that tree form which is capable of giving the maximum value of the likelihood, derived as indicated in the previous section. But because of the impossibility of investigating all the tree forms, some means of narrowing the choice is needed, and because of the very heavy numerical work involved in maximum likelihood, ways of reducing it are useful.

We have shortly described elsewhere our attempts to do this, [1, 2, 3, 4] and in fact any of the published methods of cluster analysis is likely to be informative, since well-defined clusters are robust with respect to the method used. We have, however, developed special methods with our specific aim in mind, and in particular:

1) A method of cluster analysis based on a type of generalized analysis of variance, which produces a sequence of binary splits in the group of observed populations giving at each split a maximum ratio of between to within groups variance.[4]

2) A method to determine the minimum evolutionary path connecting the observed populations, the criterion of choice between trees being here the minimum total path length.[1]

3) A method determining by least squares the amounts of evolution which must have taken place after each split to produce the observed differences between populations, the criterion of choice being here the highest additivity of these amounts.[3]

While the first method is especially useful for finding a tree form which, if not the likeliest one, is not far from it, the other two methods supply means of estimating the relative merits of different tree forms in ways which can be numerically less heavy than the full maximum likelihood treatment, and use different but still closely related, biological models.

We hope to publish a unified account of our approach to this aspect of the problem in the near future, and to present a complete account of the whole problem when, and if, the modified maximum-likelihood approach is working to our satisfaction.

REFERENCES

1. A. W. F. Edwards, L. L. Cavalli-Sforza, 1964, ***Reconstruction of evolutionary trees.*** Systematics Assoc. Publication no. 6, Phenetic and Phylogenetic Classification, 67—76.
2. L. L. Cavalli-Sforza, I. Barrai and A. W. F. Edwards, 1964, ***Analysis of human evolution under random genetic drift.*** Cold Spring Harvor Symposia on Quantitative Biology, Vol. XXIX.
3. L. L. Cavalli-Sforza and A. W. F. Edwards, 1963, ***Analysis of Human Evolution.*** Genetics Today, Proceedings of the XI International Congress of Genetics, The Hague.
4. A. W. F. Edwards and L. L. Cavalli-Sforza, 1965, ***A method for cluster analysis.*** Biometrics, Vol. 21, №. 2 362-375.

SOMMAIRE

Les caractères héréditaires d'un ensemble de populations peuvent être produits — selon une théorie évolutionniste — par un processus aléatoire fondamental (mouvement Brownien) à ramifications. L'estimation des paramètres pose des difficultés de principe, lorsqu'on essaie d'appliquer la méthode du maximum de vraisemblance. On suggère une méthode auxiliaire qui donne des résultats en accord avec le bon sens.

DISCUSSION

G. MALÉCOT. La probabilité de ramification est supposée constante, indépendante des autres ramifications. N'y-a-t-il pas des périodes où les possibilitées de ramification sont plus grandes, des périodes d'évolution explosive?

Cela ne conduit-il-pas à introduire des arbres à neouds multiples, et par suite à modifier la discussion du maximum de vraisemblance?

L. L. CAVALLI-SFORZA. Our present treatment with maximum likelihood gives equa probability to every possible tree and therefore, the likelihood computed for each will be affected by local changes in speed of evolution, not by those in the rate of branching. The methods of least squares or of minimum path are unaffected by local changes in the speed of evolution. In order to consider a constant rate of branching, the model must be specified further than we have done so far.

YU. V. LINNIK. The communication is very interesting from the mathematical point of view. The simplest case — the determination of the origin in time and space of Brownian particles can be treated systematically with the maximum likelihood method applied after eliminating the origin by means of appropriate similar statistics.

L. L. CAVALLI-SFORZA. We are looking forward to hearing more details so as to see if the treatment suggested by Professor Linnik can be extended to actual cases, and how general it can be considered.

Paper 36 (1967)

Fundamental Theorem of Natural Selection
Nature 215, 537–538.

Note from A. W. F. Edwards: R. A. Fisher published his 'Fundamental Theorem of Natural Selection' in *The Genetical Theory of Natural Selection* in 1930. The theorem and its history are covered in [140] and [238]. I first encountered it as a research student in Cambridge in 1958 when I bought the second 'Dover' edition of the book but I never discussed it with Fisher himself. It had been ignored by almost everyone except Sewall Wright until a flurry of sympathetic references by Li, Bennett, Kimura, Kempthorne, and Crow in the mid 1950s (see [238]). But unfortunately a number of writers misappropriated the name for a false theorem that the mean fitness of a population was a non-decreasing function. I was in the awkward position of arguing against this whilst remaining, in common with most people, without a complete understanding of Fisher's original theorem. This somewhat negative letter to *Nature*, however, stands the test of time. In it, reference 8 is to a draft paper of mine, considered by the co-editor of the *Annals of Human Genetics*, C. A. B. Smith, but never formally submitted. It was circulated privately at the time.

Fundamental Theorem of Natural Selection

Li[1] has pointed out, by means of a neat algebraic formulation, that Fisher's fundamental theorem of natural selection does not apply exactly in a discrete-generation random-mating situation unless there is no dominance in fitness, and that the mean fitness cannot decrease from generation to generation for a single diallelic locus.

Li writes, "Fisher obtained his result on the basis of a continuous time model with logarithmic fitness", but it should be added that Fisher[2–4] never suggested that his theorem should be applied to discrete-generation models; indeed, he most carefully stipulated, in the sentence following his original formulation[2], that "the rigour of the demonstration requires that the terms employed should be used strictly as defined". In particular, time was to be taken as continuous, and fitnesses were to be measured by Malthusian parameters of population increase. It is thus a misrepresentation to refer to any discrete-generation formulation as "Fisher's fundamental theorem".

These remarks should not be taken to imply that the interpretation of the fundamental theorem, correctly formulated, is simple. A further condition was later introduced by Fisher[3] and discussed by Kempthorne[5]. Kimura[6] has given a very general formulation, which Mode[7] has extended to the stochastic case. Several points remain to be cleared up and I shall examine them elsewhere[8].

With regard to Li's second point, that in the discrete-generation model the mean fitness for a diallelic locus cannot decrease, it should be noted that this has already been proved by Moran[9], and for any number of alleles by several investigators[10–13]. The fact that the general theorem preceded the special theorem for two alleles is due to the latter having been assumed for many years to be an obvious consequence of Wright's formulation[14]. In fact, as Moran[9] mentioned, monotonic convergence of the gene frequency must be proved, and not assumed. Cannings has pointed out to me that in the case of heterozygote advantage monotonic convergence is implicit in Fisher's original treatment[2]. This property of the mean fitness should not, at the present state of our knowledge, be taken as a manifestation of any supposed fundamental theorem in discrete time.

A. W. F. EDWARDS

Department of Statistics,
University of Aberdeen.

Received June 26, 1967.

1 Li, C. C., *Nature*, **214**, 505 (1967).

2 Fisher, R. A., *The Genetical Theory of Natural Selection* (first ed.) (Oxford Univ. Press, 1930).

3 Fisher, R. A., *Ann. Eugen.*, **11**, 53 (1941).

4 Fisher, R. A., *The Genetical Theory of Natural Selection* (second ed.) (Dover Publications, New York, 1958).

5 Kempthorne, O., *An Introduction to Genetic Statistics* (Wiley, New York, 1957).

6 Kimura, M., *Heredity*, **12**, 145 (1958).

7 Mode, C. J., *J. App. Prob.*, **3**, 327 (1966).

8 Edwards, A. W. F., *Ann. Hum. Genet.* (submitted for publication).

9 Moran, P. A. P., *The Statistical Processes of Evolutionary Theory* (Oxford Univ. Press, 1962).

10 Scheuer, P. A. G., and Mandel, S. P. H., *Heredity*, **31**, 519 (1959).

11 Mulholland, H. P., and Smith, C. A. B., *Amer. Math. Monthly*, **66**, 673 (1959).

12 Atkinson, F. V., Watterson, G. A., and Moran, P. A. P., *Quart. J. Math.*, **11**, 137 (1960).

13 Kingman, J. F. C., *Quart. J. Math.*, **12**, 78 (1961).

14 See Li, C. C., *Population Genetics* (Univ. Chicago Press, 1955).

Paper 37 (1967)

Phylogenetic analysis: models and estimation procedures

Evolution 21, 550–570.
(With L. L. Cavalli-Sforza)
Also in *American Journal of Human Genetics,* Supplement 19, 233–257.

Note from A. W. F. Edwards: All the papers reporting the work on evolutionary trees in 1963–1965 had appeared in the proceedings of conferences, often one or two years after delivery, and it was obvious that an account should be prepared for a scientific journal. This took some time because of the separation of the authors (AWFE now at the University of Aberdeen, with inadequate computer facilities, and LLCS in Pavia). Its publication in *Evolution* was somewhat delayed, but the Editor granted permission for it to be used at the workshop 'The Use of Computers in Human Genetics' at the University of Michigan, in August 1966, and simultaneously published in a Supplement of the *American Journal of Human Genetics*. The suggestion in [35] is implemented, but with only four populations.

PHYLOGENETIC ANALYSIS: MODELS AND ESTIMATION PROCEDURES[1]

L. L. Cavalli-Sforza and A. W. F. Edwards[2]
International Laboratory of Genetics and Biophysics, Naples
Pavia Section, Istituto di Genetica, Università di Pavia, Italy

Received March 30, 1966

Acceptance of the theory of evolution as the means of explaining observed similarities and differences among organisms invites the construction of trees of descent purporting to show evolutionary relationships. Whether such trees are based on fossil or living specimens, they may often be criticized for having a high subjective element. The purpose of this paper is to show how suitable evolutionary models can be constructed and applied objectively. In it we amplify and extend the methods we have given in previous communications (Edwards and Cavalli-Sforza, 1963*a*, *b*, 1964, 1965; Cavalli-Sforza and Edwards, 1964, 1966; Cavalli-Sforza, Barrai and Edwards, 1964; Cavalli-Sforza, 1966).

Considering the great variety of information provided by living organisms, it is clear that the type of data will affect both the method of treatment and the validity of the results: the higher the correlation of data and genotype, the greater is the validity likely to be. Information on nucleic acid and protein structure comes first in the scale of relevance, and that on phenotypic measurements last; discrete and continuous variation demand different treatments, and evolutionary models appropriate to both cases will therefore be required for estimation purposes. Differences which are the result of mutation are formally discrete, and evolution at the molecular level thus needs discontinuous treatment; but even in this case the limit of observation may turn the data into the continuous type, as happens, for instance, when the similarity in nucleotide sequences in the nucleic acids of two organisms is measured by hybridization techniques, or when differences between closely-related organisms are examined. In the latter case the differences may be sufficiently small to suggest that the analysis be carried out at the level of gene frequencies, semi-continuous variables which may be treated as continuous. We will be especially concerned to develop the continuous treatment, the discontinuous one being more easily obtained, in a parallel manner.

In addition to the relevance of the data, the validity of the derived evolutionary tree will be strongly dependent on the correctness of the evolutionary model used as the basis for estimation. This will be limited, first because the dynamics of evolution is not fully understood, secondly because the values of some parameters (such as selective coefficients) are unknown, or known with low accuracy, and thirdly because there are mathematical and technical limitations as to how complex the model can be.

Although data suitable for our type of evolutionary study may seem to be largely taxonomic, it should be noted that the aim of this work is not the same as that of taxonomy, as the word is normally understood (see Edwards and Cavalli-Sforza, 1964); in particular, "numerical taxonomy" (Sokal and Sneath, 1963) is not primarily concerned with phylogeny, and the fact that the techniques to be described here and those of numerical taxonomy both involve the treatment of "taxonomic" data should not be allowed to mask the differ-

[1] This paper appeared in a recent supplement of the American Journal of Human Genetics.
[2] Present address: Department of Statistics, University of Aberdeen, U.K.

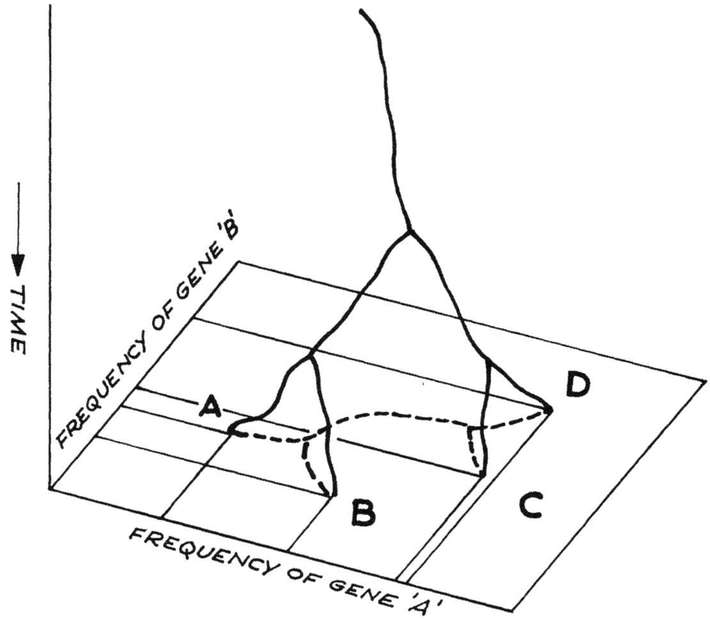

FIG. 1. An evolutionary tree and its projection onto the "now" plane.

ences between them, either at the logical or methodological levels.

EVOLUTION AS A BRANCHING PROCESS

Evolution can only be described in terms of the characters that are changing, and it is convenient to represent such changes in a multidimensional character-space in which each population occupies a position determined by the values of the characters it exhibits. In this paper the word "population" will be used to denote one of the group whose diversity is under study; it might refer to a species, a race, or even a single organism.

If a time dimension, everywhere normal to the character-space, is added, the course of evolution (were it but known) could be seen as a tree, whose branches split as populations diverge, unite as they hybridize, and end as they become extinct. Living populations would be represented by the intercept of the tree and the "now" plane (Fig. 1). In the case of discontinuous characters, the character-space would consist of a lattice of points, but to regard it as continuous will often be a good enough approximation, as indicated earlier. Data, such as gene frequencies or other measurements, will be represented by points in the space-time of Figure 1, and the problem of tracing evolutionary history will be that of fitting a suitable tree to these points. It may be noted, however, that information from the past is in practice available only for data which in other respects are less satisfactory than gene frequencies since their genetic basis is very imperfectly known, as is the case with osteometric data. But with gene frequencies data are only available, in general, for living populations, so that points will be restricted to the "now" plane. With such data we may only be able to reconstruct a projection of the tree onto the "now" plane (Fig. 1), in which case complete information on the position of the first split will not be preserved. Reconstruction of the tree in space and time will be possible, however, if we are willing to make hypotheses about the mode and speed of evolution.

Our genetic reasoning will be almost entirely confined to the analysis of gene frequencies among present-day populations, though it is clearly possible to extend it to other cases. In particular, once methods have been set up for estimating the course of evolution from present-day data, they can be extended without difficulty to include data from the past. Such an extension involves no logical jump and little increase in mathematical complexity.

The proper basis for the study of evolutionary divergence will be provided by the transformation of the space-time which makes a unit vector, in whatever direction (normal to time) and in whatever part of the space, correspond to the amount of evolutionary change expected in unit time. Such a transformed space-time will be homogeneous and isotropic with respect to evolutionary progress; in some problems it may not be Euclidean, or it may not even be possible to formulate the problem in geometrical terms, but the Euclidean representation is the simplest possible, and will suffice for the development of the argument. The correct transformation will, of course, depend on the evolutionary model and the type of data available, and in the case of gene frequencies will be treated below.

The Genetic Basis

Of the major evolutionary forces—mutation, migration, selection, and drift—we shall not incorporate the first two into our model. *Mutation pressure* is known to be usually very small compared with other pressures, so that its neglect, or its confounding with the other pressures, is reasonable: we are not here concerned with mutation in its role as the source of variation. *Migration* need not be considered at all if the evolving populations have differentiated past the specific stage; but this will not be true in most of the cases in which gene frequencies are useful, and its omission may be a source of error. We can, however, justify this practice: On the one hand, small migration rates will act essentially as almost-random disturbances partially buffering the variation due to random drift, so that the omission of migration of this magnitude is only apparent; on the other hand, large migration rates must appear only as rare accidents in a given evolutionary tree, and a "migrational accident," such as the fusion, partial or whole, of two populations, would give rise to a loop in Figure 1. The major difficulty in extending the analysis to include such loops is that they bring about an enormous increase in the number of possible trees to be examined. If, however, one or two loops are known, or assumed, to have taken place at given times and between given populations, their consideration may be practicable, although it will not be attempted in this paper.

Random genetic drift is the name given to the variation in gene frequencies which inevitably accompanies the formation of the next generation, depending, as it does, on a sample of genes from the former generation. Ignoring other sources of variation, in statistical terms this corresponds to a random walk of the gene frequencies in time. Whenever a population splits, the two branches will independently undergo random walks which will give rise to divergence between them, generally increasing with increasing time since branching. The rate of the random walk will depend on the size of the population in question (Wright's "effective breeding size") and the mating structure. The smaller the population, the greater the random variation, and thus the faster the random walk.

Under a suitable transformation this rate will be independent of the particular gene considered: the variance of a gene frequency p due to drift is approximately $p(1-p)\,(1-e^{-t/2N})$, where N is the effective size of the population and t the time elapsed in generations (Kimura, 1955). For estimation purposes, this can be made nearly independent of p by means of the angular transformation $p = \sin^2\theta$, the variance of θ then being approximately

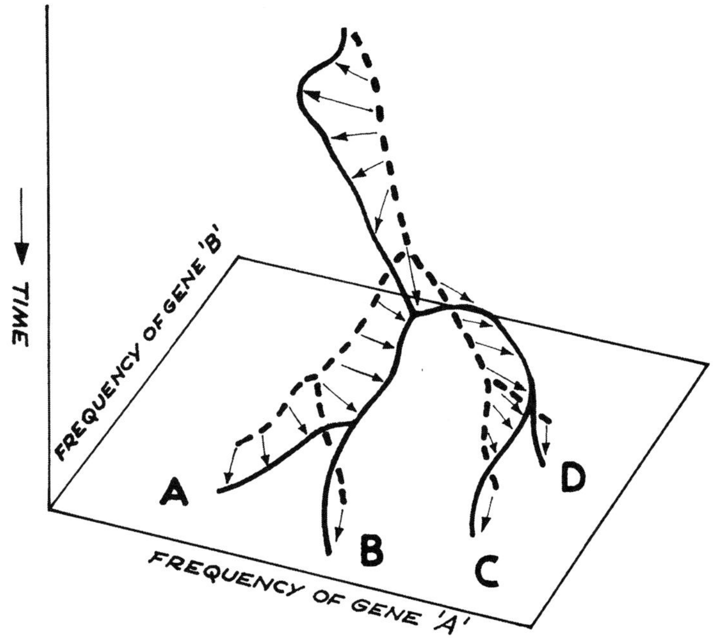

FIG. 2. The same tree as in Figure 1, but with the addition of directional selection constant in space but not time, indicated by arrows. The evolutionary paths without selection (broken lines) are translated into new paths (solid lines) without affecting the relative positions of the populations.

$(1-e^{-t/2N})/4$ (see Fisher, 1958). When $2N$ is large compared with t, this variance reduces nearly to $t/8N$.

In recent papers (Cavalli-Sforza, Barrai and Edwards, 1964; Cavalli-Sforza, 1966) we have given some reasons why we believe that random drift is likely to be more important than formerly believed in determining the variation in gene frequencies in man, and similar considerations may be applied to other organisms.

Selection may be constant in space and time, or may vary in one or the other or both. If selection is constant in space it gives rise to a shift in the gene frequencies of all the populations studied (directional selection, Fig. 2). It would be detected as a trend in the gene frequencies if data from the past were available, but it is not, by definition, a cause of differentiation between populations, and is not detectable in data solely from present-day populations.

Variable selection, however, is probably a major factor in causing divergence. If the variation is sufficiently rapid and haphazard, Kimura's (1954) model of "selective drift" will be appropriate. This model gives rise to a variation in gene frequencies which is almost indistinguishable from the effects of random genetic drift, apart from the behavior at extreme gene frequencies (0 or 1). Thus the consequence of "selective drift" will also be a fluctuation of gene frequencies akin to a random walk, with rate depending on the variability of the selective coefficients. The intensity of this variation will be a property of the gene concerned. If selection is of the stabilizing kind, such as in heterozygotic advantage, it will cause a reduction in the variance of gene frequencies, and its effect will be more or less inextricably confounded with that of random drift.

Prolonged periods of selection peculiar to individual populations will not be detectable without data from the past, and no method of phylogenetic analysis can alter the fact that any observed diversity can be explained by any evolutionary tree provided we are willing to postulate the necessary selection. Our methods invoke no such specific postulates, although if other evidence on selection in a particular population is available it should of course be taken into account. As we shall see, there are not enough degrees of freedom to estimate selection specific to individual populations.

To sum up, selective trends will be detectable only if data from the past are available. Random drift will create a random walk of the gene frequencies, to which selective drift will add, and stabilizing selection subtract, speed. If major and sudden shifts in gene frequency have occurred, because of large selective accidents or other bottlenecks in population development, they will be a source of inaccuracy in the analysis, although their presence may be detectable as departures from expectation indicated by the goodness-of-fit of the models we have developed. The whole problem of the robustness of our methods with respect to varying forms of selection, migration, and other departures from the simple model, may best be studied by Monte Carlo methods, trees being generated according to specific hypotheses, and the estimated forms compared with the known ones.

The Statistical Basis

We have given reasons above why the variation in gene frequencies of each population may be represented by a random walk. To keep the model as simple as possible, it will be supposed that no population becomes extinct, that at each split the daughter populations are both identical to their parent, and that each population is independent of every other one. We shall now examine the statistical assumptions underlying the methods of analysis.

We need hypotheses on the mode of splitting of populations and on the properties of the random walks in the individual branches of the evolutionary tree. The first type of hypothesis will affect only the form of the tree, while the second will affect its dimensions as well; both form and dimensions requiring to be estimated. As we shall see, the number of forms increases very fast with the number of populations. Depending on this number, we shall either test all possible forms or limit our analysis to a group of promising ones.

Given a particular form, the optimum tree may perhaps be estimated by maximum likelihood, or some other method, and the choice between forms will then depend on a comparison of their likelihoods or other appropriate criteria. The simplest model for splitting, which we shall use here, is that in which it occurs at random, as in a Yule process (Yule, 1924): When there are n populations each is assumed to have equal probability, in a given time interval, of generating the $(n+1)$th. On this basis the probabilities of the different forms may be written down explicitly when n is fairly small, and such probabilities will be used, where possible, as prior probabilities in the estimation procedure. The technique for enumeration is to generate the possible "topologies" (that is, tree forms irrespective of the placings of the populations on the terminal branches) for n populations from those for $n-1$ by allowing each terminal branch to split in turn, keeping track of the probabilities. The labelled populations are then distributed on the terminal branches in all possible distinct ways for each topology, each arrangement having equal probability for a given topology, and the final probability of each form thus calculated. This procedure will be clarified in the example to be given below. The calculation of the probabilities for large n has been considered by Harding (1967).

The properties to be assumed for the random walk depend directly on the biological assumptions. Using the transformation of the gene frequencies given above (which will later be generalized to the multi-allelic case), the distribution of the transformed variates will be approximately Gaussian with variance proportional to the time elapsed, while the mean will be constant in the absence of directional selection. We have already decided not to consider individual directional selection of a prolonged nature, and directional selection which is the same for all populations at any given time will not affect their relative movements, as has been indicated above, and therefore need not be considered, although it must be remembered that any inference about the gene frequencies in an earlier population will be based on its presumed absence. Thus the random walk of the gene frequencies may be regarded as a Brownian-motion or Wiener process in the transformed space.

Various models of increasing complexity can be imagined which could, under ideal conditions, be fitted in succession to the data, stopping at the lowest level necessary for a good fit:

1) The simplest model is one in which the Brownian motion has a constant rate which is the same for all characters at all times. This represents the case in which random drift alone determines the variation in gene frequencies, and population size and structure are taken to be constant.

2) The Brownian motion does not have the same rate for all characters. A simple transformation of the scale of each character to standardize the variances will allow us to include selective drift among the evolutionary forces considered, in addition to random drift, since selective drift causes the variances of characters to differ one from another. Stabilizing selection will be confounded with selective drift and thus also included. As an improvement on simply transforming the scales, the variances can themselves be estimated simultaneously with the estimation of the tree.

3) The Brownian motion is not constant in time. If it varied randomly, owing to random fluctuations in population size, this could be taken into account.

4) If nothing is known about the probability distributions of selective coefficients or population sizes, we may restrict the assumptions to that of the independence of events in individual populations, thus retaining the idea of Brownian motion due to the combined effect of random and selective drift, but with no knowledge of local variations in its rate.

We have worked out three essentially different methods. The first is estimation based on the method of maximum likelihood applied to specific models; we have fully developed this approach for model (1) and in outline for model (2), though, of the two, only the former will be described in this paper. Model (3) awaits treatment by maximum likelihood, but model (4) cannot be handled in this way owing to the lack of information on the probability distributions of selective coefficients and population sizes. Indeed, as has been mentioned above, model (4) cannot be treated rigorously by any method.

The second method is the "method of minimum evolution" (Edwards and Cavalli-Sforza, 1963*a*), which uses the intuitive idea that a plausible estimate of the projection of the evolutionary tree onto the "now" plane is given by that tree which invokes the minimum total amount of evolution (see Fig. 1). The assumptions underlying this method are not too clear; it may go some way towards handling model (4), but its success is probably due to the closeness of the solution it gives to the projection of the "maximum-likelihood" tree. The extent of the similarity merits further investigation, and experience with simulated trees should clarify its logical status. It certainly cannot be justified on the grounds that evolution *proceeds* according to some minimum principle, as recently suggested by Camin and Sokal (1965) when they applied it to the discrete case. Its success with discrete data (see also Zuckerkandl, 1965) must also be attributed to its closeness to the solution which a proper probability model would give.

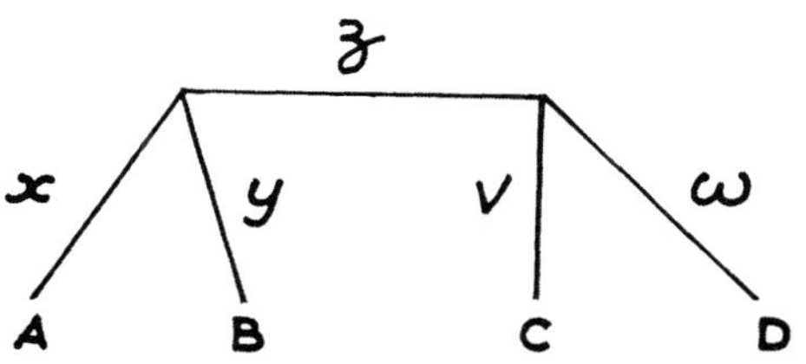

FIG. 3. Model for the "additive" tree, showing expected branch lengths.

Our third method may be called that of the "additive tree" (Cavalli-Sforza and Edwards, 1964). It assumes that distances along the tree are additive, thus implying independence of evolution in all the branches. For example, in Figure 3, the evolutionary distance between population A and the last common ancestor of populations A and B is called x, and that between population B and the same ancestor, y; it is then assumed that the observed distance between A and B, d_{AB}, is an estimate of $x + y$. For this tree we can thus set up six equations:

$$d_{AB} = x + y + \text{"error"}$$
$$d_{AC} = x + z + v + \text{"error"}$$
$$d_{AD} = x + z + w + \text{"error"}$$
$$d_{BC} = y + z + v + \text{"error"}$$
$$d_{BD} = y + z + w + \text{"error"}$$
and $$d_{CD} = v + w + \text{"error."}$$

The "error" terms represent the departures of the observed from the expected distances, and a method of estimation of v, w, x, y, and z is to minimize the sum of the squared errors—the method of least squares.

There is no trunk to the tree of Figure 3 because it is impossible to obtain information on the first split, as in the method of minimum evolution, and the branch corresponding to z is therefore represented by a single line. The tree obtained in this way cannot be represented in the character space, as may be seen by considering the case of three populations, each a distance d from the other. The three-branched least-squares tree (which has zero residual sum of squares) has branches each of length $d/2$, which do not meet if they are supposed to extend inwards from the vertices of an equilateral triangle of side d. They could be made to meet by imposing the appropriate restriction, but this would upset the simplicity of the estimation procedure, and has not been tried. The method does not, therefore, estimate the positions of nodes, either in time or space; but it has the advantage that it can be applied to distances which cannot be represented in a metric space. Apart from the assumption of independence in the separate branches, the justification of this method seems to be much the same as that of the method of minimum evolution, considered above.

DETAILS OF THE METHODOLOGY

Number of tree forms.—It has been noted above that the number of tree forms increases very rapidly with increasing n, the number of populations. The number of forms is in fact $3.5.7 \ldots (2n-3) = (2n-3)!/[2^{n-2}(n-2)!]$ when the first split can be recognized. Such a tree has $2n-1$ branches (including the trunk) so that, in progressing from a tree with $n-1$ populations to one with n, the new branch may be inserted in any one of $2(n-1)-1 = 2n-3$ places. At $n = 10$ this number is 34,459,425. When there is no information about the first split, there is no trunk, and the number of forms is reduced to $3.5.7 \ldots (2n-5)$, giving 2,027,025 for $n = 10$.

It is interesting to note, in passing, that the number of "topologies" (tree forms irrespective of the placings of the populations on the terminal branches) is given, for trees with trunks, by a_n, where

$$a_n = a_1 a_{n-1} + a_2 a_{n-2} + \ldots + a_{(n-1)/2} a_{(n+1)/2} \qquad (n \text{ odd})$$

and $$a_n = a_1 a_{n-1} + a_2 a_{n-2} + \ldots + \tfrac{1}{2} a_{n/2}(a_{n/2} + 1) \qquad (n \text{ even}).$$

This solution is derived by considering the number of ways a tree can be constructed by uniting smaller trees at their trunks, and follows Polya (1937); similar problems had been considered by Cayley (1857, 1859) in connection with the number of ways brackets can be inserted in an algebraic expression. We find $a_{10} = 98$.

With such large numbers it is thus important, when there are more than seven or eight populations, to have methods of generating promising forms for estimation. We use several somewhat intuitive methods (in the hope that they corroborate each other), and then use the forms so found, and similar ones, as a basis for metric estimation, the final choice being according to whatever criterion, such as likelihood, is appropriate.

Choice of promising forms.—Intuitively it is reasonable to suppose that present-day populations near to each other in the character space should be clustered together on the same part of the evolutionary tree, so that methods of clustering points should give some insight into the most promising form. Among the methods used, one (Edwards and Cavalli-Sforza, 1963*b*, 1965; see also Ward, 1963) divides the populations into the two clusters for which the between-clusters sum of squares is a maximum, and each cluster is then similarly treated, and so on until a complete breakdown of the original cluster has been made.

Apart from this method, we have used two others especially well-suited to the application of the method of minimum evolution. The first is based on a theorem due to Prim (1957), who showed that when branches are constrained to meet only at populations, the net of shortest length may be found by listing all the pairwise distances between populations in order of ascending magnitude and allocating branches successively to these distances, omitting any branch which completes a loop. The resulting net is the shortest net conditional on each node coinciding with one of the populations; removing the constraint, the net may be shortened step by step, as described below. The second method uses the same iterative procedure, but starting from the case in which all the nodes coincide at a single point, so chosen that the total length is then a minimum.

In the absence of a means of maximizing likelihood over different tree forms, these methods have shown promise of leading to near-optimal forms. The last two do not, however, give any indication of the position of the first split, so that several positions must be tried.

Transformation of data.—We will concentrate on populations of nonhaploid organisms, whose evolution is suitably studied by way of continuous data, be they gene frequencies or metrical characters. We have mentioned the usefulness of the angular transformation of gene frequencies, whereby the variances are standardized irrespective of the frequencies themselves (at least in the interval 0.05 to 0.95). It was pointed out by Fisher (personal communication; see also Cavalli-Sforza and Conterio, 1959) that this transformation could be generalized to multi-allelic loci, as follows.

If each of m alleles at the locus is allotted a Cartesian axis in a Euclidean space of m dimensions, and a population with gene frequencies $p_1, p_2, \ldots p_m$ is represented by the vector $(\sqrt{p_1}\ \sqrt{p_2} \ldots \sqrt{p_m})$, then the space of possible populations is the $\frac{1}{2}^m$th part of the surface of the unit hypersphere in which all the coordinates are non-negative, the population being represented by a point unit distance from the origin with direction cosines given by the above vector (Fig. 4). It follows that the angular distance between two populations with gene frequencies $p_1, p_2, \ldots p_m$ and $p_1', p_2', \ldots p_m'$ is given by θ where $\cos\theta = \sum_{i=1}^{m} \sqrt{p_i p_i'}$. Since $\theta = \pi/2$ corresponds to a complete gene substitution, it is convenient to work in terms of $2\theta/\pi$, where θ is in radians, for the unit distance is then one gene substitution.

In this representation a population may be thought of as pursuing a random walk in the curved space. Since this space is finite, with known bounds, the Gaussian

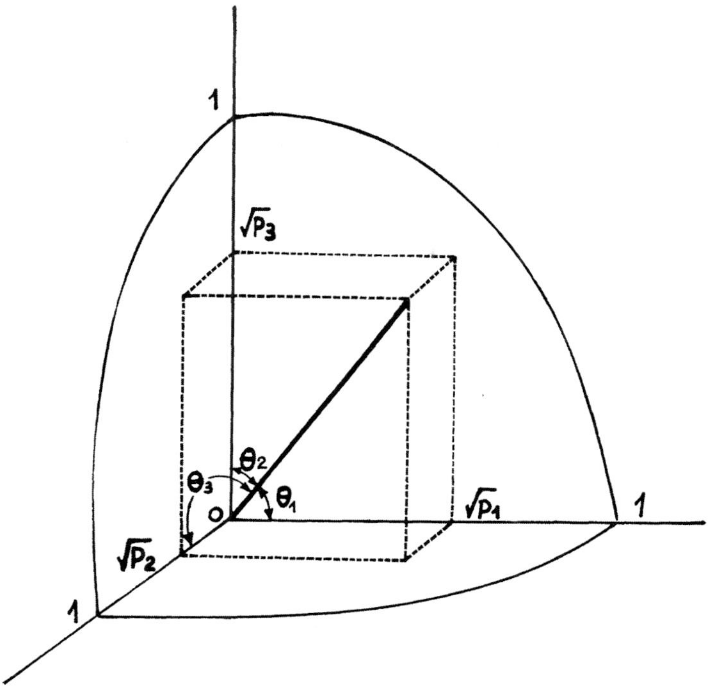

FIG. 4. Representation of a population, with gene frequencies p_1, p_2, p_3 at a single triallelic locus, on the octant of a sphere.

approximation to the gene frequency distribution will only hold if the variance is sufficiently small and the population sufficiently far removed from an edge of the space for edge effects to be neglected. The method of maximum likelihood could ideally now be applied to a model based on this transformation of the data, the log-likelihoods being summed over loci, but it turns out to be intractable owing to the curved space and difficulties with the coordinate system, so that it is necessary to approximate the curved space in the region of the populations by a Euclidean space of $(m-1)$ dimensions by means of a projection of one onto the other. An orthogonal projection onto the hyperplane tangent to the hypersphere at the centroid of the populations should suffice, although in the present work we have simply used, as the distance between two populations an arc $2\theta/\pi$ apart, the length of the chord joining them, which is $(2\sqrt{2}/\pi)$ $(\sqrt{1-\cos\theta})$. Thus the m-dimensional Euclidean space in which the hypersphere is embedded has itself been employed.

These Euclidean spaces for the separate loci (assumed unlinked) may then be combined, distances being given by Pythagoras' theorem in many dimensions, so that the square of the distance between two populations is given by the sum of the squared distances for each locus. In this way the data are represented in a Euclidean space, the scale of which is one unit per gene substitution.

Another type of continuous data of some interest is that in which measurements can only be made directly on the pairwise distances between populations. Such is the case, for instance, when "immunological" distances between populations are investigated by serological methods, or when differences in nucleotide sequences are estimated using hybridization procedures with nucleic acids. In these cases, data consist

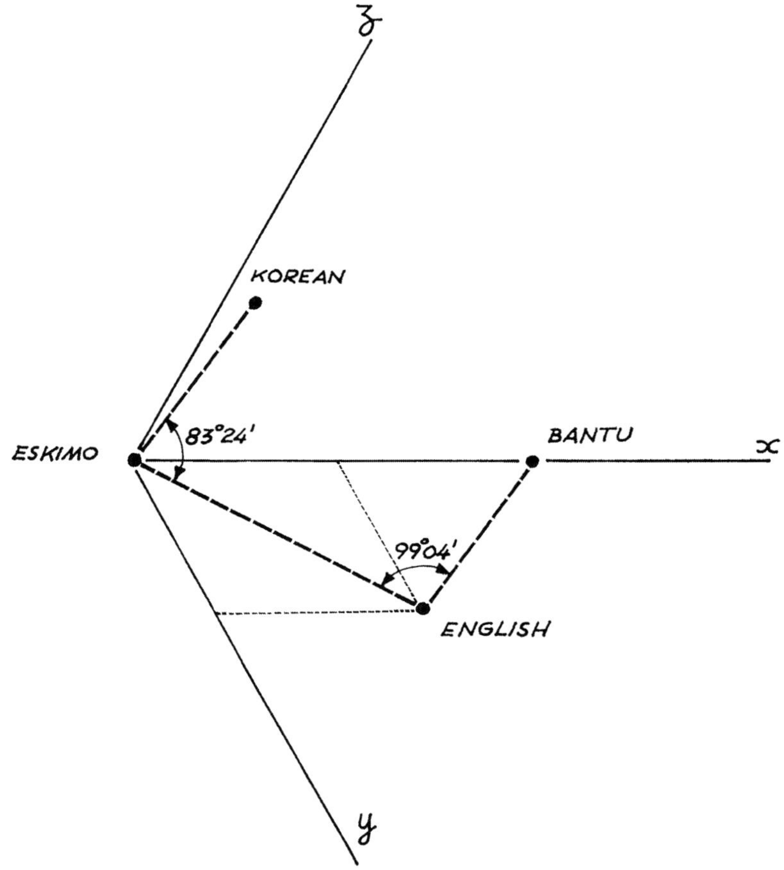

FIG. 5. Representation of the four populations in Euclidean space, showing the Prim network.

of a triangular matrix of the pairwise distances between populations, which is also the form to which multi-locus gene frequency data has been reduced by the methods described above. But whereas in the latter case the erection of a Cartesian coordinate system in Euclidean space by repeated applications of Pythagoras' theorem is bound to succeed, in the former case it will often fail, the method then generating complex numbers. But even if a nonlinear scale transformation allowing Euclidean representation cannot be found, the procedures for cluster analysis and for finding an "additive tree" by least squares may still be applied. One advantageous by-product of generating coordinates from the pairwise distances is that the maximum number of dimensions required is one less than the number of populations, however many characters (and hence dimensions) there were originally.

The case of ordinary metrical characters has been treated only to a very limited extent because of the difficulties in interpreting their genetic basis. We considered them when analyzing anthropometric data in one of the earlier papers (Cavalli-Sforza and Edwards, 1964), but the results were not very encouraging. It seems that the best procedure is to transform the original characters into a set of uncorrelated standardized variables, using the within-populations dispersion matrix (see, for example,

Rao, 1952). The lack of the complete matrix for the data just mentioned must have contributed to the unsatisfactory nature of the solution.

Estimation based on maximum likelihood.—In Euclidean space of p dimensions, the probability density a distance d from an original population after a time t has elapsed is $(1/\sigma\sqrt{2\pi t})^p \exp(-d^2/2t\sigma^2)$, owing to the Gaussian nature of the random walk, where σ^2 is the variance per unit time. The log-likelihood of a branch of length d and interval t is therefore $-(d^2/2t\sigma^2 + \frac{1}{2}p.\log 2t\sigma^2 + \frac{1}{2}p.\log \pi)$. Writing T for $2t\sigma^2$ (time and variance being confounded in their product, as is to be expected), omitting the constant and changing the sign, this becomes $d^2/T + \frac{1}{2}p.\log T$.

Let x_{ir} be the rth spatial coordinate of the ith node (or population) and t_i its time coordinate, measured backwards from the "now" plane. For the branch joining nodes i and j $(t_j > t_i)$ the above expression then becomes

$$\sum_r (x_{jr} - x_{ir})^2/(t_j - t_i) + \tfrac{1}{2}p.\log(t_j - t_i). \quad (1)$$

Summing over all the branches (denoted by (i, j)), the quantity to be minimized is

$$L = \sum_{(i,j)} [\sum_r (x_{jr} - x_{ir})^2/(t_j - t_i)] + + \tfrac{1}{2}p \sum_{(i,j)} \log(t_j - t_i). \quad (2)$$

If j, k, and l are the three nodes (or, in the case of k and l, possibly populations) connected by single branches to node i $(t_j > t_i > t_k, t_l)$, then

$$\tfrac{1}{2}\frac{\partial L}{\partial x_{ir}} = \frac{x_{ir} - x_{jr}}{t_j - t_i} + \frac{x_{ir} - x_{kr}}{t_i - t_k} + \frac{x_{ir} - x_{lr}}{t_i - t_l} = 0. \quad (3)$$

Each node yields a similar equation linear in the $x_{.r}$, the net result being a system of linear equations which enables the rth spatial coordinates of the nodes to be expressed in terms of the rth spatial coordinates of the populations and the time coordinates of the nodes, by inverting a matrix which is the same for each dimension. Assuming initial values for the time coordinates, the numerical values thus found can be inserted in the first and second partial derivatives of L with respect to the t's (these will not be quoted explicitly, but may be immediately written down from (2)), giving the scores and the information matrix. Corrections to the t's are found in the usual way, and the cycle repeated. By means of this reduction of the likelihood equations the order of the information matrix shrinks from $(n-1)(p+1)$ to $(n-1)$.

In order to begin iteration, initial estimates of the t's are required, and these may be found from cluster analysis. Considering two populations at time $T = 0$ a distance d apart, the maximum-likelihood estimate of the time coordinate of the subtending node is $d^2/2p$, while the variance of the two-population cluster is $d^2/4$. Putting the time coordinate of the subtending node equal to $2/p$ times the variance of the cluster, of however many populations, thus provides a rough estimate.

It may be noted that of the $n(p+1)$ original degrees of freedom $(n-1)(p+1)$ are used in estimating the parameters, leaving $(p+1)$. If it is not assumed that the variance of the random walks is the same in each direction, $(p-1)$ relative variances must also be estimated, leaving two degrees of freedom for a goodness-of-fit test. There are not sufficient degrees of freedom to estimate covariances as well, although these will be small if the original data have been transformed into uncorrelated variables using the between-populations dispersion matrix.

Unfortunately, except in the very simplest cases, this straightforward application of the method of maximum likelihood leads into difficulties due to the fact that the likelihood surface contains singularities, as may be seen from the following example (Cavalli-Sforza and Edwards, 1966). Since the log-likelihood of the tree is found by summing the log-likelihoods of the individual branches, each node should, in the final solution, be in the maximum-likelihood position with respect to the three adjacent nodes if these are regarded as fixed. In particular, given the positions and times of the second and third splits, those of the first should be determinable. Let us therefore consider the simple case of the determination of the origin of a Brownian-motion process when we have observed just two different populations: at time t_1 population 1 was at x_1, and at time t_2 population 2 was at x_2. Let the origin of the process, which we are required to estimate, be at (X, T). The quantity to be minimized is, from equation (2) above,

$$L = \frac{(X - x_1)^2}{T - t_1} + \frac{(X - x_2)^2}{T - t_2} + \\ + \tfrac{1}{2} \log (T - t_1) + \tfrac{1}{2} \log (T - t_2). \quad (4)$$

This surface can be plotted for varying X and T, but it is already obvious that it contains an unfortunate characteristic: suppose $t_1 > t_2$ (so that population 1 is earlier in time, which, it may be remembered, is being measured backwards), and suppose that we trace the path $X - x_1 = T - t_1$ along the surface towards (x_1, t_1). Then the first term in L will tend to zero, the second will tend to the constant $(x_1 - x_2)^2/(t_1 - t_2)$, but the third will become indefinitely large and negative as T approaches t_1, although the fourth will tend to $\tfrac{1}{2} \log (t_1 - t_2)$. Thus, provided the two observations were not made simultaneously, in which case $t_1 = t_2$, the negative log-likelihood can be made as large and negative as we please, and hence the log-likelihood as large as we please, simply by letting the origin approach the first observation along a certain path; the surface contains, in fact, a singularity at the point (x_1, t_1), and any iterative procedure, unless it is started near a well-behaved peak elsewhere in the surface, will lead to the coincidence of the origin with the first observation. It is convergence to such singularities which has led to the trivial solutions obtained when studying complete trees. Needless to say, such solutions are intuitively unacceptable, although this is not the place to consider in detail the problems of inference which they expose. Short of portraying the entire likelihood surface, which is impossible owing to the number of parameters involved, there is no fully satisfactory solution.

However, we have developed another approach which seems to us to be moderately satisfactory. Recalling that, given the time coordinates, the spatial coordinates of the nodes can be found by maximum likelihood without difficulty, the problem can be solved by finding the time coordinates by another method. One such method has already been given above, but it was intended only as an initial approximation. We now propose to fit, according to the least-squares criterion, the observed length squared of each branch to its expected squared length derived from the interval between its ends. On the Gaussian model this expected squared length is easily seen to be the interval itself.

We therefore minimize, in the notation of equations 1 to 3,

$$S = \sum_{(i,j)} \left[\sum_r (x_{jr} - x_{ir})^2 - (t_j - t_i) \right]^2, \quad (5)$$

with respect to the $t_.$, the time coordinates of the nodes. Differentiating with respect to t_i,

$$\tfrac{1}{2} \frac{\partial S}{\partial t_i} = \sum_r (x_{jr} - x_{ir})^2 - \sum_r (x_{ir} - x_{kr})^2 - \\ - \sum_r (x_{ir} - x_{lr})^2 + 3t_i - t_j - t_k - t_l \\ = 0. \quad (6)$$

Each node yields a similar equation linear in the $t_.$, enabling them to be expressed in terms of the $x_{..}$ by inverting a matrix. The values thus obtained are used in the method of maximum likelihood to generate new spatial coordinates for the nodes (equation 3 above), and the cycle repeated until convergence is attained. The procedure amounts to the iterative solution of two sets of equations exemplified by 3 and 6 above, and is a generalization of that pro-

TABLE 1. *Blood-group gene frequencies characterizing four populations.*

	Eskimo	Bantu	English	Korean
A_1	0.2914	0.1034	0.2090	0.2208
A_2	0.0	0.0866	0.0696	0.0
B	0.0316	0.1200	0.0612	0.2069
O	0.6770	0.6900	0.6602	0.5723
CDE	0.0	0.0	0.0024	0.0082
CDe	0.4985	0.1400	0.4205	0.6197
cDE	0.4906	0.0100	0.1411	0.3148
cDe	0.0109	0.6000	0.0257	0.0573
Cde	0.0	0.0200	0.0098	0.0
cdE	0.0	0.0	0.0119	0.0
cde	0.0	0.2300	0.3886	0.0
MS	0.1719	0.0900	0.2377	0.0245
Ms	0.6703	0.4800	0.3048	0.4615
NS	0.0	0.0400	0.0703	0.0646
Ns	0.1578	0.3900	0.3872	0.4494
Fy^a	0.7500	0.0600	0.4213	0.9950
Fy^b	0.2500	0.9400	0.5787	0.0050
Di^a	0.0	0.0	0.0	0.0313
Di^b	1.0	1.0	1.0	0.9687

posed by Cavalli-Sforza and Edwards (1966) to solve the two-population example considered earlier.

Having had to invoke the method of least-squares for reasons of expediency, it is necessary to defend the decision not to solve the problem entirely by this method (maximizing S with respect to the spatial as well as the time coordinates of the nodes), but rather to rely on a hybrid between least-squares and maximum-likelihood. We retain the method of maximum likelihood because

1) We prefer it as a method of estimation where it does not break down;

2) It is more tractable in the present case, as may be seen by differentiating equation 5 with respect to the $x_{..}$;

3) Likelihood provides a more useful criterion of acceptability than the residual sum of squares, as it may be directly compounded with prior probabilities.

Although it is very probable, it remains to be shown formally that the least-squares solution converges to the local maximum of the likelihood in well-behaved examples where the singularities may be discarded, and it will be worthwhile to try the full maximum-likelihood procedure using the estimates obtained by the above procedure as starting values.

It has been pointed out to us by D. G. Kendall and D. F. Kerridge (separate personal communications) that, in writing down the likelihood of a tree, we have omitted the contribution due to the "Poisson" nature of the Yule process. If we knew that n, and only n, populations *could* have been produced by the process, then the likelihood surface for the times of the splits would indeed be uniform, as we have supposed. But this is not really our model: In fact, by using a Yule process, we admit the possibility that numbers other than n could have been generated on the same model. The likelihood surface, conditional on n, is then easily seen to be proportional to $e^{-\lambda\Sigma t_i}$, where Σt_i is the sum of the times of the nodes and λ is the rate of the process. The question arises, however, of whether even this is the appropriate model, for we have not in fact observed a tree with exactly n populations, but rather we have chosen to observe n from a tree which may have had more. The likelihood surface for the nodes of a tree with n populations which have been chosen at random from a randomly-generated tree with m $(m \geqslant n)$ populations appears to be unknown. It may even be uniform.

However, even the use of the likelihood $e^{-\lambda\Sigma t_i}$ would not affect our argument, for it would weight the likelihood in favor of still shorter times for the branch lengths, thus aggravating rather than alleviating the problem of the singularities. Recourse to the least-squares modification would still be necessary, and, since the likelihood is only used to derive the spatial coordinates for given times, the addition of the quantity $-\lambda\Sigma t_i$ to the log-likelihood would not affect the estimates. It would, of course, affect the likelihood comparisons of different tree forms, though since we do not know λ, and cannot estimate it by maximum likelihood, and since we are in

TABLE 2. *Gene substitutions separating four populations, measured along the arc ($2\theta/\pi$) and along the chord ($[2\sqrt{2}/\pi]\sqrt{1-\cos\theta}$).*

		A_1A_2BO	Rh	MNSs	Fy	Di	Combined
Eskimo	arc	0.2585	0.7274	0.2222	0.5091	0.0	–
—Bantu	chord	0.2568	0.6886	0.2211	0.4957	0.0	0.9136
Eskimo	arc	0.1822	0.4718	0.2852	0.2170	0.0	–
—English	chord	0.1816	0.4611	0.2828	0.2159	0.0	0.6101
Eskimo	arc	0.1874	0.1461	0.3050	0.2883	0.1132	–
—Korean	chord	0.1868	0.1457	0.3021	0.2858	0.1131	0.4918
Bantu	arc	0.1094	0.4907	0.1554	0.2922	0.0	–
—English	chord	0.1092	0.4787	0.1549	0.2896	0.0	0.5907
Bantu	arc	0.2270	0.6398	0.1024	0.7974	0.1132	–
—Korean	chord	0.2258	0.6132	0.1023	0.6871	0.1131	1.0036
English	arc	0.2178	0.4465	0.2293	0.5052	0.1132	–
—Korean	chord	0.2167	0.4374	0.2280	0.4921	0.1131	0.7384

any case not certain of the model, we have omitted it in the following example.

It should be noted that, in the foregoing argument, if the t_i are in units of $2\sigma^2$ according to our standard notation, λ is in units of $1/2\sigma^2$, so that it is the expression $\lambda/2\sigma^2$ that is really relevant.

Estimation by the method of minimum evolution.—The method of minimum evolution (Edwards and Cavalli-Sforza, 1963*a*) uses the intuitive idea that a plausible estimate of the projection of the evolutionary tree onto the "now" plane is given by that tree uniting all the populations which invokes the minimum total amount of evolution. With suitably transformed scales, a unit quantity of evolution is a unit vector in this space, and therefore the required tree is that with the minimum total length possible. As has been mentioned, the position of the first split is undefined in the projection, and, of course, the method of minimum evolution gives no information on the time coordinates of the nodes.

The problem of finding the minimum-length tree may be referred to as the Steiner problem in p dimensions (see Courant and Robbins, 1960), and no algorism for its general solution is available, although various things are known. For example, it may be proved that, in any number of dimensions, each node must be the meet of just three branches mutually inclined at 120°, unless the node coincides with a population. It follows that, for a given tree form, a computer program may be written which will converge to the Steiner tree *for that form*, because if any two intersecting branches meet at an angle of less than 120° they may be replaced by a Y-shaped "Steiner triplet" whose inter-branch angles are 120°, the total length thus being shortened. Unless the tree form is close to the optimum, however, many branches will converge to zero length, but this fact may be used to give an indication of how the form can be improved. A Steiner tree with no branches of zero length (except those generated by nodes coinciding with populations) may be called "stable." There will usually be many stable nets for a given configuration of populations, so that stability itself is no guarantee that the minimum net has been

TABLE 3. *Coordinates of the four populations.*

	x	*y*	*z*
Eskimo	0.0	0.0	0.0
Bantu	0.9136	0.0	0.0
English	0.4695	0.3895	0.0
Korean	0.0379	0.0428	0.4885

TABLE 4. *Possible tree forms with four populations.*

	(a) rooted			
1	Esk	Ban	Eng	Kor
2	Esk	Ban	Kor	Eng
3	Esk	Eng	Ban	Kor
4	Esk	Eng	Kor	Ban
5	Esk	Kor	Ban	Eng
6	Esk	Kor	Eng	Ban
7	Ban	Eng	Esk	Kor
8	Ban	Eng	Kor	Esk
9	Ban	Kor	Esk	Eng
10	Ban	Kor	Eng	Esk
11	Eng	Kor	Esk	Ban
12	Eng	Kor	Ban	Esk
	(each with probability 1/18)			
13	Esk	Ban	Eng	Kor
14	Esk	Eng	Ban	Kor
15	Esk	Kor	Ban	Eng
	(each with probability 1/9)			
	(b) unrooted			
1′	Esk	Ban	Eng	Kor
2′	Esk	Eng	Ban	Kor
3′	Esk	Kor	Ban	Eng
	(each with probability 1/3)			

found, and, as with other methods, many different forms must be investigated.

As indicated on page 557 above, we can try the form derived from Prim's network by calculating all the interbranch angles, and replacing the two branches subtending the smallest angle by the Steiner triplet, and so on, treating the angles in order of increasing size. Alternatively, we can proceed in the same way starting with all the nodes coinciding at a central point so chosen that the total length is then a minimum.

Since the method seeks the tree of minimum length, this length is the criterion of acceptability, akin to likelihood on the full model, but the relation between these two quantities needs further study. A more detailed account of these methods for solving Steiner's problem in p dimensions is given by Edwards (1967).

Estimation on the additive tree model.—If $\mathbf{D}$ is the column vector of the $\frac{1}{2}n\cdot(n-1)$ pairwise distances between populations, $\mathbf{W}$ the column vector of the $2n-2$ branch lengths to be estimated, and $\mathbf{B}$ the $\frac{1}{2}n(n-1)\times(2n-2)$ matrix specifying the form of the tree, in which the element in the rth row and c^{th} column is 1 if the path uniting the rth pair of populations includes the cth branch, and 0 otherwise, then the least-squares estimate of $\mathbf{W}$ is given by $\hat{\mathbf{W}}=(\mathbf{B}^*\mathbf{B})^{-1}\mathbf{B}^*\mathbf{D}$, where $\mathbf{B}^*$ is the transpose of $\mathbf{B}$. Thus the corresponding expected value of $\mathbf{D}$ is $\hat{\mathbf{D}}=\mathbf{B}\hat{\mathbf{W}}$, the residual sum of squares is $\mathbf{D}^*\cdot(\mathbf{D}-\hat{\mathbf{D}})$ and it follows that the variance-covariance matrix of $\hat{\mathbf{W}}$ is $\mathbf{D}^*(\mathbf{D}-\hat{\mathbf{D}})\cdot(\mathbf{B}^*\mathbf{B})^{-1}/[\frac{1}{2}(n-1)(n-4)]$. If the errors in the observed pairwise distances are not independently distributed, or have unequal variances, the corresponding variance-covariance matrix should be incorporated in the estimation procedure, as described in Kendall and Stuart (1961); but this improvement has not been used in the present work.

Corresponding to a branch of zero length in the method of minimum evolution, a branch of negative length in the least-squares method (there is no restriction against negative branches appearing) indicates the need to change the tree form. Indeed, it is not clear that a tree with no negative branches may always be found. The residual sum of squares is the criterion for choosing between trees of different form, although the introduction of prior probabilities seems to be impossible. In comparing branch lengths found by the "additive-tree" and "minimum-evolution" methods, it should be noted that not only

TABLE 5. *Best tree found by cluster analysis, showing sums of squares removed by each split.*

0.5421	0.1210	Eskimo
		Korean
	0.1745	Bantu
		English

Total sum of squares = 0.8375

TABLE 6. *"Additive-tree" and "minimum-evolution" results for the unrooted tree forms.*

Form	Residual sum of squares ("additive tree")	Total amount of evolution ("minimum evolution")
1′	0.07054	1.7998
2′	0.08105	1.7850
3′	0.00037	1.6173

does the former method give trees unrepresentable in the character space, as mentioned earlier, but also that the longer branches are likely to be considerably shorter than their "minimum-evolution" counterparts.

AN EXAMPLE WITH FOUR POPULATIONS

In order to illustrate the use of our methods in detail, in this section we present an exhaustive treatment of a case with four populations.

The data consist of the gene frequencies for the five blood-group systems A_1A_2BO, RH (four sera), MNSs, Fy, and Di, for samples from four human populations: Eskimo, Bantu, English, and Korean. These data, which are given in Table 1, are part of a larger collection that we have used in previous communications. They are drawn from the following sources: Eskimo—Chown and Lewis (1959); Bantu—quoted by Zoutendyk, Kopeć, and Mourant (1955), with the subdivision of group *A* provided by Mourant (personal communication), and Di^a presumed absent; English—Race and Sanger (1959); Korean—Won, Shin, Kim, Swanson, and Matson (1960), with subdivision of MN by *S* and *s* according to the Japanese proportions reported by Lewis, Kaita, and Chown (1957). No special significance should be attached to this choice of samples, which was made largely for convenience.

From these data the values of $2\theta/\pi$ for the pairwise distances are calculated for each locus, and are given in Table 2. Thence the pairwise chords are found $[(2\sqrt{2}/\pi)(\sqrt{1-\cos\theta})]$, and the overall pairwise distances computed by taking the square root of the squared distances summed over loci, for each pair (Table 2). Finally (as far as the preparation of the data is concerned) a coordinate system, relative to arbitrary Cartesian axes, is erected for the four points, the coordinates being given in Table 3. The consequent arrangement of the points is shown in Figure 5.

Applying the appropriate formulae, we find there to be $5 \times 3 = 15$ possible rooted tree forms, and 3 possible forms without a root; these are listed in Table 4, together with their prior probabilities, which have been found by direct enumeration. In order to demonstrate the normal course of the analysis we first group the populations by cluster analysis (Table 5) and we also find the Prim network (Figure 5, in which the interbranch angles are also given). Cluster analysis leads to form 15, in which the initial split is into Eskimo and Korean on the one branch, and Bantu and English on the other, while, on an examination of the angles, the Prim network suggests the same form, but unrooted (3′). The second best cluster analysis leads to form 6.

The "additive-tree" and "minimum-evolution" results for the three possible unrooted forms are given in Table 6, from which it will be seen that form 3′ is, in both instances, again the best. Table 7 gives the dimensions of form 3′ on the two methods, and on the maximum-likelihood method (see below); the other two forms lead to zero (minimum-evolution) or negative (additive-tree) values for the central branch, thus indicating the need

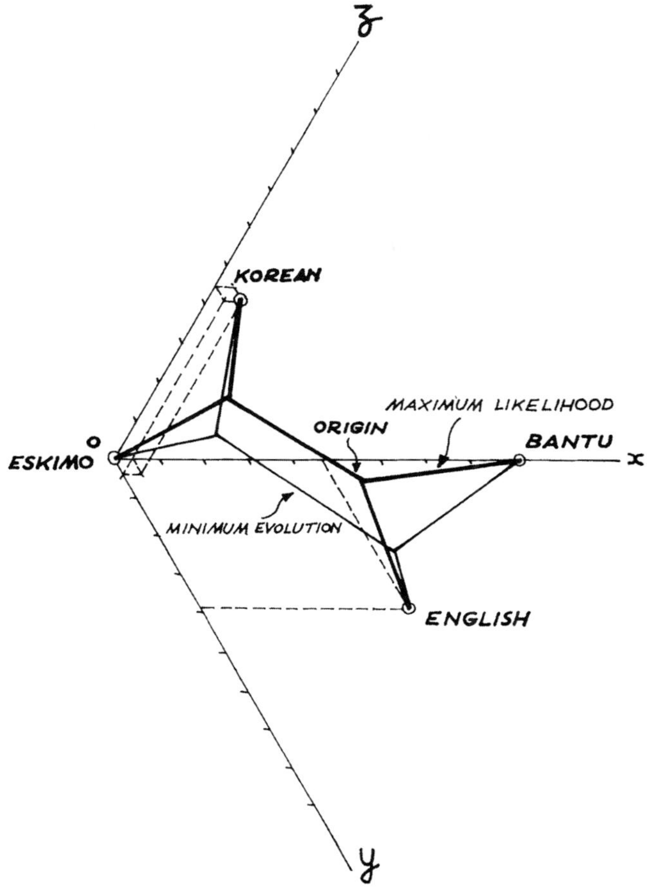

FIG. 6. The "minimum-evolution" tree compared with the projection of the "maximum-likelihood" tree into the "now" space.

for a change of form. The node coordinates for the "minimum-evolution" solution will be found in parentheses in Table 8.

Applying the modified maximum-likelihood method it transpires that only two of the rooted tree forms, numbers 6 and 15, give convergent results, the likelihood of 15 being 3.826 times that of 6. Branches of negative time interval are generated with the remaining thirteen forms. Since form 15 has twice the prior probability of form 6, the final likelihood ratio in its favor is 7.652. Its node coordinates are given in Table 8, and its spatial dimensions in Table 7. Figure 6 shows its projection in the "now" space, together with the "minimum-evolution" tree. In passing, Table 9 gives the node coordinates of the second best "maximum-likelihood" tree, of form 6.

We thus find that all the methods tried lead us to form 15 (or its unrooted equivalent, 3′), and the dimensions of this tree given by the modified method of maximum likelihood represent, we believe, the best estimates of the evolutionary paths on the given data. For the present we are not prepared to quote standard deviations for our estimates because of the complexity introduced by the number of possible forms. The problem is probably best approached by a numerical examination of the consequences of small displacements of the populations in the character space.

TABLE 7. *Dimensions of the best tree (form 3′) according to the three methods.*

Branch	Length in gene substitutions		
	"additive tree"	"minimum evolution"	"maximum likelihood"
Eskimo	0.1913	0.1998	0.2451
Korean	0.3005	0.3605	0.2946
Bantu	0.4375	0.5169	0.4895
English	0.1532	0.1270	0.2720
central	0.2752	0.4132	0.3588
Total	*	1.6173	1.6600

"Central" refers to the branch uniting the two splits.

* In comparing these lengths it must be remembered that the "additive tree" cannot be represented in the character space. The comparative shortness of the "central" branch was anticipated in the section on the "method of minimum evolution."

TABLE 8. *Node coordinates for the "maximum-likelihood" and "minimum-evolution" trees of form 15 and 3′.*

Node	x	y	z	Time
top	0.4238	0.1258	0.0972	0.1698
Esk-Kor	0.1229 (0.1221)	0.0482 (0.0747)	0.2065 (0.1394)	0.0694
Ban-Eng	0.4506 (0.4710)	0.1327 (0.2656)	0.0875 (0.0276)	0.1609

("minimum-evolution" coordinates in parentheses)

DISCUSSION

The introduction of automatic computing methods in phylogenetic analysis has the advantage of objectivity, but this is not the greatest advantage. An automatic procedure needs a clearly formulated set of rules, and discussion of the acceptability of the rules requires a knowledge of the aims. Clarification of thought is thus likely to result, and may be more important than the improvement in precision and speed brought about by the automatic procedure. In particular, by introducing a specific model the limitations of the method will be clear for all to see.

The reconstruction of evolutionary trees is a type of inductive inference which is likely to be especially weak. Relevant evidence comes from many sources, and is often conflicting and difficult to weigh; the sheer mass of evidence may be a bar to its interpretation. Under these conditions, an objective method capable, in principle, of assaying the strength of the evidence, of testing the goodness of fit of specific models, and of using various kinds of data, should be considered a step forward.

Ideally, before the use of any such method can be recommended with complete confidence, we would like to be able to state that it gives a satisfactory answer in well-known test cases. The most hopeful sources of data at the moment are gene frequency data in cattle and man, and nucleic acid hybridization and protein structure analyses. In this and previous publications we have used gene frequency data in man both on account of its suitability and its availability. Unfortunately, however, there is not much guidance from other sources as to the recent evolutionary history of man, and the material is thus not very suitable for testing our methods. But we have been encouraged by the reasonableness of results obtained on very modest amounts of data, as, for example, the form of the tree found for fifteen human populations using the same type of data as we have used in the present paper (Edwards and Cavalli-Sforza, 1964; Cavalli-Sforza, Barrai, and Edwards, 1964). In plotting the tree on a map of the world, the actual branch lengths were unavoidably, and admittedly, distorted, and only the form was undisturbed. It is therefore clear that a criticism levelled against this map—that the Maoris seem to "stem phylogenetically from the natives of Alaska" (Simpson, 1965)—is not valid. The branch leading to the Maoris could equally well have been drawn in many other ways without disturbing the basic form, and it should also be remembered that other forms with somewhat similar branching sequences were not much less likely than the one shown on the map: forms, as well as dimensions, are subject to statistical error. But given that

TABLE 9. *Node coordinates for the second best "maximum-likelihood" tree, of form 6.*

Node	x	y	z	Time
Top (Ban)	0.4325	0.0971	0.1040	0.2117
Eng	0.2940	0.1251	0.1339	0.1508
Esk-Kor	0.2620	0.1130	0.1467	0.1415

only a minute fraction of the available data was used, and that no geographical information was incorporated in the estimation procedure, the result seems encouraging.

Among the limitations of our model that have been considered in previous sections, inability to handle hybridization, convergence and parallelism (that is, the similar selective response to similar environmental stimuli in different populations) needs particular consideration. While hybridization gives rise to loops in the tree, which might possibly be detected, convergence and parallelism cause a breakdown of the assumption that evolution proceeds independently on each branch of the tree. This assumption is basic to our model (excepting the case of directional selection constant in space: see Figure 2), and probably basic to any tractable model for evolutionary divergence. In the absence of significant loops, the breakdown of this assumption is likely to be the major cause of poor fits.

However, it seems unlikely that this breakdown can occur in such a way as to cause extensive similarity at the genotypic level between organisms that have diverged greatly in the past and have since evolved in a common environment. Some convergence may be detected at a few loci, in which case a choice of loci might have to be exercised. Unfortunately, so little is known about the selective pattern of most genes that such a choice will be difficult, if not dangerous, most of the time. In man, the same or similar environments, like malaria, can bring about different selective responses, which depend on the genetic background of the exposed population, on its history, and on other environmental factors. In the case of malaria quite a variety of genetic adaptations have been observed, thalassemia, G6PD, and hemoglobins S and E being the best-known examples.

To sum up, we cannot do better than repeat an earlier warning, that "prolonged periods of selection peculiar to individual populations will not be detectable without data from the past [or other information about the selective situation], and no method of phylogenetic analysis can alter the fact than any observed diversity can be explained by any evolutionary tree provided we are willing to postulate the necessary selection." But there is reason to believe that, where enough different genes are considered, the effects of truly convergent genes will be swamped by the larger number of genes behaving independently in different populations.

In the example we have used, the best estimate for the time of the first split was 0.17 units of $2t\sigma^2$ ago, where t is the number of generations and σ^2 the variance per generation. Putting $\sigma^2 = 1/8N$, the variance due solely to drift in a population of effective size N, $t = 0.68\ N$ generations. Such estimates should serve not only to increase our knowledge of evolutionary history, but also to help us to understand the evolutionary process. For example, if independent evidence corroborates an estimate of the time taken the model is to some extent vindicated, particularly if it can be shown that the assumed population sizes are sufficient to have maintained the observed numbers of alleles.

SUMMARY

An attempt has been made to establish a procedure for estimating the course taken by evolution. The model used is that of a branching random walk, which is strictly valid only when the causes of divergence between populations are random genetic drift and variable selection. With suitable transformations of the data, evolution can then be considered as a branching Brownian-motion process. To keep the model as simple as possible it was supposed that no population becomes extinct and that each population splits, at a random time, into two daughter populations each identical to its parent. The problem was to estimate the form and dimensions of the most probable tree uniting the presently living populations. The ideal method of estimation, maximum likelihood, proved difficult and had to be replaced in part by alternative procedures. In addition to describing the available procedures in detail, a simple example is worked out fully, and the logical content and limitations of the methods are considered in depth.

ACKNOWLEDGMENTS

This work has been supported by grants from the U.S. Atomic Energy Commission and by EURATOM-CNR-CNEN Contract No. 012-61-12 BIAI.

We are particularly indebted to Dr. Laura Zonta for her painstaking cooperation in the heavy numerical work involved in the development of our methods, and to the late Professor Vittorio Galafasi and Mr. E. F. Harding for illuminating discussions on tree forms.

The final draft of this paper was prepared after A.W.F.E. had taken up an appointment in the University of Aberdeen.

LITERATURE CITED

CAMIN, J. H., AND R. R. SOKAL. 1965. A method for deducing branching sequences in phylogeny. Evolution **19**: 311–326.

CAVALLI-SFORZA, L. L. 1966. Population structure and human evolution. Proc. Roy. Soc., London, B. **164**: 362–379.

CAVALLI-SFORZA, L. L., AND F. CONTERIO. 1960. Analisi della fluttuazione di frequenze geniche nella popolazione della Val Parma. Atti Assoc. Genet. Ital. **5**: 333–344.

CAVALLI-SFORZA, L. L., AND A. W. F. EDWARDS. 1964. Analysis of human evolution. Proc. XI Int. Congr. Genet. **3**: 923–933.

——, AND ——. 1966. Estimation procedures for evolutionary branching processes. Bull. Inst. Int. Statist., 35e Session. *In press.*

CAVALLI-SFORZA, L. L., I. BARRAI, AND A. W. F. EDWARDS. 1964. Analysis of human evolution under random genetic drift. Cold Spring Harbor Symp. Quant. Biol. **29**: 9–20.

CAYLEY, A. 1857. On the theory of the analytical forms called trees. Phil. Mag. **13**: 172–176.

——. 1859. On the analytical forms called trees. Phil. Mag. **18**: 374–378.

CHOWN, B., AND M. LEWIS. 1959. The blood group genes of the Copper Eskimo. Amer. J. Phys. Anthropol. **17**: 13–18.

COURANT, R., AND H. ROBBINS. 1961. What is mathematics? Oxford Univ. Press, Oxford.

EDWARDS, A. W. F. 1967. The shortest network uniting a set of points, additional nodes being allowed. *In preparation.*

EDWARDS, A. W. F., AND L. L. CAVALLI-SFORZA. 1963*a*. The reconstruction of evolution. Abstract in: Ann. Hum. Genet., London **27**: 105 and Heredity **18**: 553.

——, AND ——. 1963*b*. A method for cluster analysis. Preprints of V Internat. Biometric Conf. Abstract in: Biometrics **20**: 383 (1964).

——, AND ——. 1964. Reconstruction of evolutionary trees. Systematics Assoc. Publ. No. 6: Phenetic and Phylogenetic Classification: 67–76.

——, AND ——. 1965. A method for cluster analysis. Biometrics **21**: 362–375.

FISHER, R. A. 1958. Statistical methods for research workers. Thirteenth ed. Edinburgh: Oliver and Boyd.

HARDING, E. F. 1967. The probabilities of root-trees generated by random bifurcation. *In preparation.*

KENDALL, M. G., AND A. STUART. 1961. The advanced theory of statistics, 2, Ch. 19. Griffin, London.

KIMURA, M. 1954. Process leading to quasi-fixation of genes in natural populations due to random fluctuation of selection intensities. Genetics **39**: 280–295.

——. 1955. Random genetic drift in multi-allelic locus. Evolution **9**: 419–435.

LEWIS, M., H. KAITA, AND B. CHOWN. 1957. The blood groups of a Japanese population. Amer. J. Human Genet. **9**: 274–283.

POLYA, G. 1937. Kombinatorische Anzahlbestimmungen für Gruppen, Graphen, und chemische Verbindungen. Acta Math. **68**: 145–253.

PRIM, R. C. 1957. Shortest connection networks and some generalizations. Bell Syst. Tech. J. **36**: 1389–1401.

RACE, R. R., AND R. SANGER. 1959. Blood groups in man. Third ed. Blackwell Scientific Publications, Oxford.

RAO, C. R. 1952. Advanced statistical methods in biometric research. Wiley, New York.

SIMPSON, G. G. 1965. Current issues in taxonomic theory (Book Review). Science **148**: 1078.

SOKAL, R. R., AND P. H. A. SNEATH. 1963. Principles of numerical taxonomy. Freeman, San Francisco.

WARD, J. H. 1963. Hierarchical grouping to optimize an objective function. J. Amer. Statist. Assoc. **58**: 236–244.

Won, C. D., H. S. Shin, S. W. Kim, J. Swanson, and G. A. Matson. 1960. Distribution of hereditary blood factors among Koreans residing in Seoul, Korea. Amer. J. Phys. Anthropol. **18:** 115–124.

Yule, G. U. 1924. A mathematical theory of evolution, based on the conclusions of Dr. J. C. Willis, F.R.S. Phil. Trans. Roy. Soc. London B, **213:** 21–87.

Zoutendyk, A., A. C. Kopeć, and A. E. Mourant. 1955. The blood groups of the Hottentots. Amer. J. Phys. Anthropol. **13:** 691–697.

Zuckerkandl, E. 1965. The evolution of hemoglobin. Sci. Amer. **212:** 110–118.

Paper 42 (1969)

Statistical methods in scientific inference
Nature 222, 1233–1237.

Note from A. W. F. Edwards: The difficulties encountered in applying the method of maximum likelihood to evolutionary trees had excited an interest in statistical inference. In October 1968, I accepted a two-year Bye-Fellowship at Gonville and Caius College, Cambridge, which gave me the opportunity for the close study of work on statistical inference, and in particular the contributions of Fisher. I rapidly formed the opinion that it was time for a book on the statistical concept of 'likelihood', and much of my Bye-Fellowship was devoted to writing one (*Likelihood*, Cambridge University Press, 1972). During the writing I gave a number of talks about likelihood, and happened to discuss my project with John Maddox, the editor of *Nature*. This paper was written at his invitation. A follow-up companion paper is my 'Science, statistics and society', *Nature* 233, 17–19, 1971.

Statistical Methods in Scientific Inference

by
A. W. F. EDWARDS
Gonville and Caius College,
Cambridge

Examination of the conflicting statistical methods currently used in scientific inference reveals an increasing awareness of the utility of likelihood. The concept of prior likelihood is introduced as a means of completing a scheme of inference which does not share the logical disadvantages of other methods.

It is remarkable that, in a century which has seen such a large growth in the application of statistics to the natural sciences, the fundamental issues of statistical inference have not been resolved. There are not many more statisticians than opinions as to how to assess rival hypotheses in the light of data, and although what follows is not a review, a short account of the more influential points of view is necessary as an introduction to a new scheme of inference which, I believe, supplies precisely those elements which many hold to be essential without any of the features to which many object.

First, it is necessary to delimit the field of enquiry by defining "statistical hypothesis", which I take to be a specific probability model for the conceptual generation of observations, adopted so that it can be compared with other hypotheses in the light of a particular set of observations. The essential statistical feature is that it is possible to attach a probability to each of the possible outcomes, and its peculiarity is that it cannot be directly verified. I thus exclude from consideration a hypothesis such as "the next ball drawn from this bag will be black", for it does not, as specified, have statistical consequences, and is directly verifiable when the draw has been made. An event such as drawing a black ball may be called an "outcome", and it has long been recognized[1] that the probability of realizing a particular outcome is—when it can be calculated—a rational measure of belief that the outcome will occur. Much of what has been written about hypotheses concerns outcomes rather than statistical hypotheses. Occasionally a hypothesis is itself an outcome of another probability model, in which case the process of inference is much simplified.

The debate on statistical inference has been conducted with more than usual vigour, perhaps because of the unconscious uncertainty of each contributor about his own position. But the polemics of the past have been replaced by the tacit acceptance of a wide variety of usage and a widespread, if muted, concern that no single position is fully satisfactory.

At one end of the spectrum is Sir Harold Jeffreys[2], who holds that prior beliefs about hypotheses, held before any data are available, can be measured in terms of probability and so can be handled by the axioms of probability and, furthermore, that there are canonical values for such probabilities, about which ideal scientists would agree. The weighing of hypotheses, or of the different parameter-values of a model, is to be achieved by applying the axioms to the prior probabilities and to the probabilities of the data conditional on the hypotheses, thereby obtaining posterior probabilities for the hypotheses. The calculation may be conveniently made using Bayes's Theorem[3]:

$$\text{Posterior probability} \propto \text{Prior probability} \times \text{Likelihood}.$$

Here the likelihood, first formally defined by Fisher[4], is equal to the probability of the data given the hypothesis. The case of continuous data may be satisfactorily handled using the concepts of probability density and the likelihood ratio (see Hacking[5]). Jeffreys weighs hypotheses by forming the ratio of their posterior probabilities on the data. Lindley[6], a modern advocate of Jeffreys's fundamental position, lays a greater emphasis on significance-testing.

Next in the spectrum come the subjectivists, who hold that although prior beliefs may properly be measured by probability, there will be such disagreement about the values of the probabilities in any one instance that the only satisfactory method is to allow everyone his own subjective value. It is often suggested that this is to be assessed through conceptual bets. The modern use of subjective probability derives from the works of Keynes[7] and de Finetti[8], and the chief advocates of its use in statistical inference today are Good[9,10] and Savage[11]. Subjectivists hold that all beliefs may be interpreted by probability, and so Bayes's theorem is used, with the understanding that the posterior probabilities have an element of subjectivity, though how much is not always clear. This point has particularly concerned Smith[12], Good[13] and Dempster[14] who, though subjectivists, favour more complex schemes involving upper and lower probabilities. Good[10] has also advanced a scheme in which there are various levels, or "types", of probability. Thus the uncertainty about a prior probability of one type can be expressed by a prior probability distribution for the prior probabilities, and so on. Lindley[15], from his point of view, regards the "prior of a prior" as the orthodox solution to the problem of uncertainty in this context. Polya[16] is prepared to use probability only to make statements of inequality about degrees of belief.

It is customary to refer to all these positions as "Bayesian", but it should be remembered that the opinions vary from Jeffreys's counsel of perfection to Polya's inequalities. Good[10] writes that "The essential defining property of a Bayesian is that he regards it as meaningful to talk about the probability P(H/E) of a hypothesis H, given evidence E". Because this definition would almost certainly have excluded Thomas Bayes, many writers have preferred to continue with the older term "Inverse probability" for the resulting type of inference. Good continues: "An extreme Bayesian believes that every intuitive probability is precise, whereas less extreme Bayesians regard intuitive probabilities as only partially ordered so that each probability merely lies in some interval of values".

In the middle of the spectrum is Bartlett[17,18], who, following Carnap, holds that probabilities can be used to describe such beliefs, but that this type of probability must not be confused with classical frequency probability. Bartlett does not, however, develop any scheme for handling the two types jointly, but seems content to adopt the methods of significance-testing.

From this point of the spectrum onwards there is a denial that beliefs about hypotheses can be measured in terms of probability. No alternative measure is suggested, except where data alone are concerned, in which case some hold that relative likelihoods may be used. The use of likelihood for this purpose has been advocated by Fisher, particularly at the beginning[4] and end[19] of his career. (In the middle[20] he was more concerned to emphasize likelihood for its role in estimation.) Modern exponents of this view are Barnard[21–23], Birnbaum[24], Hacking[5] and also myself[25]. Hypotheses are to be weighed by likelihood ratios as far as data are concerned, but any prior opinions are to be included in the final assessment in the vague intuitive way in which opinions are combined. Publication of the likelihood surface (giving the likelihood at all parameter-values, or for all the rival hypotheses) is adopted as the best method of presenting results.

Next in the spectrum come those who view the weighing of hypotheses principally in terms of tests of significance and estimation. Fisher was much concerned with this approach, and contributed greatly to both fields. He was associated with many of the early tests, and argued strongly in favour of his fiducial probability, which seems to allow, in certain circumstances, probability statements about parameter-values to be made on the basis of data. The circumstances include an acceptance of Fisher's definition of probability and the complete absence of prior information. The statistical world is somewhat bemused by the concept. In practice, the argument has not been influential because of the difficulty or impossibility of phrasing many problems in its terms. By contrast, Fisher's influence on the theory of estimation has been enormous, largely through his advocacy[27] of the "method of maximum likelihood". Although the widespread use of the standard tests of significance and of the method of maximum likelihood is a measure of Fisher's influence, since the adoption of the method has often been largely empirical, the frequency with which they are used should not be taken as a measure of their logical validity.

Finally, at the opposite end of the spectrum from Jeffreys are the many advocates of the Neyman–Pearson[28] theory of hypothesis-testing. Hacking notes that this "is very nearly the received theory". The method is unexceptionable when it is necessary to decide absolutely between two hypotheses, for it establishes a critical value (in the case of a single variable) which divides the space of possible observations into two parts: if the observation falls in one part, one hypothesis is chosen, and if in the other part, the other hypothesis is chosen. Elegant theories aim at the choice of a critical value such that some combination of the probability of rejecting the "false" hypothesis and the probability of accepting the "true" one is optimized. This is the cornerstone of Wald's[29] sequential theory. Some Bayesian enthusiasts have added inverse probability as well, thus linking the two ends of the spectrum "round the back"[30].

Importance of Likelihood

In this literature there is abundant criticism of every point of view, for each protagonist has sought support for his own position by contrasting it with others. All but the most transparent criticisms (such as the Bayesians have sometimes suffered) seem to contain convincing elements, so that it is easy to be a sceptic who feels that there has been no firm progress towards the solution of the controversy. Yet necessity forces the practising scientist to adopt certain methods. Fortunately there is general agreement that in well behaved situations the several methods are highly correlated in effect, at least when the evidence from the data is near to overwhelming the prior opinions. Perhaps the chief dichotomy in the spectrum is not between those who measure belief in hypotheses by probability, and those who do not, but between those who seek to compare rival hypotheses by some relative measure, and those who apply tests based on the concept of repeated trials. From Jeffreys to the advocates of likelihood there is a measure of agreement that hypotheses may best be weighed against some scale of relative acceptability, rather than by significance-testing. They hold[2,5] that in this context the relevance of the significance-test, with its reliance on tail-areas, has never been satisfactorily established. Furthermore, there is general agreement that the likelihood, or likelihood function in the case of continuous parameters, is very important. It enters into all Bayesian methods in addition to providing a method in its own right. It possesses some attractive properties, which Birnbaum[24] in particular has analysed in depth.

Indeed, likelihood seems to be the one rock in a shifting scene, and the only criticisms that Bayesians have consistently levelled at the likelihood method are that it does not allow for the formal incorporation of prior beliefs, and that likelihood has no "meaning". On the latter point, the root objection seems to be that likelihood has no interpretation in terms of probability, which is precisely its merit. Birnbaum[24] and Barnard[22] have suggested various "evidential" interpretations, though it seems to me that none is needed. Lindley[31] complains that arguments based on the likelihood function alone do not allow nuisance parameters to be integrated out, a restriction which prevents logical distinctions being sacrificed to expediency.

Fisher[4,19,32–35] waged continuous war on inverse probability, echoing the doubts of Cournot[26], Boole[36] and Venn[1]. All his criticisms centred on the Bayesian's use of probability as a measure of belief. Its use, he argued, to specify prior belief in the absence of any "chance set-up" (Hacking[5]) led to the error of treating a problem as though one knew that of which one was in fact ignorant; it was not invariant to transformations of the parameters; and, in particular, the uniform prior distribution used to convey complete ignorance about a parameter led to apparently gratuitous information about any new parameter functionally related to the first. I agree with these criticisms, and developments in the logic of Bayesian inference since Fisher's death seem to have done little to meet them. As noted earlier, these developments have been designed to allow for the fact that there are differing degrees of confidence in subjective probabilities, but because these degrees are themselves measured in terms of probabilities, all of which, whatever their kind in Bayesian terms, obey the axioms of probability, a confusion of probabilities results. The confusion can only be removed by maintaining a rigorous distinction between frequency probabilities on the one hand and measures of degrees of belief in statistical hypotheses on the other, as well as between subjective and objective measures. This distinction was carefully drawn by Fisher[4,20] nearly fifty years ago in his definition of likelihood. In his last book[19] he was quite explicit;

"Apart from the simple test of significance, therefore, there are to be recognized and distinguished, between the levels of certain knowledge and of total nescience, two well-defined levels of logical status for parameters lying on a continuum of possible values, namely that in which the probability is known for the parameter to lie between any assigned values, and that in which no probability statements being possible, or only statements of inequality, the Mathematical Likelihood of all possible values can be determined from the body of observations available".

"The likelihood supplies a natural order of preference among the possibilities under consideration".

Venn[1], in 1876, had ventured the same idea: "To decide this question, what we have to do is to compare the relative frequency with which the two kinds of cause would produce such a result." Bernoulli had said as much a century before[37].

Fisher's long advocacy of likelihood was overshadowed for most of his life by his work on estimation, significance-testing, and the design of experiments, and the enthusiasm with which this was greeted. Current use of likelihood can more often be traced to a Bayesian argument (as in human genetics; Smith[38]) or to the analogy with information or entropy (as in physics), than to the direct influence of Fisher. Indeed, Barnard has been alone in his continued support for likelihood on theoretical grounds, though he has now been joined by Hacking.

Testing Hypotheses

In addition to contending with inverse probability, Fisher had repeatedly to stress the irrelevance of the Neyman–Pearson method of hypothesis-testing to the weighing of scientific hypotheses. Whatever its other merits, the concept of acceptance or rejection of alternative hypotheses is alien to their assessment in pure science. What comfort is it for a scientist, who has rejected a hypothesis in favour of its rival by the slenderest of margins, to know that if he applies the method in enough cases, the frequency with which he is wrong will in some sense be a minimum? His concern is for the inference he should draw from the data he has, not from the data he might have possessed. It seems to me that Fisher's[19] indictment of the Neyman–Pearson method as applied to scientific problems is completely convincing, and anything I add would be mere repetition. But anyone inclined to discount Fisher's writings may refer to Hacking.

Nor do I regard the classical methods of significance-testing, exemplified by the use of t, χ^2 and F, which Fisher did so much to promote, as entirely satisfactory from a logical point of view. No real explanation has yet been offered as to why one is allowed to add the clause "or greater value" so as to form a region of rejection; Fisher[19] skates round this when considering fiducial limits, and in connexion with confidence limits he admits that "This feature is indeed not very defensible save as an approximation". To what? His reaction to the point is to take refuge in the comparison of likelihoods, and it is interesting to note, as Hacking does, that Gosset ran for cover in the same direction.

And so, rejecting inverse probability as erroneous, the theory of hypothesis-testing as irrelevant, and significance-testing as no more than a valuable expedient, I am left with likelihood, coupled with a desire to incorporate my prior opinions into inference in a formal manner.

Fisher evidently never felt any such desire, or he would surely have made the proposal I now make. On the contrary, any tendency towards an overformalization of scientific induction, particularly in the direction of decision theory, made him reach for his pen[19]: " . . . the Natural Sciences can only be successfully conducted by responsible and independent thinkers applying their minds and their imaginations to the detailed interpretation of verifiable observations. The idea that this responsibility can be delegated to a giant computer programmed with Decision Functions belongs to the phantasy of circles rather remote from scientific research".

Nevertheless, the numbers of those who, while rejecting this notion, feel the need for a scheme by which prior belief can be formally incorporated into statistical inference may be gauged from the strength of the Bayesian school today, many of whose adherents seem to be prepared to smother their doubts about inverse probability for the sake of introducing prior belief. To adopt alternative means for achieving this is not to assert that the whole of statistical inference can be conducted with them, but it does allow some of the simpler problems to be handled more satisfactorily than before, which, in turn, may be expected to lead to a greater understanding of statistical inference as a whole.

Prior Likelihood

These means may be provided within the framework of likelihood, and without embracing the enigmas of inverse probability, by adopting the concept of prior likelihood. For if, as Fisher asserts, we can measure our relative belief in rival hypotheses or parameter values, contributed by a set of observations, by relative likelihood, where, in the absence of a chance set-up, probability is inappropriate, then it is natural to measure our relative belief, in the absence of specific data, by the same means. Because likelihoods are not probabilities, how else are data-based or "experimental" relative likelihoods to be interpreted other than by reference to a scale of relative belief which, though arbitrary, like temperature, acquires meaning through experience and example ?

The formal treatment requires that we only contemplate relative likelihoods, or likelihood ratios, either for two contrasted hypotheses, or for a continuum of parameter-values, in which case the same end will be achieved by considering the likelihood function determined only down to a constant factor, which may be arbitrarily chosen by assigning unit value to the likelihood at some particular point (such as its maximum). Following Jeffreys's terminology[39], the natural logarithm of the likelihood ratio may be called the support for one hypothesis against another. When based exclusively on data, given the model, it may be referred to as experimental support. Loosely, the same terms may be adopted for the logarithm of the likelihood function, it being understood that true support can only be adduced for one parameter-value against another, by forming the difference of their log-likelihoods, in which case the arbitrary constant is cancelled.

I now define prior support, but only when the prior knowledge does not contain any element of a chance set-up. The prior support for one hypothesis or set of parameter-values against another hypothesis or set is S if, prior to any experiment, I support the one against the other to the same degree that I would if I had conducted an experiment which led to experimental support S in a situation in which, prior to this conceptual experiment, I was unable to express any preference for the one over the other. The defined quantity is, of course, subjective. It has, by virtue of its definition, the properties of experimental support. The conceptual experiment plays the same part as Laplace's "device of imaginary results" in Bayesian inference. Birnbaum[24] regards a "prior likelihood function" which is constant as "a natural formal representation . . . [of] the absence of prior information", but seems not to have pursued the matter. Barnard[21], similarly, has written of a "neutral result". The scheme of inference is now:

$$\text{posterior support} = \text{prior support} + \text{experimental support} \qquad (1)$$

whereas in Bayesian thinking the analogous relation may be derived by forming relative probabilities (for two hypotheses) or odds, and taking natural logarithms:

$$\text{posterior log-odds} = \text{prior log-odds} + \text{experimental support} \qquad (2)$$

for the logarithm of the ratio of the likelihoods is, of course, the experimental support, defined earlier.

Everyone agrees that, when a chance set-up exists, the application of Bayes's theorem is valid, for the prior probabilities are then determined by the set-up. Inference becomes an exercise in conditional probability, but may be formulated as equation (2). In the total absence of a chance set-up, I suggest that equation (1) is appropriate. Furthermore, I hold that any situation can be treated by one, the other, or a mixture, of these two schemes, as necessary, as long as throughout the inference log-odds and support, which are as probabilities to likelihoods, are not confused. In addition it will generally be prudent to maintain the distinction between prior and experimental support.

The differences between equations (1) and (2) are vital. If we start with a chance set-up and prior log-odds, we can state our results in probability terms, using equation (2) to give the posterior log-odds. But if we start with mere opinions, we are unable to make a terminal probability statement, but must adopt the weaker form of inference, using equation (1) to derive a posterior opinion.

Valid Bayesian prior probabilities or probability distributions can profitably be thought of as part of the model. All inferences of the kind here considered are conditional on specific statistical models, and an inference will only be generally accepted if the model on which it is based is generally accepted. The adoption of a prior chance set-up is an extension of the model, and good grounds are needed for accepting all aspects of a model before inferences based upon it are accepted. If the model is incomplete, so must the inference be. The adoption of further probabilities to specify this incompleteness, as Good[10] recommends, similarly requires a model, and is thus of no avail. The difficulty can only be overcome by adopting a measure of belief that is not subject to the axioms of probability. Likelihood, and hence support, supplies this.

The use of support avoids the inconsistencies of subjective probability. It enables situations with and without prior chance set-ups to be distinguished, so that we no longer need assume that of which we are, in fact, ignorant; it is invariant to transformations of the parameters, and, in particular, uniform prior support for a parameter implies uniform prior support for any single-valued transformation of that parameter.

It is not supposed that the weaker form of inference supplied by support can achieve, though in a more satisfactory way, as much as inverse probability, for it is precisely the "achievements" of inverse probability to which exception is taken. One important difference, mentioned by Barnard[22,23], is the inability of support to attach meaning to the disjunction of two hypotheses. Whereas in inverse probability "H_1 or H_2" is regarded as a hypothesis, provided H_1 and H_2 are mutually exclusive hypotheses, in terms of support this disjunction is meaningless. "The likelihood of A or B means no more than 'the stature of Jackson or Johnson'; you do not know what it is until you know which is meant" (Fisher[32]). And this is quite right, for although, when prior probabilities may be validly attached to H_1 and H_2 through the existence of a chance set-up, the distribution of outcomes of the hypothesis "H_1 or H_2" is defined as the weighted sum of the separate outcomes, in the absence of such a set-up no distribution of outcomes can be contemplated. The essence of a statistical hypothesis is a predicted distribution of outcomes, and so "H_1 or H_2" is not an admissible hypothesis when support is in use. Similarly, the suggestion that a continuous parameter lies in a particular interval, such as $0 < p < \frac{1}{2}$, is only a hypothesis if a prior distribution of p is admitted; otherwise it only admits of an inequality statement about the possible distributions of outcomes, which does not merit the title of hypothesis in the precise statistical sense in which the word is used here. Perhaps the lack of meaning of such statements is a clue to the obscurity of the fiducial argument.

Mendelian Example

A single illustration involving both probability and support must suffice. It is an extension of Fisher's[19] well known example. "In Mendelian theory there are black mice of two genetic kinds. Some, known as homozygotes (BB), when mated with brown yield exclusively black offspring; others, known as heterozygotes (Bb), while themselves also black, are expected to yield half black and half brown. The expectation from a mating between two heterozygotes is 1 homozygous black, to 2 heterozygotes, to 1 brown. A black mouse from such a mating has thus, prior to any test-mating in which it may be used, a known probability of 1/3 of being homozygous, and of 2/3 of being heterozygous. If, therefore, on testing with a brown mate it yields seven offspring, all being black, we have a situation perfectly analogous to that set out by Bayes in his proposition".

Fisher then computes the likelihoods of the "homozygote" and "heterozygote" hypotheses on the data, obtaining 1 and 1/128, and applies these to the prior probabilities, 1/3 and 2/3, by means of Bayes's theorem, obtaining posterior probabilities of 64/65 and 1/65. In the terms of equation (2), the prior log-odds for homozygosis against heterozygosis are log (1/2) = −0·693, the experimental support is log (128) = 4·852, and the posterior log-odds are thus 4·159, or log (64).

Now suppose there is another prior possibility: that the mouse in question is the offspring of a mating between two black mice, as before, but one is now homozygous BB, the other being Bb. The expectation amongst the offspring is one homozygous black to one heterozygous black, prior log-odds of zero. The experimental support is the same, and the posterior log-odds are therefore 4·852, or log (128).

One day we are given a black mouse to test, and are told that it is the offspring of one of these types of mating, with no prior evidence to favour one type over the other. The test-cross, as before, gave seven black mice. The experimental support is still 4·852. If the parental mating is of the first type (expectation 1 : 2) then, as we have seen, the posterior log-odds are 4·159, but if the mating is of the second type (expectation 1 : 1) the posterior log-odds are 4·852.

We must now consider the other two hypotheses, concerning the types of parental mating. We have no reason to believe in one type rather than the other, prior to the test-cross, and adopt prior support of zero. The likelihood of the first hypothesis is the probability of obtaining seven black mice from a test-cross when the tested mouse is homozygous with probability 1/3 and heterozygous with probability 2/3, or

$$\frac{1}{3} \times 1 + \frac{2}{3} \times \frac{1}{128} = \frac{65}{192}$$

and the likelihood of the second hypothesis is, similarly,

$$\frac{1}{2} \times 1 + \frac{1}{2} \times \frac{1}{128} = \frac{129}{256}$$

The likelihood ratio for the first hypothesis against the second is thus 260/387, a support of −0·398. Because the prior support was zero, the posterior support is also −0·398.

All knowledge about the hypotheses of primary interest can therefore be summarized by asserting that the probability of the mouse being homozygous is either 64/65 or 128/129, the support in favour of the second value being 0·398.

This is not a statement solely of probability or of support, but a hybrid of the two, as is to be expected from the nature of the problem. Such hybrid statements fill the continuum between support and probability. Inverse probability, by contrast, leads to the same answers for different problems. In this instance it would be argued that the prior probability of the mouse being homozygous was

$$\frac{1}{2} \times \frac{1}{3} + \frac{1}{2} \times \frac{1}{2} = \frac{5}{12}$$

and of being heterozygous, 7/12. The posterior probability of homozygosity would turn out to be 640/647. This is the answer we would obtain if we interpreted the posterior support of log (260/387) for the first type of mating against the second as if it were posterior log-odds; for we would then weight the posterior probabilities of the hypotheses of primary interest accordingly, obtaining

$$\left(260 \times \frac{64}{65} + 387 \times \frac{128}{129}\right)/(260 + 387) = 640/647$$

as before.

But this result is too strong, for it makes a probability statement where it is not valid. It draws a conclusion that should only be drawn if the mouse in question had had probability one-half of deriving from each type of mating. The earlier statement, involving support, supplies the weaker inference which is needed.

I thank Professor J. H. Edwards, Dr K. E. Machin and Dr J. H. Renwick for stimulating discussions. I also thank Professor D. G. Kendall for hospitality in his department.

Note added in revision. Through the courtesy of Mr A. D. McLaren and Dr D. Hudson I have recently seen a draft of Dr Hudson's paper which similarly advocates the use of prior likelihood.

1 Venn, J., *The Logic of Chance*, second ed. (Macmillan, London, 1876).
2 Jeffreys, H., *Theory of Probability*, second ed. (Oxford University Press, 1948).
3 Bayes, T., *Phil. Trans. Roy. Soc.*, **53**, 370 (1763) reprinted in *Biometrika*, **45**, 296 (1958).
4 Fisher, R. A., *Phil. Trans. Roy. Soc.*, A, **222**, 309 (1922).
5 Hacking, I., *Logic of Statistical Inference* (Cambridge University Press, 1965).
6 Lindley, D. V., *Introduction to Probability and Statistics from a Bayesian Viewpoint* (Cambridge University Press, 1965).
7 Keynes, J. M., *A Treatise on Probability* (Macmillan, London, 1921).
8 De Finetti, B., in *Studies in Subjective Probability* (Wiley, New York, 1964).
9 Good, I. J., *Probability and the Weighing of Evidence* (Griffin, London, 1950).
10 Good, I. J., *The Estimation of Probabilities* (MIT Press, Cambridge, Mass., 1965).
11 Savage, L. J., in *The Foundations of Statistical Inference* (Methuen, London, 1962).
12 Smith, C. A. B., *J. Roy. Statist. Soc.*, B, **23**, 1 (1961).
13 Good, I. J., *J. Roy. Statist. Soc.*, B, **23**, 1 (1961); in discussion.
14 Dempster, A. P., *J. Roy. Statist. Soc.*, B, **30**, 205 (1968).
15 Lindley, D. V., *J. Roy. Statist. Soc.*, B, **30**, 205 (1968); in discussion.
16 Polya, G., *Mathematics and Plausible Reasoning*, **2** (Princeton University Press, 1954).
17 Bartlett, M. S., *Essays on Probability and Statistics* (Methuen, London 1962).
18 Bartlett, M. S., *J. Roy. Statist. Soc.*, A, **130**, 457 (1967).
19 Fisher, R. A., *Statistical Methods and Scientific Inference* (Oliver and Boyd, Edinburgh, 1956).
20 Fisher, R. A., *Statistical Methods for Research Workers* (Oliver and Boyd, Edinburgh, 1925).
21 Barnard, G., *J. Roy. Statist. Soc.*, B, **11**, 115 (1949).
22 Barnard, G. A., *Proc. Fifth Berkeley Symp. on Math. Statist. and Probability*, **1**, 27 (1966).
23 Barnard, G., *J. Inst. Actuar.*, **93**, 229 (1967).
24 Birnbaum, A., *J. Amer. Statist. Assoc.*, **57**, 269 (1962).
25 Cavalli-Sforza, L. L., and Edwards, A. W. F., *Bull. Intern. Statist. Inst.*, **41**, 803 (1966).
26 Cournot, A. A., *Exposition de la Theorie des Chances et des Probabilités* (Hachette, Paris, 1843).
27 Fisher, R. A., *Contributions to Mathematical Statistics* (Wiley, New York, 1950).
28 Neyman, J., and Pearson, E. S., *Phil. Trans. Roy. Soc.*, A, **231**, 289 (1933). Reprinted in *Joint Statistical Papers* (Cambridge University Press, 1967).
29 Wald, A., *Sequential Analysis* (Wiley, New York, 1947).
30 Raiffa, H., and Schlaifer, R., *Applied Statistical Decision Theory* (Harvard University, 1961).
31 Lindley, D. V., *J. Amer. Statist. Assoc.*, **57**, 269 (1962); in discussion.
32 Fisher, R. A., *Proc. Camb. Phil. Soc.*, **26**, 528 (1930).
33 Fisher, R. A., *Proc. Roy. Soc.*, A, **144**, 285 (1934).
34 Fisher, R. A., *J. Roy. Statist. Soc.*, **98**, 39 (1935).
35 Fisher, R. A., *Proc. Amer. Acad. Arts and Sci.*, **71**, 245 (1936).
36 Boole, G., *An Investigation of the Laws of Thought* (Walton and Maberly, London, 1854).
37 Bernoulli, D. (1777), reprinted in *Biometrika*, **48**, 3 (1961).
38 Smith, C. A. B., *Amer. J. Human. Genet.*, **11**, 289 (1959).
39 Jeffreys, H., *Proc. Camb. Phil. Soc.*, **32**, 416 (1936).

Paper 46 (1970)

Estimation of the branch points of a branching diffusion process

Journal of the Royal Statistical Society B 32, 155–174.

Note from A. W. F. Edwards: Clarification of the difficulties encountered in applying maximum-likelihood estimation to the statistical model of an evolutionary tree was first achieved by Joseph Felsenstein in his PhD dissertation *Statistical Inference and the Estimation of Phylogenies* (University of Chicago, 1968). It bears fruit in this paper, read before the Royal Statistical Society in December 1969, as I mention in the acknowledgments. The distinction is here made between the true parameters of the modelling process (knowledge of which would enable it to be simulated) and the unknowns of a different type, the positions and times of the nodes of the tree (except the time of the first node), whose distributions could be found from such simulations. The terms 'structural parameters' and 'incidental parameters' are sometimes used in statistics for the two types, as remarked by D. V. Lindley in the Discussion (which was notable for the participation of many of the leading British statisticians of the time).

The paper explores the solution given this distinction. A likelihood can be found for the true parameters, but the remainder can only be the subject of probability statements, conditional on particular values for them (which could be their maximum-likelihood estimates). In consequence, as Felsenstein pointed out to me in early 1968, it is better not to try to estimate the positions of the nodes.

Estimation of the Branch Points of a Branching Diffusion Process

By A. W. F. Edwards

Gonville and Caius College, Cambridge

[Read before the Royal Statistical Society, at a meeting organized by the Research Section, on Wednesday, December 10th, 1969, Professor J. Durbin in the Chair]

Summary

A solution to the problem of estimating the positions and times of the branch points of a Brownian-motion/Yule process, given the positions of all the particles at a particular time, is outlined. A likelihood approach is used, and it is shown that the solution involves maintaining a clear distinction between likelihood and conditional probability if difficulties over mathematical singularities are to be avoided. Several unsolved mathematical problems are encountered, and it is concluded that some simulation studies may be required for a complete solution. Questions of scientific inference which the problem raises are discussed briefly.

1. Introduction

We define a Brownian-motion/Yule process as a stochastic process, continuous in time and space, in which an initial particle undergoes Brownian motion in a space of p dimensions and is subject to a constant risk of splitting into two particles, each of which undergoes Brownian motion and can split further according to the same process. The multiplication of particles thus follows a Yule process. Our problem is to estimate the times and positions of the splits from a knowledge of the positions of the particles at a particular time.

This process was introduced by Edwards and Cavalli-Sforza (1964) to describe the evolution of biological populations as measured by changes in gene-frequencies, and has been used in attempts to estimate the past course of evolution (see Cavalli-Sforza and Edwards, 1967). These attempts have encountered a number of difficulties that merit close examination from a statistical point of view, for they involve important points of inference.

The following treatment differs from that of Cavalli-Sforza and Edwards (1967) in that the process is here confined to a fixed time interval, arbitrarily taken as unity, and both the rate of diffusion of the Brownian-motion process and the splitting parameter of the Yule process are estimated.

2. The Model

Because of its relevance to the genetical situation, our interest in the process starts at the moment the first particle splits into two. Cavalli-Sforza and Edwards have consistently used the convention that the time of observation is taken as $t = 0$, and earlier times take positive values. We adopt this "backwards" measurement here, and assign time $t = 1$ to the moment of division of the original particle.

Let the probability of a particle splitting in an infinitesimal time interval dt be $\lambda\, dt$; λ is thus the parameter of a Yule process. Let the probability density for the position of a particle a time t after it was at a particular point be

$$(\tfrac{1}{2}\pi t\sigma^2)^{-p/2} \exp(-d^2/2t\sigma^2), \tag{1}$$

where p is the number of dimensions and d the distance of the particle from the original point; σ^2 is thus the variance (per unit time) of a p-dimensional Brownian-motion process.

Let there be n labelled particles observed in p dimensions at time $t = 0$. There will be $3.5.7.....(2n-3) = (2n-3)!/\{2^{n-2}(n-2)!\}$ possible tree forms by which these n particles could have arisen, according to the process, from the original particle (see Cavalli-Sforza and Edwards, 1967); at $n = 10$ there are 34,459,425 forms.

Let the $(n-1)$ nodes (branch points) in a particular tree form be numbered $1, 2, \ldots, (n-1)$, starting with 1 for the initial split, and numbering downwards (that is, towards $t = 0$), and let the n final particles be arbitrarily numbered $n, (n+1), \ldots, (2n-1)$. Then let (k_{im}), $1 \leqslant i \leqslant (n-1)$, $m = 1, 2$, be the matrix which lists, in ascending order in its ith row, the numbers of the two points (which may be final particles or other nodes) subtended by the ith node.

Let (ϵ_{ij}), $1 \leqslant i, j \leqslant (2n-1)$, be the matrix in which $\epsilon_{ij} = 1$ if point i subtends point j, -1 if point i is subtended by point j, and zero otherwise. Clearly, $\epsilon_{ij} = 1$ if and only if k_{i1} or $k_{i2} = j$, and $\epsilon_{ij} = -1$ if and only if k_{j1} or $k_{j2} = i$. Thus $\epsilon_{ij} = -\epsilon_{ji}$, the upper triangular non-zero elements being positive. Furthermore, the submatrix (ϵ_{ij}), $n \leqslant i$, $j \leqslant (2n-1)$, is identically zero.

Finally, let the positions of the nodes and final particles, in space and time, be given by the matrix

$$\{(x_{ik}) \,|\, (t_i)\}, \quad 1 \leqslant i \leqslant (2n-1), \quad 1 \leqslant k \leqslant p, \quad t_1 = 1, \quad t_i = 0 \quad \text{for } n \leqslant i \leqslant (2n-1),$$

where x_{ik} is the kth co-ordinate of the ith point (be it node or final particle), and t_i its time.

3. The Problem

The data consist of the positions of the particles in the space at the present time, namely

$$(x_{ik}), \quad n \leqslant i \leqslant (2n-1), \quad 1 \leqslant k \leqslant p.$$

The unknowns are the following:

(1) the form of the tree;

(2) the positions in space of all the nodes, including the first:

$$(x_{ik}), \quad 1 \leqslant i \leqslant (n-1), \quad 1 \leqslant k \leqslant p;$$

(3) the times of all the nodes except the first:

$$t_i, \quad 2 \leqslant i \leqslant (n-1);$$

(4) the parameter σ^2 of the Brownian-motion process;

and

(5) the parameter λ of the Yule process.

These unknowns are of two types which it is essential to distinguish. There are those that are fundamental to a complete specification of the branching-diffusion process, and those that refer to unknown quantities in a particular realization of that process. In the first class we have the position of the origin of the tree, at the first split $(x_{1k}, 1 \leqslant k \leqslant p)$, and the two parameters σ^2 and λ. If we knew these $(p+2)$ quantities we could proceed to simulate the process.

In the second class we have the remaining quantities, that is, the form of the tree, the positions of nodes other than the first $\{x_{ik}, 2 \leqslant i \leqslant (n-1), 1 \leqslant k \leqslant p\}$, and the times of these nodes $\{t_i, 2 \leqslant i \leqslant (n-1)\}$. Were we to simulate the process and concentrate on that subset of simulations which led to (or arbitrarily close to) the observed distribution of precisely n particles at $t = 0$, these quantities in the second class would have probability distributions.

In this way we can see how, in outline, a solution might be achieved. Let us call the unknowns in the first class the true parameters of the process. The first step must be to obtain estimates of these parameters; in view of the fact that there are $(p+2)$ of them, point estimates will have to suffice. Then, given these estimates and the data, the conditional probability distributions of the unknowns of the second class must be determined. Again, in view of the number of quantities involved, point values will have to suffice.

Elsewhere (Edwards, 1969), I have given my reasons for adopting the likelihood approach to problems of statistical inference, and I included an example which involved both likelihood and conditional probability. The present problem is the same in principle, though infinitely more complex in practice. We have a large number of unknowns, some continuous and some discrete; some are the proper subjects of probability statements, other things being equal; and others are the proper subjects of likelihood statements, other things being equal.

Ideally, if we follow the likelihood approach, we would have liked to examine all the probability distributions, multivariate in the second class of unknowns, for all the points in the space of true parameters, about which only likelihood statements can be made. But we lack the necessary large number of dimensions for such a representation, and must therefore be satisfied with determining the essential quantities of the conditional probability distributions (means, variances and covariances) at the maximum-likelihood values of the true parameters.

This, very briefly, is the basis of the solution offered in more detail in the following sections.

4. Solution for the True Parameters

The first task is to maximize the likelihood for variation in the true parameters x_{1k} $(1 \leqslant k \leqslant p)$, σ^2 and λ, for which we need to know the probability density for the particles of a Brownian-motion/Yule process, unit time after the first split and conditional on there being just n particles, as a function of these parameters. This probability density function appears to be unknown, so that even at this early stage of the solution it is necessary to resort to simulation. The approximate form of the distribution may be determined as follows.

Without loss of generality, we may set $x_{1k} = 0$, $1 \leqslant k \leqslant p$, and $\sigma^2 = 1$, since σ is obviously just a scale parameter. Simulation for various values of λ, conditional on n particles at $t = 0$, will then provide the necessary information. The magnitude of the simulation required is, however, daunting, and may turn out to be of the same order as that which would be required for a solution to the entire problem by simulation. A knowledge of the analytic form of the distribution is therefore urgently needed.

5. Solution for the Tree Form

On the assumption that we have obtained maximum-likelihood estimates for the true parameters, we can now turn to the conditional probability distributions of the other unknowns. Of these unknowns, the tree form is a discrete variable, and there can thus be no question of determining the mean tree form. A specific tree form must be adopted before anything can be said about the positions and times of nodes 2 to $(n-1)$, and it is natural to adopt the form with the highest conditional probability, given the estimates of the true parameters, and the data. We proceed to an outline of the calculation of this probability.

Considering only the Yule process, let a labelled tree form with temporally ordered nodes (see Section 2) be called a "labelled history". Let there be H_n possible labelled histories for n particles. There are $(n-1)$ nodes, and intersected by any time horizon between nodes $(r-1)$ and r there are r branch segments (including, for purposes of enumeration, the root at $r = 1$). In all, therefore, there are

$$n + \sum_{r=1}^{n-1} r = \tfrac{1}{2}n(n+1)$$

such segments in the tree, and the $(n+1)$st particle could have derived from any of them, with equal probability.

Thus $H_{n+1} = \frac{1}{2}n(n+1)\,H_n$, $H_1 = 1$, whence

$$H_n = \frac{(n-1)!\,n!}{2^{n-1}},$$

and each such labelled history is, from the point of view of the Yule process alone, equally probable, with probability $1/H_n$.

Another way of deriving this result is as follows. Let an "unlabelled history" be a tree form with temporally ordered nodes but unlabelled final particles, whose subtended branches at each node are regarded as distinct unless neither divides further. Since the first split can only occur in one place (the root), the second split in one of two places (either branch), and, in general, the rth in any of r branches, there must be $(n-1)!$ distinct unlabelled histories, each of equal probability. When we label the final particles, there are $n!$ permutations of n letters, but each unlabelled history possesses $(n-1)$ nodes, each of which generates a pair of identical labelled histories from amongst the $n!$ permutations, so that in all there must be $n!/2^{n-1}$ distinct sets, each set containing 2^{n-1} identical labelled histories. Thus there are $(n-1)!\,n!/2^{n-1}$ distinct equiprobable labelled histories.

Consider now a particular unlabelled history with node times t_i, $2 \leqslant i \leqslant n-1$, where $1 \geqslant t_2 \geqslant t_3, \ldots, t_{n-1} \geqslant 0$. Given λ, the probability density for the t_i and n, jointly, is

$$\lambda^{n-2} \exp\left\{-\lambda\left(2 + \sum_{i=2}^{n-1} t_i\right)\right\}, \tag{2}$$

because for each non-terminal branch, of which there are $(n-2)$, we have terms like $\lambda e^{-\lambda t}$ contributed, where t is the branch length in time, and for each of the n terminal branches we have terms like $e^{-\lambda t}$.

But the probability of obtaining just n populations, starting with a known split, is

$$(n-1)\,e^{-2\lambda}(1-e^{-\lambda})^{n-2},$$

so that the probability density for the node times, and a particular unlabelled history, given just n final particles, is

$$\frac{\lambda^{n-2}\exp\{-\lambda \sum_{i=2}^{n-1} t_i\}}{(n-1)(1-e^{-\lambda})^{n-2}}. \tag{3}$$

By way of a check, if we integrate the node times out of this probability density function, we should obtain the probability of an unlabelled history for n particles, which is $1/(n-1)!$. Now

$$\int_{t_2=0}^{1}\int_{t_3=0}^{t_2}\cdots\int_{t_{n-2}=0}^{t_{n-3}}\int_{t_{n-1}=0}^{t_{n-2}}\exp\left\{-\lambda\sum_{i=2}^{n-1}t_i\right\}dt_{n-1}\,dt_{n-2}\ldots dt_3\,dt_2$$

is easily shown to be equal to

$$\frac{(1-e^{-\lambda})^{n-2}}{(n-2)!\,\lambda^{n-2}},$$

so the required probability is this multiplied by $\lambda^{n-2}/\{(n-1)(1-e^{-\lambda})^{n-2}\}$, or $1/(n-1)!$, which completes the check.

The full expression for the probability density function of the node times and a labelled history with n particles, having regard only to the Yule part of the process, is therefore

$$dF = \frac{2^{n-1}\lambda^{n-2}\exp\{-\lambda\sum_{i=2}^{n-1}t_i\}}{(n-1)\,n!\,(1-e^{-\lambda})^{n-2}}\,dt_2\ldots dt_{n-1}, \tag{4}$$

which, on integrating out the node times $t_2\ldots t_{n-1}$, leads to $2^{n-1}/\{(n-1)!\,n!\}$, the probability of each distinct labelled history.

We must now consider the contribution of the Brownian-motion part of the process. Since the position of a particle does not affect its probability of splitting, the two aspects of the total process will contribute independently to the overall probability density.

Given σ^2, we know that each particle is normally distributed with mean x_{1k} and variance σ^2, in the kth dimension. Hence the distribution of the n particles in the p dimensions is multivariate normal in the variables

$$x_{ik},\quad n\leqslant i\leqslant(2n-1),\quad 1\leqslant k\leqslant p,$$

with mean vector x_{1k}, $1\leqslant k\leqslant p$, and variances all σ^2. Only the covariances are required in order to complete the specification of the distribution.

Now the covariance of any two x_{ik}'s with different k's is zero, because the Brownian motion in each dimension is, on the model, independent. But within each dimension k, if x_{jk} and $x_{j'k}$, $n\leqslant j,j'\leqslant(2n-1)$, are "united" by node l, their covariance must be $\sigma^2(1-t_l)$ (Felsenstein, 1968). In each dimension independently, therefore, the vector of values x_{ik}, $n\leqslant i\leqslant(2n-1)$, denoted by $\boldsymbol{\xi}_k$, will have probability density

$$\frac{1}{(2\pi)^{n/2}|\mathbf{T}|^{\frac{1}{2}}}\exp\{-\tfrac{1}{2}(\boldsymbol{\xi}_k-x_{1k}\mathbf{1})'\mathbf{T}^{-1}(\boldsymbol{\xi}_k-x_{1k}\mathbf{1})\}, \tag{5}$$

where $\mathbf{1}$ is the vector each of whose elements is one, and $\mathbf{T}$ is the matrix whose element in the jth row and j'th column is $\sigma^2(1-t_l)$, where l is the node at which the parts of the tree containing the jth and j'th particles branch. This formulation includes the cases $j=j'$, for then $t_l=0$ and the diagonal matrix elements are all σ^2, as required.

Since the node times are the same for each dimension, the probability density for x_{ik}, $n\leqslant i\leqslant(2n-1)$, $1\leqslant k\leqslant p$, will be simply

$$\frac{1}{(2\pi)^{np/2}|\mathbf{T}|^{p/2}}\exp\left\{\sum_{k=1}^{p}-\tfrac{1}{2}(\boldsymbol{\xi}_k-x_{1k}\mathbf{1})'\mathbf{T}^{-1}(\boldsymbol{\xi}_k-x_{1k}\mathbf{1})\right\}. \tag{6}$$

It follows that the probability density for the positions of the n particles in p dimensions and a particular labelled tree is

$$\frac{2^{n-1}\lambda^{n-2}}{(n-1)\,n!\,(1-e^{-\lambda})^{n-2}(2\pi)^{np/2}}\int_{t_2=0}^{1}\int_{t_3=0}^{t_2}\cdots\int_{t_{n-2}=0}^{t_{n-1}}\int_{t_{n-1}=0}^{t_{n-2}}$$
$$\times\frac{\exp\{-\lambda\sum_{i=2}^{n-1}t_i-\sum_{k=1}^{p}\frac{1}{2}(\xi_k-x_{1k}\mathbf{1})'\mathbf{T}^{-1}(\xi_k-x_{1k}\mathbf{1})\}}{|\mathbf{T}|^{p/2}}\,dt_{n-1}\,dt_{n-2}\ldots dt_3\,dt_2. \quad (7)$$

This is a function of λ, σ^2, the origin, the labelled history and the observations. Using the maximum-likelihood estimates of λ, σ^2 and the origin it enables the probability of each labelled history to be evaluated, in principle. Since, however, an analytic solution to this multiple integral has not been obtained, and in view of the very large number of labelled trees to be investigated with even a moderate value of n, we here encounter another obstacle to a complete solution.

In order to continue the investigation of the outline solution, we may suppose that these computational difficulties have been overcome, and that we have chosen the most probable labelled tree, conditional on the maximum-likelihood values of λ, σ^2 and the origin. We can then proceed to investigate the distribution of the remaining unknowns, the times and positions of the nodes.

6. Solution for the Node Times

Given λ, σ^2, the origin of the tree and its labelled form and the data, and using the notation for the tree form described in Section 2, it is easy to write down the probability density for the node times and positions

$$t_i,\quad 2\leqslant i\leqslant(n-1),\quad 1\geqslant t_2\geqslant t_3\ldots t_{n-1}\geqslant 0,$$
$$x_{ik},\quad 2\leqslant i\leqslant(n-1),\quad 1\leqslant k\leqslant p,$$

by compounding the contributions from each branch. We find

$$F=\lambda^{n-2}\prod_{i,j,i<j}\epsilon_{ij}\{2\pi\sigma^2(t_i-t_j)\}^{-p/2}\exp\left[-\sum_k(x_{ik}-x_{jk})^2/\{2\sigma^2(t_i-t_j)\}\right]\exp\{-\lambda(t_i-t_j)\}. \quad (8)$$

This probability density is not conditioned on the labelled tree form and the known number of particles, and there is no need to condition it, since only a constant independent of the node times and positions is involved.

Owing to the large number of variables, we are unable to examine this conditional probability distribution by graphical means, and we must resort to point "estimates". The choice lies between determining the most probable vector for the node times and positions and determining the mean vector, or possibly some combination of these approaches. In an earlier attack on this problem (Cavalli-Sforza and Edwards, 1967) the mistake was made of regarding the above probability distribution as a likelihood function of the unknowns, and in attempting to derive maximum-likelihood estimates singularities in the "likelihood" were discovered. Indeed, it is obvious that the probability distribution contains singularities, for whenever two nodes at either end of a branch coincide in space, the probability density increases indefinitely as the time interval between them tends to zero. The "maximum probability" solution is therefore trivial and meaningless.

The distribution is evidently multivariate normal in the spatial co-ordinates of the nodes for given times, so that in respect of these, finding the maximum or the mean of the distribution amounts to the same thing. We will, in fact, obtain the mean by maximization. For the node times, however, we have no alternative but to find their means.

It must be appreciated that describing a multivariate distribution by its mean vector and its covariance matrix is only fully informative if that distribution is normal. We are confronted by a dilemma because the distribution in question is normal in some directions only. It is not obvious whether the best solution will be provided by quoting the overall multivariate mean; by quoting the mean of the non-normal variables followed by the mean of the normal variables conditioned on the former mean; or by quoting the most probable value for the marginal distribution of the non-normal variables, followed by the mean of the normal variables conditioned on that. One can imagine situations in which the first method provides a very undesirable solution, for example, if the regression of the normal variables on the non-normal is very curvilinear (in a multidimensional sense). I am inclined to consider the third method the least unsatisfactory, but for any of the methods we require the marginal distribution of the node times.

This distribution will be found from (8) above by integrating out the node co-ordinates

$$x_{ik}, \quad 2 \leqslant i \leqslant n-1, \quad 1 \leqslant k \leqslant p,$$

leaving as variables the node times

$$t_i, \quad 2 \leqslant i \leqslant n-1.$$

The integration is certainly possible in principle, since it involves the repeated integration of exponential functions of a quadratic nature, but experience gained on performing the operation in simple special cases makes it seem likely that it would be a very arduous procedure. Again, it may be necessary to resort to a numerical solution.

7. Solution for the Node Positions

Once values for the node times have been adopted, finding the node positions is not difficult. Taking the logarithm of the probability density function we have

$$\log F = -\sum_{\substack{i \ j \\ i<j}}\sum \epsilon_{ij}\left[\frac{p}{2}\log\{2\pi\sigma^2(t_i-t_j)\}+\frac{\Sigma_k(x_{ik}-x_{jk})^2}{2\sigma^2(t_i-t_j)}+\lambda(t_i-t_j)\right]+(n-2)\log\lambda. \quad (9)$$

Eliminating the constant terms, and changing the sign, we will consider

$$G = \sum_{\substack{i \ j \\ i<j}}\sum \epsilon_{ij}\left[\frac{p}{2}\log\sigma^2+\frac{p}{2}\log(t_i-t_j)+\frac{\Sigma_k(x_{ik}-x_{jk})^2}{2\sigma^2(t_i-t_j)}+\lambda(t_i-t_j)\right]-(n-2)\log\lambda. \quad (10)$$

The means, variances and covariances may then be found as follows. Differentiating G with respect to the kth co-ordinate of the lth node $\{1 \leqslant l \leqslant (n-1)\}$,

$$\frac{\partial G}{\partial x_{lk}} = \frac{\partial}{\partial x_{lk}}\left\{\sum_{j=l+1}^{2n-1}\epsilon_{lj}\frac{(x_{lk}-x_{jk})^2}{2\sigma^2(t_l-t_j)}+\sum_{i=1}^{l-1}\epsilon_{il}\frac{(x_{ik}-x_{lk})^2}{2\sigma^2(t_i-t_l)}\right\}, \quad (11)$$

the first sum taking care of the contributions from the lth row, and the second taking care of those from the lth column, of the upper triangular part of the matrix (ϵ_{ij}).

Thus

$$\frac{\partial G}{\partial x_{lk}} = \frac{1}{\sigma^2}\left\{\sum_{j=l+1}^{2n-1} \epsilon_{lj} \frac{x_{lk}-x_{jk}}{t_l-t_j} - \sum_{i=1}^{l-1} \epsilon_{il} \frac{x_{ik}-x_{lk}}{t_i-t_l}\right\}$$

$$= \frac{1}{\sigma^2}\left\{\sum_{j=l+1}^{2n-1} \epsilon_{lj} \frac{x_{lk}-x_{jk}}{t_l-t_j} + \sum_{i=1}^{l-1} \epsilon_{li} \frac{x_{lk}-x_{ik}}{t_l-t_i}\right\},$$

since $\epsilon_{li} = -\epsilon_{il}$. Hence

$$\frac{\partial G}{\partial x_{lk}} = \frac{1}{\sigma^2} \sum_{j=1}^{2n-1} \epsilon_{lj} \frac{x_{lk}-x_{jk}}{t_l-t_j}, \tag{12}$$

since $\epsilon_{lj} = 0$ when $l = j$. Setting this equal to zero, we have

$$\sum_{j=1}^{2n-1} \epsilon_{lj} \frac{x_{lk}-x_{jk}}{t_l-t_j} = 0. \tag{13}$$

This states that the sum of the reciprocals of the branch gradients at each node is zero in each dimension. We note also that

$$x_{lk} = \sum_{j=1}^{2n-1} \epsilon_{lj} \frac{x_{jk}}{t_l-t_j} \Bigg/ \sum_{j=1}^{2n-1} \frac{\epsilon_{lj}}{t_l-t_j}, \tag{14}$$

and hence that in each dimension each node is at the centre of gravity of the adjacent nodes (or final particles) when these are weighted inversely as the time intervals of the respective branches.

We may re-write (14) to read

$$x_{lk} \sum_{j=1}^{2n-1} \frac{\epsilon_{lj}}{t_l-t_j} = \sum_{j=1}^{2n-1} \frac{\epsilon_{lj}}{t_l-t_j} x_{jk} = \sum_{j=1}^{n-1} \frac{\epsilon_{lj}}{t_l-t_j} x_{jk} + \sum_{j=n}^{2n-1} \frac{\epsilon_{lj}}{t_l-t_j} x_{jk}$$

for $1 \leqslant l \leqslant n-1$, whence

$$x_{lk} \sum_{j=1}^{2n-1} \frac{\epsilon_{lj}}{t_l-t_j} - \sum_{j=1}^{n-1} \frac{\epsilon_{lj}}{t_l-t_j} x_{jk} = \sum_{j=n}^{2n-1} \frac{\epsilon_{lj}}{t_l-t_j} x_{jk}. \tag{15}$$

If, now, we denote the right-hand side of (15) by y_{lk}, and define

$$\zeta_l = \sum_{j=1}^{2n-1} \frac{\epsilon_{lj}}{t_l-t_j}, \quad 1 \leqslant l \leqslant n-1, \quad \mathbf{Z} = \operatorname{diag}\{\zeta_1, \ldots, \zeta_{n-1}\},$$

and

$$\eta_{lj} = \frac{\epsilon_{lj}}{t_l-t_j}, \quad 1 \leqslant l, j \leqslant n-1, \quad \mathbf{H} = (\eta_{lj}),$$

we may write (15) in the form

$$(\mathbf{Z}-\mathbf{H})\mathbf{x}_k = \mathbf{y}_k, \tag{16}$$

where $\mathbf{x}_k$ and $\mathbf{y}_k$ are the column vectors with components x_{jk} ($j = 1, \ldots, n-1$) and y_{lk} ($l = 1, \ldots, n-1$) respectively. Inverting (16) we obtain

$$\mathbf{x}_k = (\mathbf{Z}-\mathbf{H})^{-1}\mathbf{y}_k. \tag{17}$$

This equation linearly relates the kth co-ordinates of the nodes $x_{jk}(1 \leqslant j \leqslant n-1)$ to the kth co-ordinates of the final particles $x_{jk}(n \leqslant j \leqslant 2n-1)$ by an inverted matrix which is the same for all k, and depends only on the node times.

Since the distribution is multivariate normal in the node positions given their times, this solution by maximization provides the mean vector of node positions.

Differentiation of (12) will lead to the variance–covariance matrix:

$$\frac{\partial^2 G}{\partial x_{lk}^2} = \frac{1}{\sigma^2} \sum_{j=1}^{2n-1} \epsilon_{lj}/(t_l - t_j);$$

$$\frac{\partial^2 G}{\partial x_{lk}\, \partial x_{mk}} = -\frac{1}{\sigma^2} \epsilon_{lm}/(t_l - t_m);$$

and for any two nodes, or the same node, the mixed derivative with respect to different dimensions, k, is zero.

These expressions are the elements of the "information" matrix and are the same for each dimension. The variance–covariance matrix for the node positions in each dimension is simply the inverse of this matrix.

This solution still leaves us without an error matrix for the node times and positions jointly.

8. Discussion

In the first paper describing the application of a Brownian-motion/Yule process to evolutionary trees, Edwards and Cavalli-Sforza (1964) wrote that "we do not anticipate any major difficulties in the ... computer programmes, although these are not yet working". Later in the same year (Cavalli-Sforza *et al.*, 1964) they wrote that "estimation by maximum likelihood is still causing some difficulties, because of the very high number of parameters to be estimated and some peculiarities of the system". At a meeting the following year, Cavalli-Sforza and Edwards (1966) enlarged on these "peculiarities", pointing out that the "log-likelihood" surface contains singularities (in fact, lines of singularities) wherever the two nodes at either end of a branch coincide, the branch then disappearing.

It has taken from 1964 until now to achieve a full understanding of the nature of the difficulties that have been encountered, and to see the vital distinction between likelihood and conditional probability that is essential to a solution. In the intervening years a great deal of effort has been spent on computing, and it would have been impossible to clarify the difficulties without this effort. I therefore make no apology for presenting a solution in mere outline. It is of practical interest because it demonstrates the great difficulty of attaining a formal solution; the problem is one in which the ease with which intuitive solutions may be obtained (see Cavalli-Sforza and Edwards, 1964, 1967) renders the formal solution outlined in this paper virtually worthless.

Apart from the insight which all the computing work has provided, there is one result which is of some importance even though it was obtained in connection with the "likelihood" approach. In some cases it was possible to coax the computer programs into locating the maximum of what we now know as the conditional probability distribution of the node times and positions. Generally, the distribution has no maximum other than the trivial one provided by the singularities, but with some sets of data a non-trivial one exists. It was then found that, in respect of variation in the node times, the maximum was always exceedingly flat. It appears from this that the node times may prove to be practically inestimable, as though the data are simply insufficiently informative on this score. Perhaps we shall have to settle for estimating the tree form and little more.

An obvious consequence of tackling this problem has been a greatly deepened interest in statistical inference, and I have given my own position on this controversial matter elsewhere (Edwards, 1969). It will be sufficient to state here my belief that, however much a Bayesian solution would have simplified the approach to this problem (by calling everything a conditional probability), it would be most improper for a scientist to compromise his convictions by resorting to the most expedient approach to inference for the problem at hand; and a detailed study of the present problem has, if anything, strengthened my conviction that a Bayesian approach in this instance would be a gross over-simplification. I am not prepared to give the true parameters prior probability distributions because I can see no model which would justify them. We may also note that the adoption of a Bayesian approach would not automatically resolve the dilemma of how to summarize a posterior distribution in a great many variables in terms of a few descriptive parameters, for maximizing the posterior probability would lead to the singularities, whereas quoting the means of non-normal variates might lead to values of extreme improbability.

In conclusion, it seems as though the only practicable approach based on a branching diffusion process model is through simulation, and in the meantime those geneticists and others who are concerned with specific bodies of data can feel less guilty about using intuitive methods. Professor Whittle has suggested to me how a solution might be obtained by contemplating the process reversed in time, and I hope he will be able to enlarge on this in the discussion.

Acknowledgements

This work was started whilst the author was at the University of Pavia (supported in part by Euratom–CNR–CNEN Contract 012–61–12 BIAI and by the U.S. Atomic Energy Commission), and continued at Stanford University (supported by NIH Grant GM–10452–02) and at the University of Aberdeen (supported by WHO and the Medical Research Council). The author is particularly grateful to the MRC for providing computer facilities which enabled the work to be continued, and to Professor L. L. Cavalli-Sforza, Pavia, for his collaboration and continued encouragement. I have also benefited greatly from discussions with many people, especially Dr J. H. Renwick, Dr J. Felsenstein, Mr A. D. McLaren and Mr E. F. Harding.

References

Cavalli-Sforza, L. L., Barrai, I. and Edwards, A. W. F. (1964). Analysis of human evolution under random genetic drift. *Cold Spring Harbor Symp. Quant. Biol.*, **29**, 9–20.

Cavalli-Sforza, L. L. and Edwards, A. W. F. (1964). Analysis of human evolution. *Proc. XI Intern. Congr. Genet.*, **3**, 923–933.

—— (1966). Estimation procedures for evolutionary branching processes. *Bull. Intern. Statist. Inst.*, **41**, 803–808.

—— (1967). Phylogenetic analysis. *Evolution*, **21**, 550–570; and *Am. J. Hum. Genet., Suppl.*, **19**, 233–257.

Edwards, A. W. F. (1969). Statistical methods in scientific inference. *Nature, Lond.*, **222**, 1233–1237.

Edwards, A. W. F. and Cavalli-Sforza, L. L. (1964). Reconstruction of evolutionary trees. *Phenetic and Phylogenetic Classification*, Systematics Assoc. Publ. **6**, 67–76.

Felsenstein, J. (1968). Statistical Inference and the Estimation of Phylogenesis. Ph.D. dissertation, University of Chicago.

Discussion on Dr Edwards's Paper

Professor C. A. B. Smith (University College London): Dr Anthony Edwards has presented us with what must be the simplest possible mathematical model of evolution of species and groups. He has challenged us by showing that even this problem has unsuspected depths. Even with powerful mathematical weapons and high-speed computers, anything but the most elementary approach to this problem leaves us out of our depth.

Such evolutionary problems occur in various disciplines. They are most associated with biology, but they occur in linguistics, where the process of evolution can be watched and recorded as it happens. The last 2,000 years have seen Latin change into French, Spanish and Italian; each of which is still visibly and rapidly altering. Despite the fact that the process operates quite openly, we are still very ignorant of its causes and mechanism, and why it sometimes goes faster, sometimes slower. It seems plausible that biological evolution will likewise go at different rates at different periods, and that sometimes conditions will favour the splitting up of species, and sometimes keep them uniform. If this is so, Dr Edwards's model of a uniform rate of change is a considerable over-simplification. But until we have first learnt to deal with the simplest case, we can hardly cope with further complications.

Two particular points arise. Dr Edwards is concerned particularly with the evolution of frequencies of different types in populations, and particularly gene frequencies. One may doubt whether for them the Gaussian process model is completely appropriate. If (p_1, p_2, p_3), say, represents the vectors of frequencies of three different types in a large population of N individuals and each generation is obtained by random sampling from the last (with replacement), then in the next generation p_i will change to a new value p_i+u_i. The expectation $\mathscr{E}(u_i) = 0$ and the covariance $\mathscr{E}(u_i u_j) = (p_i \delta_{ij} - p_i p_j)/N$ (where δ_{ij} is the Kronecker delta). In the p-space this is not an isotropic random drift, since the covariance matrix does not have the form $\mathscr{E}(u_i u_j) = a+b\delta_{ij}$. Cavalli-Sforza and Edwards have suggested using a square-root transformation to make the drift spherically symmetric. But this, unfortunately, introduces a steady drift into the process. Let x be a random variable with fixed mean (say $\mathscr{E}x = 0$) but variance $\sigma^2 = kt$ proportional to the time. Let $y = f(x)$ be a transform of x. Then, on expansion in a Taylor series, $y = Bx+Cx^2+\ldots$, so that the mean of y is $\mathscr{E}y = B\mathscr{E}x+C\mathscr{E}(x^2)+\ldots = Ckt+\ldots$, and this, to the first order of approximation, is directly proportional to the time unless $C = 0$. Hence no transform of p follows an isotropic Gaussian Brownian movement process with constant mean.

Dr Edwards has also raised a very simple but very fundamental point in the theory of inference. If all the factors in a statistical situation can be expressed in the terms of ordinary (frequency) probabilities, then most statisticians will accept that inferences should be expressed in terms of such probabilities. Thus if Alice's brother is colour-blind, what is the chance that the baby she is expecting will be colour-blind too? This is approximately 0·28. (I omit the details of the calculation. It is a straightforward use of the laws of probability, assuming that Alice is English, and that the frequency of colour-blindness among Englishmen is about 8 per cent.) Usually an unknown θ will not have a known initial (prior) distribution in the frequency sense. If we have a definite and agreed mathematical model, then the information provided by the sample can be conveyed to another person by telling him the likelihood function. The likelihood can be combined with a personal judgment of the initial distribution to get a final distribution for the parameter. But suppose that we have a combination of these two cases, having say "true parameters" θ, with vague or subjective initial distribution, and "other unknowns" ϕ for which a conditional distribution $P_1(\phi | \theta)$ is already specified. How do we then best summarize the information provided by a sample S? The model adopted will specify a conditional probability, $P_2(S | \phi, \theta)$, of obtaining such a sample for given values of ϕ, θ. I would suggest that it is most profitable to consider the product

$$P_3(S, \phi | \theta) = P_2(S | \phi, \theta) P_1(\phi | \theta),$$

(which will be an objective frequency probability). For if we give to θ any (possibly subjective) initial distribution, $P_4(\theta)$, we can then find the joint distribution

$$P_5(S, \phi, \theta) = P_3(S, \phi \mid \theta) P_4(\theta).$$

From this we obtain the distribution $P_6(S)$ by summing over ϕ and θ, and hence the distribution of ϕ and θ conditional on the occurrence of S, namely

$$P_7(\phi, \theta, \mid S) = P_5(S, \phi, \theta)/P_6(S).$$

This is what we want; knowing that S has occurred, it tells us how likely various values of ϕ and θ are. It can be found from a knowledge of P_3 and P_4. Here $P_3(S, \phi \mid \theta)$ accordingly represents the objective and universally agreed part of the information provided by the sample (and the mathematical model), and $P_4(\theta)$ the (possibly) personal judgment about the plausibility of different values of θ, in which different individuals could reasonably be expected to differ. This function P_3 takes the place of the likelihood in the usual theory. It seems that it is exactly this function P_3 which Dr Edwards is using in equation (8).

With regard to the problem of infinite peaks in this distribution, one may recall that analogously χ^2 with one degree of freedom has an infinite peak at zero. But it would hardly seem reasonable, given the problem of stating a typical value or estimate of χ^2, to give the value 0.

I have therefore the greatest pleasure in proposing the vote of thanks.

Mr G. J. S. Ross (Rothamsted Experimental Station): I should like first to congratulate Dr Edwards for his stimulating paper. He has constructed a most plausible model of an important process, has developed powerful computer programs for fitting the model, yet has failed to obtain any sensible answers. Now this situation is paralleled in many applications of statistical estimation. Before computers were available statisticians could propose complex models which were expected to provide excellent fits to the data, but when the computers were actually programmed to fit these models the results were inconsistent and hard to interpret. Thus we are brought right back to the question of *information*, or what complexity of model can we really expect to fit?

There are three levels of complexity of statistical models: (a) general models with parameters to explain all features of the data, (b) models retaining only statistically significant parameters, and (c) models that provide satisfactory predictions or interpolations. This categorization is familiar in the context of multiple linear regression, although since standard errors of prediction are so rarely computed the predictive advantage of simple models is not always appreciated. Polynomial regression is the simplest field for the study of this phenomenon, and data can easily be found for which a curve of the sixth degree gives a very close fit, a quartic is statistically adequate and a quadratic actually gives the minimum average confidence band width over the range of the data. In non-linear problems the computing effort is often so great that no attempt is made to evaluate critically what has been achieved by fitting a model, although the need is greater, particularly as the solution may not be unique.

In the present context there seems to be about as many parameters as observations, and so intelligible results are hardly to be expected. There are two remedies, either to increase the amount and improve the design of the data or to restrict the model in some way. The data should therefore be replicated in time, although this is no doubt impracticable for studies in human evolution, and some faster changing species might be studied. Indeed can we really expect to make inferences about evolutionary parameters from a single set of contemporaneous observations? Restrictions on the model might be achieved by setting the node times to their most probable values, but it will still be almost impossible to discriminate between different labelled histories.

It has been courageous of Dr Edwards to tell us of his experience with what must have seemed at first to be a perfectly feasible exercise. So I would like to ask one further question: if statistical methods fail to solve a problem, in what way are intuitive solutions more justified?

I have great pleasure in seconding the vote of thanks to Dr Edwards.

The vote of thanks was put to the meeting and carried unanimously.

Professor D. V. LINDLEY (University College London): I should like to make a few remarks about the principles of inference that underlie this problem. The comments are closely related to those of Professor Smith and, to avoid unnecessary confusion, I will use his notation.

The parameters in the problem are of two types. The first includes the rate of the renewal process, λ, the variance of the Brownian motion, σ^2, and the initial position of the system, $\mathbf{x}_0$. Let these collectively be denoted by $\boldsymbol{\theta}$, a vector of dimension $p+2$. Those of the second type include the times and positions of the nodes. Similarly denote this set by $\boldsymbol{\phi}$. In the language of Neyman and Scott, $\boldsymbol{\theta}$ is the set of structural parameters, $\boldsymbol{\phi}$ the set of incidental parameters.

Following Smith and writing $\mathbf{S}$ for the data, it is a simple matter to write down the probability of the data given both sets of parameters, $p(\mathbf{S} \mid \boldsymbol{\theta}, \boldsymbol{\phi})$. As a function of $\boldsymbol{\theta}$ and $\boldsymbol{\phi}$ for fixed $\mathbf{S}$, this is their joint likelihood. But we next observe that there is very considerable knowledge about the second group of parameters, $\boldsymbol{\phi}$; knowledge of a sort that might be called prior. For we know $p(\boldsymbol{\phi} \mid \boldsymbol{\theta})$ from the theory of the renewal and Brownian processes. Consequently we can write down the joint density of $\mathbf{S}$ and $\boldsymbol{\phi}$ given $\boldsymbol{\theta}$ as the product $p(\mathbf{S} \mid \boldsymbol{\theta}, \boldsymbol{\phi})\, p(\boldsymbol{\phi} \mid \boldsymbol{\theta})$. Since our primary interest centres on $\boldsymbol{\theta}$ we require to integrate this expression with respect to $\boldsymbol{\phi}$, so obtaining the probability of $\mathbf{S}$, given $\boldsymbol{\theta}$,

$$p(\mathbf{S} \mid \boldsymbol{\theta}) = \int p(\mathbf{S} \mid \boldsymbol{\theta}, \boldsymbol{\phi})\, p(\boldsymbol{\phi} \mid \boldsymbol{\theta})\, d\boldsymbol{\phi}.$$

It is this integration which is difficult to perform and hinders a satisfactory solution of the problem. Were it available, $p(\mathbf{S} \mid \boldsymbol{\theta})$ as a function of $\boldsymbol{\theta}$ for fixed $\mathbf{S}$ would be the likelihood of $\boldsymbol{\theta}$.

So far in my argument all the probabilities have been what some people call "objective". Dr Edwards suggests that we could go a stage further and introduce (subjective) probabilities for $\boldsymbol{\theta}$ and adopt a fully Bayesian argument. Personally I doubt if this helps because the real difficulty lies in the above integration. But were it to be used it is perhaps worth noting that an obvious prior for $\boldsymbol{\theta}$ is available since λ, σ^2 and $\mathbf{x}_0$ are parameters of a familiar type. The appropriate Jeffreys's prior is $d\lambda\, d\sigma^2\, d\mathbf{x}_0/\lambda\sigma^2$.

The main point that emerges is that it is quite unnecessary to distinguish different sorts of probabilities or various types of likelihood. Given any data, $\mathbf{S}$, and any parameter $\boldsymbol{\theta}$, $p(\mathbf{S} \mid \boldsymbol{\theta})$, as a function of $\mathbf{S}$ for fixed $\boldsymbol{\theta}$, is the probability of $\mathbf{S}$, given $\boldsymbol{\theta}$; as a function of $\boldsymbol{\theta}$, for fixed $\mathbf{S}$, it is the likelihood of $\boldsymbol{\theta}$, given $\mathbf{S}$. I do not see that any new principle of inference is needed.

One final remark. Dr Edwards has sensibly used the likelihood function and my comment is not addressed to him. But I would like to ask an orthodox statistician what his sample space would be in this problem. He requires one to construct unbiased estimates or make significance tests. The obvious one, indefinite repetitions of the same fixed duration of the process and observations only at the end, seems artificial. Why of fixed duration? The problem is fascinating because we only have a sample of size one.

Professor P. WHITTLE (Statistical Laboratory, Cambridge): The labour of scanning the millions of possible trees to determine which one maximizes some criterion such as likelihood seems prohibitive. It would be more natural to work back in time from the present position, determining recursively the most probable path in time, conditional on the facts that (a) co-ordinates of the lines ("branches") at the present time are known and (b) the tree is presumed to have a single root at some stage in the past, i.e. to be in fact a single tree.

The functional form of the conditional probabilities required seems to be beyond derivation, so this programme can be followed only qualitatively. Nevertheless, it does lead to a straightforward, recursive method of reconstruction. Instead of examining all trees, one lets the system evolve (or, rather, degrade) into the past, yielding, in Dr Edwards's vivid phrase, "a time machine which runs evolution backwards".

So, suppose n lines are being examined at the present moment t, the jth having vector co-ordinate x_j. The changes in the x_j over a time interval of length Δ are assumed statistically independent, and normally distributed with zero mean and covariance matrix $\mathbf{K}\Delta$: the conventional assumption that the lines follow independent vector Brownian paths, apart from branching. Let $\exp[-Q(\mathbf{x})] = \exp[-Q(x_1, x_2, \ldots, x_n)]$ be the joint density of the x_j, conditional on n and on the "single-root" assumption.

Let ξ_j denote the most probable value of $x_j(t-\Delta)$; the most probable co-ordinate of the jth line a time Δ in the past. If Δ is so small that the probability of branching in $(t-\Delta, t)$ is negligible, then $\boldsymbol{\xi} = (\xi_1, \xi_2, \ldots, \xi_n)$ will maximize

$$Q(\boldsymbol{\xi}) + \frac{1}{2\Delta} \sum_j (x_j - \xi_j)' K^{-1} (x_j - \xi_j) \tag{1}$$

whence

$$\frac{\xi_j - x_j}{\Delta} = -\mathbf{K} \frac{\partial Q}{\partial \xi_j}. \tag{2}$$

Let $\xi_j(s)$ denote the most probable value of $x_j(t-s)$: the most probable co-ordinate of the jth line a time s in the past. Then an extension of the argument leading to (2) will lead us, in the limit of small Δ, to the equations

$$\frac{\partial \xi}{\partial s} = -\mathbf{K} \frac{\partial Q(\boldsymbol{\xi})}{\partial \xi_j}, \quad (s \geqslant 0). \tag{3}$$

The dynamic equations (3) are to be integrated for $s \geqslant 0$, subject to the starting condition $\xi_j(0) = x_j(t)$, in order to determine the most probable past course of the evolutionary process. We have not considered the variation of n. However, the form of Q will presumably be such as to cause neighbouring lines to converge for small s, i.e. in the relatively recent past, and others subsequently. At each such convergence ("node") one loses a line, and carries on the system (3) with one fewer ξ_j variable. One continues this process until left with only a single line, whose ξ value will thereafter remain constant.

The function Q is unknown, but one might achieve something qualitatively correct by imagining that it has the form

$$Q(\boldsymbol{\xi}) = \tfrac{1}{2} \sum_j \sum_k H(\xi_j - \xi_k), \tag{4}$$

where $H(\xi_j - \xi_k)$ will represent an "attraction" between lines j and k. The form of the function H must be such that this attraction is strong for neighbouring lines, weak for distant lines. This will require that $H(\zeta)$ be rapidly increasing for small $|\zeta|$, slowly for large $|\zeta|$. A consideration of the case $n = 2$ leads one to an educated guess for the form of H.

Suppose that, in the case $n = 2$, the two lines separated a time T in the past, so that $x_1 - x_2$ is $N(0, 2KT)$, and has density $\exp(-R)$, where

$$R = \text{const} + \tfrac{1}{2} \log T + \frac{(x_1 - x_2)' K^{-1} (x_1 - x_2)}{4T}.$$

T is unknown, but if we estimate it by maximizing R with respect to T, and calling the result Q, we have

$$Q(x_1, x_2) \sim \text{const} + \tfrac{1}{2} \log [(x_1 - x_2)' K^{-1} (x_1 - x_2)]$$

corresponding to the choice

$$H(\zeta) = \tfrac{1}{2} \log (\zeta' K^{-1} \zeta). \tag{5}$$

Under the assumptions (4), (5) the dynamic equations (3) become

$$\frac{\partial \xi_j}{\partial s} = \sum_k \frac{\xi_k - \xi_j}{D_{jk}^2}, \tag{6}$$

where

$$D_{jk}^2 = (\xi_j - \xi_k)' K^{-1}(\xi_j - \xi_k)$$

measures the "squared distance" between lines j and k.

Thus, lines j and k converge at a rate roughly proportional to D_{jk}^{-1}, which gives a very positive meeting of two close lines as time recedes into the past. In fact, for the case of just two lines the paths in time would converge on the two arms of a common parabola, meeting at the vertex, with the common ξ value thereafter constant. For $n > 2$, the confluence of two lines will take the form of a somewhat distorted parabola.

Under the Brownian assumption, this locally parabolic form would be correct. For two lines which have just diverged (in forward time) the distance $|x_j - x_k|$ will be of the order of the square root of time elapsed since divergence; at the actual moment of divergence the two lines separate explosively. The explosive separation is mimicked in the backwards reconstruction by implosive approach of two contiguous lines, at a rate inversely proportional to separation distance.

Integration of the system (6) would give one an estimate of the tree, but with no indication of its sampling variability. A wild surmise: the random trees reconstructed by using dynamic equations (6) with a random forcing term might have sampling statistics similar to those of the original randomly evolving trees, conditional on the initial assumptions (a) and (b). However, study of this point would seem to be an order of magnitude more difficult than estimation itself.

We have assumed the matrix **K** known. The scale of **K** is intrinsically inestimable without further information. One might plausibly estimate relative values of its elements by minimizing $\sum\sum (x_j - x_k)' K^{-1}(x_j - x_k)$ subject to prescription of the value of $|\mathbf{K}|$.

Professor M. S. BARTLETT (Department of Biomathematics, Oxford): I only received a copy of the paper a day before the meeting and was somewhat disconcerted by the comparative lack of background information in it of the genetic problem which the model was supposed to represent, so that I was grateful to have Dr Edwards's much fuller verbal account. My own respect for the relevance of the likelihood function is not, while considerable, quite so extreme as his, and I would suggest that some of his difficulties may have been aggravated by too extreme a view. In particular, it would certainly seem sensible to separate out the estimation problem for the parameters of the process from incidental events of one realization (however much of interest these may be). Having done this, we must still be wary of the likelihood function on two points. The first point is largely technical, and is the complication often encountered in setting up the likelihood function for a partial set of observations like the end-products, in contrast, say, with that for the complete process. The second is more fundamental and concerns the possible danger of using a likelihood function for the wrong model. Quite apart from the technical criticisms from Professor Smith, I would note that any model which incorporates Brownian motion is certainly wrong, as the detailed data for any time-interval, however short, would in theory lead to perfect estimation of the dispersion parameter σ^2.

Let us, however, proceed to try to estimate λ and σ^2. (The estimate of the initial point is conveniently dealt with separately.) The estimate of λ from n, the number of points at time t, is a standard problem, as $p_t(n)$, or equivalently $\Pi_t(z) = E\{z^n\}$, is well known for the Yule process (given $n = 2$ at $t = 0$). Let us conjecture that, as the distance distributions are Gaussian, we should for σ^2 concentrate on $Y = \sum_{r>s} (x_r - x_s)^2$, where $x_1, \ldots, x_n$ are the n points at time t and $|x_r - x_s|$ the distance between x_r and x_s. As the joint distribution of Y and n is probably complicated, I tried to investigate the generating-function

$M = E\{Yz^n\}$, from which $E\{Y|n\}$ could be found. The differential equation for M appears to be

$$\frac{\partial M}{\partial t} = \lambda(z-1)\frac{\partial M}{\partial z}+2\lambda zM+\sigma^2\frac{\partial^2 \Pi}{\partial z^2},$$

which looks straightforward enough to solve. Unfortunately, there seems a rather awkward integrating factor in one of the auxiliary equations, so I have provisionally dealt with the unconditional mean $S = E\{Y\}$, for which (putting $z = 1$ in the above equation) we have

$$\frac{\partial S}{\partial t} = 2\lambda S+\sigma^2 m_{(2)},$$

where $m_{(2)}$ is the second factorial moment of n, viz. (for $n = 2$ at $t = 0$) $4e^{2\lambda t}-2e^{\lambda t}$. This gives

$$S/\sigma^2 = 6te^{2\lambda t}+4e^{\lambda t}(e^{\lambda t}-1)/\lambda,$$

indicating, if λ is known or estimated, how to obtain an estimate of σ^2. (As $E(n) = 2e^{\lambda t}$, the most obvious substitution is $e^{\lambda t} = \frac{1}{2}n$, though this would produce some bias in the estimate of σ^2 unless corrected for.)

Professor L. L. CAVALLI-SFORZA (University of Pavia, Italy): I do not have much to add. I would like to say that I know Dr Gomberg has been working on the Bayesian solution and his results seem rather similar to what was put on the board. The problem with this integral is that its stochastic integration takes an extremely long time on a very big computer, so that probably the Bayesian solution, if there is no simple way of solving that integral, is likely to be even less acceptable than maximum likelihood in practical terms.

Being a biologist I am interested in having solutions even if they are not perfect. I am very pleased to see that the problem offers sufficient challenge to statisticians, and that they may provide a better solution than those we are using at present. Some of the practical difficulties are seen clearly by simulation procedures, which can also offer numerical solutions to the problem of error. For instance, it is seen that trees in which divergence is early give the most difficult estimation problems. When divergence is late in the evolutionary process, estimation is satisfactory with every method, and probably also maximum likelihood gives no problem in this case. With simulation we have also the advantage of introducing any model we like, however complicated, and this can greatly help to understand what nature did. I would like to take this opportunity of thanking Dr Edwards for having contributed so greatly to this problem.

Dr E. SENETA (Statistical Laboratory, Cambridge): I wonder if it might not be a conceptual aid, as well as possibly closer to the physical situation, to consider a *discrete* time–space model, embodying the essential aspects of the model Dr Edwards discusses, in particular, a branching effect, accompanied by random motion in space. The model which I shall describe casts some light on several of the problems raised tonight. In particular, it in principle enables us to determine one of the distributions in which Dr Edwards is interested, as well as enabling us to say something about the estimation of the parameter's with which he is concerned.

Let us assume, then, that the particles of our process move in a *denumerable* state-space J, the process being defined at time instants $n = 0, 1, 2, \ldots$, there being one particle present at time 0, at some initial state of J, say i. This particle splits (reproduces), yielding a random number of "offspring", depending on its position, and hence according to a p.g.f. $f(i; s)$, $s \in [0, 1]$, say; any particular offspring, independently of the others, moves to a state (position) $j \in J$, with probability p_{ij}. At time instant one, and at successive instants, a particle at any one position reproduces analogously and independently of all other particles. At any one time, we may classify the particles present into types T_i, $i \in J$, according to their position. It is then clear that the evolution through time of the numbers of particles of various types (i.e. at various positions) follows *a multitype Galton–Watson process*, with vector-valued multivariate generating function:

$$\mathbf{f}(\mathbf{s}) = \{f^i(\mathbf{s})\},$$

where

$$f^i(\mathbf{s}) = f(i; \sum_{j \in J} p_{ij} s_j), \quad i \in J,$$

expressible, as shown, in terms of the stochastic matrix $\mathbf{P} = \{p_{ij}\}$, and the scalar p.g.f.'s $f(i; s)$, $i \in J$.

The situation pertinent to the model described by Dr Edwards is much more simple than this general model. It appears to occur when we specialize the $f(i; s)$ to be independent of i and given by

$$f(i; s) \equiv f(s) = qs + (1-q)s^2, \quad 0<q<1,$$

so that

$$m \equiv f'(1-) = 2-q>1,$$

and the matrix $\mathbf{P}$ depends only on one parameter, viz. the probability r, $0<r<1$, of transition to a nearest neighbour, in a state-space J which is the p-dimensional generalization of a square lattice. The difficulty in subsequent analysis, although the probabilistic evolution of this system depends only on the two parameters q and r and the initial state, is that in general little is known about a Galton–Watson process with an infinite number of types, although some results are known (see Moy, 1967) in certain situations, and in the present simple case it may be possible to make some deductions also.

However, if we are willing to *restrict* the state-space in some way, to make it finite, while retaining the essential features of the above model (specifically, the irreducibility and aperiodicity of $\mathbf{P}$ and the general two-parameter dependence), then we have a *standard* multitype supercritical Galton–Watson process (Harris, 1963, Chapter 2), $\{\mathbf{Z}_n\}$, of very simple form, for which in particular

$$\mathbf{Z}_n/m^n \xrightarrow{\text{a.s.}} \mathbf{W} \equiv w\mathbf{v},$$

where $\mathbf{v}$ is the (non-random) stationary distribution vector of the matrix $\mathbf{P}$, and w is a univariate random variable, depending on the branching effect only, and about whose distribution information is available. Moreover, since in an evolutionary situation an observation such as Dr Edwards desires to make on the process $\{\mathbf{Z}_n\}$ would probably be made after a *long* time n, one would effectively be observing the random variable $m^n\mathbf{W}$, whose distribution depends on q and r only (note that $m^n w$ is an approximation to the total number of particles at time n). These two parameters may then, in principle, be estimated by standard techniques.

Professor D. G. KENDALL (Statistical Laboratory, Cambridge): About twenty years ago there appeared in Series B of our Society's *Journal* a paper by Professor A. S. C. Ross on "Philological probability problems", and one of his problems, that of reconstructing the family tree of Indoeuropean languages, has something in common with that presented to us today by Dr Edwards. In the discussion on Ross's paper I sketched a solution to his problem whose details escape me now, but it may be worth looking at them again in the light of this afternoon's meeting.

I should also like to refer to a third problem which it is profitable to consider in the same context as the other two; this concerns the linking, not of peoples or of languages, but of manuscript copies of some literary work. For example, Mrs Whitteridge of Oxford has considered how one could choose from among say 100 different manuscript versions of *The Romance of the Rose* which particular one is most likely to be the primitive version (and so, the one to be edited, and published, with critical apparatus, etc.) if one believes (a) that the primitive version *is* included in the collection and (b) that medieval scribes committed copying errors, but did not try to correct the copying errors of their predecessors.

Some questions arising from this third problem formed part of a Cambridge Ph.D. dissertation by Dr Haigh of Sussex, and an account of his work will soon appear in the *Journal of Applied Probability*.

Having made these bibliographical remarks, I should like to follow them up with a word or two about a problem in the statistics of genealogy—another subject in which trees are studied statistically and probabilistically. The only thing which can be asserted with complete certainty about one's male family line of descent is that it exists and is unique. When, however, one tries to trace it, even if the enquiry can with confidence be confined to a small and isolated geographical area, one encounters the difficulty of statistical "noise" in the form of cousins (often, irritatingly enough, with the same first name and marrying wives with the same first name). With great labour one sorts the cousins out and goes back a generation or two, only to find new batches of cousins appearing as one's time-machine travels backwards. One is soon persuaded that this is a phenomenon which will go on occurring for ever, or at any rate short of Eden, and then one can take a holiday from genealogical labours and prove a theorem which makes a virtue of consanguinity. We might try to formulate the problem thus.

A sub-critical birth-and-death process has existed "for ever", and one is interested in tracing the line backwards from some specially interesting individual (one's self or one's client) and noting all the side-branches as one goes backwards. Each side-branch may be responsible for 0, 1, 2, ... male individuals alive at the present time; thus the existing *related* male population admits a canonical partition when its members are assigned to their respective side branches. Describe the statistical properties of this partition, given the number of currently existing name-bearers (supposed all ultimately to be related). I have worked out some aspects of this problem in odd moments this afternoon, but it may be more satisfactory for the moment merely to pose the problem and to produce a solution at a later stage. I should be interested to know to what extent established theories of inbreeding deal with these matters.

The author replied in writing as follows:

The points made in the discussion fall into four classes: first, concerning the validity of the model (Professor Smith mentions the genetic process; Mr Ross and Professor Bartlett have reservations which would apply to any use of the model); secondly, concerning the method of statistical inference used (Professors Smith and Lindley); thirdly, concerning approximate solutions (Professors Whittle and Bartlett); and fourthly, concerning the solution of somewhat different models (Dr Seneta and Professor Kendall).

Professor Bartlett remarked that he "was somewhat disconcerted by the comparative lack of background information in [the paper] of the genetic problem which the model was supposed to represent". I am afraid this omission was necessary because there was no time to pursue the lengthy justification of the model as applied to the genetic process, and it seemed to me that the model was of considerable statistical interest in its own right. I will therefore continue the discussion of Professor Smith's point elsewhere, and try to cover Mr Ross's and Professor Bartlett's reservations by confining my attention to the following problem.

A ciné film is taken of a physical process which is described by the Brownian-motion/Yule process to any stipulated accuracy, but unfortunately the film is destroyed except for the last frame. What, if anything, can we say about this particular realization of the process and its parameters, given only the final disposition of the particles? To cover Professor Bartlett's point, we may suppose that the exposure time was so short compared with the time required for the particles to move a detectable distance that the last frame by itself provides no information about their mean-square velocity.

It seems to me that the description of this problem in continuous terms is no less valid than the similar description of many other problems we encounter in statistics, and therefore that its solution is worth attempting. Mr Ross thinks that "intelligible results are hardly to be expected", but I feel that there are at least *some* possible dispositions of the particles which would lead me to make quite strong intuitive statements about the past history of the process.

If this is accepted, then the adequacy of the method of inference becomes of paramount interest. Both Professor Smith and Professor Lindley agree that it is necessary to focus attention on the likelihood function, and I find Professor Smith's comments at this point closely in accord with my own position. Professor Lindley would use the terms "structural parameters" and "incidental parameters" to describe my "true parameters" and "second class" of unknowns; I avoided this terminology intentionally because, as used by Neyman and Scott, an incidental parameter is a nuisance parameter, whereas my second class of unknowns are of primary interest. He suggests, in an aside, the appropriate Jeffreys's prior; I will limit myself, in the present instance, to passing the opinion that Jeffreys's prior distributions for a parameter about which we know nothing at all, and for a parameter about which we know nothing except that it is positive, are mutually inconsistent, a reflection of the logical fallacy of trying to make wholly uninformative probability statements.

Professor Lindley implies that I have found it necessary "to distinguish different sorts of probabilities or various types of likelihood". I do not think this is so; I have merely found it necessary to distinguish probability and likelihood, and have not used "any new principle of inference". This is relevant to Professor Smith's point about the infinite peaks, which, since they occur in a probability distribution rather than in a likelihood function, no longer worry me.

I will now turn to the third class of comments: suggestions for an approximate solution. I include Professor Bartlett's in this category, because I do not think one can separately estimate λ and σ^2, except as an approximation. The estimation of λ from n, the number of particles at time t in a Yule process, is indeed "a standard problem", but it is intuitively clear (and the likelihood function concurs) that in this problem there is more information about λ than n alone provides. For example, if for a given n and t the points are closely clustered in two groups, a higher value of λ is indicated than if they are widely scattered, because although in both cases the same number of fissions has occurred, in the second case it is likely that there is a greater total time during which fissions did *not* occur. Professor Bartlett's solution for σ^2 given λ is welcome; it is compatible with my statement that "σ is obviously just a scale parameter".

I am most grateful to Professor Whittle for his detailed suggestions as to how best to use a "time-machine" approach. Naturally, the idea that the process could be simulated in reverse by imagining the particles moving under some law of mutual attraction was one of the first intuitive approaches to the problem, though it appears not to have been mentioned in print until J. H. Edwards (1969) suggested it. The reason I did not pursue it was that I felt, as with so many intuitive approaches, that it could only be justified by reference to the formal "forward" solution. Professor Whittle, however, has now shown that considerable progress may be made towards isolating the appropriate gravitational function. Bearing in mind that we are, on the simplest model, assuming the Brownian motion homogeneous in space, his covariance matrix $\mathbf{K}$ becomes diagonal with elements σ^2, with a corresponding simplification in the solution. I hope to have the opportunity to do some calculations along these lines.

Fourthly, Dr Seneta proposes a discrete model for the process, which I hope he amplifies elsewhere. It does appear, however, to offer little encouragement if the form of the evolutionary tree is of primary interest.

Finally, I would like to answer a specific question posed by Mr Ross: "If statistical methods fail to solve a problem, in what way are intuitive solutions more justified?" If a statistical solution shows that the information we seek is simply not available, then of course our intuition is misguided if it offers some (as is the case, I believe, with Bayesian priors). But in the present example we have no reason to believe, from the attempted statistical analysis, that the data do not permit an inference, though there are some indications that particular sets of data might prove rather uninformative. It seems, rather, that we have not been sufficiently ingenious in devising methods for extracting the information. I therefore have no qualms about the more intuitive solutions I and others have obtained on actual data, though I would be the first to welcome a proper statistical solution with which to compare them. To respect our intuition less than this would be to deny all scientific progress before the middle of the nineteenth century.

References in the Discussion

Edwards, J. H. (1969). In *Computer Applications in Genetics*, ed. N. E. Morton, p. 146. Honolulu: University of Hawaii Press.

Harris, T. E. (1963). *The Theory of Branching Processes*. Berlin: Springer.

Moy, S.-T. C. (1967). Extensions of a limit theorem of Everett, Ulam and Harris on multi-type branching processes to a branching process with countably many types. *Ann. Math. Statist.*, **38**, 992–999.

As a result of the ballot held during the meeting, the following were elected Fellows of the Society:

Armstrong, William, M.Sc.
Baker, Geoffrey John, B.Sc.
Beese, Trefor, M.Sc.
Beesley, Ian Blake, M.A.
Bland, John Martin, M.Sc.
Branch, David Norman, B.Sc.
Chiu, Wing Kin, M.Sc.
Clapperton, Susan Mary
Crowder, Martin John, B.Sc.
Danziger, Lawrence, Bachelor-Master Bus. Admin.
Durbin, Peter Leonard, B.Sc.
England, Malcolm Stuart, B.Sc.
Evans, Ronald Ivor, M.I.E.E.E.
Fisher, Christine Mary, B.Sc.
Godfrey, Michael D., Ph.D.
Halil, Tundger
Hirji, Tajdin Natha, B.Sc.
Jones, Charles Frederick, M.A.
Nadas, Arthur, Ph.D.
Rand, Graham Kennedy, B.Sc.
Rusby, Stuart, B.Sc.
Russell, Cecil Anthony Francis, B.A.
Sleath, Martin David, M.A.
Smith, Stephen Frederick, B.Tech.
Spooner, Hilda Ann, B.Sc.
Stopher, Peter Robert, Ph.D.
Tennant-Smith, James, B.Sc.
Thomas, Robert Owen, B.Sc.
Ury, Hans Konrad, M.A.
Vitek, Pavel Jan Vitezslav
Williams, George Trevor, Sc.D.
Witt, Stephen Frank, B.A.
Yannoulis, Yannis, M.A.

Paper 49 (1971)

Review of *Evolution and the Genetics of Populations*, Volume 2, by S. Wright Heredity 26, 332–338.

Note from A. W. F. Edwards: So far as I am aware, this is the first critical look at Wright's contributions to the mathematical theory of population genetics since Fisher's 1941 paper (see [140]). It is contemporary with a paper I read to the joint meeting of the Biometric Society and the Edinburgh Group of the Royal Statistical Society in March 1971, 'The relation between Wright's "adaptive topography" and Fisher's "Fundamental Theorem of Natural Selection"' (Abstract in *Biometrics* 27, 762–763, 1971). As in 1967 ([36]), I was still handicapped through lacking a clear understanding of Fisher's theorem. The typescript has survived, and was helpful in preparing [140], 23 years later.

It is vital that we should keep population trends, ecological problems and food supply problems in the public eye. But it will not help to provoke despair.

Sir JOSEPH HUTCHINSON
St John's College, Cambridge

EVOLUTION AND THE GENETICS OF POPULATIONS. VOLUME 2: THE THEORY OF GENE FREQUENCIES. Sewall Wright. University of Chicago Press. Pp. 511. £6, 15s.

It is a platitude of mathematical population genetics that the subject's development is principally due to Fisher, Haldane and Wright. The order is alphabetic; but it is also the order in which they are leaving the scientific stage, and the order in which they have published their major books on the subject. Fisher gave us *The Genetical Theory of Natural Selection* in 1930 (Fisher, 1930*a*, 1958) and Haldane *The Causes of Evolution* in 1932 (Haldane, 1932*b*, 1966); but we have had to wait until now, full forty years after *The Genetical Theory*, for Professor Sewall Wright to distil his contribution into a major book. Happily he has been given the time and strength to perform this work; without it the perspective in which he views, and wishes us to view, mathematical population genetics would have remained indistinct.

This volume owes something to an earlier work of Wright's *Statistical Genetics in Relation to Evolution* (Wright, 1939*b*), a slim book published in Paris on the eve of war, and consequently not widely known. But it is very much larger, and includes the fruits of thirty years' more work; it is also much more partisan. The author spares no efforts to remind us of his early contributions, sometimes with startling effect: on p. 417, for example, we read that he " dealt with analysis of covariance as well as variance " in 1917. My impression had been that Fisher did not introduce the word *variance* until 1918 (Fisher, 1918); here is meat for historians.

With Haldane's contribution Wright deals generously, but this only emphasises the conflict with Fisher. Not for him Haldane's epithet " brilliant" in describing *The Genetical Theory* (Haldane, 1966, p. 13)*. The wounds inflicted in this sad battle must lie deep. Only the scientific drama, however, concerns us at present, and it is never far from the surface. Sometimes Wright avoids mentioning Fisher altogether, as in " Correlations between Relatives in the Presence of Dominance " (p. 433), a principal topic of Fisher's 1918 paper *The correlation between relatives on the supposition of Mendelian inheritance*. The " correlation array " used by Fisher is attributed correctly to Pearson (1904) (it is the only paper of Pearson quoted), and the " 1918 " paper is not cited at all, merely earning a critical mention in the next section because Fisher neglected the environmental correlations when analysing Pearson's data on human stature. But that was 53 years ago.

The old battlegrounds of adaptive surfaces, self-sterility alleles, the evolution of dominance, inbreeding, and the rate of fixation, are revisited. Naturally, the author presses his own points of view, and this is not the place to question matters of opinion. But the kernel of the book is the concept of the adaptive surface, a concept which Fisher never accepted and Haldane,

* I have since discovered that Wright once wrote of *The Genetical Theory*. "It is a book which is certain to rank as one of the major contributions to the theory of evolution". (*J. Hered.*, *21*, 349, 1930).

I think, never used. The dispute concerns a matter of fact, not opinion, and we must pursue it. The adaptive surface is the hero—or villain—of the piece; it reminds us of the scientific drama at each appearance.

The problem is as follows. On p. 39, Wright reveals his famous formula for the change in the frequency q_x of the xth of a series of alleles in one generation under natural selection:

$$\Delta q_x = q_x(1-q_x)\frac{\partial \bar{W}}{\partial q_x} / 2\bar{W}.$$

$\bar{W}$ is the " mean fitness ", and the partial derivative is the slope of the function $\bar{W}$ in the direction of the corner of the simplex* corresponding to a population all A_x, the xth allele. No one, as far as I know, disputes the formula, though many feel that the introduction of the derivative of $\bar{W}$ is somewhat artificial; but Wright, of course, is going to hang a whole theory on $\bar{W}$ and its derivatives.

Now if all the Δq_x are zero, we have an equilibrium point, by definition; that is, the gene frequencies all remain unchanged from one generation to the next. Similarly, if the gene frequencies are such that $\bar{W}$ is a stationary point with respect to them, all the $\frac{\partial \bar{W}}{\partial q_x}$ are zero, and hence all the Δq_x zero. Thus the non-degenerate equilibrium gene frequencies and the gene frequencies at the stationary point of $\bar{W}$ *with respect to variation in the* q_x are one and the same thing. So far so good.

Now (p. 41) Wright finds the conditions under which the stationary point of $\bar{W}$ is a maximum *with respect to variation in the* q_x. He does so correctly, though it might be neater to use the theory of constrained derivatives. But what has this to do with changes in the q_x *with respect to time* under natural selection? All that the above equation tells us is that, since Δq_x and $\frac{\partial \bar{W}}{\partial q_x}$ are of the same sign, the *direction* of gene-frequency change is such that a very small step in that direction would lead to an increase in $\bar{W}$.† The direction of the change is not even necessarily along the line of steepest ascent on the $\bar{W}$ surface; but, more important, there is nothing in the equation itself which makes it obvious that the *magnitude* of the gene-frequency change is going to be sufficiently small for it to count as a " very small step ".

Imagine yourself on the side of a smooth hill in a fog. Your " equation of motion " guarantees that (1) the size of your step is proportional to the gradient, reducing to zero at the top, and (2) the direction of each step is such that, were you to proceed a very small distance—but not a whole step—in that direction, you would certainly move to a higher point. Would you necessarily arrive at the top of the hill? No, because the actual size of any

* A simplex is the generic name for the sequence of geometrical figures " point, line, triangle, tetrahedron . . ." on or in which we may represent the frequencies of genes at a locus with 1, 2, 3, 4 . . . alleles.

† On reflection, even this generally accepted statement requires proof. Though it is obviously true for two alleles, it is not obviously true for three or more, because the element " ∂q_x " refers to the direction towards the " x "-apex of the simplex, whilst the element " Δq_x " refers to the direction perpendicular to the flat opposite the " x "-apex. These directions only coincide at the centre of the simplex, or when there are but two alleles. However, Mr E. F. Harding and I have obtained a proof, which we propose to publish shortly.

step might be such that it would take you over a shoulder of the hill to a *lower* point. The ony certain way to arrive at the top under this scheme would be to make the steps infinitesimally small, for then each shuffle would necessarily take you a little higher. Alternatively, you might examine the equation of motion in more detail to see whether, hidden in it, was the assurance that even with man-sized steps you would necessarily gain height with each step.

So it is with Wright's " surface of mean fitness " or " adaptive topography ". In order to identify the problem of whether the equilibrium gene-frequency points are stable or unstable with the problem of whether the stationary points of $\bar{W}$ are at a maximum or a minimum of the $\bar{W}$ surface, we must either assume such weak selection that each step can be shown to be small enough for $\bar{W}$ to certainly increase, or we must prove that, for any magnitude of selection, and hence for large step sizes, $\bar{W}$ does in fact increase. Wright does neither: in connection with the multiple-allele case with a particular fitness matrix, he simply jumps to the statement (p. 43), " It is fairly obvious and easily verified that equilibrium is stable ($\bar{W}$ maximum) ". But what is " fairly obvious " in this case is that $\bar{W}$ is a maximum with respect to variations in gene frequency, from which it does not necessarily follow that the equilibrium is stable.

With just two alleles we can easily prove that $\bar{W}$ always increases (or, to be precise, never decreases), for we merely have to show that there is a single stationary point of $\bar{W}$, which is a trivial matter, and that the direction of gene frequency change is monotonic towards, or away from, the stationary point, which is clear from Fisher's (1930*a*) treatment. But with three or more alleles, although $\bar{W}$ still has only one stationary point, being a quadratic function of the gene frequencies, one cannot appeal to monotonicity in gene frequency change.

Wright refers to the work of Kimura (1956*b*) and Penrose, Smith and Sprott (1956) for the establishment of the multiple-allele stability conditions, and also to Mandel (1959), but in fact Mandel was at pains to point out that both Kimura and Penrose *et al.*, had assumed a continuous-time model which " is justified only in the case of infinitesimally slow selection ". As we have seen, with such very small steps in gene-frequency change, $\bar{W}$ must increase, as it will whenever selection is so slow that we can effectively treat populations of genes rather than of genotypes.

The nettle of what happens with selection of any magnitude and discrete generations, which is precisely Wright's model, was first grasped by Owen (1954) who correctly gave the stability conditions for the three-allele case, though without proof. The proof, actually for any number of alleles, was supplied by Mandel and Hughes (1958), who arrived at the conditions by showing that, sufficiently near the equilibrium point, $\bar{W}$ is in fact a non-decreasing function. They also " conjecture that the mean viability $\bar{W}$ increases each time the population undergoes the process of mating, but to prove this statement generally appears to be a matter of considerable mathematical difficulty ". This difficulty was overcome, and the increasing nature of $\bar{W}$ finally established, by Scheuer and Mandel (1959), Mulholland and Smith (1959), and Atkinson, Watterson and Moran (1960), and in 1961 Kingman sealed the matter by giving a greatly simplified proof.

One might have supposed that Wright would have eagerly siezed on the above work as in part validating the concept of an adaptive topography, at least for the case of a single autosomal locus in a diploid organism undergoing random mating, with any number of alleles and constant genotypic fitnesses. But in fact he completely ignores all the above papers, in spite of them being cited in Moran's (1962) book. This omission is significant, because if Wright were to admit the existence of the problem, he would have to prove $\bar{W}$ an increasing function in every case to which he applied the concept of an adaptive topography, and so far a proof has only been obtained for a single autosomal locus. Indeed, as soon as one proceeds to the next most complex case, two loci with epistacy, a counter-example is known (Moran, 1964), though Wright dismisses Moran's criticism as " based on a misunderstanding of the concept " of an adaptive surface.

In the case of selection with inbreeding, Li (1967) notes that " the peak-and-valley theory of natural selection no longer holds ", but adds that " Wright is well aware of the fact ". It appears that at equilibrium the gene frequencies are at a stationary point of the " modified average fitness $\bar{W}^*$"; but that does not justify a new " peak-and-valley " representation, because $\bar{W}^*$ may not be an increasing function. Weir (1970) is not hopeful on the point. In the case of autotetraploids " the average fitness does not necessarily increase under selection " (Li, 1967); nor does it with linked loci. For a multi-allelic sex-linked locus neither a proof nor a counter-example is known to me, though Cannings (1968) is of the opinion that the use of a potential function " would probably be misleading ". The situation is similar when the fitnesses differ between the sexes, where there may be two distinct stable equilibria (Owen, 1953), whilst with frequency-dependent selection " the concept of adaptive peak ceases to apply " (Li, 1967).

This is not an auspicious foundation for a general theory. By ignoring the fundamental problem, Wright inevitably forfeits our sympathy in his attempts, later in the book, to treat more complex situations by fitness functions, sometimes specially constructed. He tries to sustain a general theory by distinguishing between the peaks of the special " fitness functions " and the peaks of the mean fitness functions, which do not, in general coincide. But by this stage in the argument I am forced to conclude that the concepts merely obscure the issues, and are best abandoned. Fisher (1941), in a paper which all who seek to resolve this controversy for themselves should study, says " Wright's conception embodied in equation (6) [that quoted above] that selective intensities are derivable, like forces in a conservative system, from a simple potential function dependent on the gene ratios of the species as a whole, has led him to extensive but untenable speculations ". Fisher was no doubt familiar with potential theory, since he read mathematics in Cambridge before the first war, but his " Fundamental theorem of natural selection " (Fisher, 1930*a*) was as far as he would travel along that road. This theorem, which is a very restricted continuous-time theorem, tells us all that we can hope to learn about the increase in fitness in a general way, and serves to emphasise, rather than to obscure, the essential peculiarities of selection in *diploid* populations. It preceded Wright's (1932) concept of an adaptive topography; Wright devotes three lines to it (p. 121).

To sum up this discussion on the core of Wright's book, the general theory of gene-frequency change cannot be satisfactorily treated by appealing to the concept of an adaptive surface, because the mean fitness is not necessarily an increasing function (unless selection is so slow that we can no longer call the theory general); in the single case in which we know the mean fitness does increase, the concept is hardly necessary, and misleading to the extent that populations do not climb the surface by the steepest route; in more complex cases the stationary points of the adaptive surface may not correspond to the equilibrium gene frequencies, so that for a potential theory representation other " fitness functions " have to be found. These appear to offer no conceptual advantage, and the demonstration that they are in fact potential functions requires the solution of the problem of interest in any case, unless selection is so slow that, again, the theory is no longer general. Finally, the concept obscures the fact that in a diploid population the interests of the individual and of the population do not usually coincide.

The book is refreshingly well-written, direct and to the point. The introduction is an admirably concise précis; Chapter 2 opens at such a pace that the author does not even mention that he is dealing with the large-population theory initially, and it is not long before the reader is confronted with some samples of an unfortunate characteristic of Wright's mathematical work: the proliferation of subscripts. This " underwriters' disease " reaches epidemic proportions at times, a peak being reached on page 469, with the subscript WW(E)W(E1)W(E2). Subscripted subscripts, and even superscripted subscripts, occur. Indeed, the use of path coefficients as a method of analysis often seems to relegate the essential difficulties to the subscripts, rather than to illuminate the problem.

Chapter 3—Systematic change of gene frequency: single loci—has already drawn extensive comment above. We may additionally note that the section " Unequal selection in males and females " does not mention Owen's (1953) interesting work; and that the conditions for stable equilibrium at a sex-linked locus appear to have been first given by Bennett (1958), and proved by Mandel (1959*a*), Kimura (1960) and Edwards (1961), by different methods. None of this work is mentioned; nor does Wright consider genes which affect the sex-ratio in the way suggested by Fisher (1930*a*).

In Chapter 4 the concept of an adaptive surface is extended to multiple loci, and in Chapter 5 to frequency-dependent selective values. A general " fitness function " is introduced in order to handle those cases in which the stationary points of the mean fitness are not the same as the points of genetic equilibrium. In effect the fitness function is found by determining the expression for Δq in the particular case under consideration, equating it to $q(1-q)\frac{\partial F}{\partial q}\,/\,2F$, and then solving for F. "The set of formulas for the Δq's has the important property that these show by their form that within their range of approximate validity the set of gene frequencies tends to move up the gradient of the ' surface ' $\log \bar{w}$, except as qualified by the positive factor, $0{\cdot}5q(1-q)$, and thus tends to approach a ' peak ' in this surface " (p. 121). The " important property " is the one mentioned earlier in this review, which Wright does not prove; it is interesting to find him suggesting that the factor $0{\cdot}5q(1-q)$ might upset it (in fact it does not).

Thus for slow selection F plays the role of $\bar{W}$ in simple models, and may be regarded as an adaptive surface which " would determine the course of evolution of the system ". But the truth of the matter is that what happens to F, or $\bar{W}$, is a consequence of evolution and not a determinant. The general theory for which Wright seeks is a will-o'-the-wisp, and solving for F is simply a mathematical juggle designed to rescue the concept of an adaptive topography from oblivion. That may by now be impossible, but at least the demise of $\bar{W}$ draws attention to the fact that the *average effect* and the *average excess* are not, in general, the same (Fisher, 1930*a*).

Chapter 6 deals briefly with cytoplasmic inheritance, and in Chapters 7-11 Wright treats the theory of inbreeding, to which he has contributed so much. Naturally, the treatment is by path coefficients. There need be no controversy about the correctness of the approach, though it is fruitful to reflect on ways in which the presentation might be improved. The major advance since Wright's early work on the subject is undoubtedly the replacement of the definition of the inbreeding coefficient as the " correlation between uniting gametes " by the " probability that the two uniting genes are identical by descent ". It is understandable that Wright should not adopt the more modern treatment, but his treatment suffers correspondingly. He barely mentions Malécot's work, and Cotterman (1941), to whom is due the concept of gene-identity by descent, is not mentioned at all. Kempthorne's (1957) contributions are not discussed, though Fisher is frequently (and justifiably) taken to task for his 1949 book. But on the whole we may share the reaction of Rupert Brooke:

> " So shall I curb, so baffle, so suppress
> This too avuncular officiousness,
> Intolerable consanguinity ".

Chapter 12—Population structure—shows the use of the inbreeding coefficient at its best, and Chapters 13 and 14 deal with the stochastic distribution of gene frequencies, with a long digression on self-incompatibility alleles. Chapters 15 and 16 are on components of variability, path coefficients being used freely. Chapter 17 is a summary of the entire book.

One may readily sympathise with Wright in his desire to give us a lasting account of his contributions to the subject of mathematical population genetics, but in this volume he renders himself a disservice. By offering us a general treatise which adheres so closely to his own earlier work he runs the risk of having this work too harshly criticised. His coverage of the literature is extensive but also, as we have seen, selective. His treatment of some topics is not along lines which are generally adjudged the most illuminating, and his insistence on relying so heavily on the concept of an adaptive surface creates more problems than it solves.

There is no doubt that his contribution to the subject has been great, but if his wish is that future generations should fully understand and appreciate this contribution, it would have been better to have presented us with a historical account than to offer us a treatise so biassed in favour of his own particular approaches. For it is generally true that the study of the foundations of a subject enhance one's opinions of the founders, whereas the realisation that earlier views are being too closely followed has the reverse effect. We must view this book in a historical setting, for it would be a disservice to the subject and to Professor Wright if we were to regard it as an account of the contemporary state of mathematical population genetics.

A. W. F. Edwards
Gonville and Caius College, Cambridge

References

(Only those references made in the review but not in the book are given.)

ATKINSON, F. V., WATTERSON, G. A., AND MORAN, P. A. F. 1960. A matrix inequality. *Quart. J. Math.*, *11*, 137-40.

BENNETT, J. H. 1958. The existence and stability of a selectively balanced polymorphism at a sex-linked locus. *Austral. J. Biol. Sci.*, *11*, 598-602.

CANNINGS, C. 1968. Equilibrium under selection at a multi-allelic sex-linked locus. *Biometrics*, *24*, 187-189.

COTTERMAN, C. 1941. The correlation between relatives in a random-mating population. *Sci. Month*, *53*, 227-234.

EDWARDS, A. W. F. 1961. The population genetics of sex-ratio in *Drosophila pseudoobscura*. *Heredity*, *16*, 291-304.

HALDANE, J. B. S. 1966. *The Causes of Evolution.* Cornell University Press.

KIMURA, M. 1960. *Population Genetics* (Published in Japanese).

KINGMAN, J. F. C. 1961. On an inequality in partial averages. *Quart. J. Math.*, *12*, 78-80.

MANDEL, S. P. H. 1959*a*. Stable equilibrium at a sex-linked locus. *Nature*, *183*, 1347-1348.

MANDEL, S. P. H., AND HUGHES, I. M. 1958. Change in mean viability at a multi-allelic locus in a population under random mating. *Nature*, *182*, 63-64.

MULHOLLAND, H. P., AND SMITH, C. A. B. 1959. An inequality arising in genetical theory. *Amer. Math. Monthly*, *66*, 673-683.

SCHEUER, P. A. G., AND MANDEL, S. P. H. 1959. An inequality in population genetics. *Heredity*, *31*, 519-524.

OWEN, A. R. G. 1953. A genetical system admitting of two distinct stable equilibria under Natural Selection. *Heredity*, *7*, 97-102.

OWEN, A. R. G. 1954. Balanced polymorphism of a multiple allelic series. *Caryologia*, *Supp. 6*, 1240-1241.

WEIR, B. S. 1970. Equilibria under inbreeding and selection. *Genetics*, *65*, 371-378.

Paper 53 (1972)

Affinity as revealed by differences in gene frequencies
In *The Assessment of Population Affinities in Man*, eds.
J. S. Weiner and J. Huizinga, Oxford: Clarendon Press,
37–47.
(With L. L. Cavalli-Sforza)

3

AFFINITY AS REVEALED BY DIFFERENCES IN GENE FREQUENCIES

A. W. F. EDWARDS *and* L. L. CAVALLI-SFORZA

Gonville and Caius College, Cambridge
and
Istituto di Genetica, University of Pavia

Introduction

In 1964 (Edwards and Cavalli-Sforza) we gave details of a method of measuring the distance between two populations on the basis of the differences in gene frequencies between them. The method, which is described below, was suggested to one of us by R. A. Fisher, and was used by Cavalli-Sforza and Conterio (1960). Mr. J. C. Gower, of Rothamsted, has kindly drawn our attention to a much earlier account of the method by Bhattacharyya (1946) which both we and our chief critics (Balakrishnan and Sanghvi 1968) have overlooked. Bhattacharyya's work was received by the journal *Sankhyā* as early as 1941, the delay in publication presumably being caused by war-time conditions. Since Fisher had a paper published in the same volume of *Sankhyā* it seems likely that he was familiar with Bhattacharyya's contribution, though in advising us many years later he may well have forgotten the precise reference, particularly as the transformation used was given by him for the case of two alleles in early editions of *Statistical methods for research workers* (see Fisher 1958).

The method has been criticized by Balakrishnan and Sanghvi (1968) on grounds which seem to us to be erroneous for reasons which are given by Edwards (1971), who confirms Bhattacharyya's results by a different method. Having wrongly rejected Bhattacharyya's transformation, Balakrishnan and Sanghvi promote other methods based on linear transformations, which, as Edwards (1971) shows, cannot satisfy the agreed requirements for a measure of distance. The present paper is principally devoted to an account of Bhattacharyya's method as we have developed and used it.

Stochastic distance

The concept of stochastic distance is an attempt to appropriate to the stochastic case some of the attributes of distance as used in deterministic situations. We are all familiar with the ordinary linear measures of distance in a Euclidean space, though, as Ellis (1966) points out, our reasons for using them are really ones of mathematical expedience rather than compelling logic. If we can transfer some of the attributes of deterministic distance to the stochastic case we may, through this familiarity, be able to interpret it more easily. To retain as much similarity as possible it is important to try and continue to work in a Euclidean space, even if more than three dimensions should be required, in spite of the fact that some measures will fit more naturally into a non-Euclidean space (see Johnson 1968).

If in the deterministic case two points are a distance d apart, we can say something about their mutual relationship. For example, if we travel along the straight line from one to the other at a constant velocity v, the journey will take time d/v, and if we travel with constant acceleration f, starting from rest, we will arrive at a velocity of $\sqrt{2fd}$. Distance is evidently a very useful concept, and summarizes much of the information about the relationship of two points.

In a stochastic case, however, the information that distance summarizes must be probabilistic in kind. Two points a distance d apart must lie in some defined probabilistic relationship to one another for the distance to convey any meaning, and in order to define the relationship we must define the stochastic law which governs movement in the space. There are many such diffusion processes from which to choose, but until we agree on a standard we will be unable to converse freely about stochastic distance. One of the desirable properties of deterministic distance is its universality throughout space: for many purposes we do not need to know where the points are, but only their distance apart. For this to be true of a stochastic distance it is clear that the diffusion process employed must be identical throughout the space and in all directions. This will be so if the probability distribution which characterizes the process (by describing the probability with which a point will have moved to another position in time t) is radially symmetrical and has a form independent of the original position of the point. Furthermore, a consistent interpretation of distance will only be achieved if the process is time-homogeneous, so that the same expected distribution is obtained after an elapsed time t however t is subdivided. This requirement implies a distribution whose variance is proportional to time, and the other requirements are most naturally and simply satisfied by a radially-symmetrical multivariate normal distribution. The simplest satisfactory diffusion process is therefore that which is governed by a normal distribution whose variance is proportional to time, and this is well known as the spatially homogeneous Brownian motion process. The probability density for the position of a point a distance d from its starting place, in any specified direction, is

$$(1/\sigma\sqrt{2\pi t})^p \exp(-d^2/2t\sigma^2),$$

where t is the elapsed time, p the number of dimensions of the Euclidean space of movement, and σ^2 the variance of the process (in every direction) per unit time.

With this convention we can interpret any distance in terms of the probability of that distance being travelled in a certain time, and from a knowledge of distance we can calculate many quantities of interest, since we know the characteristic distribution. Of course, many other diffusion processes could be used, but each results in its own interpretation of distance, and confusion will only be avoided if we adopt a particular convention, such as using the spatially homogeneous Brownian motion process.

In applying distance thus defined to a particular situation it will be necessary to transform the space which is natural to the problem at hand so that the Brownian motion model obtains, at least approximately. In some cases, of course, such a transformation will not be possible, and it will not then be possible to talk about conventional stochastic distance.

The genetic model

In our efforts to reconstruct evolutionary pathways from a knowledge of present gene frequencies amongst a group of populations, we have argued (Cavalli-Sforza and Edwards 1967) that it is essential to adopt a specific model for the processes of evolutionary divergence in gene frequency. After considering various possibilities we concluded that for fairly closely related populations (such as the races of man) it was reasonable to suppose that the process of gene frequency change was brought about by random genetic drift or by forces which closely simulate random genetic drift. Kimura (1955) has shown that this implies a distribution of gene frequencies, approximately multinomial in form, in which the variance of a gene frequency initially p is $p(1-p)(1-e^{-t/2N})$, where N is the effective population size and t is the time elapsed in generations. If t is small compared with N the variance of the ith gene frequency will be approximately $p_i(1-p_i)t/2N$, and (because of the multinomial form) the covariance of the ith and jth gene frequencies $-p_i p_j t/2N$.

Thus the characteristic distribution of this particular stochastic process is not that of the spatially homogeneous Brownian motion process. We have, however, two motives for transforming the process into the standard form: first, by doing so we will be able to talk meaningfully about distance, and, secondly, we will be able to apply statistical estimation procedures for reconstructing the evolutionary tree which are based on the Brownian motion process (Cavalli-Sforza and Edwards 1967; Edwards 1970). The required transformation for a single locus is that suggested by Bhattacharyya, and is described in the next section.

The angular transformation

If the gene frequencies p_i, $1 \leq i \leq k$, of k alleles are plotted along k Cartesian axes, any point representing a population will lie in the $(k-1)$-flat $\sum p_i = 1$. Fig. 1 shows this for the case of $k = 3$ alleles. But, on the model of random genetic drift, the characteristic distribution for the stochastic process in this space will not be of the required form. Bhattacharyya suggested the transformation to new variables θ_i, $1 \leq i \leq k$, given by

$$\sin^2 \theta_i = p_i, \quad \text{for all } i.$$

This transformation is equivalent to the plotting of the square roots of the gene frequencies, $\sqrt{p_i}$, along k Cartesian axes, resulting in the population space being the $(1/2^k)$th part of the surface of the unit hypersphere in k dimensions, since the sum of the squares of the coordinates is then simply the sum of the gene frequencies, which is unity. The angle between the radius vector of a population and the ith axis is $(\pi/2-\theta_i)$, and $\sin \theta_i$ is the ith direction cosine of the vector. For $k = 3$ alleles the new population space is the octant of a sphere, shown in Fig. 2.

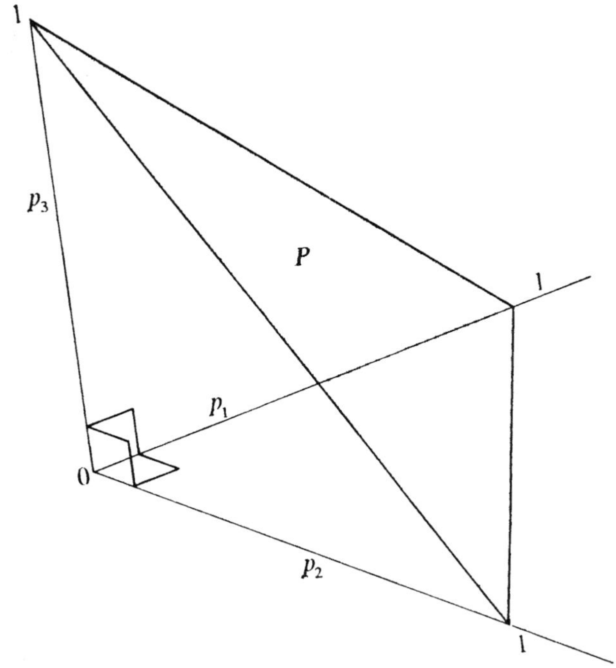

FIG. 1. Cartesian representation of a population P with gene frequencies p_1, p_2, and p_3.

Bhattacharyya showed that to a first approximation the distribution in the transformed space which corresponds to the multinomial distribution in the original space is (without reference to any particular system of coordinates) normal, with radial symmetry, and dispersion (or variance in any direction) independent of its mean. If the multinomial had mean p_i, $1 \leq i \leq k$, variances $p_i(1-p_i)t/2N$, and covariances $-p_ip_jt/2N$, then the normal has mean θ_i and variance in any direction $t/8N$. Naturally, the new variables θ_i will be correlated among themselves, since a functional relation exists among them, but, notwithstanding the comments of Balakrishnan and Sanghvi, these correlations are irrelevant to a consideration of distance because distance is a concept which does not depend on the particular coordinate system in use. Edwards (1971) has derived the variances and covariances of the θ_i and confirmed the radial symmetry of the transformed distribution by showing that the correlations among the θ_i are precisely those expected between these non-orthogonal variables.

This angular transformation thus provides us with a population space in which to measure distance which conforms to the Brownian motion standard over small areas, but which suffers the disadvantage over large areas of being curved rather than Euclidean. Bearing in mind the above comments on the desirability of a Euclidean space, the next step must be to find a projection of the curved space into a Euclidean one which will best preserve the properties already achieved. Indeed, a Euclidean space is essential if we are to be able to combine distances from a number of different loci. The projection required is described in the next section.

Before leaving the curved space of Bhattacharyya we should note that in it the measure of distance between two populations with gene frequencies p_i and p'_i is given by the great-circle distance separating them, which is readily found from the direction cosines to be α, where

$$\cos\alpha = \sum_{i=1}^{k} \sqrt{p_i p'_i}.$$

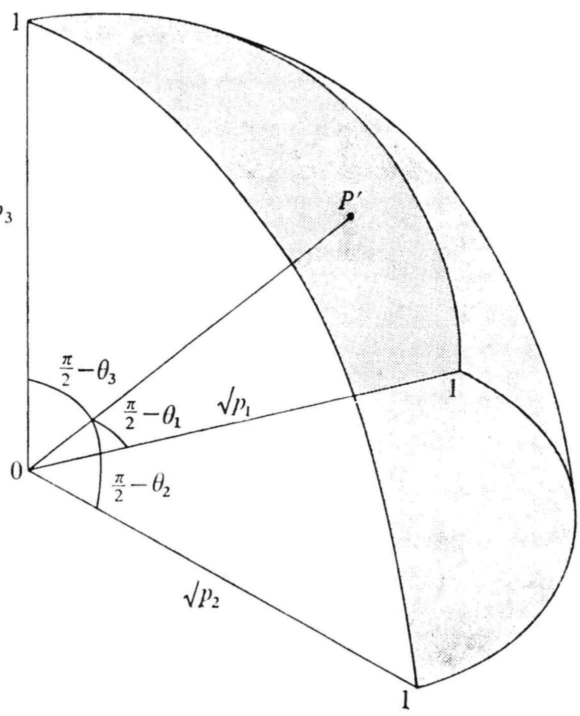

FIG. 2. Representation of the population P by a point P' on the octant of a sphere, achieved by plotting the square roots of the gene frequencies along Cartesian axes.

Stereographic projection into a Euclidean space

Edwards (1971) has shown that a stereographic projection of the surface of the hypersphere onto the tangent plane, along an axis equally inclined to all the original Cartesian axes, preserves the essential properties of the curved space over small areas, whilst admitting only a modest scale distortion at the vertices. The geometry of the projection for the case of three alleles is shown in Fig. 3,

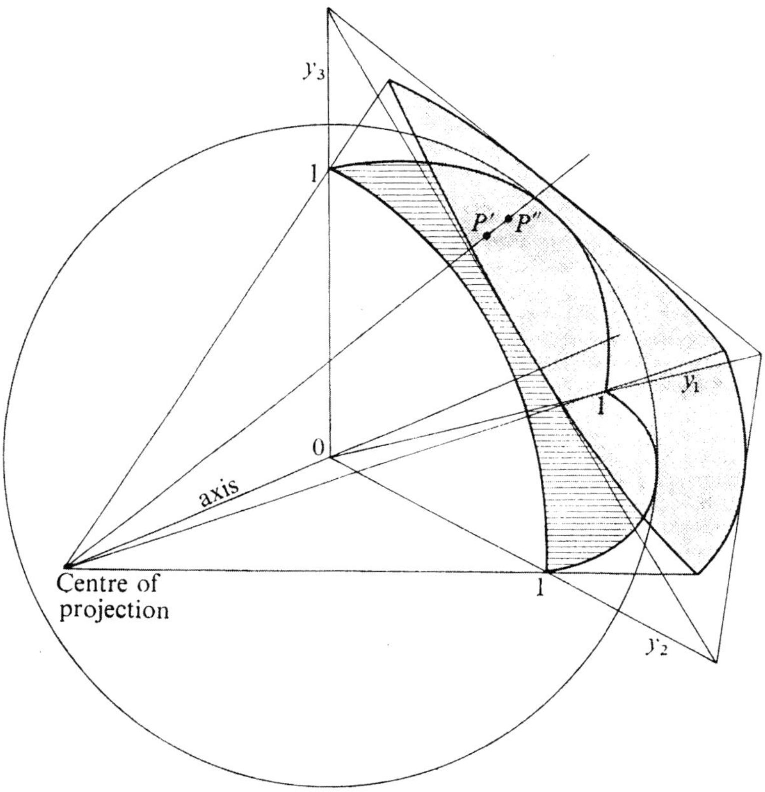

FIG. 3. Stereographic projection of the point P' into the point P'' in the plane tangent to the octant of the sphere at the centre of the octant.

and the resulting population space, graduated in units of the original gene frequencies, in Fig. 4. Referring to the original Cartesian axes, the coordinates of the new point representing a population are

$$y_i = \frac{2(\sqrt{p_i}+\sqrt{1/k})}{1+\sum\sqrt{p_i/k}} - \frac{1}{\sqrt{k}}, \quad \text{all } i,$$

and populations thus displayed will lie in a Euclidean space of $k-1$ dimensions, since $\sum y_i = \sqrt{k}$.

In this space the distance between two populations with gene frequencies p_i and p'_i is given by E where

$$E^2 = \frac{8\{1-\sum\surd(p_i p'_i)\}}{\{1+\sum\surd(p_i/k)\}\{1+\sum\surd(p'_i/k)\}}.$$

Near the centre of the space the scale is $\pi/2$ units to one gene substitution, but at a vertex it is greater than this by a factor of $2\surd k/(1 \times \surd k)$, which amounts to 1·27 for $k = 3$. The distance between two vertices is evidently $2\surd 2/(1 + 1/\surd k)$, or about 1·79 for $k = 3$, whereas $\pi/2 = 1{\cdot}57$. It seems best to adopt a unit of measurement which is correct at the centre, and we therefore express the distance between two populations as

$$E = \frac{4\surd 2}{\pi}\cdot\sqrt{\frac{\{1-\sum\surd(p_i p'_i)\}}{\{1+\sum\surd(p_i/k)\}\{1+\sum\surd(p'_i/k)\}}}\ \text{gene substitutions.}$$

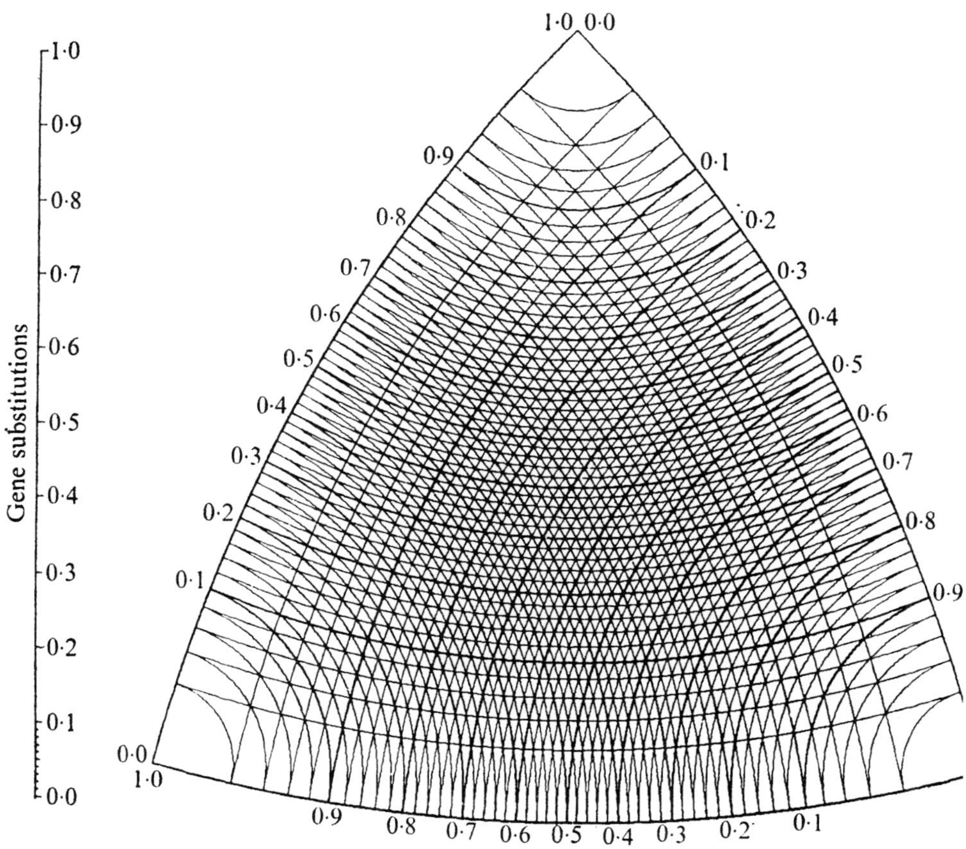

FIG. 4. Stereographic projection of an octant of a sphere for plotting populations by their gene frequencies at a triallelic locus. The grid lines are at intervals of 0·1.

Comparison with other measures of distance

Apart from E, two other distance measures have been suggested as approximations to the fundamental $\cos^{-1}\sum\surd(p_i p_i')$ of Bhattacharyya. Edwards and Cavalli-Sforza (1964) took the length of the chord as an approximation to the length of the arc, and have used this in subsequent work (Cavalli-Sforza and Edwards 1967). This has the advantage that the points representing populations are arrayed in a Euclidean space (of k dimensions), but this space is not isotropic for Brownian motion, and long distances are underestimated.

If the chord distance is expressed in terms of the p_i and p_i' and a further approximation made (see Bhattacharyya) we obtain a measure suggested by Mahalanobis:

$$G^2 = \sum\frac{(p_i - p_i')^2}{(p_i + p_i')}.$$

Bhattacharyya seems to have omitted a factor of one half in deriving this, but throughout the present discussion all such scale factors may be disregarded. Sanghvi (see Balakrishnan and Sanghvi 1968) later rediscovered this measure, but since it involves a double approximation to the fundamental arc measure, and is non-Euclidean, it has little but simplicity to commend it.

Other methods mentioned by Balakrishnan and Sanghvi involve only linear transformations of the original points, and therefore make no contribution towards stabilizing the multinomial variance. They also depend, for their transformations, on the data in question (as does Mahalanobis's D^2), and hence will not provide distances which are universally comparable. It was for this reason that we never followed up our suggestion (Cavalli-Sforza and Edwards 1967) that one might use a projection of the curved space centred on the centroid of the actual populations.

Naturally, over small areas near the centre of the population space all these measures amount to the same thing, in which event, as Mr. J. C. Gower has pointed out in Chapter 1, one might as well abandon all pretence of sophistication and use the straightforward distance in the original $(k-1)$-flat given by

$$X^2 = \sum(p_i - p_i')^2.$$

But if an attempt is to be made to overcome the vagaries of the multinomial distribution there can be no doubt that Bhattacharyya has described the optimal transformation, and little doubt that the stereographic projection described by Edwards (1971) is the best subsequent transformation, independent of the actual data, which ensures a Euclidean representation.

Combination of distances over loci

In the space in which E is measured the variance in any direction is $t/8N$, and is independent of the number of alleles k. It immediately follows that squared distances may be summed over loci to achieve an overall squared distance.

Another way of seeing that the individual squared distances should not be weighted in the combination (as has sometimes been suggested) is to observe that for small distances the squared distance between two populations both of which are supposed to have derived from a single ancestral population is, apart from a constant involving $t/8N$, distributed as χ^2_{k-1} (Bhattacharyya). The squared distances may therefore be summed directly, just as are independent χ^2 values.

Since, for a constant elapsed time, the expected squared distance at a locus is proportional to $k-1$, the expected squared distance summed over loci is proportional to $\sum_{\text{loci}} (k-1)$. In applications in which squared distances are being used to estimate times, therefore, a standard measure will be obtained for a single locus by dividing E^2 by $k-1$, and for several loci by dividing $\sum_{\text{loci}} E^2$ by $\sum_{\text{loci}} (k-1)$.

Distance on other genetic models

Cavalli-Sforza (1969) has recently pointed out the advantages of relating genetic distance to the kinship coefficient, and has shown how it may be interpreted on a model of differentiation by directed evolution. His paper should be consulted for further details.

REFERENCES

Balakrishnan, V. and Sanghvi, L. D. (1968) Distance between populations on the basis of attribute data. *Biometrics* **24**, 859–65.

Bhattacharyya, A. (1946) On a measure of divergence between two multinomial populations. *Sankhyā* **7**, 401–6.

CAVALLI-SFORZA, L. L. (1969) Human diversity. *Proc. XII Int. Congr Genetics, Tokyo.* **3,** 405–17.

—— and CONTERIO, F. (1960) Analisi della fluttuazione di frequenz geniche nella popolazione della Val Parma. *Atti Ass. genet. ital.* **5** 333–44.

—— and EDWARDS, A. W. F. (1967) Phylogenetic analysis: models an estimation procedures. *Evolution, Lancaster, Pa.* **21,** 550–70.

EDWARDS, A. W. F. (1970) Estimation of the branch points of a branchin diffusion process. *Jl R. statist. Soc. B.* **32,** 155–74.

—— Distances between populations on the basis of gene frequencies. *Biometrics.* In press.

—— and CAVALLI-SFORZA, L. L. (1964) Reconstruction of evolutionary trees. *Publs Syst. Ass.* **6,** 67–76.

ELLIS, B. (1966) *Basic concepts of measurement.* Cambridge University Press.

FISHER, R. A. (1958) *Statistical methods for research workers,* 13th ed. Oliver and Boyd, Edinburgh.

JOHNSON, L. A. S. (1968) Rainbow's end: the quest for an optimal taxonomy. *Proc. Linn. Soc. N.S.W.* **93,** 8–45.

KIMURA, M. (1955) Random genetic drift in multiallelic locus. *Evolution, Lancaster, Pa.* **9,** 419–35.

Note added in proof

Whilst looking through the stock of graph paper in the Department of Genetics of the University of Cambridge, where Sir Ronald Fisher was Professor, for some ordinary 'triangular' graph paper, I chanced upon some paper identical to that shown in Fig. 4. It is headed 'Trilinear square-root chart: polar projection of an octant of the surface of a sphere', and contains no other means of identification. In spite of the most widespread enquiries I have been unable to ascertain its origin, but it seems likely that it was drawn at Fisher's suggestion in the early 1950s. I am not aware that it has been given or used in any publication.

A.W.F.E.

Paper 56 (1972)

Likelihood

Bulletin of the Institute of Mathematics and its Applications 8, 329–331.

Reprinted by permission of the Institute of Mathematics and its Applications.

Likelihood

A. W. F. EDWARDS

Gonville and Caius College, Cambridge

IT is a natural consequence of mathematical training that theoretical statisticians should wish to develop their subject axiomatically, but equally it is a consequence of the applied background of statistics as a whole that the axiomatic basis should be chosen in accordance with the canons of scientific inference. But herein lies our first difficulty, because the philosophy of science itself is a field with many different points of view. A dominant one, however, is that it is the business of science to explain, or to "make plain" and I shall assume this to be the business of statistical inference in science also. I restrict my remarks to inference in science and am not concerned with extra-scientific activities which also fall within the province of theoretical statistics, such as decision theory.

A suitable axiomatic basis will only be achieved, I believe, through a dialogue between scientist and statistician, in which each must exercise some patience in explaining his field to the other. My training is wholly in science and not at all in mathematical statistics, so I hope you will be patient over my deficiencies in your field. In the last 10 years I have learnt some of the techniques of statistics and it is because, as a scientist, I have found some of these techniques to be of doubtful relevance to scientific inference, that I have become interested in their axiomatic basis.

If I had an infinite time at my disposal I should first examine the reasons for the inadequacy of the usual schemes of inference, taken one by one; then I should examine the historical development of the subject in some detail and finally I should develop the scheme that I find most appropriate, based on R. A. Fisher's concept of Likelihood. But I am unable to do this adequately even within the 200 pages of my book,[1] so I can hardly scratch the surface of the subject in this paper. I will have to treat each of these three topics superficially; I will certainly be unable to claim that likelihood solves all our problems, but I hope I will leave you with the feeling that an axiomatic approach to statistical inference based on likelihood is well worth exploring in depth over the next few years. Such studies may reveal that the approach is inadequate, but I believe they are more likely to show that likelihood is a concept which mathematical statisticians have underrated for far too long.

In my third year as an undergraduate in Cambridge I elected to read Genetics. R. A. Fisher was in his last year as Professor of Genetics and in his first lecture that year, on the estimation of recombination fractions, he introduced us to the method of maximum likelihood. In those days I was in the habit of taking rough notes during lectures and laboriously producing a "fair" copy in the evening. It reads "p [the recombination fraction] may be estimated by the Method of Maximum Likelihood. This method gives the most probable value of p. . . . It can be shown that, for large samples, this method gives the estimate of p with the smallest variance." Here, in a nutshell, is the confusion which runs through so much of modern statistics: the note says "most probable" when it means "most likely" and it fails to make clear that the two criteria by which the method is judged (that it gives the most "probable" value and that it gives the estimate with the smallest variance) are totally different *in kind*.

Let us examine the second point first. I call the distinction "Hacking's divide" because it has been most clearly made by Ian Hacking, first in his book "Logic of Statistical Inference,"[2] and more recently in a paper "Jacques Bernoulli's Art of Conjecturing."[3] Like so many important distinctions in statistics, it first may have been made a long time ago, but it is still not widely appreciated. On one side of the divide is the criterion that any estimator we use should be usually right. That is, in a long series of trials we require that an estimator, say an interval estimator, should include the true value with high probability. For a given probability, the narrower the interval, the better we like the estimator. This is the same kind of criterion as wanting the distribution of an estimator, in repeated sampling from the same population, to have as small a variance as possible, or perhaps to be "minimum variance unbiased."

On the other side of the divide is the criterion that the estimator should be credible on each occasion of use. Hacking emphasises the difference between these two *desiderata* by means of an example. Suppose p is the probability of success in a Bernoulli trial, to be estimated from the number a of successes in n, an even number of such trials. Let $f(a,n)$ be an interval estimator in [0,1] that is "usually right." That is, in repeated usage $f(a,n)$ will usually include the value p and thus meet the first criterion. Now consider the estimator $f^*(a,n)$, which i just the same as $f(a,n)$ except that when $a = n/2$ precisely it is replaced by zero. Thus should a happen to be 500 in 1000 trials, the estimator of p is 0. f^* still meets th first criterion, because it is usually right, the probabilit of obtaining $a = n/2$ being negligible for large n, what ever p, but it is most certainly not credible on eac occasion of use. When $a = n/2$, nobody accepts that is a good estimator of p.

On one side of Hacking's divide, then, criteria c excellence in repeated sampling, appropriate to befor trial betting; on the other side, criteria appropriate t after-trial evaluation. Elsewhere Hacking has spoken c the problem of prediction and the problem of explanatio as the two sides of the coin.

Now it is generally agreed that if you need a theory

statistics based on repeated-sampling criteria, the Neyman-Pearson theory supplies it. True, one may interest oneself in point estimators as an alternative to interval estimators, or in "classical" tests of significance as an alternative to the Neyman-Pearson tests, but by-and-large the "usually right" theory is the received theory of statistics, spiced with some "usually close" or "not usually wrong." Neyman, indeed, holds that inductive inference is impossible: we may indulge in "inductive behaviour" by adopting rules that make us usually right, but we may not measure the credibility of our answer on any particular occasion. This may be a reasonable procedure for an industrialist maximising his profits in the long run, but is it any use for the scientist interested in explanation? That it is not what the scientist requires may be concluded from the simple fact that scientists insist on interpreting confidence intervals as though they were measures of credibility on each occasion of use. And sometimes they are not: if $a = n/2$ precisely, is $p = 0$ a good explanation?

The answer, of course, is a resounding NO. Now I do not for one moment assert that repeated sampling theories have led to the acceptance of many bad explanations; on the contrary, they usually lead to good explanations because in most situations the two criteria are similar. This is part of the confusion of our subject. Thus confidence intervals, significance levels and the other paraphernalia of the repeated-sampling approach may often be interpreted as measures of credibility on particular occasions. But should we not be more direct and ask specifically what properties we require of a measure of credibility? For before-trial betting everybody agrees we should contemplate the probability of being right; but what shall we use for after-trial evaluation?

The reason that $p = 0$ is a bad explanation of $a = n/2$ is that, if p were 0, this value of a would be impossible; and the reason that $p = 0 \cdot 1$, say, would not be a good explanation is that the probability of $a = n/2$ would still be much less on this hypothesis than on the hypothesis $p = 0 \cdot 5$. We are led to prefer that hypothesis which maximises the probability of what we observed, or, in other words, the most likely hypothesis. The concept of likelihood forces itself to our attention.

Before exploring the concept, let us make an historical digression. For the evidence from history is that the 250 year old discussion of statistical inference shows no signs of abating.

Bernoulli, in "Ars Conjectandi" (1713) posed the problem of making statements about an unknown binomial parameter and offered an interval estimator which was "usually right." De Moivre, in the preface to the "Doctrine of Chances" (1718), advocated adopting hypotheses of high likelihood: "Further, the same Arguments which explode the Notion of Luck, may, on the other side, be useful in some Cases to establish a due comparison between Chance and Design: We may imagine Chance and Design to be, as it were, in Competition with each other, for the production of some sorts of Events, and may calculate what probability there is, that those Events should be rather owing to the one than to the other." Bayes (1763) introduced the postulate that hypotheses were the proper objects of probability statements, which enabled him to adopt an "after-trial evaluation" approach, to which I will return. Lambert (1760) and Daniel Bernoulli (1777) both maximised likelihoods, as we should now say, but Gauss (1809) was more explicit: he adopted a Bayesian position and then chose the estimate with the highest posterior probability. But later he preferred repeating-sampling criteria as a justification for least-squares estimation. Laplace (1820) advocated methods on both sides of Hacking's divide.

And so the discussion continued. By the beginning of this century some statisticians advocated least squares, some the method of moments and some Bayesian estimation. But in 1912 an undergraduate at Gonville and Caius College, Cambridge, wrote a paper which was to have revolutionary effects. His name was R. A. Fisher. He argued that least squares methods were arbitrary because they depended on the measurement scale of the variables; that the method of moments was arbitrary because it depended on an arbitrary choice of moments to equate in the population and the sample; and that Bayesian methods of estimation were arbitrary because they depended on the parametric form adopted. His proposal was the method of maximum likelihood, which suffered none of these objections. At first he did not use the word "likelihood" but wrote of "inverse" or "relative" probability. His intention, however, is quite clear. "P is a relative probability only, suitable to compare point with point, but incapable of being interpreted as a probability distribution, or of giving any estimate of absolute probability." Fisher offered no further justification of his proposal, but in 1921, in response to a misunderstanding of it by Karl Pearson and his school, who thought it was a Bayesian proposal, he emphasised that it was not by using the word "likelihood." He wrote "What we can find from a sample is the *likelihood* of any particular value of ρ, if we define the likelihood as a quantity proportional to the probability that, from a population having that particular value of ρ, a sample having the observed value r should be obtained. So defined, probability and likelihood are quantities of an entirely different nature."

The introduction of a new term is seldom greeted with universal acclaim; but here Fisher was seen to be making a subtle and vital distinction. I believe that the new terms I propose and which I will mention presently may similarly be justified. In his first papers Fisher seemed to regard likelihood as a measure of relative credibility, but in 1922 and the following years he sought to establish the method of maximum likelihood by appealing to the criteria of repeated-sampling. However, his researches soon led him to return across Hacking's divide: "Thus, basing our theory entirely on considerations independent of the possible relevance of mathematical likelihood to inductive inferences in problems of estimation, we seem inevitably led to recognise in this quantity the medium by which all such information as we possess may be appropriately conveyed." By 1938 he is writing, somewhat wistfully, "A worker with more intuitive insight than I might perhaps have recognised that likelihood must play in inductive reasoning a part analogous to that of probability in deductive problems. My own procedure has been more pedestrian!"

Meanwhile, Neyman and Pearson were busy crossing Hacking's divide in the opposite direction. They started from the realisation that "there is little doubt that the criterion of likelihood is one which will assist the investigator in reaching his final judgement," but found it necessary to introduce repeated-sampling considerations in order to develop their approach further.

In the background to all these discussions the Bayesians have been hard at work, like double-bass players in an orchestra. Taking their cue from Bayes and Laplace they maintain not only that after-trial evaluation is the right approach, but that the proper axiomatic basis for credibility is provided by the axioms of probability. Jeffreys, de Finetti, Savage and Lindley are recent exponents of this point of view.

I cannot here cover all the grounds of dispute between the "repeated samplers" and the Bayesians. The former have declined to accept that the probability axioms are universally appropriate for the problem of induction. The most telling point, in my view, is that the Bayesian scheme is incapable of achieving a satisfactory representation of ignorance. Furthermore it insists on answering questions about the plausibility of sets when the scientist is asking questions about the plausibility of points. The Bayesians, for their part, have produced devastating criticism of the repeated-sampling approach. But they have been too sure of themselves to notice that the territory on one side of Hacking's divide is not all Bayesian. May not likelihood rather than probability be the proper measure of credibility? Must we not distinguish carefully these two concepts which I confused, long ago, in my lecture notes?

Since much of the confusion of thought in statistics has arisen because similar, but not identical, concepts have the same name, the first task of someone who wishes to draw attention to the differences is to adopt or adapt new words. We even need to find a new phrase for the likelihood method, for otherwise, you can be sure, most statisticians will think we are talking about maximum likelihood. I suggest the word "support" for log-likelihood, and hence the "Method of Support."

I use evaluation as the analogue of estimation, as did Fisher in 1921. Hence I refer to evaluates and evaluators rather than estimates and estimators when I wish to consider best-supported parameter values. I use two techniques to communicate information from support functions when the whole function is not displayed: first, m-unit support limits or regions to define the part of the parameter space where the support is within m units of the maximum; and, secondly, minus the second derivative of the support function taken at the maximum. This I call the observed information, to distinguish it from the expected information of the repeated-sampling theory. Its reciprocal is the formation, because it tells us about the form of the support function near the maximum. Similarly we find the coformation of two parameters and we speak of the formation matrix. The square-root of the observed formation, corresponding to the standard error in the conventional theory, I have christened the span.

Next, we may state the basic axiom of our position, the Likelihood axiom. "Within the framework of a statistical model, all the information which the data provide concerning the relative merits of two hypotheses is continued in the likelihood ratio of those hypotheses on the data, and the likelihood ratio is to be interpreted as the degree to which the data support the one hypothesis against the other."

It is taking surprisingly long for this axiom to emerge explicitly. After all, it only says that, other things being equal, the scientist will prefer the hypothesis which best fits the data and the relative excellence of the fit is to be measured by the probability of obtaining the data.

Finally, let us look to the future. We must examine closely the treatment of nuisance parameters and the concept of conditional inference. We must find a replacement for degrees of freedom which does not depend on repeated sampling. We must resolve the demarcation dispute between probability and likelihood, making up our minds about fiducial probability in the process. But when all is said and done I feel sure that we shall appreciate the wise choice of statistical procedure made so many years ago by Shakespeare's Earl of Westmorland: "by shape of likelihood the news was told."

References

1. Edwards, A. W. F., "Likelihood," Cambridge University Press, 1972.
2. Hacking, I., "Logic of Statistical Inference," Cambridge University Press, 1965.
3. Hacking, I., "Jacques Bernoulli's Art of Conjecturing," *Brit. J. Phil. Sci.*, 1971, **22**, 209-229.

Paper 60 (1974)

The history of likelihood

International Statistical Review 42, 9–15.
Reprinted in *Likelihood* (expanded edition),
Baltimore, MD: Johns Hopkins University
Press (1992).

The History of Likelihood

A. W. F. Edwards
Gonville and Caius College, Cambridge

Introduction

One of R. A. Fisher's most influential papers was "On the mathematical foundations of theoretical statistics" (1922), in which he propounded the *method of maximum likelihood* as a means of point estimation, and hence established a whole branch of statistical reasoning. Yet this paper does not contain his original statement of the method, which was published in 1912, nor does it contain his original definition of *likelihood*, which appeared in 1921. The great innovation of the 1922 paper was, rather, the clear specification of one approach to the problem of estimation, and the elucidation of the properties of maximum-likelihood estimators.

Methods similar to the method of maximum likelihood have a history prior to the work of Fisher, but the definition of likelihood itself, and the realization that it could be used independently as a measure of relative support for hypotheses, appears to be entirely his own, and the main purpose of the present paper is to investigate the background to his introduction of the new concept.

With the decline in the esteem with which repeated-sampling methods are held, in the face of the Bayesian revival, the concept of likelihood has come to the fore as offering an approach to statistical inference that is neither repeated-sampling nor Bayesian (see Edwards, 1972). Although it is too early to predict the extent to which pure likelihood methods can supersede other approaches, the concept of likelihood is, in addition, so fundamental to both Bayesian inference and Neyman-Pearson methods that an account of its origins needs no further justification. In the former, the likelihood is the vehicle which carries the observational or experimental results in Bayes' Theorem, whilst in the latter, likelihood was one of the key concepts in the original formulation by Neyman and Pearson (1928). The present paper traces the history of both *likelihood* and the *method of maximum likelihood*; it is essential to keep the distinction between the two clearly in mind.

The Method of Maximum Likelihood

The problem of the estimation of a parameter was first put into something like its modern form in the fourth part of James Bernoulli's posthumous work *Ars Conjectandi*, published in 1713. Bernoulli considered the estimation of a binomial parameter, and after propounding his famous theorem, was able to use it to make "confidence interval" statements which, however, depended upon the unknown parameter (see Hacking, 1971). Fifty years later Bayes' *Essay* was published, also posthumously, containing the solution to Bernoulli's estimation problem in terms of inverse probability, and making quite explicit the postulate required. It was not, however, in connection with the binomial that alternative methods were first advanced, for in science a more pressing problem was to find the " best" value for a parameter whose determination was subject to "error". It was appreciated that the assumed error distribution would be effectively continuous, and some of the early suggestions for estimating the parameter (as we should now say) of a continuous distribution were close to the method of maximum likelihood.

It is well known that if a uniform prior distribution is adopted in a Bayesian formulation of the estimation problem, maximizing the posterior probability is equivalent to maximizing the likelihood. This circumstance has tended to obscure the distinction between the Bayesian and the non-Bayesian justification of the method, and in order to emphasize the distinction I shall refer to the method of maximum *likelihood* when no Bayesian connection is offered, and otherwise to the method of maximum *probability*.

In 1760 Lambert proposed a method of estimation, in his *Photometria*, which is analytically identical to the method of maximum likelihood. The work is not directly mentioned in Todhunter's *History*, but Sheynin (1971) gives a full account. Lambert finds the multinomial probability for four observations from a continuous distribution. Then, dropping the irrelevant multinomial coefficient, he proceeds to maximize the "likelihood" by differentiating the log-likelihood and equating it to zero. Lambert's method was referred to by John Bernoulli in his article on the mean ("milieu") in the second of the volumes on "Mathématiques" of the *Encyclopédie Méthodique*, published in Paris in 1785 (but presumably written much earlier). John Bernoulli did not there describe it, on the grounds that it was already published, but instead devoted most of his article to the unpublished methods of Daniel Bernoulli and Lagrange. John refers to the former as "un petit écrit latin de M. Daniel Bernoulli, qu'il me communiqua, en 1769". It was later published (1778) and is given in English translation by Kendall (1961). Todhunter devotes Articles 424–427 to it, but does not appreciate that the method proposed by Bernoulli may be interpreted as being free of a Bayesian justification. Bernoulli simply states (in translation): "I think that of all the innumerable ways of dealing with errors of observation one should choose the one that has the highest degree of probability for the complex of observations as a whole". Unfortunately he then proposes a semi-elliptic error distribution, which results in an intractable maximization problem. Euler (1778) was less than enthusiastic about Bernoulli's admittedly "purely metaphysical speculation". His paper follows Bernoulli's, both in the original and in translation.

Lagrange's method was also subsequently published (1770–73). As Plackett (1958) observes, Lagrange "purports to show that the mode of the distribution of sample means is the same as the population mean". But he is in difficulty over the fact that in the discrete cases with which he worked, the most probable sample, being integral, might have a mean which could not match that of the population. However, in Problem VI Lagrange tackles the practical difficulty that one does not know the population distribution, by *estimating* the proportions with which the various errors occur; and this he does by maximizing the likelihood, finding the maximum-likelihood estimators of multinomial proportions. Todhunter (Art. 562) says that Lagrange "[takes] it as evident that the most probable values" are these, but there seems to be no justification for implying that Lagrange was using an inverse probability argument, though it is true, as Todhunter also remarked, that in the subsequent investigations into the probability that the estimates do not differ from the true values by more than assigned quantities, the inverse argument is used; for without its use this probability must generally involve the unknown true values, which, in Lagrange's treatment, it does not.

In passing we may note that the copy of Todhunter in the Library of Gonville and Caius College, Cambridge, contains a correction in Fisher's handwriting to Art. 561: where Todhunter writes "It appears from common Algebra that the greatest value of μ is when . . .", in connection with Lagrange's determination of the mode, Fisher has (quite correctly) replaced "greatest" by "most frequent".

There are hardly any references to these contributions until modern times, though Daniel Bernoulli's paper is mentioned by Keynes (1921) and discussed briefly by Levy and Roth (1936), who thought he was using an inverse probability argument. Lambert's contribution seems to have vanished for 200 years, partly, no doubt, because Ostwald's *Klassiker* translation omitted the part of interest (Sheynin, 1971). They seem to have been without influence on the subsequent development of estimation theory, probably because the method of least squares, which was shortly to be introduced, provided a practical procedure of such utility that logical qualms were reserved until later.

The Method of Maximum Probability

In *Theoria Motus Corporum Coelestium* (1809), Gauss introduced the method of least squares as a necessary consequence of (1) accepting the arithmetic mean as the optimum estimator, and (2) applying the method of maximum probability, using a uniform prior distribution. He shows that the error distribution, for which the mean is the maximum-probability estimator, is the normal, and that in the normal case maximizing the posterior probability is equivalent to minimizing the sum of squares. The subsequent history of least squares is not our present concern; but Gauss's original treatment is explicitly Bayesian, and therefore does not constitute an application of the method of maximum likelihood.

Similarly Laplace, in *Théorie Analytique des Probabilités* (1820), frequently used maximum-probability values for parameters, having adopted a uniform prior probability distribution. It is true that he writes of "values which render the observed event the most probable", but only in connection with a prior distribution; furthermore, when considering transformation to a new parameter, the differential element is taken into account, indicating the use of a probability density rather than a likelihood. Nor does Laplace always use the maximum probability argument: for example, when considering the classical problem of errors he thinks that the median of the posterior probability distribution should be used: "Some celebrated mathematicians", he writes, "have suggested that one should choose, for the mean, that which renders the observed result the most probable, and hence the abscissa which corresponds to the greatest ordinate of the [posterior probability] curve; but our mean is evidently the one indicated by probability theory".

Of particular interest is Laplace's approximate evaluation of the posterior probability that the true parameter value should lie in a certain interval about the most probable value, for he uses an expansion whose first and most important term is none other than minus the second differential of the log-likelihood evaluated at the maximum, or the *observed information* in my terminology (Edwards, 1972). At the end of the century Karl Pearson and L. N. G. Filon (1898) offered a very similar calculation, finding "the frequency distribution for errors in the values of the frequency constants". I conclude from this phrase that the authors were thinking in terms of inverse probability; E. S. Pearson (1967) concurs. In 1896 Karl Pearson had written: "Thus, it appears that the observed result is the most probable, when r (the correlation coefficient as a parameter) is given the value $S(xy)/(n\sigma_1\sigma_2)$. This value presents no practical difficulty in calculation, and therefore we shall adopt it". Pearson adds the footnote: "It seems desirable to draw special attention to this best value of the correlation coefficient . . .".

That Pearson was following Gauss's method of maximum probability is clear from a letter he wrote to Fisher twenty years later (26 June 1916; quoted by E. S. Pearson, 1968). Fisher had drafted a note pointing out that a contributor to the May 1916 issue of *Biometrika* had used a method of estimation (minimum χ^2) which was arbitrary to the extent that it depended on the way the data were grouped: Pearson replied, declining to publish the note; though "If you will write me a defence of the Gaussian method, I will certainly consider its publication. I frankly confess I approved the Gaussian method in 1897, but I think it logically at fault now". The reason that Pearson found fault with it at that time seems to have been that he could not accept the arbitrary nature of the prior probability distribution needed for completion of the method. Later, after Fisher had sought justification for the method of maximum likelihood in the repeated sampling properties of the estimators, Pearson, by then clear that inverse probability was not involved, attacked it as lacking justification in small samples.

Before recording Pearson's criticism of "the indiscriminate use of Bayes' Theorem" we must sketch the situation that led him to make it. Fisher advanced the method of maximum likelihood in 1912, but unfortunately he used the phrase "inverse probability", that same phrase by which Todhunter had described the Bayesian or Laplacian method. It turned out to be one of the most influential errors of terminology in statistics, for it led directly to his first quarrel with Pearson, who did not look beyond the phrase to Fisher's account and subsequent use of the method, which was non-Bayesian. We shall examine Fisher's 1912 account in detail below; his first *use* of the method was in 1915, in his paper on the correlation coefficient: "I have given elsewhere a criterion, independent of scaling, suitable for obtaining the relation between an observed correlation of a sample and the most probable value of the correlation of the whole population" (p. 520). On the following page he wrote not of the most *probable* value, but of the most *likely* value. Was this the germ of the later definition of *likelihood*? That Fisher's criterion was *independent of scaling* should perhaps have given a clue to Pearson, but it did not. Fisher's paper, in which the distribution of r in samples from a normal bivariate population was first obtained, sparked off an extensive investigation by Pearson and his collaborators. The results appeared in *Biometrika* the following year under the title *On the distribution of the correlation coefficient in small samples. Appendix II to the papers of "Student" and R. A. Fisher. A Cooperative Study*. The authors were Soper, Young, Cave, Lee and Pearson. There is a section "On the determination of the 'most likely' value of the correlation in the sampled population" in which the authors ask "What is the most reasonable value?" It seems likely that Fisher knew nothing of this section until the paper was in proof (E. S. Pearson, 1968), if then, in spite of the description of the paper as an appendix to one of his own. It contained a criticism of what the authors supposed was Fisher's use of a prior distribution for ρ, and concluded "It will thus be evident that in problems like the present the indiscriminate use of Bayes' Theorem is to be deprecated. It has unfortunately been made into a fetish by certain purely mathematical writers on the theory of probability, who have not adequately appreciated the limits of Edgeworth's justification of the theorem by appealing to *general* experience".

Pearson, then, did not use maximum likelihood; nor did he originally appreciate its difference from maximum probability; and his use of maximum probability was short-lived. Thus Haldane's (1957) suggestion that Pearson's was the method "which was developed by Edgeworth as 'the method of maximum credibility', and by Fisher as 'the method of maximum likelihood'" is a little disingenuous. Elsewhere Haldane also found it difficult to dissociate Fisher's method from inverse probability (see Haldane, 1932, and Fisher's reply, 1932). Wright (1968) repeats Haldane's suggestion, though attributing the method to Gauss. Edgeworth was clearly appealing to inverse probability (Pearson, 1967), though latterly he wished to justify prior distributions by general experience rather than to assume them as representations of ignorance: "I submit that very generally we are justified in assuming an equal distribution of *a priori* probabilities over that tract of the measurable with which we are concerned" (1908).

Likelihood-ratio Arguments

It has long been the practice to argue that we may distinguish between hypotheses by comparing the probability with which each would lead to the observed result. Thus de Moivre (1756), writing in 1717: "Further, the same Arguments which explode the Notion of Luck, may, on the other side, be useful in some Cases to establish a due comparison between Chance and Design: We may imagine Chance and Design to be, as it were, in Competition with each other, for the production of some sorts of Events, and may calculate what Probability there is, that those Events should be rather owing to one than to the other".

Laplace (1820), in the sixth principle enunciated in the *Essai philosophique*, writes: "Each of the causes to which an observed event may be attributed is indicated with just as much likelihood as there is probability that the event will take place, supposing the event to be constant". The translation is that of Truscott and Emery (1951), but in the original we find that the French word is "vraisemblance", and that the above sentence is only the first clause of a longer sentence, being concluded by a semi-colon and not a full stop. In translation Laplace continues: "The probability of the existence of any one of these causes is then a fraction whose numerator is the probability of the event resulting from this cause and whose denominator is the sum of the similar probabilities relative to all the causes; if these various causes, considered *a priori*, are unequally probable, it is necessary, in place of the probability of the event resulting from each cause, to employ the product of this probability by the possibility of the cause itself". The word "possibility" is also a loose translation, for Laplace wrote "celle", referring to *probability*. Laplace was evidently being purely " Bayesian".

Perhaps the first example of a likelihood-ratio argument (as we now call such comparisons of probabilities) to come from a writer who explicitly rejected inverse probability is that of Venn (1876): "To decide this question, what we have to do is to compare the relative frequency with which the two kinds of cause would produce such a result".

Fisher and the Definition of Likelihood

On 1 October 1909, R. A. Fisher was admitted to Gonville and Caius College, Cambridge, as an entrance scholar. John Venn was President of the College, and F. J. M. Stratton had been a Fellow for three years. In his third year as an undergraduate (1912) Fisher published a paper "On an absolute criterion for fitting frequency curves". He observes that the similarity of the results obtained by various methods "is harmful from the theoretical standpoint as tending to obscure the practical discrepancies, and the theoretical indefiniteness which actually exist". He mentions the method of least squares, but dismisses it as "obviously inapplicable to frequency curves" because "an arbitrariness arises in the scaling of the abscissa line, for if ξ, any function of x, were substituted for x, the criterion would be modified". Of the method of moments, it "is possibly of more value, though its arbitrary nature is more apparent", for "a choice has been made without theoretical justification in selecting [this] set of r equations of the general form $\Sigma x^p f = \sum_1^n x^p$". Thus does Fisher clear the decks.

"But we may solve the real problem directly." "The most probable set of values for the θ's will make P a maximum", where P is the probability of the observations, given the parameter set θ, less the differential elements dx. Here Fisher seems to propose the method of maximum probability, but without specific reference to a prior distribution. He then considers the case of estimating the mean and *modulus of precision* of a normal distribution, and observes that "the inverse probability system may be represented by the surface traced out by a point at a height P above the point on a plane, of which m [the mean] and h [the modulus] are co-ordinates". Again, Fisher seems to be relying on inverse probability, but this impression is soon to be dispelled: "We shall see (in § 6) that the integration with respect to m is illegitimate and has no definite meaning with respect to inverse probability". After some discussion of bias, we come to § 6, which I quote in full: "We have now obtained an absolute criterion for finding the relative probabilities of different sets of values for the elements of a probability system of known form. It would now seem natural to obtain an expression for the probability that the true values of the elements should lie within any given range. Unfortunately we cannot do so. The quantity P must be considered as the relative probability of the set of values $\theta_1, \theta_2, \ldots, \theta_r$; but it would be illegitimate to multiply this quantity by the variations $d\theta_1, d\theta_2, \ldots, d\theta_r$, and integrate through a region, and to compare the integral over this region with the integral over all possible values of the θ's. P is a relative probability only, suitable to compare point with point, but incapable of being interpreted as a probability distribution over a region, or of giving any estimate of absolute probability.

"This may be easily seen, since the same frequency curve might equally be specified by any r independent functions of the θ's, say $\phi_1, \phi_2, \ldots, \phi_r$, and the relative values of P would be unchanged by such a transformation; but the probability that the true values lie within a region must be the same whether it is expressed in terms of θ or ϕ, so that we could have for all values

$$\frac{\partial(\theta_1, \theta_2, \ldots, \theta_r)}{\partial(\phi_1, \phi_2, \ldots, \phi_r)} = 1,$$

a condition which is manifestly not satisfied by the general transformation.

"In conclusion I should like to acknowledge the great kindness of Mr J. F. M. Stratton [*sic*], to whose criticism and encouragement the present form of this note is due."

In view of § 6 we must conclude that Fisher was using the phrase "inverse probability" incorrectly, as he later admitted (1922). He was manifestly describing the method of maximum likelihood, and pointing out its invariance to parameter transformation. It would have been a remarkable achievement for an undergraduate if Fisher had done no more than criticize Gauss's method of least squares for its dependence on the choice of a scale for the variate, Pearson's method of moments for its arbitrariness, and, by implication, Laplace's method of maximum probability for its dependence on the choice of parametric form; but Fisher went further and gave us a concept "suitable to compare point with point, but incapable of being interpreted as a probability distribution over a region". He introduced the concept of a likelihood surface for two parameters, and, by being prepared to use his "relative probability" to compare "point with point", gave likelihood a precise meaning, though not yet a name.

The confusion that was caused by Fisher introducing a new concept without a new name has been described above, and we will not pursue the matter much further here. It suffices to add that Fisher, seeing the importance of distinguishing his new concept from probability, introduced the word *likelihood* in a paper written in October 1920 as his reply to the "Cooperative Study" and published the following year. The last section of this is entitled "Note on the confusion between BAYES' Rule and my method of the evaluation of the optimum". *Likelihood* is defined, and, "so defined, probability and likelihood are quantities of an entirely different nature". Fisher discusses the invariance of likelihood to parameter transformation, in contrast to probability, and clearly indicates that likelihood may be used not only as a quantity to be maximized, but also as a relative measure applied to different hypotheses or parameter values.

From § 6 of Fisher (1912) one might feel that Stratton suggested the emphasis in this section to Fisher. I can find no support for this idea in the earlier statistical writings of Stratton (Stratton and Compton, 1910; Wood and Stratton, 1910), and we may conclude that Stratton did not regard Fisher's method as being of particular importance because Brunt (1917) does not mention it, although he acknowledges his "indebtedness to Mr F. J. M. Stratton, of Gonville and Caius College, Cambridge, to whose University lectures I owe most of my knowledge of the subjects discussed in this book, and upon whose notes I have drawn freely". E. S. Pearson, who attended lectures by Stratton, informs me that neither did they provide any evidence of Stratton thinking Fisher's contribution important.

References

Bayes, T. (1763). An essay towards solving a problem in the doctrine of chances. *Phil. Trans. Roy. Soc.* **53**, 370–418. Reprinted in *Biometrika*, **45**, 293–315 (1958), and by Pearson and Kendall (1970).

Bernoulli, D. (1778). The most probable choice between several discrepant observations and the formation therefrom of the most likely induction (with a comment by Euler). *Acta Acad. Petrop.* 3–33. English translation in *Biometrika*, **48**, 3–13 (1961), reprinted by Pearson and Kendall (1970).

Bernoulli, J. (1713). *Ars Conjectandi*. Basel.

Brunt, D. (1917). *The Combination of Observations*. Cambridge University Press.
de Moivre, A. (1756). *Doctrine of Chances*, 3rd ed. London.
Euler, L. (1778). See Bernoulli, D. (1778).
Edgeworth, F. Y. (1908). On the probable errors of frequency constants. *J. Roy. statist. Soc.* **71**, 381–397, 499–512, 651–678.
Edwards, A. W. F. (1972). *Likelihood*. Cambridge University Press.
Fisher, R. A. (1912). On an absolute criterion for fitting frequency curves. *Mess. math.* **41**, 155–160. Reprinted in Fisher (1971).
Fisher, R. A. (1915). Frequency distribution of the values of the correlation coefficient in samples from an indefinitely large population. *Biometrika*, **10**, 507–521. Reprinted in Fisher (1971).
Fisher, R. A. (1921). On the "Probable Error" of a coefficient of correlation deduced from a small sample. *Metron*, **1**, part **4**, 3–32. Reprinted in Fisher (1971).
Fisher, R. A. (1922). On the mathematical foundations of theoretical statistics. *Phil. Trans. Roy. Soc.* A, 222, 309–368. Reprinted in Fisher (1950) and Fisher (1971).
Fisher, R. A. (1932). Inverse probability and the use of likelihood. *Proc. Camb. Phil. Soc.* **28**, 257–261.
Fisher, R. A. (1950). *Contributions to Mathematical Statistics*. New York: Wiley.
Fisher, R. A. (1971). *Collected Papers*, Vol. 1 (1912–24). University of Adelaide.
Gauss, C. F. (1809). *Theoria Motus Corporum Coelestium*. Hamburg. Werke, **7**, Göttingen (1906). English translation by C. H. Davis. Boston: Little, Brown (1857).
Hacking, I. (1971). Jacques Bernoulli's *Art of Conjecturing*. *Brit. J. Phil. Sci.* **22**, 209–229.
Haldane, J. B. S. (1932). A note on inverse probability. *Proc. Camb. Phil. Soc.* **28**, 55–61.
Haldane, J. B. S. (1957). Karl Pearson, 1857–1957. *Biometrika*, **44**, 303–313.
Kendall, M. G. (1961). Translation of Bernoulli (1778), q.v.
Keynes, J. M. (1921). *A Treatise on Probability*. London: Macmillan.
Lagrange, J. L. (1770–73). Mémoire sur l'utilité de la méthode de prendre le milieu entre les résultats de plusieurs observations. *Miscellanea Taurinensia*, **5**, 167–232. *Œuvres* **2**, 171–234. Paris: Gauthier-Villars (1868).
Lambert, J. H. (1760). *Photometria*. Augustae Vindelicorum.
Laplace, Marquis de (1820). *Théorie analytique des probabilités*, 3rd ed. Paris: Courcier. The introduction is: *Essai philosophique sur les probabilités*. Reprinted in English as *A Philosophical Essay on Probabilities*; New York: Dover Publications (1951).
Levy, H., Roth, L. (1936). *Elements of Probability*. Oxford: Clarendon Press.
Neyman, J., Pearson, E. S. (1928). On the use and interpretation of certain test criteria for purposes of statistical inference. *Biometrika*, **20**a, part I, 175–240; part II, 263–294. Reprinted in *Joint Statistical Papers*, Cambridge University Press (1967).
Pearson, E. S. (1967). Some reflexions on continuity in the development of mathematical statistics, 1885–1920. *Biometrika*, **54**, 341–355. Reprinted by Pearson and Kendall (1970).
Pearson, E. S. (1968). Some early correspondence between W. S. Gosset, R. A. Fisher and Karl Pearson, with notes and comments. *Biometrika*, **55**, 445–457. Reprinted by Pearson and Kendall (1970).
Pearson, E. S., Kendall, M. G. (1970). *Studies in the History of Statistics and Probability*. London: Griffin.
Pearson, K. (1896). Mathematical contributions to the theory of evolution – III. Regression, Heredity and Panmixia. *Phil. Trans. Roy. Soc.* A, **187**, 253–318. Reprinted in *Karl Pearson; Early Statistical Papers*, Cambridge University Press (1948).
Pearson, K., Filon, L. N. G. (1898). Mathematical contributions to the theory of evolution – IV. On the probable errors of frequency constants and on the influence of random selection on variation and correlation. *Phil. Trans. Roy. Soc.* A. **191**, 229–311. Reprinted in *Karl Pearson; Early Statistical Papers*, Cambridge University Press (1948).
Plackett, R. L. (1958). The principle of the arithmetic mean. *Biometrika*, **45**, 130–135. Reprinted by Pearson and Kendall (1970).
Sheynin, O. B. (1971). J. H. Lambert's work on probability. *Arch. Hist. Exact. Sci.* **7**, 244–256.
Soper, H. E., Young, A. W., Cave, B. M., Lee, A., Pearson, K. (1916). On the distribution of the correlation coefficient in small samples. Appendix II to the papers of "Student" and R. A. Fisher. A cooperative study. *Biometrika*, **11**, 328–413.
Stratton, F. J. M., Compton, R. H. (1910). On accident in heredity with special reference to right- and left-handedness. *Proc. Camb. Phil. Soc.* **15**, 507–512.
Todhunter, I. (1865). *A History of the Mathematical Theory of Probability*. Cambridge and London: Macmillan.
Truscott, F. W., Emery, F. L. (1951). Translation of Laplace's *Essai Philosophique*: see Laplace (1820).
Venn, J. (1876). *The Logic of Chance*, 2nd ed. London: Macmillan.
Wood, J. B., Stratton, F. J. M. (1910). The interpretation of experimental results. *J. Agric. Sci.* **3**, 417–440.
Wright, S. (1968). *Evolution and the Genetics of Populations*. Vol. 1. *Genetic and Biometric Foundations*. Chicago University Press.

Résumé

Le présent article trace l'histoire de l'idée générale de la vraisemblance, de ses origines dans les œuvres de de Moivre (1717), de Lambert (1760) et de D. Bernoulli (1778), jusquà sa formulation explicite par Fisher (1912). On considère les parties des œuvres de Gauss et de Laplace qui s'y rapportent, et en s'occupe surtout de l'évolution des idées de Fisher entre 1912 et 1922.

Paper 96 (1986)

Are Mendel's results really too close?
Biological Reviews 61, 295–312.

Note from A. W. F. Edwards: As a student I was made aware of Fisher's 1936 paper on this subject, though I never discussed it with him. But he did repeat some of Mendel's experiments each year in the departmental garden for demonstration purposes, and I discovered for myself the difficulty of deciding whether peas of intermediate form should be classed as 'round' or 'wrinkled'. By the time I wrote *Likelihood* I was harbouring doubts about tests of significance in general, and on page 190 remarked 'It would be interesting to rework Fisher's analysis using the justification of χ^2 offered by the Method of Support [i.e. likelihood]'. This paper is the result. It contains some thoughtful remarks about χ^2 goodness-of-fit tests.

ARE MENDEL'S RESULTS REALLY TOO CLOSE?

By A. W. F. EDWARDS
Gonville and Caius College, Cambridge

(*Received* 26 *March* 1985, *accepted* 29 *May* 1986)

CONTENTS

I. INTRODUCTION

Gregor Mendel read his famous paper 'Experiments in plant hybridization' to the Brünn Natural History Society in 1865 (Mendel, 1866). It attracted wide attention only from 1900 onwards, and in 1911 R. A. Fisher, then starting his third year as an undergraduate in Gonville and Caius College, Cambridge, noted in a talk to the Cambridge University Eugenics Society:

> It is interesting that Mendel's original results all fall within the limits of probable error; if his experiments were repeated the odds against getting such good results is about 16 to one. It may have been just luck; or it may be that the worthy German abbot, in his ignorance of probable error, unconsciously placed doubtful plants on the side which favoured his hypothesis (Norton & Pearson, 1976; Bennett, 1983).

Twenty-five years later Fisher published his analysis of Mendel's data (Fisher, 1936), admitting, in a covering letter to the editor of the journal, 'I had not expected to find the strong evidence which has appeared that the data had been cooked' (Bennett, 1965). Nearly 20 years later still he prepared an introduction and commentary to Mendel's paper which was published posthumously in 1965 (Bennett, 1965), and his final verdict was, 'The data have evidently been sophisticated systematically, and after examining various possibilities, I have no doubt that Mendel was deceived by a gardening assistant, who knew too well what his principal expected from each trial made.' A letter from Fisher to E. B. Ford dated 2 January 1936 (Bennett, 1983) amplifies what Fisher called his 'abominable discovery': 'I cannot conceive that Mendel himself had any hand in it'. From his comments in 1936 it seems that Fisher had by then forgotten what he had written in 1911.

Fisher's remark about the odds of 16 to 1 was probably taken directly from Weldon (1902), but that is not to suggest that Weldon himself harboured any suspicions about Mendel's results. He subjected them to a statistical analysis using probable errors but concluded only that Mendel's statements were 'admirably in accord with his experiment'. Fisher had encountered Mendel's paper itself in the English translation in Bateson's newly published *Mendel's Principles of Heredity* (Bateson, 1909) which he had bought as a freshman (Fisher, 1952). In 1936 he referred to a footnote of Bateson's as raising 'a point of great interest' because it hinted that Mendel's paper should not be taken as a literal account of his experiments, and this may be said to have been the first such suggestion.

Fisher's 1936 conclusion slowly became the received wisdom, but his painstaking analysis and his defence of Mendel's integrity have sometimes been incorrectly reported as having exposed a scientific fraud of major proportions, and the name of Mendel is in danger of acquiring the connotations of Piltdown or Burt. Recently there has been a spate of papers seeking to defend Mendel against the imaginary charges. Thus Pilgrim (1984) writes 'The purpose of this [i.e. his own] paper is to demonstrate that Fisher's reasoning was faulty and to clear the name of an honest man', and 'There is no evidence that Mendel did anything but report his data with impeccable fidelity. It is to the discredit of science that it did not recognize him during his lifetime. It is a disgrace to slander him now'. An answer to Pilgrim is Edwards (1986). A similar example of missionary zeal is afforded by Monaghan & Corcos (1985), who write that 'suggestions that Mendel's data were biased in support of an hypothesis about inheritance in peas are not supported'. Their paper is, however, completely untrustworthy in its statistical analysis (see Section III below). A recent biography of Mendel (Orel, 1984) concludes, on the basis of Weiling's analysis to which I refer below, 'This would seem to have removed the last shadow of doubt as to the credibility of Mendel's experiments and his published results'.

As I will show, such views are not supported by the evidence. In the present paper, after reviewing the many 'explanations' of Mendel's data, I shall let the facts speak for themselves as plainly as possible: there is no point in addressing the issue with statistical techniques of such complexity that they are more difficult to interpret than the original data. I shall assume that readers are familiar with Mendel's paper and that they have an English translation to hand, preferably that printed in Bennett (1965), to which my page references will apply (thus: M, p. 20). I cannot myself comment on the quality of this translation from the German, but it certainly reproduces Mendel's numerical material without error. Others have felt the need for an improved translation (see Stern in the foreword to Stern & Sherwood, 1966) and one has been made (Sherwood, *loc. cit.*).

II. A SURVEY OF 'EXPLANATIONS'

The critical re-evaluation of experiments and observations is an important part of scientific endeavour. ('There is no substitute for a careful, or even meticulous, examination of all original papers purporting to establish new facts', Fisher, 1936.) Though in certain notorious cases this has led to the exposure of fraudulent work, it generally results in an increased appreciation of the original material or even a fresh discovery from it. Occasionally some puzzling aspects of the data surface on a close examination, and may even lead to the suggestion that the experiments did not take place exactly as the author reported them. Thus Galileo's unrepeatable results in his experiments on acceleration down an inclined plane mentioned in the *Dialogo* led

Mersenne (1637) to doubt whether he had actually performed them. In that example it was the incomprehensibly bad data which raised doubts, but long before the era of modern statistics it was understood that suprisingly good data could equally well create justifiable suspicion. Babbage (1830), writing before the time of Mendel, classified some of the methods used to produce data too close to expectation:

> *Trimming* consists in clipping off little bits here and there from those observations which differ most in excess from the mean, and in sticking them on to those which are too small.
>
> *Cooking*... One of its numerous processes is to make multitudes of observations, and out of these to select those only which agree, or very nearly agree.

Many commentators on Mendel's results share Fisher's belief that some kind of 'cooking' has taken place (for example: de Beer, 1964; Sturtevant, 1965; Crew, 1966), but most have sought alternative explanations for the surprisingly good fit of the data to the hypothesis.

A popular one is that the data are the result of an imperfectly understood sampling process. An early example of such a misunderstanding occurs in the work of Francis Galton, who, by the standards of his time, had a remarkably clear head for statistical enquiry. In *Hereditary Genius* (Galton, 1892), however, he concluded that since judges were all men and came from families of average size five, each must have (on average) $2\frac{1}{2}$ sisters and $1\frac{1}{2}$ brothers. Galton is here wrongly assuming binomial sampling, with probability one half, *including* the judge; might not we be assuming the wrong sampling process for Mendel's data? (Galton, 1904, later realized his error.)

The natural suggestion is that Mendel stopped counting the peas in each experiment when he was satisfied with the result, and this has been put forward numerous times (Dunn, 1965*a*, *b*; Olby, 1966; Beadle, 1967; van der Waerden, 1968; C. D. Darlington, reported to me by J. H. Edwards; J. Kříženecký, reported by Orel, 1968). The difficulty is that there is no evidence whatsoever that Mendel did this, and in many of the experiments his description clearly excludes the possibility, so that to accept it as a means of exonerating Mendel from having reported data which had been adjusted in some way is to saddle him with the charge of not having reported his experimental method accurately. 'The style throughout suggests that he expects to be taken entirely literally; if his facts have suffered much manipulation the style of his report must be judged disingenuous' (Fisher, 1936). Van der Waerden (1968) is the only one of the authors quoted above who pursued the suggestion in detail, and although he found it possible to account for some of the earlier instances of surprisingly good fits in the experiments if he assumed sequential sampling, he concluded 'All in all my impression is that some data from later years were probably biased.'

Another favourite suggestion has been that the data exhibit non-binomial variability not because of the sampling process, but because of the reproductive physiology of the pea (Sturtevant, 1965; Weiling, 1966; Thoday, 1966; Beadle, 1967). Sturtevant and Beadle 'explored this possibility to see if it was sufficient to account for the apparent bias. It works in the right direction but is not sufficient' (Beadle, 1967). Weiling (1966, 1971, 1985) made a detailed comparison of Mendel's data with data from other authors and concluded that the results were all consistent with the hypothesis that the variance of the segregation was smaller than the binomial variance by a factor of between 0·6 and 1·0. In the next section I give my reasons for rejecting this suggestion: the evidence is that the pea is in fact an excellent randomizer!

To help explain the infra-binomial variance in the experiments in which Mendel progeny-tested F_2 plants which displayed one of the dominant plant-characteristics to see if they were homozygous or heterozygous (M, pp. 20, 26), Weiling (1966) suggested that on average only 8 of the 10 seeds sown to test each parent for heterozygosity will have produced a scorable plant, thus making the probability $(3/4)^8$ that no recessive was raised, rather than the $(3/4)^{10}$ assumed by Fisher (1936). Taking into account the resulting misclassification of some heterozygotes as homozygotes, the expectations are then even further from Mendel's data, and the fit consequently *worsened*. By contrast, Sturtevant (1965) thought that 'if the experiment included at least 10 seeds but often more than 10, then the correction to the 2:1 expectation will be less, and Fisher's most telling point will be weakened', and Wright (1966; also reported by Dunn, 1965*b*) advanced the same idea: 'if the average was 12, the probability of misclassifying [the heterozygote] falls from ·056 [$(3/4)^{10}$] to ·031 [$(3/4)^{12}$]'. This would *improve* the fit of Mendel's results. As a matter of fact this suggestion had already been made by Fisher (1936) in connection with the 473 plants with coloured flowers from the trifactorial experiment which needed progeny-testing to determine whether they were homozygous or heterozygous, for which Mendel did not restate his procedure. 'It was presumably the same as before', wrote Fisher, adding:

> if we could suppose that larger progenies, say fifteen plants, were grown on this occasion, the greater part of the discrepancy would be removed. However, even using families of 10 plants the number required is more than Mendel had assigned to any previous experiment... Such explanations, moreover, could not explain the discrepancy observed in the first group of experiments, in which the procedure is specified,....

Nothing could illustrate better than these two opposing theories the ingenuity that has been expended on accounting for Mendel's surprisingly good data. Here are two explanations, one postulating that *fewer* than ten plants were scored, and the other that *more* than ten were scored, both with a view to accounting for results judged too close to a 2:1 ratio. In the next section I shall comment further on Fisher's handling of this question, and show that Weiling's suggestion is of little importance statistically, but once again we should note that there is no evidence that Mendel did either of the things suggested. He stated explicitly (M, p. 20) that 'ten seeds of each were cultivated', implying that there were exactly ten which grew. In Sherwood's translation (Stern & Sherwood, 1966) '*angebaut*' – 'cultivated' – is rendered as 'sown', but I am advised (K. Martin, personal communication) that this is an unwarranted change.

One puzzling aspect of the problem of Mendel having apparently overlooked the fact that progeny-testing with only ten plants will have led to some heterozygotes being misclassified as homozygotes is that towards the end of his paper (M, p. 49) he was completely clear that a form with low probability (one-eighth in the case he was treating) might by chance not appear in a particular experiment.

Wright (1966) also suggested that Mendel might often have been able to distinguish a 'segregating group from a nonsegregating one even in the absence of any recessives'.

In other words, for some characters Mendel might have been able to spot the heterozygote, and Wright mentioned seed-coat colour especially. This was indeed often true for seed-coat colour (M, p. 16), though Mendel never stated he used the fact. Indeed, he explicitly employed progeny-testing (M, p. 20) or 'further investigation' (M, p. 26), which all commentators have assumed to mean progeny-testing, in the case of seed-coat colour as with the other plant characters. Mendel (M, p. 15) also mentioned that F_1 hybrids from short × tall parents were usually taller than the tall parent, but he thought this possibly due to the 'greater luxuriance which appears in all parts of plants when stems of very different length are crossed' and did not mention its occurrence in the F_2. It is one thing to notice that tall offspring are generally taller than their tall parent, but another thing altogether to notice that a tall population of plants actually consists of two sub-populations. Even if there were some truth in the suggestion that in respect of seed-coat colour Mendel could distinguish between a mixed group of ten plants and a homozygous group, the explanation would only affect one character and the statistical consequences would be small. But the most telling objection to this hypothesis is, again, the fact that Mendel explicitly described a quite different procedure (M, p. 20).

In addition to the papers cited above, there have been many relevant reviews, though mostly of a somewhat inconclusive nature, and all lacking any extensive statistical analysis of the scope of Fisher (1936) or even Weiling (1966). Mention may, however, be made of the wide-ranging commentaries by Piegorsch (1983), who studied Weiling's analysis with care, and Root-Bernstein (1983), who concluded his review of earlier work by saying 'The issue remains unresolved,' and then, accepting the existence of bias, suggested that some plants were difficult to classify and on that account were tallied last in a direction to fit the ratios (cf. Fisher, 1911, quoted above in Section I). In respect of the others, I believe that I have read all the published material on this question, and that no useful purpose would be served by referencing papers which, though they may have fulfilled a valuable educational role at the time, are of no lasting value. Similarly, I make no reference to the many excellent commentaries on Mendel's paper that do not touch on the goodness-of-fit issue.

III. PREVIOUS STATISTICAL ANALYSES

Fisher (1936) gave the results of the most extensive statistical analysis of the data so far. On recomputing his χ^2 values (1936, table V) which he calculated to four places of decimals, I find them all correct, though Piegorsch (1983) noted that the famous 'P-value' for $\chi^2 = 41{\cdot}6056$ on 84 d.f. should be ·99997 and not ·99993!

There are, however, two minor criticisms of Fisher's analysis. First, the 20 d.f. *Illustrations of plant variation* are not independent of the 2 d.f. *3:1 ratios – seed characters* since the data for the former are included in the latter. Subtracting them out increases χ^2_2 from 0·2779 to 0·6232. Secondly, in those experiments in which, as Fisher was the first to point out, 2:1 ratios are not to be expected (see Section II above), he did in fact use them in his calculations of χ^2, presumably because they were what he thought Mendel expected. Root-Bernstein (1983) regarded this as a 'ploy' of Fisher's (a choice of word more revealing of Root-Bernstein's thought than Fisher's), but there are arguments both ways.

Fisher was testing the divergence between Mendel's counts and their expectations, and the problem is *which* expectations, Mendel's or nature's? The difficulty is that any goodness-of-fit test confounds two sources of departure from expectation, the systematic (due to using the incorrect expectation) and the random (due to binomial variability). Usually one is assuming binomial variability and testing an expectation, in which case the null hypothesis under test clearly involves that expectation, but if one is assuming an expectation and testing the variability against the binomial model, as when the implied alternative is that the data have been adjusted towards the expectation, it is arguable that one should then assume *that* expectation. However, we need not explore this statistical by-way further (for which see Edwards, 1972, chapter 9) since even assuming the alternative expectations on Mendel's data, Fisher's overall conclusion is barely affected: Weiling (1966) computed that the χ^2_{84} is increased from 41·6056 to 48·910 and remarked correctly, 'This value is also highly significant.' Finally, on Fisher's statistical analysis, we should note that van der Waerden's (1968) suggestion of a numerical error seems to be a mistake.

After Fisher, Weiling (1966) has given the next most extensive statistical treatment. He squarely addressed the question of whether the variance might not be binomial by postulating that it is infra-binomial by a factor *c*, for the reasons given in Section II above, and estimating *c*. Exactly the same thought occurred to me at a later date (Edwards, 1972, chapter 9) in a general discussion of χ^2. My k^2 is his *c*, my figure 25 highly reminiscent of Weiling's figures, and my final comment indicative of an intention to do what, unbeknown to me, Weiling had already done:

> Since the alternative hypothesis [in considering the goodness-of-fit of Mendel's data] is that the variance of the results is less than expected, rather than that the means are in doubt, the use of χ^2 seems entirely appropriate. It would be interesting to rework Fisher's analysis using the justification of χ^2 offered by the Method of Support.

Of course, once one has estimated *c* (for which Weiling found the broad limits 0·6–1·0) there is nothing left to test, and Weiling persuaded himself that Mendel's results were compatible with those of other workers who repeated his experiments around the turn of the century. However, the facts indicate otherwise. The overwhelming majority of the post-Mendelian data on segregation in the pea comes from two sources, Bateson & Killby (1905) and Darbishire (1908, 1909), for which Weiling calculated $\chi^2_{408} = 411·101$ and $\chi^2_{654} = 597·689$ respectively, or $\chi^2_{1062} = 1008·79$ in aggregate. The estimate of *c* is therefore 1008·79/1062 = 0·9500. In other words these massive data exhibit a standard deviation of about 97·5 % of that expected on a binomial model, and far from this lending support to the biological hypothesis that the pea does not segregated randomly, it fills one with admiration for the perfection of the randomizing mechanism! The compatibility which impressed Weiling is but a reflection of the wide interval estimate he obtained for *c* on Mendel's data.

Moreover, the conclusion that the pea is a good randomiser is actually supported by an alternative analysis Weiling gave which he misinterpreted. If the data really do depart from the binomial model in the way he suggested, then not only should the observed χ^2 values be less than their expectations (= their degrees of freedom), but more than half of the component χ^2_1's should be less than their theoretical median value of 0·4549. He found (his table 4) that for Bateson & Killby's data 116 were less and 120 more, and for Darbishire's 347 were less and 307 more (I assume that only 236 d.f. in Bateson & Killby's data refer to single-factor segregations). These figures, 463 *vs.* 427, lend no real support to the infra-binomial-variance hypothesis, but Weiling did not appreciate this because he wrote:

> In spite of the consequent results of our investigations into all the analyzed pea crosses (yielding an appreciable deviation of the value *c* from 1), the χ^2-test may, without hesitation, at least be applied to the evaluation of the individual segregations. Then the test of frequency distribution of all other inquiries into the underlying individual probabilities shows no noticeable deviation in the direction of P = 1·0, as [his] Table 4 illustrates.

Weiling here conceded 'no noticeable deviation' but thought this consistent with an appreciable overall departure of *c* from 1. He thus assumed that a collective of individual χ^2_1's can exhibit no systematic departure from their theoretical distribution even when their sum does, but this is a misunderstanding of the nature of χ^2. I consider the logic of χ^2 further in the next section.

Weiling also analysed the much less extensive data of Correns and Tschermak, and the fact that I have concentrated my remarks on the data of Bateson & Killby and Darbishire is explained by the circumstance that I had already chosen these data-sets for a complete re-analysis because of their extent and their ready availability to me. When I started the re-analysis I was not aware of the existence of the other smaller data-sets in the German literature, and since I do not read German I must leave their reconsideration to others. I only labour this point in order to deflect any suggestion that I have deliberately ignored data which supports this or that hypothesis. I hope to publish a full analysis of the very large English data sets in due course.

Weiling is to be congratulated for his determined attempt to rescue the Mendelian experiments from the Fisherian conclusion. He left no stone unturned in his search for explanations, increasing the value of χ^2 by using the true instead of the 2:1 ratio and by postulating that not all the 10 seeds sown in some of the experiments yielded mature plants, and then accounting for the too-small χ^2 still remaining by the hypothesis of a diminished variance. But in the end the attempt fails: there is too much to explain. The diminished-variance hypothesis is not supported by other more extensive data, and, as I show in my own analysis below, simply reducing the variance is in any case not enough to explain the peculiarities of Mendel's data.

No such congratulations can be offered to Monaghan & Corcos (1985). All their four tables contain substantial errors. At the trivial level, many of the χ^2 values they give are simply wrong to the accuracy claimed. Even the simple addition of these values is wrong in table I, and this departure alone from Fisher's value should have altered the authors to the fact that either they or Fisher were in error. Table III contains the astonishing assertion that $0{\cdot}95 < P < 0{\cdot}99$ for $\chi^2_{67} = 32{\cdot}57$, on which obvious error the authors hang their conclusions. Table IV *Results of other workers compared to those of Mendel* leads to the observation 'that the results of Bateson for cotyledon shape [*sic*] and Tschermak for cotyledon color deviate very widely from the other workers', which is not surprising given that Monaghan & Corcos have copied them both down wrongly from the literature. The list of references is a bibliographic disaster area. I spare the reader further criticism.

My overall impression from reading all the commentaries since Fisher (1936) is that a good deal of special pleading, not to mention downright advocacy, has failed to make any substantial impact on Fisher's conclusion. Moreover, as Fisher was the first to point out, much of the special pleading requires that Mendel's account itself would have to be judged 'disingenuous'. Certainly, there is every reason for each student of Mendel to keep an open mind and to make his own analysis and form his own conclusion. I give my own attempt in Section V, preceded in Section IV by a discussion of the precise nature of goodness-of-fit χ^2, the statistical procedure on which the whole issue hinges and which has often been misunderstood.

IV. GOODNESS-OF-FIT χ^2

Karl Pearson published his famous χ^2 goodness-of-fit test in the year that Mendel's paper was rediscovered (Pearson, 1900), and it is of interest that he warned against its use with large numbers of degrees of freedom:

> Thus, if we take a very great number of groups our test becomes illusory. We must confine our attention in calculating P to a finite [*sic*] number of groups, and this is undoubtedly what happens in actual statistics. *n* will rarely exceed 30, often not be greater than 12.

The basis of Pearson's concern seems to have been that in a large number of dimensions practically the whole of the probability in a multivariate normal distribution is to be found in the 'tails', or regions remote from the central region of high probability *density*. He seems to have sensed the dilemma facing anyone who ponders the question of what a 'typical' sample-point might be from a standardized multivariate normal distribution in, say, 84 dimensions (the number of degrees of freedom in Mendel's data). If, for example, the choice falls somewhere in the region of high probability density within one unit (= standard deviation in one dimension) of the centre, or mean, then the corresponding χ^2_{84} is less than one and, by this criterion, the point is surprisingly close to the centre. Yet if the choice falls about nine units from the centre, to make χ^2_{84} about 81 and thus near its expectation, it is in a region where the probability density has fallen to about $2{\cdot}6 \times 10^{-18}$ of its maximum value.

It is interesting to find this implicit concern for the logic of 'tail-area' significance tests so early in their history. More recent explicit criticisms are well-known, often in the writings of Bayesians (for example, Jeffreys, 1939) but not always (for example, Edwards, 1972). What is the 'rejection region' to be? Taking the 5% significance level as an example, is it to be the region delimited by the line (or surface, etc.) of constant probability density which divides the sample space into probabilities of 0·95 and 0·05 such that the probability density everywhere in the latter part is less than the probability density everywhere in the former part? Or is it to be the line which delimits that part of the sample space beyond a certain distance from the mean, the distance chosen so that just 5% of the probability is in it? In either case, what transformations of the sample space are allowed (since they manifestly affect both criteria)?

The χ^2 distribution is a transformation of the multivariate normal sample space, many dimensions being mapped into one by combining all points of equal probability density. On either of the above criteria the rejection region for the standardized multivariate normal will be the outer skirts delimited by a hypersphere, but in the case of χ^2, which is not a symmetrical univariate distribution, it will be the two tails, consisting either of equal probability content 0·025 or cut off at equal probability *densities* and totalling 0·05 in probability. The corresponding region in the multivariate normal is not that described above, but rather two disjoint regions delimited by hyperspheres, one surrounding the centre (the image of the left-hand tail of χ^2) and one far removed from it (the image of the right-hand tail). A χ^2 value much below expectation (such as 41·6056 on 84 d.f.) is represented in the multivariate normal space by a point within the hyperspherical region surrounding the centre, the region of highest probability density, yet in the χ^2 space it is in a region of low probability density. The criticism of Mendel's results is that they are too close to expectation in the normal space as judged by being too far from expectation in the χ^2 space. *Quis custodiet ipsos custodes?*

We may emphasize this *Paradox of the Probable* in another way. Suppose the χ^2 test had antedated Mendel, and that in his paper he had reported a value of 84·0000 on 84 d.f. The reaction of a latter-day Fisher might well have been to conclude that Mendel's assistant had known that what Mendel really needed for his paper was not good mendelian ratios but a good value of χ^2.

I have discussed these questions at length in my book *Likelihood*: '*The quantification of surprise in terms of probability is likely to tell only half the story*' (Edwards, 1972). The other half depends on the fact that the more improbable a result on a particular hypothesis as judged on *some* criterion, the greater the possibility of an alternative hypothesis leading to the result with a substantially higher probability. In other words there is a greater possibility of a hypothesis of higher likelihood, and hence one which, other things being equal, is more satisfactory. In particular, any *pattern* which we recognize in some data and which is unexplained on the current hypothesis is a signal that we should seek an alternative hypothesis, because an alternative which accounts for it is almost bound to have a higher likelihood. This may be the case even where the actual data are more probable than any other that might have occurred given the hypothesis, for we can then often more readily think of an alternative explanation than would be the case for some more arbitrary outcome. Thus numerical equality of the sexes at a children's party (see *Likelihood*, p. 190) is readily explained although, being the most probable outcome of a random selection of children, it might be thought to be the least in need of explanation. It is surely for this reason that attempts to base a theory of statistical estimation on maximising probabilities (Barndorff-Nielsen, 1976) have failed.

To sum up, we must not allow our judgment to be dominated by tests of significance and other calculations of probability which are at best pointers for further thought and at worst misleading. This is especially important when, as in the case of Mendel's experiments, the suggestion is that the data are 'too good' in the sense of being *too probable*, too close to the target. If it were just a question of having hit the bull's eye with a single shot we might conclude, as some commentators have (e.g. Pilgrim, 1984), that Mendel was simply lucky, but when a whole succession of shots comes close to the bull's eye we are entitled to invoke skill or some other factor. A χ^2 of 41·6056 on 84 d.f. may look like a single lucky shot, but in reality it is a succession of 84 shots which must be manifesting some kind of pattern calling out for investigation. This will be the basis of my own analysis of Mendel's data in the next section.

Finally, there is one technical point about χ^2 which should be noted. It has been implied (Weiling, 1966) that one of the reasons for doubting the χ^2 analysis is that it is based on the assumption of normally distributed data, whereas genetic segregations are (at least on the simplest hypothesis) binomial. But it is a fact that the *expectation* of χ^2 on a binomial model is *exactly* equal to its degrees of freedom (Pearson, 1900). This is true *by definition*, for Pearson obtained his goodness-of-fit criterion by assuming that the normal distribution in question had a mean and variance given by the binomial distribution. In modern notation:

Let x be normally distributed with mean μ and variance σ^2. Then by definition $\chi_1^2 = (x-\mu)^2/\sigma^2$. Now put $\mu = np$ and $\sigma^2 = npq$, the mean and variance of a binomial distribution, and let the observed number of successes x be a and of failures be $b = n\text{-}a$. Then, substituting for μ and σ^2,

$$\begin{aligned}\chi^2_1 &= (a\text{-}np)^2/npq,\\ &= (a\text{-}np)^2/np+(a\text{-}np)^2/nq,\\ &= (a\text{-}np)^2/np+(b\text{-}nq)^2/nq,\\ &= \Sigma[\,(\text{obs}-\text{exp})^2/\text{exp}]\text{ exactly.}\end{aligned}$$

It is obvious that the expectation is exactly 1 because $E(a\text{-}np)^2 = npq$ by the definition of variance, and if all the χ_1^2's are independent then the expectation of their sum will exactly equal the degrees of freedom.

It follows that although the fine detail of the χ^2 distribution may be wrong its mean and variance are right, and there can be no doubt over its applicability save when sample sizes are very small, much smaller than Mendel's, and even then the expectation is still correct.

V. A FRESH ANALYSIS

The 84 d.f. in Mendel's data may be individually identified in a number of different ways depending on the order in which the simultaneous segregations are considered. I shall arbitrarily order the loci according to Mendel's own classification. Thus in analysing the results of selfing AaBb where A, B is the order of the loci in Mendel's list, I shall take the segregation at the A locus first and then the two segregations at the B locus for each of the two A-locus phenotypes, making three independent comparisons in all. With this arrangement we will then have 84 independent variates to analyse.

For our purposes it will be better to work with χ rather than χ^2, partly because it preserves the information about the direction in which the segregation departs from expectation, but also because χ is expected to have a normal distribution to an excellent approximation, and methods for analysing a sequence of normal variables are commonplace. By the argument of the previous section the expectation of each χ on the binomial hypothesis is exactly 0 and its variance exactly 1, so that its distribution is $N(0, 1)$ for all practical purposes. It may also be noted that since in the binomial case χ is given exactly by dividing the observed deviation from the hypothetical binomial mean by the hypothetical standard deviation, whether we derive it as the square-root of Pearson's χ^2 criterion or directly as a standardized deviation is immaterial.

There is, of course, also an arbitrariness as to which departures from expectation we regard as positive and which as negative, but we will consistently regard a deviation in favour of the first of the two segregating classes as positive, and the first class will always

Table 1. *Notation for Mendel's seven characters (M, p 17).*

	Seed characters
A, a	Form of seed
B, b	Colour of albumen
	Plant characters
C, c	Colour of the seed-coats
D, d	Form of pods
E, e	Colour of the unripe pods
F, f	Position of flowers
G, g	Length of stem

be taken to be the one with the larger expectation. In the case of the experiments on gametic ratios, where 1:1 is expected, the first class will always be the one with the larger number of dominant genes, so the order will be either AA:Aa or Aa:aa as the case may be.

The notation for the characters is given in Table 1, and agrees with Mendel's for the major part of his paper. Table 2 gives the 84 values of χ and their origins. The page references are to Mendel's paper in Bennett (1965), and 'A' stands for AA+Aa and 'a' for aa in accordance with the usual convention. As mentioned at the beginning of our Section III, the data for the illustration of plant variation (M, p. 17) should be subtracted out from the data in the first two experiments (M, pp. 16–17), and this has been done. Similarly, modified expectations based on the ratio 0·629124:0·370876 have been used instead of 2:1 where appropriate. All calculations have been rounded to four significant figures after completion.

It will be seen that of the fifteen χ values for the segregation ratio 0·6291:0·3709 (rows 30–35 and 61–69), 13 are positive and only 2 negative, suggesting something of a bias towards the larger class. The total segregation is 720:353, or 0·6710:0·3290, with an associated χ value of +2·8408, indicating a very poor fit indeed. Mendel (M, pp. 20, 21) explicitly accepted this type of experiment as establishing the 2:1 ratio, and as an aid to judging the extent to which the data conform to this rather than the true ratio we can compute the likelihood ratio for the 2:1 hypothesis *vs.* the 0·6291:0·3709 hypothesis on the assumption of binomial variation (for likelihood techniques, see Edwards, 1972). The result is 57·90, or a support value (taking logs) of just over 4, which is large enough to substantiate what Sturtevant (1965) called 'Fisher's most telling point' in his argument that the data have been biased in the direction of agreement with what Mendel expected.

Having concluded from an examination of the *means* for these fifteen segregations that possibly all is not well, we now turn to an examination of the *variances* of the other 69 segregations. For we know from the analyses considered in Section III that it is the infra-binomial variance that raises doubts about the data. As far as the means are concerned, of their 69 χ values, 38 are positive, 30 negative, and 1 zero. Their mean is 4·0880 which, dividing by $\sqrt{69}$ to make it a standard normal deviate, is equal to 0·4291, as unexceptionable a value as one could wish for. By contrast, the variance of the χ values is certainly exceptional. χ^2 is now the appropriate statistic, and the sum of the 69 χ^2 values from Table 2 is 30·8138, highly remarkable on any interpretation of tests of significance.

Table 2. *Mendel's segregations and their χ^2 values.*

	Character	Segregation: Expected	Segregation: Observed		Total	χ
			F_2, Seed characters, pp. 16–17			
1	A	3:1	5138	1749	6887	−0·7583
2	B	3:1	5667	1878	7545	+0·2193
			F_2, Illustrations of plant variability, p. 17			
3	A	3:1	45	12	57	+0·6882
4	A	3:1	27	8	35	+0·2928
5	A	3:1	24	7	31	+0·3111
6	A	3:1	19	10	29	−1·1793
7	A	3:1	32	11	43	−0·0880
8	A	3:1	26	6	32	+0·8165
9	A	3:1	88	24	112	+0·8729
10	A	3:1	22	10	32	−0·8165
11	A	3:1	28	6	34	+0·9901
12	A	3:1	25	7	32	+0·4082
13	B	3:1	25	11	36	−0·7698
14	B	3:1	32	7	39	+1·0170
15	B	3:1	14	5	19	−0·1325
16	B	3:1	70	27	97	−0·6448
17	B	3:1	24	13	37	−1·4237
18	B	3:1	20	6	26	+0·2265
19	B	3:1	32	13	45	−0·6025
20	B	3:1	44	9	53	+1·3482
21	B	3:1	50	14	64	+0·5774
22	B	3:1	44	18	62	−0·7332
			F_2, Plant characters, p. 18			
23	C	3:1	705	224	929	+0·6251
24	D	3:1	882	299	1181	−0·2520
25	E	3:1	428	152	580	−0·6712
26	F	3:1	651	207	858	+0·5913
27	G	3:1	787	277	1064	−0·7788
			F_3, Seed characters, pp. 19, 20			
28	A	2:1	372	193	565	−0·4165
29	B	2:1	353	166	519	+0·6518
			F_3, Plant characters, p. 20			
30	C	0·63:0·37	64	36	100	+0·2252
31	D	0·63:0·37	71	29	100	+1·6743
32	E	0·63:0·37	60	40	100	−0·6029
33	F	0·63:0·37	67	33	100	+0·8462
34	G	0·63:0·37	72	28	100	+1·8813
35	E	0·63:0·37	65	35	100	+0·4322
			F_2, Bifactorial experiment, p. 23			
36	A	3:1	423	133	556	+0·5876
37	B, among 'A'	3:1	315	108	423	−0·2526
38	B, among 'a'	3:1	101	32	133	+0·2503

Table 2. (*cont.*)

	Character	Segregation: Expected	Segregation: Observed		Total	χ
			F_3, Bifactorial experiment, p. 24			
39	A, among 'AB'	2:1	198	103	301	−0·3261
40	A, among 'Ab'	2:1	67	35	102	−0·2100
41	B, among 'aB'	2:1	68	28	96	+0·8660
42	B, among Aa 'B'	2:1	138	60	198	+0·9045
43	B, among AA 'B'	2:1	65	38	103	−0·7664
			F_2, Trifactorial experiment, p. 26, seed characters			
44	A	3:1	480	159	639	+0·0685
45	B, among 'A'	3:1	367	113	480	+0·7379
46	B, among 'a'	3:1	122	37	159	+0·5037
			F_3, Trifactorial experiment, p. 26, seed characters			
47	A, among 'AB'	2:1	245	122	367	+0·0369
48	A, among 'Ab'	2:1	76	37	113	+0·1330
49	B, among 'aB'	2:1	79	43	122	−0·4481
50	B, among Aa 'B'	2:1	175	70	245	+1·5811
51	B, among AA 'B'	2:1	78	44	122	−0·6402
			F_2, Trifactorial experiment, p. 26, plant character			
52	C, among AaBb	3:1	127	48	175	−0·7419
53	C, among AaBB	3:1	52	18	70	−0·1380
54	C, among AABb	3:1	60	18	78	+0·3922
55	C, among AABB	3:1	30	14	44	−1·0445
56	C, among Aabb	3:1	60	16	76	+0·7947
57	C, among AAbb	3:1	26	11	37	−0·6644
58	C, among aaBb	3:1	55	24	79	−1·1043
59	C, among aaBB	3:1	33	10	43	+0·2641
60	C, among aabb	3:1	30	7	37	+0·8542
			F_3, Trifactorial experiment, p. 26, plant character			
61	C, among AaBb	0·63:0·37	78	49	127	−0·3488
62	C, among AaBB	0·63:0·37	38	14	52	+1·5174
63	C, among AABb	0·63:0·37	45	15	60	+1·9384
64	C, among AABB	0·63:0·37	22	8	30	+1·1816
65	C, among Aabb	0·63:0·37	40	20	60	+0·6020
66	C, among AAbb	0·63:0·37	17	9	26	+0·2610
67	C, among aaBb	0·63:0·37	36	19	55	+0·3903
68	C, among aaBB	0·63:0·37	25	8	33	+1·5276
69	C, among aabb	0·63:0·37	20	10	30	+0·4257
			Gametic ratios, p. 32, seed characters, first experiment			
70	A	1:1	43	47	90	−0·4216
71	B, among AA	1:1	20	23	43	−0·4575
72	B, among Aa	1:1	25	22	47	+0·4376
			Gametic ratios, p. 32, seed characters, second experiment			
73	A	1:1	57	53	110	+0·3814
74	B, among Aa	1:1	31	26	57	+0·6623
75	B, among aa	1:1	27	26	53	+0·1374

Table 2. (*cont.*)

	Character	Expected	Segregation Observed		Total	χ
	Gametic ratios, p. 32, seed characters, third experiment					
76	A	1:1	44	43	87	+0·1072
77	B, among AA	1:1	25	19	44	+0·9045
78	B, among Aa	1:1	22	21	43	+0·1525
	Gametic ratios, p. 32, seed characters, fourth experiment					
79	A	1:1	49	49	98	0·0000
80	B, among Aa	1:1	24	25	49	−0·1429
81	B, among aa	1:1	22	27	49	−0·7143
	Gametic ratios, p. 32, plant characters					
82	G	1:1	87	79	166	+0·6209
83	C, among Gg	1:1	47	40	87	+0·7505
84	C, among gg	1:1	38	41	79	−0·3375

We now go further than any previous analysis of Mendel's data by looking at the distribution of the 69 χ values themselves to see whether the unexpectedly low variance is a reflection of any particular anomaly. Fig. 1 displays the values on normal probability paper. It is immediately obvious that the reduced variance is not characteristic of the whole data, as Weiling's theory would require, but is confined to the tails of the distribution, where the extreme variates are not extreme enough to conform to expectation. Between the upper and lower quartiles (the 25-percentiles) of the expected distribution the slope of the plot is acceptable, but outside these points it is not. It must be remembered that a normal probability plot is not a linear regression, so that intuitive judgments of the departure from expectation may be misleading; for comparison, a simulated plot from a standard normal distribution is given in Fig. 2, using the first 69 random normal deviates from Lindley & Scott (1984), reading columnwise.

Another way to judge the apparent lack of extreme deviates is by examining the range, which is from −1·4237 (row 17) to +1·5811 (row 50), a total of 3·0048, whereas the expected range for a sample of 69 random normal deviates is 4·7440. Indeed, the expected absolute values of the first three most extreme deviates out of 69 are 2·6216, 2·2755, and 2·0828, and nothing on Mendel's data approaches these. The inescapable conclusion is that some segregations beyond the outer 5-percentiles (approximately) have been systematically biased towards their expectations so as to fall between the 5-percentiles and the 25-percentiles. Further analysis shows that the effect is not confined to particular sample sizes or segregation ratios, but is quite general.

It is relevant to recall here that Mendel thought it worthwhile to repeat an experiment in which a segregation of 60:40 had appeared where he had expected 2:1 (M, p. 20) a segregation with a corresponding χ-value of only −1·4142. As Fisher (1936) observed, this is not a significant value; yet only twice out of 69 times does Mendel record more extreme segregations (those quoted above). We should not forget, however, that when Mendel actually singled out the most extreme segregations in the first two experiments for comment (M, p. 17) they were no different from what one might expect (Fisher, 1936).

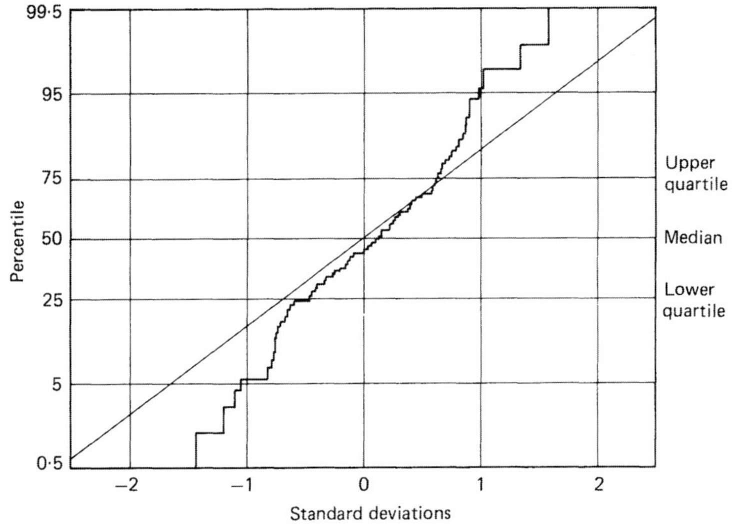

Fig. 1. The values of χ for the 69 segregations with undisputed expectations plotted on normal probability paper.

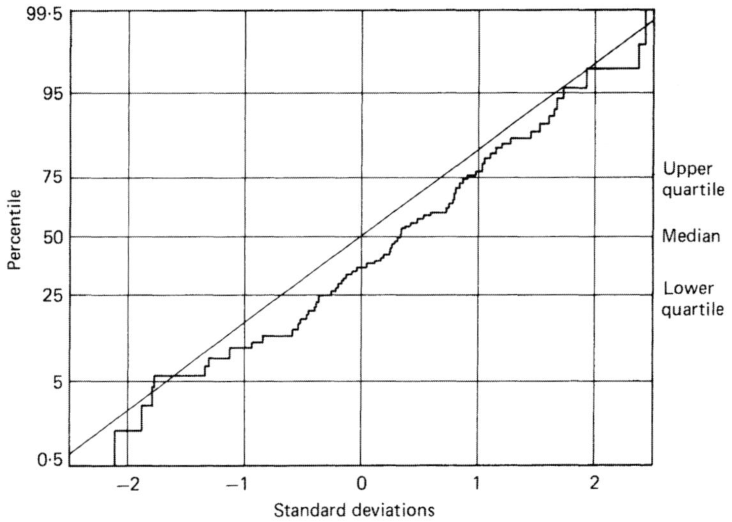

Fig. 2. The values of 69 simulated standard normal deviates plotted for comparison with Fig. 1.

VI. CONCLUSIONS

We thus see that Mendel's data exhibit two independent peculiarities both of which point to the same conclusion, namely, that in general the segregations agree more closely with what Mendel expected than chance would dictate. Where Mendel expected a 2 : 1

ratio even though the true expectation must have been different it was the former ratio which was achieved, and throughout the rest of his results there is a persistent lack of extreme segregations.

In Sections II and III above I reviewed earlier analyses and explanations and concluded that only Fisher (1936) could withstand detailed criticism. It may be helpful if I admit at this point that for many years I supposed that Fisher's analysis was going to be able to be faulted because of its total reliance on the 'repeated sampling' logic of the χ^2 goodness-of-fit test which I had come to mistrust, but a complete review of the whole problem has now persuaded me that his 'abominable discovery' must stand. I agree with Sturtevant (1965): 'In summary, then, Fisher's analysis of Mendel's data must stand essentially as he stated it.' As to the precise method by which the data came to be adjusted I would rather not speculate, though it does seem to me that any criticism of Mendel himself is quite unwarranted. Even if he were personally responsible for the biased counts his actions should not be judged by today's standards of data recording. In 1967 Dobzhansky reviewed six separate volumes celebrating the centenary of Mendel's paper and had this to say:

> I believe that a far simpler explanation [than Fisher's suggestion of an over-zealous assistant] is at least as plausible. Few experimenters are lucky enough to have no mistakes or accidents happen in any of their experiments, and it is only common sense to have such failures discarded. The evident danger is ascribing to mistakes and expunging from the record perfectly authentic experimental results which do not fit one's expectations. Not having been familiar with chi-squares and other statistical tests, Mendel may have, in perfect conscience, thrown out some crosses which he suspected to involve contamination with foreign pollen or other accident.

With the reservation that in order to explain some of the results misclassification rather than simple throwing away of plants will have to be invoked, I agree with Dobzhansky. Perhaps after all the undergraduate Fisher came close to the correct solution 75 years ago: 'It may be that the worthy German abbot, in his ignorance of probable error, unconsciously placed doubtful plants on the side which favoured his hypothesis.'

VII. SUMMARY

1. All the analyses and discussions of Mendel's data in his 1866 paper 'Experiments in plant hybridization' are reviewed, with special reference to the suggestion by Fisher (1936) that the segregations are in general closer to Mendel's expectations than chance would dictate. It is concluded that in spite of many attempts to find an explanation, Fisher's suggestion that the data have been subjected to some kind of adjustment must stand.

2. A fresh analysis based on the individual consideration of the 84 independent segregations reported by Mendel confirms this conclusion in two separate ways. In the words of my title, Mendel's results really are too close.

VIII. REFERENCES

BABBAGE, C. (1830). *Reflections on the Decline of Science in England and on Some of its Causes*. Fellowes, London.

BATESON, W. (1909). *Mendel's Principles of Heredity*. Cambridge University Press.

BATESON, W. & KILLBY, H. (1905). Experimental studies in the physiology of heredity: Peas (*Pisum sativum*). In *Royal Society Reports to the Evolution Committee* **2**, 55–80.

BARNDORFF-NIELSEN, O. (1976). Plausibility inference. *Journal of the Royal Statistical Society* B**38**, 103–131.

BEADLE, G. W. (1967). 'Mendelism, 1965'. In *Heritage from Mendel* (ed. R. A. Brink), pp. 335–350. University of Wisconsin Press, Madison.

BENNETT, J. H. (ed.). (1965). *Experiments in Plant Hybridisation*, by Gregor Mendel, with a Commentary and Assessment by R. A. Fisher. Oliver & Boyd, Edinburgh.

BENNETT, J. H. (ed.). (1983). *Natural Selection, Heredity, and Eugenics*. Clarendon Press, Oxford.

CREW, F. A. E. (1966). *The Foundations of Genetics*. Pergamon, Oxford.

DARBISHIRE, A. D. (1908). On the result of crossing round with wrinkled peas, with especial reference to their starch-grains. *Proceedings of the Royal Society* B**80**, 122–135.

DARBISHIRE, A. D. (1909). An experimental estimation of the theory of ancestral contributions in heredity. *Proceedings of the Royal Society* B **81**, 61–79.

DE BEER, G. (1964). Mendel, Darwin, and Fisher (1865–1965). *Note and Records of the Royal Society* **19**, 192–226.

DOBZHANSKY, T. (1967). Looking back at Mendel's discovery. *Science* **156**, 1588–1589.

DUNN, L. C. (1965*a*). *A Short History of Genetics*. McGraw-Hill, New York.

DUNN, L. C. (1965*b*). Mendel, his work and place in history. *Proceedings of the American Philosophical Society* **109**, 189–198.

EDWARDS, A. W. F. (1972). *Likelihood*. Cambridge University Press. (Reprinted 1984.)

EDWARDS, A. W. F. (1986). More on the too-good-to-be-true paradox and Gregor Mendel. *Journal of Heredity* **77**, 138.

FISHER, R. A. (1936). Has Mendel's work been rediscovered? *Annals of Science* **1**, 115–137. (Reprinted in Bennett, 1965, pp. 59–87, and Stern & Sherwood, 1966, pp. 139–172.)

FISHER, R. A. (1952). Statistical methods in genetics. *Heredity* **6**, 1–12.

GALTON, F. (1892). *Hereditary Genius*, 2nd ed., Macmillan, London.

GALTON, F. (1904). Average number of kinsfolk in each degree. *Nature* **70**, 529 and 626.

JEFFREYS, H. (1939). *Theory of Probability*. Clarendon Press, Oxford. (Reprinted 1983.)

LINDLEY, D. V. & SCOTT, W. F. (1984). *New Cambridge Elementary Statistical Tables*. Cambridge University Press.

MENDEL, G. (1866). Versuche über Pflanzen-Hybriden. *Verhandlungen des naturforschenden Vereines in Brünn* **4**, 3–47. (English translations in Bateson, 1909 [reprinted in Bennett, 1965, pp. 7–51] and Stern & Sherwood, 1966, pp. 1–48.)

MERSENNE, M. (1637). *L'Harmonie Universelle, Seconde Partie*. Ballard, Paris.

MONAGHAN, F. & CORCOS, A. (1985). Chi-square and Mendel's experiments: where's the bias? *Journal of Heredity* **76**, 307–309.

NORTON, B. & PEARSON, E. S. (1976). A note on the background to, and refereeing of, R. A. Fisher's paper '[On] The correlation between relatives on the supposition of Mendelian inheritance'. *Notes and Records of the Royal Society* **31**, 151–162.

OLBY, R. C. (1966). *Origins of Mendelism*, Schocken, New York, (2nd ed. University of Chicago Press, 1985).

OREL, V. (1968). Will the story on 'too good' results of Mendel's data continue? *BioScience* **18**, 776–778.

OREL, V. (1984). *Mendel*. Oxford University Press.

PEARSON, K. (1900). On the criterion that a given system of deviations from the probable in the case of a correlated system of variables is such that it can be reasonably supposed to have arisen from random sampling. *London, Edinburgh, and Dublin Philosophical Magazine, Fifth series* **50**, 157–175.

PIEGORSCH, W. W. (1983). The questions of fit in the Gregor Mendel controversy. *Communications in Statistics: Theory and Methods* **12**, 2289–2304.

PILGRIM, I. (1984). The too-good-to-be-true paradox and Gregor Mendel. *Journal of Heredity* **75**, 501–502.

ROOT-BERNSTEIN, R. S. (1983). Mendel and methodology. *History of Science* **21**, 275–295.

STERN, C. & SHERWOOD, E. R. (1966). *The Origin of Genetics: A Mendel Source Book*. Freeman, San Francisco.

STURTEVANT, A. H. (1965). *A History of Genetics*. Harper & Row, New York.

THODAY, J. M. (1966). Mendel's work as an introduction to genetics. *Advancement of Science* **23**, 120–124.

VAN DER WAERDEN, B. L. (1968). Mendel's experiments. *Centaurus* **12**, 275–288.

WEILING, F. (1966). Hat J. G. Mendel bei seinen Versuchen 'zu genau' gearbeitet? – Der χ^2-Test und seine Bedeutung für die Beurteilung genetischer Spaltungsverhältnisse. *Der Züchter* **36**, 359–365. (An English translation by W. W. Piegorsch is available as Paper BU-718-M of the Biometrics Unit, Cornell University, Ithaca, New York.)

WEILING, F. (1971). Mendel's 'too good' data in *Pisum* experiments. *Folia Mendeliana* **6**, 75–77.

WEILING, F. (1985). What about R. A. FISHER's statement of the 'too good' data of J. G. MENDEL's *Pisum*-paper? *45th Session of the International Statistical Institute*. Amsterdam.

WELDON, W. F. R. (1902). Mendel's laws of alternative inheritance in peas. *Biometrika* **1**, 228–254.

WRIGHT, S. (1966). Mendel's ratios. In Stern & Sherwood (1966), 173–175.

Paper 117 (1989)

Probability and likelihood in genetic counselling
Clinical Genetics 36, 209–216.

Note from A. W. F. Edwards: The purpose of the paper is stated in the opening abstract.

Probability and likelihood in genetic counselling

A. W. F. Edwards
Department of Community Medicine, University of Cambridge, Cambridge, England

The aim of this paper is to clarify some of the concepts used in the calculation of risk in genetic counselling. The use of probability as a measure of risk in a particular case is reviewed, with special reference to the Fisher – Neyman controversy. The technical concept of likelihood is then introduced and applied to risk probabilities themselves. The difficulties of interpretation are discussed, with a distinction drawn between calculating the risk for a child and inferring the genotype of his mother, even though the risk is wholly determined by the genotype. Fiducial inference is briefly mentioned, and the paper ends with a discussion of the problem which arises when the functional relationship between an unknown parameter and the unknown risk is not 1:1. The conclusion is tentatively reached that risk can only be gauged by probability and thus that prior probabilities must sometimes be assumed, but the paper is primarily intended as a guide and catalyst for the informed discussion of some of the difficulties, which may involve ethical dilemmas.

Received 19 March, revised 10 April, accepted for publication 22 April 1989

Key words: counselling; likelihood; probability; risk; support.

Probability as a Proportion

Probabilistic reasoning is an area of continuing controversy in philosophy, and even the geneticist's simple use of it may not be straightforward. Suppose that a child homozygous for a recessive disease gene has been born to a normal couple and we are asked to advise on the risk for a future pregnancy. The advice will undoubtedly be 'the probability that another child will be affected is 1/4', but what *exactly* does this mean?

The conventional, or 'frequency', interpretation is that the putative child would belong to a class of children, namely those born to parents both of whom are heterozygous, of which one quarter are homozygous for the disease gene. We will not concern ourselves with the philosophical question of the extent of this class – whether, for example, it is infinite – nor will we be drawn into a discussion of the role played by our knowledge that there is a cellular explanation for the Mendelian proportion of one quarter (it is sufficient to recall that historically the observation of the proportion long preceded the physical explanation). But one question must concern us: what is it that allows us to associate the probability of one quarter with this particular putative child? It is not enough that the child should be a member of a class of which a known proportion is homozygous for the recessive gene: he must also *not* be a member of any *recognizable* subset of that class in which the proportion is different (Fisher 1956, 1958, 1959). An example will make this clear.

Suppose that the disease is not fully penetrant, and that it is expressed in only half the children homozygous for the gene. There are now two subclasses of children of heterozygous parents, one in which the double-recessive manifests the disease and another, equal in frequency, in which it does not. The putative child will be a member of one of these subsets, but they are not *recognizable* subsets because there is no way of telling to which subset the child will belong. The probability of the child being affected is therefore $1/2 \times 1/4 = 1/8$. But if the subsets were recognizable we should certainly use the appropriate one: perhaps the disease manifests only in the male, in which case we might determine the sex of the foetus and state that the risk was 0 or 1/4, as appropriate.

It has long been known that there is a difficulty with the frequency interpretation of probability. Venn's original example (1866) is a good one: 'Let us assume, for example, that nine out of ten Englishmen are injured by residence in Madeira, but that nine out of ten consumptive persons are benefited by such a residence. ... John Smith is a consumptive Englishman; are we to recommend a visit to Madeira in his case or not?' Here we have a recognizable subset to which our individual belongs, but we can only guess the appropriate probability from the two partially-relevant class proportions. I discuss below the important question of whether it is meaningful to be uncertain about such a risk probability.

Probability of the Single Case

In the scheme discussed so far it has been implicitly assumed (because it is generally agreed) that, provided the conditions mentioned are met, the probability derived from the class proportion can be applied as a risk to the specific future case. But what if the case exists already, though nothing is known about it? Suppose the pregnancy is already in progress and the foetus a reality. Is the analysis still valid?

It may come as a surprise to learn that in the theory of mathematical statistics which achieved mid-century dominance, Neyman's theory, the validity of a probability statement applied to a particular existing object, such as *this* foetus, or *that* playing card, is denied. Neyman would allow only that the probability that a particular foetus was affected was zero or one, and we did not know which. Fisher strenuously disagreed, and I will here do no more than give references which provide an entry into the literature of the controversy: Fisher (1955), Neyman (1961). In practice all scientists seem to share Fisher's opinion that such single-case statements, if properly derived, are valid, and whatever philosophical developments ultimately take place on this issue there is no doubt that from an operational point of view such probability statements are effective and valuable. But once again, as at the end of the last section, the issue raises the question as to what is meant by uncertainty about a risk probability.

Likelihood

In 1921, Fisher realised that there was a need to identify one of the elements which occur in probability calculations by a separate name, *likelihood*. Most geneticists first encounter this word in its technical sense in the phrase *method of maximum likelihood*, Fisher's method of statistical parameter-estimation in which the value of the parameter or parameters which maximise the likelihood is preferred to all others. So far in our discussion unknown parameters have not intruded into the argument, but suppose we were to ask our original question about the putative child of *any* parents, not just ones who have already produced an affected child. The answer q^2 would then involve the population frequency q of the recessive gene, which might be an unknown or imperfectly-known *parameter*.

This is not the place for a discussion of statistical estimation, but the point has to be made that whereas probabilities refer to events or classes of events, likelihoods refer to parameters or hypotheses. The likelihood of a parameter value or of a hypothesis, given some data, is defined to be the probability of the data assuming the parameter value or hypothesis. Thus whilst with probabilities the alternatives are other events, with likelihoods the alternatives are other parameters values. For a fuller account the reader is referred to the book *Likelihood* (Edwards 1972, 1984). The short account in Appendix 3 of Thompson (1986) may also be found helpful.

A consequence of this definition is that whereas probabilities are additive, likelihoods are not, and they only have a *relative* interpretation. Thus one value of a gene frequency or recombination fraction may be three times more likely than another on some data, but it cannot have an absolute likelihood. Although likelihood is specific to particular data, it is perhaps misleading to use the notation for conditioning as with probability, P(data|parameter), and I now prefer a double bar for clarity, L(parameter||data) (Edwards 1987a).

The separation of likelihood from probability has led to a great clarification throughout statistics, and it is important to maintain the terminological distinction in all applications. Thus in dealing with a parameter such as a gene frequency which can take any value in a continuum (here $0 \leq q \leq 1$) the likelihood is a function of q – the *likelihood function* – but just as point likelihoods cannot be added, likelihood functions cannot be integrated. In this they are quite different from probability distributions, and the phrase 'likelihood distribution' should not be used. For the same reason a likelihood function does not possess a mean or a median like a probability distribution does. It may, however, possess a mode, but since this is then the maximum-likelihood value of the parameter this standard terminology is to be preferred.

A further important difference between likelihood functions and probability distributions lies in their behaviour when the parameter or variable is transformed. In the case of a likelihood function the transformation is achieved by simple substitution, but with a probability distribution this is not the case because of the presence of the differential element. An example of a transformed likelihood function will be given below, where the likelihood function for a risk is derived from the likelihood function for a recombination fraction, the risk being itself a function of the recombination fraction. I shall also discuss the problem which arises if the transformation is not single-valued (1:1).

The definition of likelihood means that likelihoods obtained from separate sets of data may be multiplied together to obtain an overall likelihood. It is thus convenient to work with the logarithm of the likelihood, for then the log-likelihoods may be summed to achieve the same end. Normally the natural logarithm (logarithm to the base e) of the likelihood is used in statistics, and indeed when Barnard (1949) introduced the word *lod* for the *logarithm* of the *odds* he meant the natural logarithm, but in 1955 Morton introduced the idiosyncratic phrase *lod score* to mean the log-likelihood to the base 10, normalized to be zero at recombination fraction one-half. However, the pressure to conform to standard practice is likely to be felt even by recombination fractions in due course (see Edwards 1988), and lods may be expected to give way to units of log-likelihood to the base e, with a rate of exchange of 0.4343 lods to one unit of log-likelihood ($\log_{10} e$).

In the case of a parameter such as a gene

frequency or recombination fraction an *interval* estimate is often quoted, being an interval within which it is likely the parameter lies. If this is obtained by finding the parameter values at which the log-likelihood or *support* is two units (say) below the maximum, such values may be called the 2-unit *support limits* (*Likelihood*, p. 71) and the corresponding interval the *support interval*. It is misleading to refer to such an interval as a 'confidence' interval, because the word 'confidence' has been preempted by Neyman's theory and has a very different technical meaning, even though in simple cases involving the normal distribution numerical equivalence may occur.

Finally in this section on likelihood terminology, I stress the importance of being careful about the use of the synonyms *probably* and *likely*, for having so pointedly distinguished likelihood from probability it is a pity not to extend the distinction to these words as well when writing or speaking technically, using the former where the implied measure is probability and the latter where likelihood is implied. Thus: 'the child will probably be normal since it is likely that the recombination fraction is very low'.

Likelihoods for Risk Probabilities

I now discuss the calculation of likelihoods for a risk probability, using an example shown to me by J. Clayton of Toulouse. This example leads directly to a discussion of the question referred to twice already, as to whether it is meaningful to attach uncertainty to a risk.

Suppose an X-linked recessive disease locus has alleles + and *d*, and is linked with a known recombination fraction θ ($0 \leq \theta \leq \frac{1}{2}$) to a marker locus with alleles *1* and *2*. A boy is born to a normal couple and has genotype *1d*. A second pregnancy is diagnosed as being a boy with marker genotype *2*. What is the risk that he is affected (disease genotype *d*)?

Clearly the mother is *12+d*, but there is no indication as to whether she is *1+/2d* or *1d/2+*. The likelihood of the former is the corresponding probability of the genotype of her first son, which is $\frac{1}{2}\theta$ since a recombinant is then involved, and the likelihood of the latter is $\frac{1}{2}(1-\theta)$ since a recombinant is not then involved. The likelihood ratio for the two maternal phases is thus $\theta{:}1-\theta$.

Now if the first phase (*1+/2d*) is the true one the probability that the second son is affected (*d*) is $1-\theta$, but if the second phase (*1d/2+*) is true this probability is θ. Thus the risk for the second son is either $1-\theta$ or θ, with a likelihood ratio $\theta{:}1-\theta$ for the former versus the latter. For example, if $\theta = 0.02$ the risk is either 0.98 or 0.02, the latter being 49 times more likely than the former. Note how, in this argument, the likelihood ratio for the phases has been transferred to the risks, because there is a 1:1 correspondence between the two phases and the two risk probabilities.

Here, then, is an example of an uncertain risk probability. There are two subsets with known probabilities, and although we do not know to which our own case belongs, one is more likely than the other, to a degree depending on the recombination fraction θ.

If the two phases are known to be equally represented in the population, then the overall risk is computed as:

$$\theta \times (1-\theta) + (1-\theta) \times \theta = 2\theta(1-\theta).$$

The frequency interpretation is that this is the proportion of second sons of marker genotype *2* who are affected and who have elder brothers of genotype *1d*. The difficulty arises when we know nothing about the frequencies of the two phases, and may be brought sharply into focus by considering what happens when θ is very small.

In this case if we assume the two phases equally probable, then the risk is approximately 2θ, by the above formula, but without this assumption the more likely of the two risk probabilities is θ, and the likelihood ratio in favour of this is approximately $1{:}\theta$. In other words, if θ is very small the evidence for the maternal phase *1d*/*2*+ is overwhelming and the risk is thus θ.

The difference between the two risks, 2θ and θ, arises because the former continues to allow for the remote possibility that the phase is really *1*+/*2d* even when the evidence is overwhelmingly against it. Generally in science we accept a hypothesis (in the sense of dismissing its competitors) if the evidence is overwhelming, and this example raises the question as to just how strong the evidence, in the form of the likelihood ratio, has to be for us to ignore the less-likely phase completely. I do not know the answer to this question, and put it forward for discussion.

But is it meaningful to talk about two different risk probabilities at all? Are we not constrained to quoting a single probability for the risk, and therefore to guessing the population frequencies of the phases if we do not know them in order to compute this probability? In particular, if we are totally ignorant are we not constrained to using the principle of indifference (*Likelihood*, p. 55) to justify the assumption of equal probabilities, even though there is a strong aversion to this Bayesian practice?

I believe the dilemma may be resolved by recognizing that two different problems are involved. One is the direct question, What is the risk that the second son is affected?, whilst the other is the indirect question, What is the evidence that the mother has a high risk of her second son being affected? The first is a question about the son, the second a question about the mother.

In the first case a single probability is appropriate, and on the necessary assumption (in the absence of contrary evidence) that the phases are equally probable, it is $2\theta(1-\theta)$ as we have already seen. Without a single probability there is no secure basis for a decision, such as a decision to terminate the pregnancy, and decision theory tells us that in taking decisions we are implicitly adopting values for the unknown probabilities which appear in our formulations. The conclusion seems inevitable that when considering the son it is *not* meaningful to speak of uncertainty in the risk probability. Any feeling of unease in this conclusion is, I submit, due to confusing prediction about the son with inference about the mother.

For in the second case, that of the mother, she is characterized either by a high risk $1-\theta$ or low risk θ of her second son being affected when he is known to carry marker gene *2*, and although it may not be of practical value it is perfectly logical to make a likelihood statement about which risk is the more likely on the evidence of the genotype of her first son.

Such likelihood statements about probabilities have been made before (*Likelihood*, pp. 61–65). A recent example (Edwards 1986, 1987b) involved the estimation of the risk of another Chernobyl-like nuclear accident.

Fiducial inference

This section is intended only for those who enjoy the finer points of theories of statistical inference, and continues the discussion of the example of the last section in terms of fiducial probability. For a valid fiducial inference to be made a problem has to possess a particular structure. In all uncontroversial applications of fiducial probability so far, that structure has involved continuous parameter and sample spaces (for reviews of fiducial inference, see the articles in the *Encyclopedia of Statistical Science* by Buehler, Edwards, and Stone 1983). But to exhibit the fiducial argument in the simplest possible case I invented the 'sentry-box' example with discrete parameter- and sample-spaces (Edwards 1976; see also *Likelihood*, 1984 edition, pp. 207 & 217). By a strange coincidence it turns out that our genetical example has exactly the same formal structure as the sentry-box example, and estimating the mother's phase from a knowledge of her first son's phenotype is equivalent to estimating from which of two sentry-boxes a sentry came if he is observed at a particular one of two guard positions.

In the terminology of the genetical example, the argument runs as follows. It is supposed that at the outset nothing is known about the mother's phase or about the frequency of the two phases in the population. Now consider the following inference rule for the case of an only son: 'If the son is *1d* or *2+* say the mother is *1d/2+*, and if the son is *1+* or *2d* say the mother is *1+/2d*'. The peculiar property of this rule is that whatever the mother's phase the statement to which it leads is correct with probability $1-\theta$, because in each case the statement entails and is entailed by the statement 'there has been no recombination'. This property is known technically as the *confidence property* and is necessary, but not sufficient, for the fiducial argument to hold. The other required property is that the son's genotype be uninformative about whether a recombination has in fact occurred. This property, which seems to hold in this case, ensures that the probability $1-\theta$, which applies to the long run, may be validly applied to the particular case once the son's genotype is known. We may therefore conclude, according to the fiducial argument, that if the son is *1d* or *2+* the mother is *1d/2+* with probability $1-\theta$, and if the son is *1+* or *2d* the mother is *1+/2d* with probability $1-\theta$.

It will be noticed that the fiducial probability $1-\theta$ is the same as the Bayesian posterior probability would be on the assumption of equal prior probabilities for the two phases, but the two arguments are entirely different in ways which the interested reader can pursue through the literature on fiducial probability (for references see the encyclopedia articles quoted above). In particular, the strength of the fiducial argument would have been the same even if no-one had yet produced a theory of population genetics, or empirical data, in support of the phases being equally frequent in the population.

A final peculiarity of this example is that it allows fiducial prediction (Fisher 1956), a fact which was not noticed in connection with the sentry-box example, but we will not investigate this here. It is sufficient to have observed that, for the first time, an actual scientific example involving discrete hypotheses has yielded to the fiducial argument. This is, however, likely to be of greater interest to students of statistical inference than practitioners of human genetics.

Likelihoods for Risk Probabilities Involving the Recombination Fraction

So far the unknown which has entered into the calculation of risk has been the mother's phase, a discrete parameter which can only take two values, but it might equally well be a continuous parameter such as the recombination fraction θ. We can continue with the same example if we suppose the mother's phase is known to be *1+/2d* but now the recombination fraction is not known exactly but has an uncertainty expressed by a likelihood function $L(\theta)$. The risk R of a son who is marker-genotype *1* being affected is simply $R=\theta$, and by the rule which transforms likelihood functions by direct substitution, the likelihood function for the risk is simply $L(R)$.

Exactly the same problems of meaning arise here as when the mother's phase was unknown: is it meaningful to have a likelihood function for a risk? The answer now, as then, hangs on whether we are interested in theorizing about the risk or predicting about the son. In the latter case there again seems to be no alternative to adopting a prior distribution for the recombination fraction and proceeding according to the laws of probability.

We can (somewhat artificially) extend the example to illustrate a more complicated relationship between R and θ by asking for the risk that male twins of marker genotypes *1* and *2* (and thus non-identical) should both be affected, which is $R=\theta(1-\theta)$, a 1:1 relationship since $0\leq\theta\leq\frac{1}{2}$. Thus $\theta=(1-\sqrt{(1-4R)})/2$, and $L(R)=L((1-\sqrt{(1-4R)})/2)$.

It may happen that the risk R and the recombination fraction θ are not in a 1:1 relationship, as has been pointed out by Krawczak (1987). The likelihood function for R is then folded back on itself, some values of R possessing more than one likelihood. If we are making a prediction about a child then once again there seems to be no alternative but to adopt a prior probability distribution for θ. With reference to Krawczak's example it will thus be seen that I do not fully share his concern, or that expressed by Ott (1985). It is not strictly possible to derive a single likelihood function from a folded one – technically speaking the fold is a sort of nuisance parameter which cannot be eliminated – but taking the largest of the likelihoods at each point would give the profile likelihood as an acceptable solution. However, in Krawczak's example, and the one in the previous paragraph, to express uncertainty about the risk does not seem to me to be the appropriate thing to do.

Acknowledgements

The problems considered in this paper arose out of discussions with Dr M.Krawczak, Dr J. Clayton, Dr J. Ott, and Professor J. H. Edwards, to all of whom I am grateful. They do not share all the views here expressed, however, and the paper is an attempt to clarify the issues for further discussion rather than to resolve them.

Note Added In Proof

In his recent paper 'Evaluating the uncertainty of risk prediction in genetic counseling: a Bayesian approach' (*Am. J. Med. Genet.* **31**, 513–519; 1988), A. Rogatko applies Bayesian analysis to the problem, resulting in a probability distribution for a risk probability. He assumes without discussion that this is a meaningful concept.

References

Barnard,G.A. (1949). Statistical inference. *J. Roy. Statist. Soc. B*, **11**, 115–149.

Buehler, R. J. (1983). Fiducial inference. *Encyclopedia of Statistical Sciences*, Vol. 3. New York, Wiley, pp. 76–81.

Edwards, A. W. F. (1972). *Likelihood.* Cambridge University Press.

Edwards, A. W. F. (1976). Fiducial probability. *The Statistician* **25**, 15–35.

Edwards, A. W. F. (1983). Fiducial distributions. *Encyclopedia of Statistical Sciences*, Vol. 3. New York, Wiley, pp. 62–63.

Edwards, A. W. F. (1984). *Likelihood.* (paperback edition) Cambridge University Press.

Edwards, A. W. F. (1986). How many reactor accidents? *Nature* **324**, 417–418.

Edwards, A. W. F. (1987a). Likelihood. *The New Palgrave: A Dictionary of Economics*, Vol. 3. London, Macmillan, pp. 184–185.

Edwards, A. W. F. (1987b). How many nuclear reactor accidents? *Nature* **327**, 663.

Edwards, A. W. F. (1988). Lod score or log-likelihood? *Nature* **333**, 308.

Fisher, R. A. (1921). On the 'Probable Error' of a coefficient of correlation deduced from a small sample. *Metron* **1**, 3–32.

Fisher, R. A. (1955). Statistical methods and scientific induction. *J. Roy. Statist. Soc. B*, **17**, 69–78.

Fisher, R. A. (1956). *Statistical Methods and Scientific Inference.* Edinburgh, Oliver & Boyd.

Fisher, R. A. (1958). The nature of probability. *Centennial Rev.* **2**, 261–274.

Fisher, R. A. (1959). Mathematical probability in the natural sciences. *Technometrics* **1**, 21–29.
Krawczak, M. (1987). Genetic risk and recombination fraction – an example of non-monotonic dependency. *Hum. Genet.* **75**, 189–90.
Morton, N. E. (1955). Sequential tests for the detection of linkage. *Am. J. Hum. Genet.* **7**, 277–318.
Neyman, J. (1961). Silver Jubilee of my dispute with Fisher. *J. Oper. Res. Soc. Japan* **3**, 145–154.
Ott, J. (1985). *Analysis of Genetic Linkage in Human Families*. Baltimore, Johns Hopkins University Press.
Stone, M. (1983). Fiducial probability. *Encyclopedia of Statistical Sciences*, Vol. 3. New York, Wiley, 81–86.
Thompson, E. A. (1986). *Pedigree Analysis in Human Genetics*. Baltimore, Johns Hopkins University Press.
Venn, J. (1866). *The Logic of Chance*. London and Cambridge, Macmillan.

Address:

A. W. F. Edwards
Department of Community Medicine
University of Cambridge
Fenner's, Gresham Road
Cambridge, CB1 2ES, England

Paper 140 (1994)

The Fundamental Theorem of Natural Selection

Biological Reviews 69, 443–474.

Note from A. W. F. Edwards: This long review paper was made possible by the publication of Ewens (1989) in which Fisher's fundamental theorem was presented through the construction of a discrete-generation analogue of the continuous-time original. For the first time Fisher's intentions for his theorem became clear to me. Although Price (1972) had attempted a clarification it did not bear fruit until Ewens presented Price's view in a more comprehensible way, stressing the *partial* nature of the fitness change involved. When Price's paper appeared in the *Annals of Human Genetics* I had noted it with interest (at the time I was subscribing privately to the journal) but was then too preoccupied with statistical inference and the publication of my *Likelihood* to give it close attention. I did, however, point it out as important to Alan Robertson of Edinburgh, whom I knew to be interested, and to Ewens himself. For a later view of the theorem, taking into account the continuous-time nature of Fisher's 1930 development and its original concept of fitness, see [238].

THE FUNDAMENTAL THEOREM OF NATURAL SELECTION

By A. W. F. EDWARDS

Department of Community Medicine, University of Cambridge

(*Received* 11 *February* 1994; *accepted* 25 *February* 1994)

CONTENTS

The fundamental theorem of natural selection was a valiant attempt by R. A. Fisher to fit a quart into a pint pot.

The present author reviewing *Population Genetics* by Ewens, 1969*a* (Edwards, 1969).

Many authors have maintained that the theorem holds only under very special conditions, while only a few (*e.g. Edwards*, 1967) *have thought that Fisher may have been correct – if only we could understand what he meant*! *It will be shown here that this latter view is correct.*

Price, 1972.

My own reason for not including a planned chapter on the fundamental theorem is… [*that*] *in spite of many efforts I am still not satisfied that I am able to provide an account that does the theorem justice.*

The present author in the Epilogue to *Foundations of Mathematical Genetics* (Edwards, 1977).

I. INTRODUCTION

R. A. Fisher enunciated his Fundamental Theorem of Natural Selection in Chapter II of The Genetical Theory of Natural Selection (1930):

The rate of increase in fitness of any organism at any time is equal to its genetic variance in fitness at that time.

For 42 years after its publication the theorem was universally misunderstood and its name often attached either to a simpler approximate theorem (which was true) or to a general theorem about the mean fitness never decreasing (which was not). Then in 1972 the late G. R. Price (who was also jointly responsible for the notion of an Evolutionarily

R. A. FISHER in 1929

Stable Strategy; Maynard Smith and Price, 1973) published a detailed analysis and explanation (Price, 1972) which led in due course to Ewens's clearer formulation of the theorem and hence to its wider acceptance (Ewens, 1989; for some background discussion, see Edwards, 1990).

The purpose of the present review is not only to give an account of the Fundamental Theorem as now understood, but to lay to rest the misunderstandings and misnamings of earlier years. In order to achieve this it will be necessary to review some statements which are now best forgotten, since this paradoxical activity is the only way to clear away the confusion which has surrounded the theorem. The authors of the recalled statements can rest assured that not only have they kept distinguished company, but also that to change one's view of the Fundamental Theorem has notable precedents. However, I shall not review the arguments of Price and Ewens in favour of the present interpretation of the theorem, except in so far as they are implicit in what follows. I believe them to be correct, and the reader who still harbours lingering doubts should study Price (1972) and Ewens (1989) for himself.

It will be helpful at the outset if we keep in mind the following simple 'growth-rate' theorem:

In a subdivided population the rate of change in the overall growth-rate is proportional to the variance in growth rates.

This elementary theorem is applicable wherever there are differential growth-rates, in economics for example (Edwards 1987*a*) as well as in biology. It opens a window on to the Fundamental Theorem, yet seems never to have been published in its own right, by Fisher or anyone else. Fisher clearly possessed it in 1930, but it was not until many years later that it came to be given explicitly by writers commenting on the Fundamental Theorem. I first saw it in the text-book by C. C. Li (1955) as a 'simplified version of Fisher's *fundamental theorem of natural selection*'. Simplified it certainly was, for it omitted mating; in a subsequent edition (1976) Li replaced this version by another which, as we shall see, has been one of the two common misinterpretations of the Fundamental Theorem. Somewhat paradoxically, another source for this elementary theorem is the section entitled 'An extension of Fisher's Fundamental Theorem of Natural Selection to include interpopulation selection' in Crow (1955), for if the genic variances are zero all that remains is the interpopulation variance.

The growth-rate theorem is easily proved: Let a population be subdivided into k parts in proportions p_i ($i = 1, 2, \ldots, k$) and let the ith part change in size by a factor w_i in unit time (its 'growth-rate'). At the end of the unit time interval the new proportion of the ith part will be $p'_i = p_i w_i / w$, where $w = \Sigma p_i w_i$ is the overall growth rate. The new overall growth rate $w' = \Sigma p'_i w_i$ is therefore equal to $\Sigma p_i w_i^2 / w$ and the change in the overall growth-rate, $w' - w$, is $(\Sigma p_i w_i^2 - w^2)/w$. But the numerator of this expression is simply the variance of the w_i by definition (say v), being their expected squared value minus their squared expected value. Thus $w' - w = v/w$ and the theorem is proved. The constant of proportionality, $1/w$, is of course just a scaling factor: dividing both sides by w leads to $w'/w - 1 = v/w^2$, showing that the proportionate change is equal to the square of the coefficient of variation. A continuous version of the theorem is easily obtained by proceeding to a limit, the growth in each sub-population then being exponential.

The growth-rate theorem exactly captures the vague notion that the more variable are the growth-rates the more quickly will the most rapidly-growing parts of the population (or sectors of the economy) come to dominate the rest. Eventually dominance by the fastest will be complete, all variability in growth-rates will have vanished, and no further increase in the overall growth-rate can occur. It is as though the process of change, or evolution, consumes the variability. It was the intrinsic difficulty of adapting this simple theorem to the complexities of Mendelian genetics, coupled with Fisher's cryptic exposition in *The Genetical Theory*, which caused his successful attempt to be misunderstood for so long.

II. BACKGROUND

It is to Adam Smith's *An Inquiry into the Nature and Causes of the Wealth of Nations* (1776) that we must turn for the germ of the idea behind the Fundamental Theorem. In Book IV: *Of Systems of Political Economy* he wrote:

> As every individual, therefore, endeavours as much as he can both to employ his capital in the support of domestic industry, and so to direct that industry that its produce may be of the greatest value; every individual necessarily labours to render the annual revenue of the society as great as he can. He generally, indeed, neither intends to promote the public interest, nor knows how much he is promoting it. By preferring the support of domestic to that of foreign industry, he intends only his own security; and by directing that industry in such a manner as its produce may be of the greatest value, he intends only his own gain, and he is in this, as in many other cases, led by an invisible hand to promote an end which was no part of his intention. Nor is it always the worse for the society that it was no part of it. By pursuing his own interest he frequently promotes that of the society more effectually than when he really intends to promote it.

From this idea it is but a small step to the realization that an economy progresses not merely because the efforts of individual producers to better themselves raise the average level of production, but also because the 'invisible hand' will drive inefficient producers out of business and thus raise the average still further. As Sober (1984) remarked, 'The Scottish economists offered a non-biological model in which a selection process improves a population as an unintended consequence of individual optimization', and he saw this as one of the influences in the formation of Darwin's views on the effects of natural selection. We can only speculate on whether Fisher was familiar with these ideas in economics in the 1920s; if so, a possible source would have been the economist J. Maynard Keynes, whose *A Treatise on Probability* was published in 1921 and whom Fisher had known since his undergraduate days (Box, 1978). He will, however, certainly have read Darwin's comment: 'The larger and more dominant groups thus tend to go on increasing in size; and they consequently supplant many smaller and feebler groups' (*The Origin of Species*, p. 428; Darwin, 1859).

At the time when Fisher entered on his researches the importance of heritable variation as the raw material on which natural selection acts was not well understood, and the prevailing intellectual atmosphere was one of doubt as to whether Darwinian natural selection could account fully for evolutionary change. Furthermore, no one had yet successfully quantified this variation mathematically. Indeed, not until Francis Galton had statisticians learnt to regard biological variability as intrinsically interesting, in contrast to variability in physics and astronomy which they had been accustomed to think of as 'error'. Karl Pearson, in his many papers on evolution, had pursued the mathematical study of biological variation, but it was the young Fisher who saw the overwhelming advantages of working not with the standard deviation but its square, which he christened the variance in his paper *The correlation between relatives on the supposition of Mendelian inheritance* (Fisher, 1918*a*). As he explained in a contemporary article in the *Eugenics Review*: 'This mean square deviation I term the variance, and use it as a measure of variability, by reason of this important property, namely, that two independent causes of variability acting together produce a variance which is the sum of the variances produced by either separately' (Fisher, 1918*b*). Indeed, he went so far as to say, in the opening sentence of this article, 'The great service which the modern development of statistics has rendered to eugenics is that it supplies a definite method of measuring and analysing variability'.

Throughout the 1920s, during the gestation of *The Genetical Theory of Natural Selection*, Fisher was fond of quoting Darwin's dictum, from Chapter II of *The Origin of Species*, that 'wide ranging, much diffused, and common species vary most'. 'The present note', he wrote in his major 1922 paper (Fisher, 1922*a*), 'is designed to discuss the distribution of the frequency ratio of the allelomorphs of dimorphic factors, and the conditions under which the variance of the population may be maintained'. It was a remarkable 'note', for it introduced to the world not only the concept of stable polymorphism through heterozygotic advantage but also the chain-binomial model for gene-frequency change and the associated concept of the distribution of gene frequencies. To handle the chain-binomial model Fisher introduced the diffusion approximation, thus inaugurating stochastic diffusion theory (see Kendall, 1990). In addition, he applied the Galton–Watson branching process to the survival of an individual gene and handled it by functional iteration. (For further comments on this paper and its impact, see Edwards, 1978, 1990; the branching-process model is also described in Fisher's simultaneous publication 1922*b* which carries the date 'Feb. 2nd, 1921'.)

Fisher's 1922 investigations were thus motivated by a desire to investigate the importance of the various factors affecting the variance of a population. Stochastic losses due to finite population size will cause a slow decay, but the major contributions will be the gain due to mutation and the loss due to selection: 'The decay in the variance of a species breeding at random without selection, and without mutation, is almost inconceivably slow: a moderate supply of fresh mutations will be sufficient to maintain the variability. When selection is at work even to the most trifling extent, the new mutations must be much more numerous to maintain equilibrium... Thus a numerous species, with the same frequency of mutation, will maintain a higher variability than will a less numerous species: in connection with this fact we cannot fail to remember the dictum of Charles Darwin [quoted above]'.

Armed with his appropriate measure of variability and inspired by Darwin's work, it was thus natural that Fisher should proceed to quantify the relationship between the variability in the genetic contributions of individuals to subsequent generations, and the rate at which the population changes in consequence. In 1926, in a note in *Nature* with E. B. Ford, he wrote 'it is easily demonstrable that in species in which a higher proportion of the total variance is ascribable to genetic causes, the effective selection will be more intense than in species in which the variance is to a larger extent ascribable to environmental variations' (Fisher & Ford, 1926; repeated in Fisher & Ford 1928). As many have remarked, when Fisher wrote 'it is easily demonstrable' you knew you were in for a long haul.

On 13 November 1928 Fisher wrote to Leonard Darwin, Charles Darwin's fourth son and Fisher's early mentor, telling him how *The Genetical Theory of Natural Selection* was progressing: 'I have finished drafts of Chapters II [*The Fundamental Theorem of Natural Selection*] and III, but not yet started IV. They are at your disposal but I do not want to press a lot of heavy reading on you. Chapter II is heavy. ... [P.S.] Thank you ever so much for real encouragement' (Bennett, 1983). On the only occasion on which I met Lady Fisher (24 June 1974) I asked her about *The Genetical Theory*. She said that except for Chapters IV and V Fisher dictated it to her as he paced up and down their living room, she taking it down longhand. She added that the Fundamental Theorem 'came to him as he dictated'. Fisher's daughter and biographer, Mrs Joan Fisher Box, had earlier told me that he had been dissatisfied with its exposition, but could not see how to improve it (Edwards, 1990; Box, 1978, adds that the period of dictation was between October 1928 and June 1929).

Calling the relative genetic contributions of individuals their fitnesses, Fisher sought a relationship between the variance in fitness in the population and the fitness of the whole population (as measured by the mean of the individual fitnesses). It was presumably at this stage that he discovered the exact growth-rate theorem given in the Introduction above. He will instantly have seen that as it stood it was incapable of direct application to the evolutionary problem for the simple but highly complicating reason that the Mendelian mechanism ensures that although a population may be said to have a continuing existence, the individuals which constitute it do not. The variability which passes from one generation to the next through the medium of reproduction is related to, but not identical to, the phenotypic variability manifest in the reproducing generation.

Appreciation of this point will have come naturally to Fisher. In *The correlation between relatives* he had already shown how the total genotypic variance at a single locus in a population, that is, the variance contributed by the variability of genotypes in an environment assumed uniform, could be considered as the sum of two components. The first component, called the genetic variance, is due to the additive effects of the genes ('an additive part which reflects the genetic nature without distortion'), whilst the second, residual component, the dominance variance, is due to the non-additive effects of the genes. The basic idea comes from linear regression: regress the genotypic values on the number (0, 1 or 2) of genes of a particular allelic type and the resultant analysis of variance separates the linear ('genetic') and the residual ('dominance') effects. (For more information consult, for example, the books by Falconer, 1989, or Edwards, 1977).

Here I digress for a moment to introduce a uniform terminology, for there has been a change of usage since Fisher first wrote. He used the word 'genetic' straightforwardly as the adjective from the word 'gene', so that by 'genetic variance' he meant the variance contributed by the genes acting additively. It would have been consistent with his 1918 usage to have written, for example, 'The genotypic variance (in a uniform environment) is the sum of the genetic and dominance variances'. Unfortunately, in an attempt at clarification, some writers started to refer to the 'additive genetic variance', which carried the implication that some of the genetic variance was non-additive.

Before long, therefore, 'genetic' replaced 'genotypic' as in 'The total genetic variance is the sum of the additive genetic variance and the dominance variance'. Crow and Kimura (1970), surveying the resultant confusion, recommended using the word 'genic' as the adjective from 'gene', with the statement thus becoming 'The genotypic variance is the sum of the genic variance and the dominance variance'. I shall adhere to this usage; as it happens, the adjective 'genic' was used by Fisher (1932), though it was actually coined by C. B. Bridges ten years earlier.

III. THE FUNDAMENTAL THEOREM OF NATURAL SELECTION

Price's (1972) masterly analysis of the Fundamental Theorem is wholly convincing, but it is not the simplest approach to the theorem, which was provided by Ewens (1989). Ewens pointed out that 'the Fundamental Theorem as properly understood... is exact in discrete time and thus not specifically a continuous-time result'. Both Fisher's original derivation and Price's commentary are couched in the language of continuous time and the differential calculus, and part of the lack of understanding of the theorem has been the erroneous view that whilst a continuous-time version might be correct if properly interpreted, no exact discrete-generation version exists (Edwards, 1967, 1977). I now give Ewens's proof of the theorem for a single multiallelic locus. In his view 'the discrete-time calculation... was clearly well-known to [Fisher] and possibly formed the basis of [the] continuous-time result'. I shall refer to my *Foundations of Mathematical Genetics* ('FMG'; Edwards, 1977) for background information, but using capital A rather than lower-case a for the genes so as to conform to Ewens's notation and release a for another purpose.

The essence of Price's interpretation is the recognition of the fact that the Fundamental Theorem refers to the *partial change* in mean fitness, not the total change. This will be defined and derived in the third part of the proof which follows.

Consider a discrete-generation model and let P_{ii} be the frequency of genotype $A_i A_i$, and $2P_{ij}$ the frequency of genotype $A_i A_j$. (Hardy–Weinberg equilibrium is not assumed.) The frequency of A_i is then $p_i = \Sigma_j P_{ij}$ and if the fitness of $A_i A_j$ is w_{ij} then the mean fitness is $w = \Sigma_{i,j} P_{ij} w_{ij}$. In the adult population the frequency of A_i will therefore be $p'_i = \Sigma_j P_{ij} w_{ij}/w$. We now define the *average excess* a_i of gene A_i in conformity with Fisher (1930, 1958), but in the words of FMG p. 12:

With respect to any character, the average excess of a gene is the difference between the mean for that sub-population consisting of all the homozygotes and half the heterozygotes which carry that gene, and the mean of the whole population.

Hence in respect of the character 'fitness'

$$a_i = (P_{ii} w_{ii} + \sum_{j \neq i} P_{ij} w_{ij})/\sum_j P_{ij} - w = \sum_j P_{ij} w_{ij}/p_i - w. \qquad (1)$$

Notice that the change in the frequency of A_i through the action of the differential fitnesses, $\Delta p_i = p'_i - p_i$, may now be expressed in terms of the average excess:

$$\Delta p_i = \sum_j P_{ij} w_{ij}/w - p_i = (p_i/w)(\sum_j P_{ij} w_{ij}/p_i - w) = p_i a_i/w. \qquad (2)$$

In passing, this proves the obvious fact that the weighted sum of the average excesses is zero, $\Sigma_i p_i a_i = 0$, since necessarily $\Sigma_i \Delta p_i = 0$.

The second of the new concepts which Fisher developed for the Fundamental Theorem is that of the *average effect* α_i of the gene A_i on the deviation of the fitness of a genotype from the mean fitness w. Even in the simplest single-locus model, the effect of a gene on the fitness of a genotype depends not only on which allele it is, but on which allele accompanies it in the genotype, except in the special case of no dominance in fitness, when the deviation from the mean fitness is determined by the simple sum of the effects of the two genes: $w_{ij}-w = \alpha_i+\alpha_j$. 'Then the effect on the fitness of an individual whose genotype contains an A_j gene of replacing that gene by an A_i gene is $\alpha_i-\alpha_j$' (Ewens, 1989). In the presence of dominance, however, we may not speak unambiguously of the effect of the gene A_i, but we can compute its average effect by appealing to the idea of regression, regressing the fitness of genotypes on the numbers of each allele present to find the 'additive effects' of the alleles (Fisher, 1918; see above, Section II). We may use Fisher's own words from 1941, which was the only occasion on which he was mathematically explicit (his attempts at verbal definition have not always corresponded exactly with this (Falconer, 1985)):

> The direct mathematical measure of the average effect of a... gene substitution is the partial regression... of the genotypic measurement on the numbers 0, 1 or 2 of the allelomorphic genes [alleles] in each genotype. This is the natural measure of the average effect of a gene substitution.

We now implement this definition by the standard regression procedure of finding the values of the α_i which minimize the residual sum of squares $\sum_{i,j} P_{ij}(w-w_{ij}+\alpha_i+\alpha_j)^2$. The normal equations are, for all i, $\sum_j P_{ij}(w-w_{ij}+\alpha_i+\alpha_j) = 0$, and on expanding this and recalling that $\sum_j P_{ij} w_{ij} = p_i a_i + p_i w$ from the expression (1) for the average excess a_i, we have

$$p_i \alpha_i + \sum_j P_{ij} \alpha_j = p_i a_i. \tag{3}$$

The genic variance is the sum of squares removed by the regression:

$$\begin{aligned} v &= \sum_{i,j} P_{ij}(\alpha_i+\alpha_j)^2 = 2\sum_{i,j} P_{ij}\alpha_i^2 + 2\sum_{i,j} P_{ij}\alpha_i\alpha_j \\ &= 2\sum_i p_i\alpha_i^2 + 2\sum_{i,j} P_{ij}\alpha_i\alpha_j. \end{aligned}$$

Now

$$\sum_j P_{ij}\alpha_i\alpha_j = p_i(a_i-\alpha_i)\alpha_i \quad \text{from (3)}$$

so

$$2\sum_{i,j} P_{ij}\alpha_i\alpha_j = 2\sum_i p_i(a_i-\alpha_i)\alpha_i$$

and we arrive at

$$v = 2\sum_i p_i \alpha_i a_i, \tag{4}$$

Fisher's expression for the genic variance in terms of the average effects α_i and average excesses a_i of the genes (Fisher, 1930, though the formulation involving multiple alleles is 1958).

The third and final part of the proof of the theorem is to show that the partial change in the mean fitness w due to changes in the genotypic frequencies P_{ij} from one generation to the next (P'_{ij}) is precisely equal to v/w. For it is a standard result of the regression model that the weighted mean departure of the genotypic fitnesses from their regression values is zero, or $0 = \Sigma_{i,j} P_{ij}(\alpha_i + \alpha_j)$, and adding the mean fitness w to both sides leads to $w = \Sigma_{i,j} P_{ij}(w + \alpha_i + \alpha_j)$ since $\Sigma_{i,j} P_{ij} = 1$. Thus the change in w due only to the change in the genotype frequencies P_{ij} and not including changes in w and the α_i is

$$\begin{aligned}
\delta w &= \sum_{i,j} (P'_{ij} - P_{ij})(w + \alpha_i + \alpha_j) \\
&= \sum_{i,j} (P'_{ij} - P_{ij})(\alpha_i + \alpha_j) \\
&= 2\sum_{i} (\alpha_i \sum_{j} (P'_{ij} - P_{ij})) \\
&= 2\sum_{i} \alpha_i \Delta p_i \quad \text{since} \quad p_i = \sum_{j} P_{ij} \\
&= 2\sum_{i} p_i a_i \alpha_i / w \text{ by (2)} \\
&= v/w \text{ by (4).}
\end{aligned}$$

Thus the partial change in the mean fitness w due to changes in the genotypic frequencies P_{ij} from one generation to the next is exactly equal to v/w. Ewens (1989) makes detailed comments on the mathematical idiosyncracy of computing δw whilst holding w constant on the right-hand side.

(Readers unfamiliar with the elementary mathematics of the least squares fit of a linear model should minimize the residual sum of squares for themselves and prove that the variance removed by the fitted regression is the genic variance as given above. The remainder of the results can be made to follow in a variety of different ways. FMG pp. 12–16 may be helpful, but note that the treatment there is only for the special case of a population in Hardy–Weinberg equilibrium.)

If fitnesses are standardized so that $w = 1$ then $\delta w = v$, and if we take a generation to be the unit of time, then we may present our modern interpretation of the Fundamental Theorem in words as

> The rate of increase in the mean fitness of any organism at any time ascribable to natural selection acting through changes in gene frequencies is exactly equal to its genic variance in fitness at that time.

This is as close as I can get to Fisher's wording (1930, 1958) whilst following the interpretation of Price and Ewens. 'Genic' has replaced 'genetic' for the reasons given in Section II and 'mean fitness' has replaced 'fitness of any organism'; these are uncontroversial rewordings (Fisher himself wrote of 'the average fitness' in his letter to Sewall Wright dated 19 January 1931; see below). I have added 'ascribable to natural selection acting through changes in gene frequencies' from Fisher's own words:

'ascribable to natural selection' is from his 1941 explanation of the Theorem (p. 57) and '[due to all] changes in gene frequencies' is from the sentence preceding his statement of the Theorem (1930, 1958), modified as suggested by Price. Price and Ewens repeatedly emphasize that Fisher thought of the immediate effect of natural selection as being only through the changes in gene frequencies. For the sake of minimizing the changes I have kept Fisher's word 'organism', though it would not now be the first choice; in fact Fisher himself replaced it by 'species' in the Summary at the end of Chapter II of *The Genetical Theory*. 'Population' might now be the preferred word.

Although I have only treated the case of a single locus in this account, Fisher clearly viewed the Fundamental Theorem as making a statement covering all loci contributing to the determination of fitness. Ewens (1989) has supplied the multilocus mathematics.

IV. FISHER'S OWN VIEW OF HIS THEOREM (1930)

Throughout the rest of his life Fisher made comments which throw light on his own attitude to the Fundamental Theorem, and these will be reviewed below in the course of describing how others saw it, but his comments in the 1930 edition of *The Genetical Theory* (Chapter II: *The Fundamental Theorem of Natural Selection*) need to be considered first. These are best studied in the original, and here I only give a brief summary. Of course, the name of the Theorem itself tells us the great importance that Fisher attached to it, as well as his comment that it 'hold[s] the supreme position among the biological sciences'.

Immediately after his statement of the Theorem Fisher writes 'The rigour of the demonstration requires that the terms employed should be used strictly as defined', which should have served as a warning to those who have supposed that the Theorem is approximate only. Then he considers the impact of finite population size on the genic variance for the special case of a random-mating population, making the important statement that with random mating the average excess a and the average effect α are equal.

Next Fisher likens the Fundamental Theorem to the second law of thermodynamics in some important respects, especially that 'each requires the constant increase of a measurable quantity, in the one case the entropy of a physical system and in the other the fitness, measured by m, of a biological population'. The m refers to 'The Malthusian parameter of population increase' which Fisher had introduced in an earlier section of the chapter (II) on the Fundamental Theorem and which, as he explicitly recognized, could be negative as well as positive. It seems that he was here carried away by his enthusiasm, but the qualification follows quickly amongst a list of 'profound differences': '(3) Fitness may be increased or decreased by changes in the environment...'. Moreover, '(4) Entropy changes are exceptional in the physical world in being irreversible, while irreversible evolutionary changes form no exception among biological phenomena'. 'Finally, (5) entropy changes lead to a progressive disorganization of the physical world... while evolutionary changes are generally recognized as producing progressively higher organization in the organic world'. I do not know if this is the first time that such an explicit statement was made.

In making these comparisons Fisher is relying on his training in mathematical physics. He took the Cambridge Mathematical Tripos in 1912 and then spent a year of graduate study in physics under (Sir) James Jeans. He was also influenced by (Sir)

Arthur Eddington's views; he and Eddington had been in communication over a statistical question in 1920 (Fisher, 1920), and Eddington had been a referee of Fisher's 1922(*c*) classic paper *On the mathematical foundations of theoretical statistics*. In *The Genetical Theory* Fisher quotes Eddington in the law of entropy, as well as quoting him approvingly in the Preface: 'We need scarcely add that the contemplation in natural science of a wider domain than the actual leads to a far better understanding of the actual' (Eddington, 1928).

The remaining sections of Chapter II are entitled *The nature of adaptation*, *Deterioration of the environment*, and *Changes in population*. In the first Fisher gives his well-known geometrical analogy to explain why, when many characters are considered simultaneously, large undirected changes will in all probability be deleterious, whilst in the second and third sections he considers the impact of his approach on the study of population growth. Once again we are reminded that it is a partial change in mean fitness to which the Theorem applies:

> Against the action of Natural Selection in constantly increasing the [mean] fitness of every organism, at a rate equal to the [genic] variance in fitness which that population maintains, is to be set off the very considerable item of the deterioration of its inorganic and organic environment. It is only if the former of these agencies exceeds the latter that there can be any actual increase in population, while in the reverse case the population will certainly decrease.

V. SEWALL WRIGHT'S VIEW OF THE THEOREM

The American evolutionary biologist Sewall Wright was a contemporary of Fisher's, born just two months before him, and by 1930 had established himself as a leading figure. In 1922 he had invented the inbreeding coefficient and path coefficients and in 1929 he and Fisher had had an exchange in the *American Naturalist* on the theory of dominance (see Provine, 1986). They had met for the first time in 1924 (Bennett, 1983). From June 1929 until June 1931 they were in friendly correspondence (printed in part in Bennett, 1983, and almost in its entirety in Provine, 1986), initially in connection with the evolution of dominance but widening into a more general discussion, especially after Wright had sent Fisher the manuscript of his major paper *Evolution in Mendelian populations* (Wright, 1931) in August 1929.

Fisher arranged for complimentary copies of *The Genetical Theory of Natural Selection* to be sent to a number of American scientists, including Wright, who also received a review copy from the *Journal of Heredity*. The long review which resulted (Wright, 1930) was the only major one the book was to receive in the United States (Bennett, 1983). In his accompanying letter to Wright, Fisher commented (19 March 1930) 'In some ways the first chapter is the most important, and in some the second'. Wright thanked him on 10 June: 'I wish to thank you very much for sending me a copy of your recent book. I have found it extremely interesting and stimulating'. He does not mention the Fundamental Theorem in this letter or the subsequent one, but he does so in his review.

Wright's review of *The Genetical Theory* is eight pages long. 'It is a book which is certain to take rank as one of the major contributions to the theory of evolution'. Then: 'The core of Dr Fisher's theory of selection is given in Chapter II. He reaches a formula on which he lays great emphasis as "the fundamental theorem of natural selection".'

One's first impression is that the genetic variance in fitness must in general be large and that hence if the theorem is correct the rate of advance must be rapid. As Dr Fisher insists, however, the statement must be considered in connection with the precise definition which he gives of the terms. He uses "genetic variance" in a special sense [i.e. the genic variance].

Wright had thus understood the importance of the genic variance, but he went on, in the review, to remark that the Fundamental Theorem needed revision in order to incorporate dominance, epistasis, and mutation. He, like many others in the next forty-two years, had not taken Fisher's point in full, but, more than any of his contemporaries, he had studied *The Genetical Theory* with care.

Fisher expressed delight with the review when he wrote on 19 January 1931, but added:

I had not intended to take up any special point in this letter, but I am tempted to mention this one, 'The formula itself seems to need revision...'. The point here is that the average fitness is continually being increased by selection, at exactly the same rate as it is being decreased by mutation. This cause of deterioration of adaptedness, due to mutations of the organism, is, in my treatment, classed with the parallel deterioration due to changes in the environments.

Fisher added that Wright will find this point emphasized should he 'happen to re-read p. 41' (*Deterioration of the environment*).

Wright replied on 3 February: 'I was very glad to hear that you were pleased with my review', and

I am very sorry that I overlooked your treatment of mutations on page 41. The essential difficulty which I felt with your conclusions on pages 34–37 still remains, but I should have worded it differently.... I assumed that all internal factors...were intended to be taken into account. I do not know how I managed to overlook your clear statement to the contrary on page 41. I think that if the theorem on page 35 had been stated with its qualifications: "The rate of increase in fitness of any population at any time is equal to its genetic [genic] variance in fitness at that time, except as affected by mutation, migration, change of environment and the effects of random sampling," I would not have been confused by the discussion in the following pages.

This was indeed a noble effort to understand Fisher's obscure theorem, but since it did not include a qualification equivalent to 'ascribable to natural selection acting through changes in gene frequencies' it was not what Fisher had meant. Wright's interpretation was that the theorem was true in respect of total fitness, not partial fitness, provided mutation and other factors were not taken into account.

In 1939 Wright published a short monograph *Statistical Genetics in relation to Evolution* (Wright, 1939) in which he did not mention the Fundamental Theorem explicitly, and in his much larger *The Theory of Gene Frequencies* (Wright, 1969) it only rated a passing reference (p. 121). In 1955, however, he was clear that 'Fisher's formula is more general [than his own] in that it does not involve the assumption of random combination within and among loci (long continued random mating) which we have made' and that 'Fisher's form of statement indicates that it is intended to apply only momentarily rather than to define "fitness" as a quantity that governs the course of evolution, analogously to a potential function'.

Yet from a few remarks in 1988 (see below; Wright 1988) it might seem that he never changed his view of the theorem after all, for he then revealed that he still thought that the assumption of panmixia was basic to its derivation.

VI. THE RISE AND FALL OF THE ADAPTIVE TOPOGRAPHY

Wright may have been the first to misunderstand the nuances of the Fundamental Theorem, but he was certainly not to be the last. The irony is that although he came closer to divining Fisher's meaning than anyone else was to do for many years, and certainly made a much greater effort, his misinterpretation led him to develop the concept of an adaptive topography, which was to foster further misunderstandings in due course. For his letter of 3 February 1931 continued, after a few intervening lines:

> Some aspects of the ideas which I tried to express in pages 353 to 355 of my review might be visualised as follows: Think of the field of visible joint frequencies of all genes as spread out in a multidimensional space. Add another dimension measuring degree of fitness. The field would be very humpy in relation to the latter because of epistatic relations, groups of mutations which were deleterious individually producing a harmonious result in combination.

The letter contains a drawing of a continuous line with three humps of different height (fitness) followed by a description of how a species might be seen to evolve in this representation. It would 'tend to move steadily' up the slope towards the nearest maximum 'under the influence of selection' but if this maximum was not the absolute maximum the species would only be able to escape from it and make further progress in fitness through 'something other than the steady pressure of selection'. Wright suggested four factors, (1) a changing environment would continually change the system of humps, (2) new mutations would add further dimensions to the field, allowing new paths of advance, (3) random genetic drift in a small species would allow stochastic jumps from one hump to another, as would (4) a subdivided large species. Wright's image was that of a potential function expressing the selective effect, accompanied by stochastic variation: in modern computer jargon, he has invented 'simulated annealing'.

The image of a 'field of joint frequencies of all genes spread out in a multidimensional space' will not have been new to Fisher, for in November 1930 he had received a letter from J. B. S. Haldane in Cambridge, written on the 9th, which said:

> I have...got a paper on the 2-gene [i.e. 2-locus] case, showing that in the case of dominance when *AB* and *aabb* are more viable than *Abb* or *aaB* there is no stable equilibrium apart from homozygosis. In this case if we plot the proportion of recessive genes we get trajectories like: [here follows a diagram] representing the change in the composition of the population. The general case of *m* genes and several (>2) stable equilibria involves problems in *m*-dimensional analysis situs which are too much for me. If heterozygotes have a pull as such we may get stable equilibria of another type, but these may not be stable in the long run, allowing for the appearance of modifiers, as you point out.

The diagram in the letter is identical to the one which eventually appeared in the paper (Haldane, 1931*a*; reproduced in Haldane 1932, 1966, 1990) except for the labelling. Although Bennett (1983) published the first paragraph of this short letter, he omitted the second, which is published here for the first time with the permission of the Barr-Smith Library of the University of Adelaide.

Thus within the span of three months Fisher had received letters describing multidimensional gene-frequency spaces, first from Haldane and then from Wright. Now Fisher was already famous in statistics for the facility with which he used mathematical arguments in multidimensional space, first displayed in his derivation of the distribution of the correlation coefficient (Fisher, 1915): 'The problem of the frequency distribution of the correlation coefficient *r*, derived from a sample of *n* pairs, taken at random from an infinite population, may be solved, when that population can be represented by a normal surface, with the aid of certain very general conceptions derived from the geometry of *n* dimensional space'. No one was better placed than Fisher, who had had a distinguished undergraduate mathematical career at Cambridge, to understand why untutored excursions into hyperspace were likely to be misleading. Fisher cautioned Wright in his letter of 31 May 1931:

Your letter of February 3rd contains a point about non-optimal points of genetic stability which I should like to take up with you. In one dimension, a curve gives a series of alternate maxima and minima, but in two dimensions two inequalities must be satisfied for a true maximum, and I suppose that only about one fourth of the stationary points would satisfy both. Roughly I would guess that with *n* factors only 2^{-n} of the stationary points would be stable for all types of displacement, and any new mutation will have half a chance of destroying the stability.

When Wright replied on 5 June he mentioned that he had received an offprint of Haldane's paper, and on Fisher's point he commented '2^{1000} is an infinity of such a high order that the maxima may be very widely scattered and still permit a practically infinite number of them'.

Fisher, who was spending the summer at Ames, Iowa, visited Wright in Chicago at the end of June. In later life Wright could recall little of their discussions (Provine, 1986); all we have to go on is the thank-you note from Fisher:

This is just a note to thank you and Mrs Wright for your kindness and hospitality to me in Chicago. I wish I could better understand your view on those points on which I differ from you, but on the points I have discussed with Lush [J. L. Lush of Iowa State College, Ames], I see little chance that I shall ever do so. However, there is a substantial body of theory on which I think we do agree and that after all is of infinitely more interest to the world at large than the very obscure points still in dispute.

It is easy to surmise that Wright's enthusiasm for his image of an adaptive topography, based on his deep biological knowledge and intuition, came up against the rapier-like quality of Fisher's analytical mind, trained in mathematics and mathematical physics and thus thoroughly familiar with the geometry of *n*-dimensional Euclidean space and potential theory. I conjecture that the very obscure points still in dispute were about *n*-dimensional surfaces. This appears to have been the last letter between Fisher and Wright.

Wright's image of an adaptive topography proved to be compulsive viewing, and came to dominate thinking in evolutionary biology for a long time. Confusingly, however, in his first publication on it (Wright, 1932) he departed from the mental picture of a multidimensional field of gene frequencies which he had described in his letter to Fisher and instead described a field of gene combinations: 'If the entire field of possible gene combinations be graded with respect to adaptive value under a particular set of conditions, what would be its nature?' In this new image, a point in the space represented a particular gene combination, not a particular set of gene frequencies, though fitness (adaptive value) was still added in a further dimension as previously. A species was no longer characterized by a point representing its gene frequencies, but by a cluster of points representing the individuals, a sort of bounded cloud. Wright then based his evolutionary arguments on a two-dimensional drawing of this multidimensional space, with the third dimension, for fitness, represented by contours as on a map. 'A species whose individuals are clustered about some combination other than the highest would move up the steepest gradient toward the peak', as if fitness were a potential function. Wright then drew the same conclusions about the importance of random genetic drift from his new analogical image as he had from the old.

Setting aside for a moment the question of whether the properties of the multidimensional space could really be thought about in only three dimensions, Wright's new model is severely restricted by being *haploid*. It is based on *gene combinations*, that is to say, *haplotypes* in modern terminology. Even in as many dimensions as there are loci, a point represents not an individual, as Wright asserts, but half an individual. Yet there is no doubt that he is thinking about diploid individuals, for the opening sentence of his paper is 'The enormous importance of bi-parental reproduction as a factor in evolution was brought about a good many years ago by East' and then, immediately before introducing 'the field of possible gene combinations', 'There is no difficulty in accounting for the probable uniqueness of each individual human being or other organism which is the product of biparental reproduction'. It is to the haplotype that Wright attaches a fitness, not an individual. It is (to paraphrase Dawkins, 1976) the *selfish haplotype* which flourishes on Wright's metaphorical mountains. Consequently the model cannot take into account dominance in fitness, which was precisely the phenomenon which Fisher struggled successfully to allow for in the Fundamental Theorem.

Returning to the question of the two-dimensional representation of a multidimensional space, Wright's diagrams showing 'The combinations of from 2 to 5 paired allelomorphs' are technically *Boolean cubes* for from 2 to 5 factors, starting with the square (2 factors) and cube (3); in his 'diagrammatic representation of the field of gene combinations in two dimensions instead of many thousands', 'it is assumed that the genotypes are packed, side by side, in a two dimensional space in such a way that each is surrounded by genotypes that differ by only one gene replacement' (Wright, 1988). Now there is a fundamental mathematical reason why one cannot represent a Boolean cube in two dimensions so that all adjacent corners in n dimensions are adjacent to each other in the two dimensions. The reason is to do with the structure of a Venn diagram for n sets.

Each of the 2^n segments of an n-set Venn diagram corresponds logically to each of the 2^n corners of an n-dimensional Boolean cube, and the general importance of Venn diagrams may be said to stem from the fact that they are, in a sense, the best possible 2-dimensional representations of n-dimensional Boolean cubes (see, for example, my popular article *How to iron a hypercube*; Edwards, 1991). Wright's two-dimensional field may be thought of as an attempt to construct a Venn diagram for a huge number n of sets. A satisfactory solution to this problem has only recently been found (Edwards, 1989), from which it may be shown that the mean number of segments surrounding a segment (corresponding to Wright's 'genotypes that differ by only one gene replacement') is $4\text{-}2^{3-n}$, which tends to 4 as n tends to infinity. In other words the number of immediate neighbours to a corner of a Boolean n-cube, which is of course n (corresponding to the n loci, at each of which a single gene substitution may occur) is represented in the best possible two-dimensional version, on average for large n, by just 4. There is absolutely no way in which the proximity structure of a large-dimensioned Boolean cube can be modelled in two dimensions, and any conclusions drawn from the assumption that it can must be treated with reserve. In 1932 Wright did indeed say 'The two dimensions of figure 2 are a very inadequate representation of such a field' (i.e. of 'the entire field of possible gene combinations'), but with the enthusiastic reception of his graphical arguments he threw caution to the wind.

Both Fisher and Haldane were present when Wright gave this paper at the 6th International Congress of Genetics at Cornell in August 1932. There is no record of any discussion, but Fisher will probably have felt that Wright's paper confirmed his own view that excursions into n-dimensional space were naive and unhelpful. For his part, Wright will have heard Fisher warn not to expect too much from the application of mathematics to biology, as if Fisher foresaw the dangers into which he thought Wright was drifting:

> I believe that no one who is familiar, either with mathematical advances in other fields, or with the range of special biological conditions to be considered, would ever conceive that everything could be summed up in a single mathematical formula, however complex. (Fisher, 1932)

What he may not have foreseen is that T. Dobzhansky was to promote Wright's adaptive topography with such enthusiasm that it captivated a whole generation of evolutionary biologists. 'Many others besides Dobzhansky were much impressed by Wright's 1932 paper' writes Provine (1986). 'His adaptive surface diagrams appeared and reappeared frequently in papers and books and were constantly referenced'.

In 1935 Wright reverted to his original idea of an adaptive topography, the space once again becoming a space of gene frequencies ('cf. Haldane, 1931' he writes) with the addition of a dimension for 'the average adaptive value' which he identifies with the quantity in Fisher's Fundamental Theorem.

> If evolution were controlled only by selection, the locus of a population characterised by any given set of gene frequencies would move up the steepest gradient in the field.

Once again Wright introduces the misleading notion of a potential function, this time on the basis of a different graphical representation. Fisher was less than impressed; at a Royal Society meeting on the theory of natural selection the following year he remarked causticly 'It has been proposed ... that the environment, picturesquely renamed the landscape, governs the course of evolutionary change, much as the field of force determines the trajectory of a comet' (Fisher, 1936). In 1949 Wright was to accept that 'steepest ascent' did not apply (Wright, 1949):

> With constant genotypic selective values, the species tends to move toward one of the peaks, though not, as carelessly stated previously, up the steepest gradient in the surface $\bar{W}$ from the point at which it is located.

However, this admission that even in the simplest case there was in fact no potential function did nothing to prevent the spread of the contrary belief. The idea that the average adaptive value was a potential function of the gene frequencies which determined gene-frequency changes was to dominate much thinking in evolutionary biology during the next four decades. A single example, from an influential population geneticist addressing a Zoological Congress (Lewontin, 1965), will suffice:

If we then know these [genotype] fitnesses and the frequencies of the different genotypes we can calculate a parameter $\bar{W}$, the average fitness of the population. As the frequencies of the different genotypes change, the average fitness will change... This can be represented in a topographical diagram... [which] shows an "adaptive topography".... By connecting points of equal $\bar{W}$ we get equipotential lines forming a kind of topographic map.

We can now state a fundamental theorem of intrademe evolution. Provided that fitnesses of genotypes are not functions of the commonness or rarity of those genotypes and provided the population is very large in size, then the frequencies of the genotypes [*sic*] in the population will change under natural selection so as to maximise $\bar{W}$.

As Provine (1986) has related, after 1935 the two versions of an adaptive topography travelled together:

Wright's concept of the surface of selective value has become famous because of its apparently great heuristic value in conveying graphically the relationship between organisms, mechanisms of evolution, and adaptation. Despite its great attractiveness and apparent ease of interpretation, the surface of selective value is one of Wright's most confusing and misunderstood contributions to evolutionary biology. Wright himself has contributed to the confusion.

In the first place, Wright contributed not just one conception of his surface but two very different versions that mathematically were wholly incompatible. Worse, he frequently treated these two incommensurable conceptions as if they were together a unitary conception, and he switched back and forth between them often.

In 1971 I had expressed precisely these views in a paper 'Fisher's Fundamental Theorem of Natural Selection and Wright's Adaptive Topography: a controversy revisited' read at a joint meeting of the Biometric Society and the Royal Statistical Society (Edinburgh Group) (Edwards, 1971*a*, abstract only). The paper remains unpublished because it was rejected for publication on more than one occasion. However, at the same time I wrote a long review of Volume 2 of Wright's great work *Evolution and the Genetics of Populations – The Theory of Gene Frequencies* (Wright, 1969) in which, after much detailed discussion of the gene-frequency version of the adaptive topography, I remarked 'The general theory for which Wright seeks is a will-o'-the-wisp' (Edwards, 1971*b*). In my book *Foundations of Mathematical Genetics* (1977) I explained the omission from it of

all treatment of gene-frequency change by the use of differentials.... This is a major departure from current text-book practice (see for example Crow and Kimura, 1970; Elandt-Johnson, 1971; and Li, 1976). Wright (1969) gives the fullest account of the treatment, with its associated general 'fitness function'. It may only be a matter of notation, but I believe it to have been the source of misunderstandings to which I refer in my review of Wright's book (Edwards, 1971*b*).

Provine's book on Sewall Wright contains no references to these criticisms nor, much more remarkably, to the devastating criticism of the adaptive topography published by Fisher in 1941, to which I now turn.

In 1938 E. B. Ford, of Oxford, sent Fisher the page proofs of his forthcoming book *The Study of Heredity* (Ford, 1938). Fisher read it 'with the greatest pleasure and interest' but wishes 'to take up argumentatively' Ford's statement, on pages 174–5, of the views developed by Sewall Wright: 'either the statement or the original views seem to be confused'. In his letter (2 May) Fisher then clearly distinguishes Wright's two types of adaptive topography, the gene-combinations one and the gene-frequencies one, and we may be sure that Ford was not able to correct his pages 174–5 before his book went to press, because they do indeed completely confuse the two types by describing the first kind and then arguing as if it were the second.

Fisher was evidently uncertain whether the confusion originated with Wright himself or with Ford, which shows that he had not been following the American developments closely, but he must have been sufficiently interested to follow the matter up, perhaps goaded by the fact that Ford had not amended his book and by an essay of Wright's in *The New Systematics* (Wright, 1940). The result was his second publication on the Fundamental Theorem (Fisher, 1941) in which, with startling clarity, he drew attention to Wright's confusion:

> Wright's conception...that selective intensities are derivable, like forces in a conservative system, from a simple potential function dependent on the gene ratios as a whole, has led him into extensive but untenable speculations. For example, in the New Systematics, p. 170 [Wright, 1940], we find: 'As already noted, $\bar{W}$ [the mean fitness] is a function of all gene frequencies. In the practically infinite field of gene combinations, possible from differences in only a few thousands, or even hundreds of loci, there are likely to be an enormous number of different harmonious combinations of characters. These would appear as peak values of $\bar{W}$, separated by valleys or saddles in a multidimensional surface'.
>
> Prof. Wright here confuses the number of genotypes, e.g. 3^{1000}, which may be distinguished among individuals, with the continuous field of variation of gene frequencies. Even if a potential function, such as $\bar{W}$ is supposed to be, really existed, the large number of genotypes supplies no reason for thinking that even one peak, maximal for variations of *all* gene ratios should occur in this field of variation.

Fisher ended his 1941 paper with a fully-worked numerical example of 'a model involving powerful selection, in which the fitness of the species as a whole, judged by external criteria is entirely inoperative'. Many years later, as we shall see, others were to exhibit models in which the mean fitness did not increase, but, with supreme irony, these were held to be counterexamples to Fisher's Fundamental Theorem rather than to Wright's adaptive topopgraphy. Fisher went on to make one or two other comments on the concept of an adaptive topography in 1958 (see the next Section, and Edwards, 1990).

In 1988 Wright, well-advanced in years, chose to defend himself against Provine's severe criticism (Wright, 1988), and did so by ignoring the existence of the gene-frequency-based adaptive topography altogether, thus deflecting Provine's remarks on the incompatibility of the two versions. Against Provine's extensive mathematical criticism of the 'gene-combinations' adaptive topography Wright offered no defence, preferring to concentrate on defending his 'shifting-balance' theory of evolution verbally as opposed to graphically. But Provine (and I) had not argued that the theory was wrong; only that the graphical metaphor was inappropriate. Wright had defended the defensible but ignored the criticism. As to my own criticisms of the gene-frequency-based adaptive topography mentioned above (Edwards, 1971*b*), Wright did not respond, though I know he read the review because J. F. Crow informed me privately at the time that he had not been upset by it, for which I am grateful. P. A. P. Moran (author of Moran, 1962, 1964) wrote to me in 1974 'I was very interested in your review of Wright in Heredity which I had missed. I thoroughly agree with it'.

It was only human of Wright to have basked in the enthusiasm with which his 1932 diagrams were adopted by population biologists who were not technically competent to see their mathematical naivety, but it was strange that the metaphor was powerful enough to ensure that even the criticism of Fisher (1941) had no impact. In 1988 Wright pleaded that 'The concept of "surfaces of selective value" was introduced in 1932 merely in connection with a nonmathematical comparison of elementary evolutionary processes' and suggested that Provine 'was looking for something more mathematical than was intended'. To hold that view one would have to ignore both the gene-frequency surface of selective value and Wright's many papers on the topic after 1932, which is what Wright did, for in 1988 he referred to nothing between then and 1977.

VII. J. B. S. HALDANE AND THE THEOREM

Bennett (1983) has described the British reception of *The Genetical Theory*. It seems that no one noticed the Fundamental Theorem, not even Haldane. In a letter dated 17 December 1930 Fisher expressed his disappointment that Haldane was not to review the book in the *Eugenics Review*, and wondered whether even so Haldane would allow what he had already written to appear later. But Haldane had lost his manuscript, to Fisher's distress: 'I should greatly have liked to read it'. Happily, a couple of months later it turned up and Haldane sent it to Fisher, who found it 'extremely interesting'. It was subsequently published as an 'essay-review' under the title 'Mathematical Darwinism' (Haldane, 1931*b*).

> The book before us is of the latter class [books which, though often inaccurate in detail, state a new point of view, and lay the foundations of new branches of science]. It is extremely difficult and highly controversial. ... No serious future discussion either of evolution or eugenics can possibly ignore it.

Haldane did not mention the Fundamental Theorem explicitly either in this essay or in his review in the *Mathematical Gazette* (Haldane, 1931*c*).

Haldane's own book *The Causes of Evolution* came out in 1932 (Haldane, 1932, 1966, 1990). In the Introduction he referred to *The Genetical Theory* as 'a brilliant book', a compliment which Fisher returned in his own unpublished review of *The Causes of Evolution* (since published by Bennett, 1983). However, Haldane 'ignored the theorem' in his book (the expression is from Leigh, 1990), and I am not aware that he ever alluded to it, though he did embrace the adaptive topography at least as a metaphor: 'A large population of any given composition will inevitably evolve towards one particular maximum of W' (Haldane, 1958).

The 'official' review which the *Eugenics Review* commissioned was contributed by C. G. Darwin (1930), brother of Leonard, who similarly made no mention of the theorem.

VIII. 1955–72: THE THEOREM MISUNDERSTOOD

For twenty-five years after its publication the Fundamental Theorem had no impact on the literature other than through its influence on Sewall Wright. It seems not to have been noticed by either of the two principal writers on mathematical genetics of the day besides Wright and Haldane, Hogben (1933, 1946) and Dahlberg (1947). But then Hogben and Dahlberg were more interested in genetical statistics than evolution. More remarkably, Ford (1938) does not mention it, even though, as we saw in Section II, he was Fisher's co-author (Fisher & Ford, 1926, 1928) when the germ of the idea was first published. Malécot's *Les mathématiques de l'hérédité* (1948; English translation 1969) is silent on the theorem. But in 1955 Li's *Population Genetics* appeared and remained the principal text until Moran (1962), Ewens (1969*a*), Jacquard (1970), and Crow & Kimura (1970). As mentioned in Section I, Li produced a theorem which he claimed was 'a simplified version of Fisher's *fundamental theorem of natural selection*', though he did not specify in what respect it was simplified.

Li's theorem was the growth-rate theorem of Section I applied to genotypes, and it is indeed true that the mean fitness of the genotypes in a particular generation is increased through the action of selection by an amount precisely equal to the *genotypic* variance in fitness, if we standardize the fitnesses so that the mean is unity before selection acts. But Li claimed that 'As the individuals with low viability will be relatively less numerous after the operation of selection than before it, so the *average* fitness of the selected population will be higher than that of the preceding generation'. This does not follow, because the preceding generation and the present one are separated by mating, which ensures that in general the genotypic frequencies amongst the young are not the same as those amongst their parents. In his review of Li's book Bennett (1956) remarked:

> ...the author makes a very serious error as the result of confusing genotypic with genetic variance. In what he claims is a simplified version of Fisher's fundamental theorem of natural selection, he asserts that the increase in the average selective value of the population per generation is equal to what is, in fact, the genotypic variance of the selective value.

Li's terminology was indeed unfortunate, for in his book he used 'genetic variance' in Fisher's original sense (now 'genic variance') but called the genotypic variance the 'actual genetic variance', thus inadequately separating the two notions. Bennett's review, however, had no effect on the subsequent literature. The only non-trivial case where the round of mating has no effect on the mean fitness is when fitness is exactly determined by the genes acting multiplicatively, and the population is mating at random, for then it will be in Hardy–Weinberg equilibrium both before and after mating.

In the same year as Li's book appeared, Crow (1955) discussed the theorem, and concerted attempts to understand it soon followed. Crow & Kimura (1956) and Kempthorne (1957) struggled to come to terms with Fisher (1941), but misinterpreted Fisher's 'simple example' (which had been only intended to demonstrate that the average effect and the average excess need not be the same) as leading to an additional condition required for the truth of the theorem.

[As a critic enjoying the advantage of having read Price (1972) I may seem harsh in my assessment of those who wrote before Price, but I can mete out similar treatment to my own 1967/68 typescript *On the Fundamental Theorem of Natural Selection* referred to by Turner (1970), which contained the same mistake. Because of my doubts, however, it was never published, and on reading Price I scribbled on it 'nb this is mostly nonsense – 2 Apr 73' (Edwards, 1990). But a brief note in *Nature* (Edwards, 1967) revealed my mistake.]

Fisher had examined a case of non-random mating in which departure from Hardy–Weinberg equilibrium was measured by a parameter λ:

> If λ remains constant the actual change in the mean or total measurement in constant environment will be that due to change of gene ratio [frequency] only.... If, in fact, the value of λ changes, the change in the population mean will differ from that ascribable merely to the change in gene ratio,....

But as he had intended the Theorem to refer only to this 'change in the population mean ascribable merely to the change in gene ratio', a changing value of λ did not invalidate it, and constancy of λ was not an additional condition as Crow, Kimura and Kempthorne had concluded.

Kempthorne had corresponded with Fisher on just this point two years earlier, sending him a typescript which had prompted Fisher to reply (31 January 1955; Bennett, 1983) 'I do not at all agree with the last sentence of the opening paragraph of your introduction, "later in 1941 Fisher showed that this is true only if the quantity [λ] remains constant"', and he gave his reasons at length. Again, on 18 February, Fisher had written 'I should be entirely satisfied if you cared to use the two quotations from page 56 of my 1941 paper, which seemed to express just what I mean' (I believe these to encompass the quotation above), and 'the only evolutionary effect, either in increased fitness or in anything else, that I can recognize as such, is constituted by the changes in gene ratio'. But Kempthorne was unmoved, and included the apparent condition in his account of the Theorem in his book (1957), covering himself by commenting 'I have found Fisher (1930, 1941) entirely obscure with regard to the derivation of the result'.

In fact Fisher, Kempthorne, and Kimura were engaged in a three-way correspondence about the Fundamental Theorem in 1955–56 (Kempthorne, 1957; Bennett, 1983; Edwards, 1990), for Kimura had also sent Fisher the draft of a paper, which eventually appeared as Kimura (1958). In this paper Kimura obtained a very general formula for the total change in mean fitness for the continuous-time model which he considered to be a 'reformulation of the results published by Fisher in 1930 under the title *The Fundamental Theorem of Natural Selection*, and in no way differs from it in substance'. Mathematically Kimura's result was indeed true, but since it treated the total change in mean fitness, and not the partial change, it was not really Fisher's theorem: it differed from it by the addition of two equal components, one to each side of the equation, thus obscuring Fisher's intention (Ewens, 1989). Moran (1962) in his book the *Statistical Processes of Evolutionary Theory* did not mention the Fundamental Theorem by name, but followed essentially the treatment by Kimura (1958).

Kimura went on to enunciate his related 'Maximum principle in the genetical theory of natural selection', according to which it should be possible to derive the changes in gene frequencies just as the laws of motion can be derived from the principle of least action in physics. I have given my reasons elsewhere for thinking that, on the contrary, it 'simply reflects a feature of the mathematical structure, but adds nothing to an understanding of that structure' (Edwards, 1974; 1977).

The attempts of Crow & Kimura (1956) and Kimura (1958) to interpret the Fundamental Theorem, coupled with Wright's occasional references to it (e.g. Wright, 1955), had an undesirable side-effect in that the theorem began to acquire a wider audience without its true nature having been elucidated. Consequently its name came to be used rather loosely for the notion of increasing mean fitness, on the basis of Fisher's wording of the theorem taken without consideration of all his qualifying statements. Thus a widely-read paper by Kojima & Kelleher (1960) opened as follows:

> In 1930 R. A. Fisher formulated the fundamental theorem of natural selection which states that "the rate of increase in (mean) fitness of any organism at any time is equal to its genetic variance in fitness at that time" (Fisher 1930). Later, some modifications and variations of this

theorem were considered by Wright (1949, 1955), Crow and Kimura (1956) and Kimura (1958) in attempts to interpret the theorem over wider genetic circumstances.

When the members of a population are mating at random and their fitness values do not vary over time, Fisher's theorem is usually understood to imply that the mean fitness will always increase with time and that this increase is equal to the additive genetic variance in fitness.

There was, however, nothing in Kimura's paper to support such an interpretation unless one ignored everything except the additive effects of the genes.

In another widely-quoted paper from the same year Lewontin & White (1960), treating the case of two loci by Wright's adaptive topography without mentioning the Fundamental Theorem, wrote:

> If the fitnesses of the various genotypes are constants and do not depend upon the frequencies of the genotypes, it can be demonstrated that the gene frequencies will change in such a way as to increase the value of $\bar{W}$. That is, the population will tend to "ascend" the slopes of the topography, coming to rest on a "peak".

This prompted Moran (1964) to write a paper entitled *On the nonexistence of adaptive topographies* in which he demonstrated that the mean fitness did not necessarily increase in the two-locus model and concluded that 'Wright's conception of an 'adaptive topography' is untenable'.

The 1960 papers of Kojima & Kelleher and Lewontin & White are good examples of the uncertainty which at that time surrounded the distinction between Fisher's formulation and Wright's, for the first pair of authors were using Fisher's terminology and the second pair Wright's. This was all the more remarkable because Lewontin and Kojima were in communication at the time, being joint authors of another paper (Lewontin & Kojima, 1960) in which they wrote 'These mean adaptive values may then be put in the form of adaptive topography, a concept introduced by Wright (1932)'. (In fact, as we saw in Section VI, Wright introduced the gene-frequency adaptive topography in 1935, not 1932.)

The uncertainty was also evident in Kimura's (1965) response to Moran (1964), in which he showed that subject to the side condition of unchanging Robbins coefficient of linkage disequilibrium the mean fitness increased: Kimura mentioned Wright when considering the equilibria and Fisher when considering the increase of mean fitness.

As a result of these developments the wheel turned full circle and the name of Fisher's Theorem became attached to the idea of increasing mean fitness in the discrete-generation model. Thus Ewens (1969*b*) opened his *A generalized fundamental theorem of natural selection* as follows:

> A classical result (Fisher 1930) of population genetics is that if, in a random-mating population, the fitness of any individual depends on his genetic constitution at a *single* locus, then the mean fitness of that population increases (or at worst remains stable) with time. This is true for an arbitrary number of alleles at the locus and for arbitrary fitness values; (for the best proof of this theorem see Kingman 1961).

Ewens was here quoting Fisher (1930) in connection with a theorem about increasing mean viability at a single multi-allelic locus in the discrete-generation model, which was not proved until 1959, as we shall see in the next section. He went on 'Unfortunately, if it is supposed that fitness depends on the genetic constitution at two loci, Fisher's result no longer holds (Moran 1964)', though what Moran had actually claimed to have demonstrated was that 'Wright's conception' of an adaptive topography was untenable; he had not mentioned Fisher's Theorem.

A consequence of this shift in terminology was that when other writers began to contribute to the literature they inherited a false certainty about the theorem. Thus Karlin & Feldman (1970*a*), writing about the two-locus model, had no doubt what the theorem was:

[The Fundamental Theorem of Natural Selection] states that when, in a random mating population, the fitness of any individual depends on his genotype at a single locus then the mean fitness of the population increases from generation to generation with constancy occurring only at the equilibria of the gene frequencies.

Moreover, after making some remarks about attempts to find conditions under which this 'theorem' might be true for two loci, these authors wrote (Karlin & Feldman, 1970*b*) 'it appears that these attempts to save the fundamental theorem are quite pointless'. Many other instances of the misappropriation of the name of Fisher's theorem for something that, though related in a way not then clearly understood, was certainly different, may be found in Volume 1 of *Theoretical Population Biology* (1970).

Whilst these developments prompted by consideration of the two-locus model were taking place, Li had by 1967(*a*, *b*) come to regard the Fundamental Theorem for a discrete-generation model as asserting that the mean fitness (standardized) would increase by an amount equal to the genic variance, and he considered any departure from this as vitiating the theorem. Thus in the case of a single locus with two alleles 'Fisher's theorem, if applied to discrete-generation model, is true only when the three [genotypic] fitness values are additive' (Li, 1967*a*). In the only piece I have hitherto published on the Fundamental Theorem (Edwards, 1967) I protested uncertainly and ineffectively.

1969 and 1970 saw the publication of three major textbooks in mathematical genetics, Ewens's (1969*a*) *Population Genetics*, Jacquard's (1970) *Structures génétiques des populations*, and Crow & Kimura's (1970) *An Introduction to Population Genetics Theory*.

Chapter 3 of Ewens (1969*a*) was entitled *The Fundamental Theorem of Natural Selection*, and discussed the properties of the mean fitness for the standard discrete-generation model. Not until the last section was the Fundamental Theorem itself mentioned:

The results of the previous sections indicate that the fitness of any population will usually increase, the rate of increase bearing a close relationship to the additive part of the genetic variance. Such a result has been called (Fisher, 1930) the 'Fundamental Theorem of Natural Selection', and has had considerable influence.

It is interesting to see how Ewens, who twenty years later was to explain Price's interpretations of the Theorem so clearly (Ewens, 1989), is here cautious in his choice of words, a caution that was to evaporate by the time of his next book (Ewens, 1979; see below).

Jacquard (1970) arrived at the 'Fundamental Theorem' by missing the second-order term in his account of the change in mean fitness at a multiallelic locus assuming random mating, as I pointed out in my review (Edwards, 1972*a*); in the English edition (Jacquard, 1974) the error was corrected and the Fundamental Theorem considered to be an approximation only.

Crow and Kimura's (1970) textbook naturally relied heavily on their earlier papers described above. They too showed caution in discussing Fisher's precise meaning (having found his treatment 'recondite'), but they did not make the advance in interpretations involving partial fitness which Price was to make two years later.

A fourth major textbook of the era was Elandt-Johnson's (1971) ***Probability Models and Statistical Methods in Genetics*** which, though it concentrated more on statistics than evolution, ventured a formal statement of the Fundamental Theorem:

> THEOREM. If a population satisfies conditions (10.18) [one locus, multiple alleles, constant fitnesses, random mating, discrete generations], and additionally the fitnesses of the genotypes... are additive functions of the genes..., then the rate of increase in mean fitness per generation is proportional to the additive genetic variance with respect to fitness.

This corresponded to Li's view described above; in my review (Edwards, 1972*b*) I protested at Elandt-Johnson's 'misappropriation of the name [of the Fundamental Theorem]' and added for good measure that this was 'an unfortunate characteristic of much modern writing in mathematical genetics'.

Thus in the decade 1960–70 the name 'Fundamental Theorem of Natural Selection' was usually used for a false 'theorem' which had grown out of Wright's concept of an adaptive topography but which the literature now associated with the name of Fisher, whose reputation was thereby adversely affected. Fisher died in 1962, and his last published comment on the matter was one of protest (Fisher, 1958*a*; see Edwards 1990):

> I have never indeed written about $\overline{W}$ and its relationships, and now that the alleged relationship has been brought to my attention, I must point out that the existence of such a "potential function" as that which Wright designates by $\overline{W}$, is not a general property of natural populations, but arises only from the special and restricted cases which Wright has chosen to consider. Selective tendencies are not, in general, analogous to what mechanicians describe as a conservative system of forces. To assume this property is one of the gravest faults of Wright's formulation.

But his protest, like the earlier one (1941), was in vain, and mathematical geneticists impressed by the discovery of fresh examples in which the mean fitness did not increase held them against 'the Fundamental Theorem' in spite of the fact that the first such example had been constructed by Fisher himself (1941) for the specific purpose of criticizing Wright's idea of an adaptive topography (see above). Sometimes, however, as in the quotation from Elandt-Johnson above, the name 'Fundamental Theorem' was used solely for the case of additive fitnesses, a theorem which at least had the virtue of being true even if very restricted.

As already mentioned in Section VI, in a paper delivered in 1971 (Edwards, 1971*a*; abstract only) I tried at least to disentangle the adaptive topography and the Fundamental Theorem, but it was a Herculean task in the absence of a clear understanding of the latter. That understanding was not available until Price (1972).

Perhaps the last major account of the theorem which did not have the possibility of relying on Price was that of Turner (1970). In a review *Changes in mean fitness under natural selection* he took Li's 1967 view about when the Theorem was exact, and regarded Kimura's (1958) treatment of the Theorem as 'brilliant' since it 'eliminates the flaws in Fisher's theorem' ('which is lacking mathematical rigour and [in which] the concept of fitness is hazy'). But at least Turner's last words were accurate:

The simple view of evolution that mean fitness will increase because the less fit genotypes die is not adequate because sexual reproduction shuffles the genes and the fitnesses of genotypes may change with changes in the population.

Since this was just the difficulty Fisher had intended his Fundamental Theorem to surmount forty years earlier, there was still room for progress.

With the advantage of hindsight we can now see that it should perhaps have been obvious that if (1) the Fundamental Theorem was exact, (2) the definition of genic variance on the 'right-hand side' of the theorem was not in dispute, and (3) the total increase in mean fitness (the 'left-hand side') in the simplest discrete models did not exactly equal the genic variance in general, then Fisher could not have meant the left-hand side to refer to the total increase in mean fitness. Perhaps it was the force of this logic which persuaded Price of the truth of the theorem (he does not relate), in which case he succeeded where so many had failed because he was the first to combine adequate technique with an absolute confidence that Fisher was, in some sense worth capturing, correct.

IX. INCREASING MEAN FITNESS IN A MULTIALLELIC MODEL

Whilst Crow, Kimura, and Kempthorne in the United States were making the attempts to understand the Fundamental Theorem described in the last section, a separate development stemming from the theorem was taking place in Fisher's Department of Genetics at Cambridge University (see Edwards, 1990).

Owen (1954), a colleague of Fisher's, had considered the conditions for equilibrium in the discrete-generation, random-mating, three-allele model, and his graduate student Mandel together with a fellow-student Hughes (Mandel & Hughes, 1958) conjectured that for an arbitrary number of alleles 'the mean viability [fitness] increases each time the population undergoes the process of mating, but to prove this statement generally appears to be a matter of considerable mathematical difficulty'. (Their mathematics shows that they really meant 'increases each generation' and not just as a result of mating.) They were only able to prove the result for gene frequencies arbitrarily close to an equilibrium state. They held that the Fundamental Theorem did not constitute a proof since 'its derivation depends on a continuous time argument' (but see Ewens, 1989).

I have related the history of the proof of the conjecture elsewhere (Edwards, 1977, Section 4.8). In brief, the conjecture was stated to be true by Mandel (1959) and proved by Scheuer & Mandel (1959). Simultaneously it was under consideration by Mulholland & Smith (1959), a proof being obtained by Mulholland, and in the following year a third independent proof was supplied by Atkinson, Watterson & Moran (1960). Subsequently a simplified proof was given by Kingman (1961), a matrix formulation of which will be found in Edwards (1977).

None of these authors supposed that the result constituted a proof of the Fundamental Theorem, and Kingman in his definitive account did not even mention the Theorem. Nor was there any anticipation that a result so difficult to prove could be extended to more complex models.

There is unfortunately no record of any comment by Fisher on precisely how this work related to his Fundamental Theorem (Edwards, 1990); in 1971 Owen was adopting the conventional wisdom by using the name of the theorem to refer to straightforward increasing mean fitness (Arunachalam & Owen, 1971).

Li (1969) gave a neat formulation of the total change in the mean fitness at a multiallelie locus under random mating. In the same notation as Section III, he found this to be

$$\Delta w = v/w + \sum_{i,j} \Delta p_i \Delta p_j w_{ij},$$

thus identifying the difference between the total change Δw and the partial change δw. His derivation will also be found in Edwards (1977).

X. THE SECOND EDITION OF *THE GENETICAL THEORY* (1958)

The second edition of *The Genetical Theory* was published in 1958 (Fisher, 1958*b*), but Fisher had been thinking about a reprint since 1955 ('I do not like to call it a new edition', he wrote to E. B. Ford on 25 November; Bennett 1983). Price (1972) has described in detail the difficulties introduced in Chapter II through poor editing and proof-correction. The wording of the Fundamental Theorem itself remained unchanged, but Fisher added three paragraphs emphasizing the point about dominance at the end of the section containing the theorem, paragraphs which Wright, incidentally, never remarked on (nor did Price, 1972, though Bennett, 1983, thought it 'should not be overlooked' since it 'explains how, in [the] Fundamental Theorem of Natural Selection, the evolutionary effects ascribable to the dominance component of the genotypic variance are credited to gene substitutions at other loci'):

> In addition to the genetic variance of any measurable character there exists, as has been seen, a second element comprised in the total genotypic variance, due to the heterozygote being in general not equal to the means of the two corresponding homozygotes. This component, ascribable to dominance, is also in a sense capable of exerting evolutionary effects....
>
> This indirect and conditional factor in selection seems to have been able to produce effects of considerable importance....
>
> Such effects ascribable to the dominance component of the genotypic variation are not in reality additional to the evolutionary changes accounted for by the fundamental theorem; for in that theorem they are credited to the gene-substitutions needed, for example, to develop bigger or brighter flowers....

XI. THE RECEPTION OF PRICE'S VIEW OF THE THEOREM (1972)

In this review I have repeatedly referred to G. R. Price's account of the Fundamental Theorem (1972), but at the time it was almost completely ignored. Indeed, from personal knowledge I can only name four people who looked favourably on Price's interpretation prior to Ewens (1989). The first was C. A. B. Smith, the joint editor of the journal *Annals of Human Genetics* in which it was published, whom Price thanks for his help. Smith had just published a related paper with Price (Price & Smith, 1972) whose 'secondary purpose' was 'to prepare the way for an explanation of Fisher's "fundamental theorem" in a future paper'. Professor Smith informs me that he 'half accepted' Price's explanation. The second was J. H. Bennett who wrote (Bennett, 1983) 'In an interesting article... Price came close, I believe, to Fisher's meaning in a number of respects...'. The third was D. S. Falconer, who referred to it in the second edition of his book *Introduction to Quantitative Genetics* (1981) which I discuss below, and the fourth was me. In April 1973 I wrote to A. Robertson, of Edinburgh, 'After thinking about Price's *Ann. Hum. Genet.* paper I believe we have all been too naive in our interpretations. I now see dimly what Fisher was driving at, and his 1941 paper in particular is at last beginning to make sense'. 'Price has got closer than anyone in divining Fisher's thoughts' I wrote to Ewens in 1976, enclosing a typescript in which I had written that Price's account 'must be the starting point for any future discussion of the theorem' (see Edwards, 1990).

This was not an auspicious start for a reappraisal of the Fundamental Theorem.

For the most part population genetics theory carried on as if nothing had happened. Karlin (1975) continued his earlier use of the name 'Fundamental Theorem' (quoted above in Section VIII from Karlin & Feldman, 1970*a*), complete with a section entitled 'The Nonvalidity of the Fundamental Theorem of Natural Selection'. Li updated his 1955 book into *First Course in Population Genetics* (Li, 1976) without mentioning Price's paper (though he did refer to other papers in the *Annals of Human Genetics* by Price). His 1967 interpretation of the Fundamental Theorem (see Section VIII) replaced that of 1955. In my *Foundations of Mathematical Genetics* (Edwards, 1977) I declined to discuss the Theorem (see the third quotation at the head of this review). Ewens replaced his 1969 book with *Mathematical Population Genetics* (Ewens, 1979) but did not change his view of the Fundamental Theorem, and did not mention Price. He wrote:

> We make a final remark on the Fundamental Theorem, taken to be the statement that the rate of increase of mean fitness is equal to the additive genetic variance in fitness. Over the years many purported general "proofs" of this theorem have been given, some claiming the status of an authorized version, many mutually contradictory. In the previous section we have shown that for continuous-time models the theorem is exactly true only in certain special cases. For the discrete-time models we have so far considered the increase in mean fitness is not generally identical to the additive genetic variance. We show in the next section that for discrete-time models involving two loci the mean fitness can decrease. The value of such "proofs" is thus somewhat in doubt.

Happily, ten years later Ewens (1989) was to be the catalyst to the acceptance of Price's 'authorized version'.

Other books published between 1972 and 1989 which mentioned the Fundamental Theorem and might have noted Price's interpretation, but did not, include Jameson (1977), Nagylaki (1977), Spiess (1977), Roughgarden (1979), Hartl (1980), Mayo (1983), Sober (1984), and Hofbauer & Sigmund (1988). Charlesworth (1980), however, did at least refer to Price (1972), amongst others, for 'Other interpretations of the Malthusian parameter approach to age-structured models'. Bulmer's (1980) *The Mathematical Theory of Quantitative Genetics* does not mention the Fundamental Theorem explicitly, but derives the approximate expression for the total increase in mean fitness.

Of the above books, Roughgarden's *Theory of Population Genetics and Evolutionary Ecology* (1979) may serve as an example of an influential text from which students will continue to learn a view of history which should be discarded. Not only does it not refer to Price (1972), but even more remarkably it does not even refer to *The Genetical Theory of Natural Selection* itself. Instead, it states that 'The first general connection between natural selection and $\bar{w}$ was established by Sewall Wright', and later that 'Another general theorem about natural selection is Fisher's "fundamental theorem of natural selection"'. This is held to be the continuous-time version of the growth-rate theorem mentioned in Section I. Finally, with weak selection 'we can extend Fisher's fundamental theorem of natural selection to random mating diploid populations'.

The notable exception to these strictures is Falconer (1981) who wrote 'A proof of the theorem and an explanation of why it has caused so much difficulty is given by Price (1972)' and then proceeded to use standard selection-differential mathematics to derive the result that the change in mean fitness in one generation is equal to the genic variance, apparently exactly. Since the result is certainly not exact even with only two alleles at a locus and random mating (except in the special case of no dominance) an explanation is required, and it turns out that the change in mean fitness involved in the selection mathematics is precisely the partial change envisaged by Price. The reason for this is best seen from Falconer's later account (1985): the change in the mean fitness is computed from the regression of fitness on gene frequency, but the dependence of this regression itself on the gene frequency is ignored. Falconer has evidently provided what Price had in mind when he wrote:

> (It may be noted that if our aim were merely to derive the theorem rather than to explain how Fisher derived it, the derivation can be accomplished far more simply if we work entirely with regression coefficients and covariances without using Fisher's special 'average excess' and 'average effect' variables.)

Professor Falconer tells me (December 1993) that he now thinks his treatment is probably the same as one he heard from I. M. Lerner in about 1948.

In a recent long commentary to a new edition of Haldane's *The Causes of Evolution* (1990), E. G. Leigh, Jr mentions both Price (1972) and Ewens (1989) but still computes the change in mean fitness without recognizing that it is not only the gene frequency which, in his formulation, changes from one generation to the next. Like Falconer, he thus appears to obtain an exact expression for the total change in mean fitness which is in fact only the partial change as obtained by Ewens.

XII. CONCLUSION

Fisher's Fundamental Theorem of Natural Selection is important for three reasons. First is the historical fact (Sections V and VI) that it led Wright to the idea of an adaptive topography in gene-frequency space which has dominated so much thinking in evolutionary biology. We have seen how this involved a maximisation principle of doubtful validity with which Fisher's name became associated in spite of his protests. In *Nature* in 1986 Felsenstein reviewed Provine's biography of Wright (Provine, 1986) and I challenged his view that it was Fisher, rather than Wright, who had espoused the idea of searching for the quantity evolution would maximize (Edwards, 1987). Later Felsenstein (1989), in reviewing another book this time in *Science*, associated both men with the idea, and we submitted a further interchange of views to *Science*, but they were never published. Part of Felsenstein's reply to my remarks, which I quote with his permission, seems admirably to sum up the state of affairs:

> Edwards is no doubt right that Fisher [and Haldane] were fine mathematicians, and that Fisher did not draw from his Fundamental Theorem of Natural Selection the implication that natural selection would maximise mean population fitness. But historically the more important point is what lessons readers drew from their work....

Whatever Fisher thought of fitness maximization, I can testify that population geneticists of the 1960s thought that his Fundamental Theorem was a theorem that guaranteed that change of mean fitness, being expressed in terms of a variance, could not be negative. No doubt we misinterpreted his work, but the point is that this is how we interpreted it. Fisher's influence on subsequent generations of researchers was not to inspire skepticism of fitness maximisation; they had to rediscover its limitations for themselves. The counterexample found in Fisher's 1941 paper is relevant to a discussion of his intellectual system, but it happens to have been of no influence on subsequent researchers. My review invoked Fisher as the patron saint of this belief; this he was, even if he may not have been a believer himself.

Nagylaki (1991) went further: 'Indeed, it is fortunate that Fisher has apparently been misinterpreted, for the canonical interpretation has led to a rich and illuminating literature'.

Secondly, and of more permanent value, is the fact that the theorem gives mathematical precision to the previously vague notion 'that in species in which a higher proportion of the total variance is ascribable to genetic causes, the effective selection will be more intense than in species in which the variance is to a larger extent ascribable to environmental variations' (Section II). There is no point in criticizing Fisher for not having achieved more, for we now understand (what he had always understood) that an exact statement is only possible in respect of the partial change in mean fitness. As Price (1972) remarked, 'We may note that the 'fundamental theorem' is very probably the most that anyone has yet been able to say correctly about evolutionary increase in fitness under general and realistic natural conditions. Thus the theorem is by no means a trivial, uninteresting result'.

In a very clear account which can profitably be read as a sequel to the present review, Frank & Slatkin (1992) explain the implications of the new interpretation of the Fundamental Theorem for evolutionary biology. They ask: 'What did Fisher really mean by the Fundamental Theorem? Is Fisher's interpretation of the Fundamental Theorem useful? and Why was Fisher misinterpreted, even though he stated on many occasions that he was not talking about the average fitness of a population?'. In answer to their second question they write:

> It is a matter of taste whether a particular partition of evolutionary change is useful; all of the component parts must ultimately sum to the same total change. We believe that Fisher's partition is useful – indeed fundamental. To support this conclusion, we analyse an interesting biological problem, the evolution of clutch size, and show that Fisher's view leads to a significantly deeper understanding of the problem.

They add, in conclusion, 'What one gains from the Fundamental Theorem ... is a stronger ecological sense of the relationship between natural selection and evolution'. Not everyone agrees, however: Nagylaki (1991), whilst accepting that Price and Ewens 'have argued cogently', believed that 'the biological significance and analytical usefulness of the partial change in the mean fitness remain to be demonstrated. Thus, under the Price-Ewens interpretation, Fisher's assertion is a theorem but does not seem to provide any fundamental insight into evolution under natural selection' [the quotation earlier in this Section follows at this point]. Though Frank and Slatkin thank Nagylaki, amongst others, for his comments on their paper, they do not record what they were. Ewens (1989) also recorded a 'negative assessment of the theorem as a biological statement'. Crow, whose record of thoughtful comment on the theorem now extends over forty years, is not wholly persuaded by Ewens (1989): 'We'll never know exactly what Fisher had in mind', but 'To me, even if Fisher's fundamental theorem is inexact and incomplete, this succinct statement captures the essence of the way selection works, and encapsulates a great deal of evolutionary insight in a simple expression' (Crow, 1990).

The third reason why the theorem is important is that correctly interpreted it has a considerable potential for future developments in mathematical population genetics. To discuss these would be beyond the scope of the present review, and the reader is referred to Ewens (1992) *An optimizing principle of natural selection in evolutionary population genetics* for an entry into the recent literature.

It will probably be many years before the name of the Fundamental Theorem of Natural Selection is returned to its originator and ceases to be attached to other related formulations. The magnitude of the task can be gauged by the appearance in the very journal which carried Ewens (1989) of a paper (Findlay, 1990) which opens: 'Fisher's fundamental theorem of natural selection states that the rate of change in the mean biological fitness of a population is equal to the additive genetic variance in fitness. Since variances are always positive, this implies that mean fitness never decreases,...'. Although the author quotes Ewens (1989) – as Ewens (1990) – his treatment of the theorem completely ignores the content of Ewens's paper.

Nagylaki (1993) has tentatively suggested that the theorem should be called the Fisher–Price–Ewens theorem, but that seems likely to confuse future generations: let the theorem (and only the theorem) carry the name its originator gave it.

I hope the present review will contribute to the process of reassessment of the Fundamental Theorem started by Price and Ewens, and that Fisher's original intentions will soon find their way into the textbooks to be discussed on their own merits.

XIII. REFERENCES

ARUNACHALAM, V. & OWEN, A. R. G. (1971). *Polymorphisms with Linked Loci.* Chapman & Hall, London.

ATKINSON, F. V., WATTERSON, G. A. & MORAN, P. A. P. (1960). A matrix inequality. *Quarterly Journal of Mathematics* (2) **11**, 137–140.

BENNETT, J. H. (1956). Review of Li (1955). *Heredity* **10**, 275–278.

BENNETT, J. H. (1983). (Ed.) *Natural Selection, Heredity, and Eugenics. Including selected correspondence of R. A. Fisher with Leonard Darwin and others.* Clarendon Press, Oxford.

BOX, J. F. (1978). *R. A. Fisher: the Life of a Scientist.* Wiley, New York.

BULMER, M. G. (1980). *The Mathematical Theory of Quantitative Genetics.* Clarendon Press, Oxford.

CHARLESWORTH, B. (1980). *Evolution in Age-Structured Populations.* Cambridge University Press.

CROW, J. F. (1955). General theory of population genetics: synthesis. *Cold Spring Harbor Symposium on Quantitative Biology* **20**, 54–59.

CROW, J. F. (1990). Fisher's contributions to genetics and evolution. *Theoretical Population of Biology* **38**, 263–275.

CROW, J. & KIMURA, M. (1956). Some genetic problems in natural populations. *Proceedings of the Third Berkeley Symposium of Mathematical Statistics and Probability* **4**, 1–22.

CROW, J. & KIMURA, M. (1970). *An Introduction to Population Genetics Theory.* Harper & Row, New York.

DAHLBERG, G. (1947). *Mathematical Methods for Population Genetics.* Karger, Basle.

DARWIN, C. (1859). *On the Origin of Species by means of Natural Selection.* Murray, London.

DARWIN, C. G. (1930). Natural selection. (Review of Fisher, 1930.) *Eugenics Review* **22**, 127–130.

DAWKINS, R. (1976). *The Selfish Gene.* Oxford University Press.

EDDINGTON, A. S. (1928). *The Nature of the Physical World.* Cambridge University Press.

EDWARDS, A. W. F. (1967). Fundamental theorem of natural selection. *Nature* **215**, 537–538.

EDWARDS, A. W. F. (1969). Review of Ewens (1969). *Heredity* **24**, 672–673.

EDWARDS, A. W. F. (1971*a*). Fisher's Fundamental Theorem of Natural Selection and Wright's Adaptive Topography: a controversy revisited. (Abstract) *Biometrics* **27**, 762–763.

Edwards, A. W. F. (1971*b*). Review of Wright (1969). *Heredity* **26**, 332–338.
Edwards, A. W. F. (1972*a*). Review of Jacquard (1970). *Heredity* **28**, 151–152.
Edwards, A. W. F. (1972*b*). Review of Elandt-Johnson (1971). *Heredity* **29**, 109–111.
Edwards, A. W. F. (1974). On Kimura's maximum principle in the genetical theory of natural selection. *Advances in Applied Probability* **6**, 1–3.
Edwards, A. W. F. (1977). *Foundations of Mathematical Genetics*. Cambridge University Press.
Edwards, A. W. F. (1978). Review of *Stochastic Models in Population Genetics* (ed. W-H. Li). *Heredity* **41**, 239–247.
Edwards, A. W. F. (1987*a*). Fisher, Ronald Aylmer. *The New Palgrave : A Dictionary of Economics*, Macmillan, London. Vol. II, pp. 376–377.
Edwards, A. W. F. (1987*b*). Evolution and optimisation. *Nature* **326**, 10.
Edwards, A. W. F. (1989). Venn diagrams for many sets. *New Scientist* **121**, 51–56.
Edwards, A. W. F. (1990). Fisher, $\bar{W}$, and the Fundamental Theorem. *Theoretical Population Biology* **38**, 276–84.
Edwards, A. W. F. (1991). How to iron a hypercube. *Mathematical Gazette* **75**, 433–436.
Elandt-Johnson, R. C. (1971). *Probability Models and Statistical Methods in Genetics*. Wiley, New York.
Ewens, W. J. (1969*a*). *Population Genetics*. Methuen, London.
Ewens, W. J. (1969*b*). A generalized fundamental theorem of natural selection. *Genetics* **63**, 531–537.
Ewens, W. J. (1979). *Mathematical Population Genetics*. Springer, Berlin.
Ewens, W. J. (1989). An interpretation and proof of the Fundamental Theorem of Natural Selection. *Theoretical Population Biology* **36**, 167–180.
Ewens, W. J. (1992). An optimizing principle of natural selection in evolutionary population genetics. *Theoretical Population Biology* **42**, 333–346.
Falconer, D. S. (1981). *Introduction to Quantitative Genetics* (2nd ed.). Longman, London.
Falconer, D. S. (1985). A note on Fisher's 'average effect' and 'average excess'. *Genetical Research* **46**, 337–347.
Falconer, D. S. (1989). *Introduction to Quantitative Genetics* (3rd ed.). Longman, London.
Felsenstein, J. (1986). Evolution in the twentieth century (Review of Provine, 1986, and Wright, 1986). *Nature* **324**, 175.
Felsenstein, J. (1989). Mathematics vs. evolution (book review). *Science* **246**, 941–942.
Findlay, S. (1990). Fundamental theorem of natural selection in biocultural populations. *Theoretical Population Biology* **38**, 367–384.
Fisher, R. A. (1915). Frequency distribution of the values of the correlation coefficient in samples from an indefinitely large population. *Biometrika* **9**, 507–521.
Fisher, R. A. (1918*a*). The correlation between relatives on the supposition of Mendelian inheritance. *Transactions of the Royal Society of Edinburgh* **52**, 399–433.
Fisher, R. A. (1918*b*). The causes of human variability. *Eugenics Review* **10**, 213–20.
Fisher, R. A. (1920). A mathematical examination of the methods of determining the accuracy of an observation by the mean error, and by the mean square error. *Monthly Notices of the Royal Astronomical Society* **80**, 758–770.
Fisher, R. A. (1922*a*). On the dominance ratio. *Proceedings of the Royal Society of Edinburgh* **42**, 321–41.
Fisher, R. A. (1922*b*). Darwinian evolution of mutations. *Eugenics Review* **14**, 31–34.
Fisher, R. A. (1922*c*). On the mathematical foundations of theoretical statistics. *Philosophical Transactions of the Royal Society*, A **222**, 309–368.
Fisher, R. A. (1930). *The Genetical Theory of Natural Selection*. Clarendon Press, Oxford. 2nd ed: Dover Publications, New York, 1958.
Fisher, R. A. (1932). The evolutionary modification of genetic phenomena. *Proceedings of the Sixth International Congress of Genetics* **1**, 165–172.
Fisher, R. A. (1936). The measurement of selective intensity. *Proceedings of the Royal Society of London* B, **121**, 58–62.
Fisher, R. A. (1941). Average excess and average effect of a gene substitution. *Annals of Eugenics* **11**, 53–63.
Fisher, R. A. (1958*a*). Polymorphism and natural selection. *Bulletin of the International Statistical Institute* **36**, 284–288; *Journal of Ecology* **46**, 289–293.
Fisher, R. A. (1958*b*). *The Genetical Theory of Natural Selection* (2nd ed.). Dover, New York.
Fisher, R. A. (1971, 1972, 1973, 1974). *Collected Papers* (5 vols.), Bennett, J. H. (ed.). University of Adelaide.
Fisher, R. A. & Ford, E. B. (1926). Variability of species. *Nature* **118**, 515–516.
Fisher, R. A. & Ford, E. B. (1928). The variability of species in the Lepidoptera, with reference to abundance and sex. *Transactions of the Royal Entomological Society of London* **76**, 367–379.
Ford, E. B. (1938). *The Study of Heredity*. Thornton Butterworth, London.
Frank, S. A. & Slatkin, M. (1992). Fisher's fundamental theorem of natural selection. *Trends in Ecology and Evolution* **7**, 92–95.

HALDANE, J. B. S. (1931*a*). A mathematical theory of natural selection. Part VIII. Metastable populations. *Proceedings of the Cambridge Philosophical Society* **27**, 137–142.

HALDANE, J. B. S. (1931*b*). Mathematical Darwinism: A discussion of *The Genetical Theory of Natural Selection*. *Eugenics Review* **23**, 115–117.

HALDANE, J. B. S. (1931*c*). Review of Fisher (1930). *Mathematical Gazette* **15**, 474–475.

HALDANE, J. B. S. (1932, 1966, 1990). *The Causes of Evolution*. Longmans, London (1932); Cornell University Press, Ithaca (1966); Princeton University Press (1990).

HALDANE, J. B. S. (1966). *See* Haldane (1932).

HALDANE, J. B. S. (1985). The mathematical theory of natural selection. Calcutta Mathematical Society: *The Golden Jubilee Commemoration Volume* (*1958-59*), 99–103.

HALDANE, J. B. S. (1990). *See* Haldane (1932).

HARTL, D. L. (1980). *Principles of Population Genetics*. Sunderland, Massachusetts: Sinauer.

HOFBAUER, J. & SIGMUND, K. (1988). *The Theory of Evolution and Dynamical Systems*. Cambridge University Press.

HOGBEN, L. (1933). *Nature and Nurture*. Allen & Unwin, London.

HOGBEN, L. (1946). *An Introduction to Mathematical Genetics*. Norton, New York.

JACQUARD, A. (1970). *Structures génétiques des populations*. Masson, Paris.

JACQUARD, A. (1974). *The Genetic Structure of Populations*. Springer, Berlin.

JAMESON, D. L. (ed.) (1977). *Evolutionary Genetics*. Dowden, Hutchinson & Ross, Stroudsburg, Pennsylvania.

KARLIN, S. (1975). General two-locus selection models: some objectives, results and interpretations. *Theoretical Population Biology* **7**, 364–398.

KARLIN, S. & FELDMAN, M. W. (1970*a*). Convergence to equilibrium of the two locus additive viability model. *Journal of Applied Probability* **7**, 262–271.

KARLIN, S. & FELDMAN, M. W. (1970*b*). Linkage and selection: Two locus symmetric viability model. *Theoretical Population Biology* **1**, 39–71.

KEMPTHORNE, O. (1957). *An Introduction to Genetic Statistics*. Wiley, New York.

KENDALL, D. G. (1990). Obituary: Andrei Nikolaevich Kolmogorov (1903–1987). *Bulletin of the London Mathematical Society* **22**, 31–100.

KEYNES, J. M. (1921). *A Treatise on Probability*. Macmillan, London.

KIMURA, M. (1958). On the change of population fitness by natural selection. *Heredity* **12**, 145–167.

KIMURA, M. (1965). Attainment of quasi linkage equilibrium when gene frequencies are changing by natural selection. *Genetics* **52**, 875–890.

KINGMAN, J. F. C. (1961). A mathematical problem in population genetics. *Proceedings of the Cambridge Philosophical Society* **57**, 574–582.

KOJIMA, K-I. & KELLEHER, T. M. (1961). Changes in mean fitness in random mating populations when epistasis and linkage are present. *Genetics* **46**, 527–540.

LEIGH, E. G., JR. (1990). Afterword to Haldane (1990).

LEWONTIN, R. C. (1965). Selection in and of populations. In Moore, J. A. (ed.). *Ideas in Modern Biology. Proceedings of the XVI International Congress of Zoology* **6**, 299–311.

LEWONTIN, R. C. & KOJIMA, K-I. (1960). The evolutionary dynamics of complex polymorphisms. *Evolution* **14**, 458–472.

LEWONTIN, R. C. & WHITE, M. J. D. (1960). Interaction between inversion polymorphisms of two chromosome pairs in the grasshopper, *Moraba scurra*. *Evolution* **14**, 116–129.

LI, C. C. (1955). *Population Genetics*. University of Chicago Press.

LI, C. C. (1967*a*). Fundamental theorem of natural selection. *Nature* **214**, 505–506.

LI, C. C. (1967*b*). Genetic equilibrium under selection. *Biometrics* **23**, 397–484.

LI, C. C. (1969). Increment of average fitness for multiple alleles. *Proceedings of the National Academy of Science, Washington* **62**, 395–398.

LI, C. C. (1976). *First Course in Population Genetics*. Boxwood, Pacific Grove, California.

MALÉCOT, G. (1948). *Les mathématiques de l'hérédité*. Masson, Paris.

MALÉCOT, G. (1969). *The Mathematics of Heredity*. Freeman, San Francisco.

MANDEL, S. P. H. (1959). The stability of a multiple allelic system. *Heredity* **13**, 289–302.

MANDEL, S. P. H. & HUGHES, I. M. (1958). Change in mean viability at a multiallelic locus in a population under random mating. *Nature* **182**, 63–64.

MAYNARD SMITH, J. & PRICE, G. R. (1973). The logic of animal conflict. *Nature* **246**, 15–18.

MAYO, O. (1983). *Natural Selection and its Constraints*. Academic Press, London.

MORAN, P. A. P. (1962). *The Statistical Processes of Evolutionary Theory*. Clarendon Press, Oxford.

MORAN, P. A. P. (1964). On the nonexistence of adaptive topographies. *Annals of Human Genetics* **27**, 383–393.

MULHOLLAND, H. P. & SMITH, C. A. B. (1959). An inequality arising in genetical theory. *American Mathematical Monthly* **66**, 674–683.

NAGYLAKI, T. (1977). *Selection in One- and Two-Locus Systems*. Springer, Berlin.

NAGYLAKI, T. (1991). Error bounds for the fundamental and secondary theorems of natural selection. *Proceedings of the National Academy of Science, Washington* **88**, 2402–2406.

NAGYLAKI, T. (1993). The evolution of multilocus systems under weak selection. *Genetics* **134**, 627–647.

OWEN, A. R. G. (1954). Balanced polymorphism of a multiple allelic series. *Proceedings of the Ninth International Congress of Genetics, Caryologia*, Suppl. to **6**, 1240–1241.

PRICE, G. R. (1972). Fisher's 'fundamental theorem' made clear. *Annals of Human Genetics* **36**, 129–140.

PRICE, G. R. & SMITH, C. A. B. (1972). Fisher's Malthusian parameter and reproductive value. *Annals of Human Genetics* **36**, 1–7.

PROVINE, W. B. (1985). The R. A. Fisher – Sewall Wright controversy. *Oxford Surveys in Evolutionary Biology* **2** (ed. R. Dawkins & M. Ridley), pp. 197–219. Oxford University Press.

PROVINE, W. B. (1986). *Sewall Wright and Evolutionary Biology*. University of Chicago Press.

ROUGHGARDEN, J. (1979). *Theory of Population Genetics and Evolutionary Ecology: An Introduction*. Macmillan, New York.

SCHEUER, P. A. G. & MANDEL, S. P. H. (1959). An inequality in population genetics. *Heredity* **13**, 519–524.

SMITH, A. (1776). *An Inquiry into the Nature and Causes of the Wealth of Nations*. Strahan & Cadell, London.

SOBER, E. (1984). *The Nature of Selection*. MIT Press, Cambridge, Massachusetts.

SPIESS, E. B. (1977). *Genes in Populations*. Wiley, New York.

TURNER, J. R. G. (1970). Changes in mean fitness under selection. In *Mathematical Topics in Population Genetics* (ed. K-I. Kojima), pp. 32–78. Springer, Berlin.

WRIGHT, S. (1930). Review of Fisher (1930). *Journal of Heredity* **21**, 349–356.

WRIGHT, S. (1931). Evolution in Mendelian populations. *Genetics* **16**, 97–159.

WRIGHT, S. (1932). The roles of mutation, inbreeding, crossbreeding and selection in evolution. *Proceedings of the Sixth International Congress of Genetics* **1**, 356–366.

WRIGHT, S. (1935). Evolution of populations in approximate equilibrium. *Journal of Genetics* **30**, 257–266.

WRIGHT, S. (1939). *Statistical Genetics in Relation to Evolution*. Hermann, Paris.

WRIGHT, S. (1940). The statistical consequences of Mendelian heredity in relation to speciation. *The New Systematics* (ed. J. S. Huxley), pp. 161–183. Oxford University Press.

WRIGHT, S. (1949). Adaptation and selection. In *Genetics, Paleontology, and Evolution* (ed. G. L. Jepson, G. G. Simpson and E. Mayr). Princeton University Press.

WRIGHT, S. (1955). Classification of the factors of evolution. *Cold Spring Harbor Symposium on Quantitative Biology* **20**, 16–24D.

WRIGHT, S. (1969). *Evolution and the Genetics of Populations, Vol. 2: The Theory of Gene Frequencies*. University of Chicago Press.

WRIGHT, S. (1986). *Evolution: Selected Papers* (ed. W. B. Provine). Chicago University Press.

WRIGHT, S. (1988). Surfaces of selective value revisited. *American Naturalist* **131**, 115–123.

REFERENCE ADDED IN PROOF

BENNETT, J. H. (1956). Population genetics and natural selection. *Genetica* **28**, 297–307.

This paper, which seems to have been entirely overlooked since its publication, includes a section *The fundamental theorem of natural selection* in which Professor Bennett gives the clearest account of Fisher's theorem prior to the work of Price (1972).

Paper 141 (1995)

Assessing molecular phylogenies

Science 267, 253.

Assessing Molecular Phylogenies

David Hillis *et al.* attempt (*1*) to assess the performance of methods of phylogenetic analysis by numerical simulation studies. They compare five methods using a simple criterion—the probability of obtaining the correct evolutionary tree with a given amount of simulated data. In a related paper (*2*) Hillis and colleagues increased to ten the number of methods compared.

This approach is known to be unsatisfactory. It is easy to construct elementary examples in which such a "confidence" method, though giving the correct answer with a specified probability (usually 95%), is wrong in a particular case. Coining some memorable terms for the purpose, Hacking (*3*) contrasted good "before-trial evaluation" methods that are "usually right" with good "after-trial evaluation" methods that are "credible on each occasion of use." The estimation of evolutionary trees, like most estimation problems in biology, is an after-trial evaluation problem, and the use of before-trial criteria is an inappropriate way of judging the various methods. For more than 70 years the backbone of the appropriate after-trial statistical estimation theory has been known to be Fisher's method of maximum likelihood (*4*).

The sound logical basis of maximum likelihood (*5*) sets it apart from the other algorithmic methods tested. If the probability model is specified, maximum likelihood would be the preferred method in principle even if it did not lead to the correct answer with the highest probability (though there are theoretical reasons for not being surprised that it scores well on this criterion).

It was for these reasons that Cavalli-Sforza and I strove from the beginning to improve on our least-squares distance-matrix approach (*6*) by applying maximum-likelihood to a well-defined evolutionary model for continuous characters (*7*), and why I suggested using maximum-likelihood for the discrete-character case as well (*8*). Moreover, when we discussed the justification of our "method of minimum evolution" or "parsimony" (*9*), we did so not on the grounds which have since been advanced (*10*), but because we correctly expected it to be a good approximation to the maximum-likelihood solution (*7*).

The message from statistical theory is simple: You cannot simulate evolutionary trees without a probability model, and if you possess a probability model you should be performing efficient after-trial evaluation using the method of maximum likelihood. Simulation is valuable in assessing the robustness of a model, but not the suitability of an estimation procedure; modern statistical theory has solved that problem already.

A. W. F. Edwards
Gonville and Caius College, Cambridge, CB2 1TA, United Kingdom

REFERENCES

1. D. M. Hillis, J. P. Huelsenbeck, C. W. Cunningham, *Science* **264**, 671 (1994).
2. D. M. Hillis, J. P. Huelsenbeck, D. L. Swofford, *Nature* **369**, 363 (1994).
3. I. Hacking, *Brit. J. Philos. Sci.* **22**, 209 (1971).
4. R. A. Fisher, *Philos. Trans. R. Soc. London Ser. A* **222**, 309 (1922).
5. A. W. F. Edwards, *Likelihood* (Cambridge Univ. Press, Cambridge, 1972; Johns Hopkins Univ. Press, Baltimore, MD, 1992).
6. L. L. Cavalli-Sforza and A. W. F. Edwards, *Genet. Today: Proc. 11th Int. Congr. Genet. The Hague 1963* **3**, 923 (1964).
7. A. W. F. Edwards and L. L. Cavalli-Sforza, "Phenetic and phylogenetic classification," *Systematics Assoc. Publ. No. 6* (1964), p. 67, reprinted in *Cladistic Theory and Methodology*, T. Duncan and T. F. Stuessy, Eds. (Van Nostrand Reinhold, New York, 1985); L. L. Cavalli-Sforza and A. W. F. Edwards, *Evolution* **21**, 550 (1967); *Am. J. Hum. Genet. Suppl.* **19**, 233 (1967).
8. A. W. F. Edwards, *New Sci.* **30**, 438 (1966).
9. ——— and L. L. Cavalli-Sforza, *Heredity* **18**, 553 (1963); *Ann. Hum. Genet.* **27**, 104 (1963).
10. C.-B. Stewart, *Nature* **361**, 603 (1993).

16 June 1994; accepted 12 September 1994

Paper 146 (1996)

The origin and early development of the method of minimum evolution for the reconstruction of phylogenetic trees

Systematic Biology 45, 79–91.

THE ORIGIN AND EARLY DEVELOPMENT OF THE METHOD OF MINIMUM EVOLUTION FOR THE RECONSTRUCTION OF PHYLOGENETIC TREES

A. W. F. EDWARDS

Department of Community Medicine, University of Cambridge, Institute of Public Health, Forvie, Robinson Way, Cambridge CB2 2SR, England

Abstract.—The method of minimum evolution was introduced by Edwards and Cavalli-Sforza (1963, Heredity 18:553, Ann. Hum. Genet. 27:104–105) for the reconstruction of phylogenetic trees. Its relationship to the subsequently developed parsimony methods and the logical basis of the methods are discussed, with special reference to probability models. The minimum evolution method did not derive from Hennig's phylogenetic systematics but rather as an approximation to the maximum-likelihood solution for a model of random evolution. [Phylogenetic trees; minimum evolution; Darwin principle; Ockham's razor; parsimony.]

In *The Origin of Species*, Darwin (1859) replaced the hypothesis of the individual creation of each species with the hypothesis of a single origin of life followed by the evolution of new species. He did not appeal to Ockham's razor in support of the change and would probably have been surprised to have been told that his achievement was an example—perhaps the example *par excellence*—of the application of Ockham's razor. His argument, rather, was one of analogy with the way lineages in genealogies could be traced back to common ancestors.

Ockham's razor, sometimes now called the principle of parsimony, is traditionally associated with William of Ockham (ca. 1280–1349): "*Pluralitas non est ponenda sine necessitate*" (multiplicity ought not to be posited unnecessarily). In recent years this principle has sometimes been invoked by systematists to justify particular methods of phylogenetic systematics (for reviews, see Sober, 1988, and Stewart, 1993). In this context, the principle manifests itself as what we might call *the hypothesis of the single origin*, whether of life itself or of a particular character. Darwin drew his first evolutionary tree with a single root in 1837, and the principle that the hypothesis of a single origin is to be preferred where possible can be called *the Darwin principle*. Thus, if a particular character appears the same or very similar in two species, the hypothesis of a common origin for this character shold be preferred over the hypothesis of two independent origins and parallel selection, unless of course other evidence overwhelms it. In *The Origin* (Darwin, 1859:13), Darwin wrote, "Perhaps the correct way of viewing the whole subject would be, to look at the inheritance of every character whatever as the rule, and non-inheritance as the anomaly." In 1862, he wrote, "Is it not a more simple and intelligible view that all Orchids owe what they have in common to descent . . . ?" (in Cronin, 1991:29). The Darwin principle seems so natural that phylogenetic systematists have usually assumed it without discussion, but occasionally there have been explicit appeals to Ockham's razor.

In 1963, L. L. Cavalli-Sforza and I suggested the following principle of minimum evolution for estimating the shape and dimensions of a phylogenetic tree: "The most plausible estimate of the evolutionary tree is that which invokes the minimum nett [*sic*] amount of evolution" (Edwards and Cavalli-Sforza, 1963b:553).

Stewart (1993:604) claimed that the principle of minimum evolution was derived from the principal of parsimony; she stated that "By the criterion of parsimony, the 'best' or most parsimonious tree is the one that requires the fewest total events (Cavalli-Sforza and Edwards, 1967; Farris, 1970)."

The purpose of the present review is to correct the impression that Cavalli-Sforza and I based our minimum-evolution method on the principle of parsimony and to recall the actual basis of our argument, which was statistical. The tendency to suppose that the principle of minimum evolution is based on Ockham's razor has been exacerbated by the habit of referring to them both as parsimony. There has even been a recent tendency to refer to any minimizing principle (even the method of least squares) as parsimonious, which is to destroy the usefulness of the word by giving it too wide an application. To avoid confusing such diverse ideas, I shall refrain from using these words as far as possible.

The Darwin Principle

Ockham's razor invites us to prefer the hypothesis of the single origin, but the rise of probabilistic thinking in the 18th century rendered this explanation for the preference obsolete. The probabilistic revolution applied the calculus of probability, invented in connection with games of chance, to the problem of the observational errors of astronomy and geodesy. The failure of a hypothesis to account exactly for observational data was no longer seen as a fault of either the hypothesis or the data but as a measurable reflection of the random component in observations. "*Anceps fortuna aequitate rationis reprimitur,*" Pascal wrote in 1654: "Proper calculation masters fickle fortune" (Edwards, 1987:79). People still wanted simple hypotheses, but they were now probabilistic ones and needed to be judged not only by simplicity but by a definite measure of how well they explained the data, i.e., by likelihood (Fisher, 1921).

Probabilistic reasoning leads naturally to the Darwin principle. Suppose we hear that a friend has borne male twins. Are they identical? If so, the probability that they are both boys is 0.5, but if they are not identical the probability is 0.25 (assuming that boys occur with probability 0.5). Setting aside for the moment any question of the prior probabilities, the likelihood ratio in favor of the twins being identical is therefore 2:1. The hypothesis of the single origin of the male sex is better supported than the hypothesis of two separate origins. Now suppose that the character about which we are told is not sex but a much rarer genetic character segregating in families with low frequency, p. The twins both have this character. Are they identical?

This time the probability that they both possess the character is p if they are identical but p^2 if they are not. The likelihood ratio in favor of identical twins is therefore $p{:}p^2$ or $1{:}p$, which is large and convincing if p is small. The more improbable the event that has been observed, the greater the support for the hypothesis of a single origin. The word *likelihood* is here used in its technical statistical sense as defined by Fisher (1921) (see Edwards [1972, 1992] for a full account).

In this example, it would have been possible to conduct the argument entirely in terms of probability by virtue of the fact that there is a known prior probability that twins are identical (monozygotic). This probability is about 0.25, so that even without evidence of the sexes or other genetic characteristics of the newborn twins, it is possible to hypothesize that they are not identical and to be correct about 75% of the time. Incorporating the data is then a simple exercise in conditional probability. With probability p for the genetic character, the probability ratio in favor of the hypothesis of the single origin (monozygosity) is found by multiplying 1:3 (the prior probability ratio) by $1{:}p$ (the likelihood ratio) to give $1{:}3p$ (the posterior probability ratio). The conditional probability of monozygosity is thus $1/(1 + 3p)$, which approaches 1 as p approaches zero. Such a manipulation of probabilities is a particularly simple example of the use of Bayes's theorem.

In the case of evolutionary trees, however, one cannot attach prior probabilities to the creationist hypothesis and to Darwin's alternative hypothesis of a single origin. The hypotheses must be judged by likelihood alone, and, as with the twins, the single origin of a character will have the higher likelihood, the magnitude increasing as the rarity of the character increases (and its frequency decreases). What I have called the Darwin principle thus arises directly from long-accepted ideas about statistical inference. Jefferys and Berger (1992:71) reviewed the Bayesian explication of Ockham's razor, but they conceded that the likelihood ratio (their "Bayes factor" *B*) is the linchpin of the argument: "The point here is that it may suffice to consider only *B*; the Bayes factor may well answer the question without the need to formally involve the prior odds."

The Darwin principle was, of course, in use in connection with human similarities millenia before Darwin. No one ever suggested that each individual was a special creation like a pre-Darwinian species (souls were another matter), and inferring relationship from similarity of appearance must always have reflected a subconscious desire to avoid having to account independently for similar facial and other features. Darwin himself, in a rare mention of the "doctrine of chances," was more explicit (1859:12–13), using the likelihood argument for the complementary problem of inferring inheritance given a known relationship:

> When a deviation appears not unfrequently, and we see it in the father and child, we cannot tell whether it may not be due to the same original cause acting on both; but when amongst individuals, apparently exposed to the same conditions, any very rare deviation, due to some extraordinary combination of circumstances, appears in the parent—say, once amongst several million individuals—and it reappears in the child, the mere doctrine of chances almost compels us to attribute its reappearance to inheritance.

Nowadays, human geneticists implicitly use the same argument to infer relationship given the mode of inheritance: if a rare gene is found in two people in a small community the instant reaction is to inquire whether they are related. So important to modern mathematical genetics is the distinction between genes identical in state and genes identical by descent that these two concepts are kept well apart, the latter playing a particularly vital role in probability calculations of genealogies concerning the similarities of relatives (Cotterman, 1941, 1975; see Cannings and Thompson, 1981).

None of the above arguments and examples are new. The link between the method of minimum evolution and likelihood first advanced in connection with phylogenetic trees by Edwards and Cavalli-Sforza (1964) has been examined by Felsenstein (1981b) and was the central theme of Sober (1988), who guided his readers through the philosophical literature from Ockham's razor, through the argument from design, to the principle of the common cause, and finally to likelihood. I differ from him only in that I think probabilistic reasoning implicitly replaced Ockham's razor long ago as a scientific tool. To see likelihood as the justification of Ockham's razor, although good philosophy, does not lead to good science because, as the example of phylogenetic trees shows, in all but a restricted set of probability models the naive application of parsimony does not lead to the solution with greatest likelihood. Parsimony as a philosophical notion (Ockham's razor) justifiable by likelihood arguments in simple cases needs to be kept distinct from "parsimony" as an algorithm (the method of minimum evolution) for reconstructing phylogenetic trees.

It is surprising to find the philosophy of systematics ensnared in prestatistical arguments, as though Darwin had lived before Pascal, Bernoulli, Gauss, and Laplace and that therefore the implications of Darwin's revolutionary hypothesis have to be studied without reference to modern theories of scientific inference.

Methods of Least Squares, Maximum Likelihood, and Minimum Evolution

Distance Matrix Method and Least Squares

Before we suggested the branching Brownian motion process as a model, Cavalli-Sforza and I (Edwards and Cavalli-Sforza, 1963a:9) had introduced least-squares estimation based on the matrix of pairwise distances:

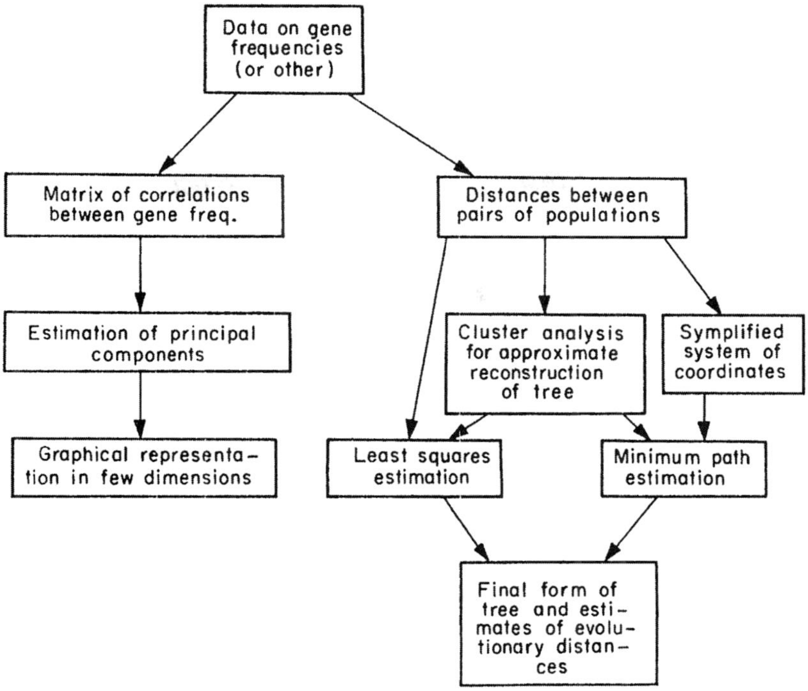

FIGURE 1. Reconstruction of phylogenetic trees: flow chart of the analysis. [Reproduced from Cavalli-Sforza and Edwards (1964).]

> An additive model may be appropriate, in which the lengths of the segments are so estimated that the observed distance between two points is closely approximated by the expected distance along the tree, formed by the sum of the connecting segments. Least squares provides an elegant estimation procedure, and we are currently engaged in applying this approach to taxonomic data. Our results will be published elsewhere.

The first computations (for a human evolutionary tree based on blood-group data) were reported in the same year and described by Cavalli-Sforza and Edwards (1964). Figure 1 (reproduced from that paper) shows the arrangement of the analysis, with least squares and minimum evolution (=minimum path) as alternative procedures and principal component analysis as an ancillary. The least-squares or distance matrix method became well known after its use by Fitch and Margoliash (1967), who employed a weighted least-squares criterion. Confusion has however been caused by describing even the use of least squares as a parsimony procedure.

Brownian Motion Model and Maximum Likelihood

The additive model with least-squares estimation suffers from the disadvantage of not providing a solution representable in evolutionary space–time, so that there can be no estimates of ancestral states. Realizing this, Cavalli-Sforza and I sought alternative methods for estimating the positions of the nodes of the evolutionary tree—the ancestral states—in space–time. We proposed the branching Brownian motion process for the evolution of continuous characters (Cavalli-Sforza et al., 1964; Edwards and Cavalli-Sforza, 1964; elaborated on by Cavalli-Sforza and Edwards, 1966, 1967). A Brownian motion process is an elementary stochastic process in which movement in a space is characterized by the normal distribution whose variance is proportional to the time elapsed. In our case, the space was the multidimensional gene-frequency space. Such a model could be criticized for its unrealistic simplicity, but one has to start somewhere, and we wrote computer programs in an attempt to find the maximum-likelihood solution. These programs were difficult to develop for a variety of reasons. Progress was slow partly because our first high-level language, FORTRAN II, became available to us only in 1963.

Ultimately, a full mathematical analysis of the maximum-likelihood approach was given by Edwards (1970), and further work and programs were reported by Felsenstein (1973b, 1981a) and Thompson (1975). Goldman (1990) reviewed the entire field.

Method of Minimum Evolution

To obtain some approximate results, we reasoned that if in the evolutionary space statistical theory estimated the expected position of any intermediate point between two known points to be on the line joining them, thus minimizing the amount of evolution invoked (as it seemed to do), perhaps the maximum-likelihood solution for the point intermediate with respect to three known points might at least be approximated by the point that minimized the sum of the three branch lengths. In 1963, I had just acquired *What is Mathematics?* by Courant and Robbins (1961), which described this geometrical problem and its solution. The first page I wrote on the method of minimum evolution, some time in the spring of 1963, contained the following:

> The data for *n* species may be exactly represented as points in a space of ($n - 1$) dimensions.
> *Antonio's theorem*
> The best taxonomic tree is formed by that string network which connects all the points using the least total length of string.

With commendable caution, the first note I added was "(1) In what sense it is 'best' has yet to be determined."

As is well known, there is no complete method of solution to the minimum-length tree problem, but an iterative program called EVOMIN with a good prospect of leading to the minimum solution was easily written in FORTRAN II. The first published analysis using it (Cavalli-Sforza and Edwards, 1964; Cavalli-Sforza et al., 1964; Edwards and Cavalli-Sforza, 1964) was sufficiently well received that it is still reappearing in books many years later (e.g., Harvey and Pagel, 1991). Figure 2, the first published minimum-evolution tree, is reproduced from Cavalli-Sforza and Edwards (1964). The EVOMIN program was circulated on both sides of the Atlantic from 1966 onwards with documentation, *Minimum evolution: Program specification.* It was used, for example, by Fitch and Neel (1969) and Kidd and Sgaramella-Zonta (1971). The procedure was described by Cavalli-Sforza and Edwards (1967), but the first full account was by Thompson (1973).

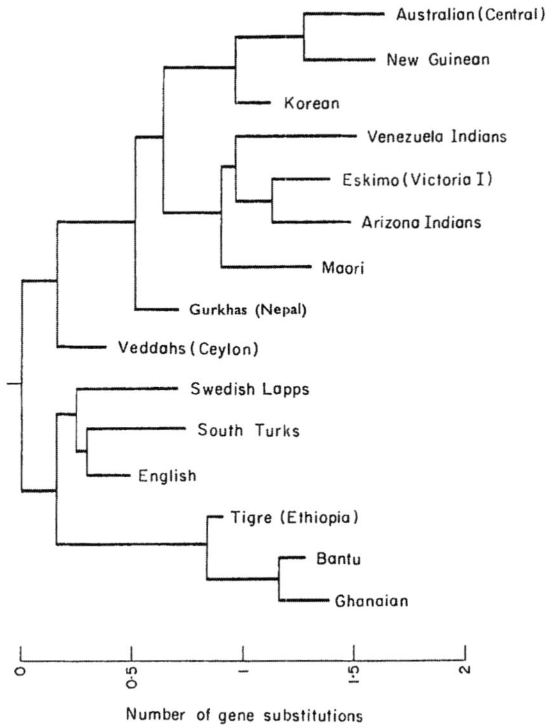

FIGURE 2. The first minimum-evolution phylogenetic tree, computed from blood-group gene frequencies. [Reproduced from Cavalli-Sforza and Edwards (1964).]

The idea of the method of minimum evolution arose solely from a desire to approximate the maximum-likelihood solution.

> It is probable that this method gives a tree which is approximately the same as the projection of the maximum-likelihood tree onto the "now" character space. (Edwards and Cavalli-Sforza, 1964:75)

However, approximate methods have been used

> with success and we anticipate that the results thus obtained are very close to those obtainable by maximum likelihood. (Cavalli-Sforza et al., 1964:10)

> Its success is probably due to the closeness of the solution it gives to the projection of the "maximum-likelihood" tree. (Cavalli-Sforza and Edwards, 1967:555)

It soon became clear from simple cases that there was no question of minimum evolution providing the maximum-likelihood solution exactly on the Brownian motion model (Cavalli-Sforza and Edwards, 1967; Edwards, 1969). It thus came as no surprise that a similar situation obtains with discrete models (Felsenstein, 1973a). (For further discussion of the relationship between maximum likelihood and minimum evolution, see Felsenstein [1988] and Goldman [1990].)

Apparent Linearity of Evolution

Statistical theory applied to the Brownian motion model estimates any intermediate point on the evolutionary path between two points to lie on the straight line joining them. This solution arises from the unimodal nature of the spherically symmetrical multivariate normal distribution.

Imagine the multidimensional character space with the positions of the two endpoints marked. Now consider a definite intermediate time. For each hypothetical position of the corresponding intermediate point in the space, there is an associated probability density defined by the model, found by multiplying together the two probability densities associated with the two parts into which the time interval has been divided. The symmetrical multivariate normal distribution with any given variance parameter is unimodal, and its probability density depends only on the distance from the mean, decreasing with increasing distance. It follows that the joint probability density for any point not on the straight line joining the two endpoints will be less than the density at the base of the perpendicular from such a point to the straight line, because the new point will be closer to both endpoints and both of the contributing probability densities will therefore be greater. Thus, the point of highest density, or the most probable position for the intermediate point, must lie on the straight line. Its whereabouts on this line depend, of course, on the intermediate time in question. However many time intervals the total interval is divided into, this result will hold for all of them, so that in this sense the most probable path of evolution will be the straight line.

Too much should not be read into this result. It is similar to "In a million tosses of a fair coin the most probable result is 500,000 heads." The straight line may indeed be more probable than any other path specified, but the probability of any particular path will always be vanishingly small. If we ask a slightly different question, such as "At a particular intermediate time, what is the most probable distance of the intermediate point from the straight line?" the answer will not be zero (because the distribution of distance from the mean of a multivariate normal distribution does not, for sufficient dimensions, have its mode at the origin). Nevertheless, the fact that the shortest path is the most probable one is suggestive, but simple examples show that as soon as three points are considered the equivalence breaks down.

Minimum Evolution for Discrete Characters

The original wording of the principle of minimum evolution in 1963 was deliberately general, designed to cover both continuous and discrete characters. The argument in the previous section is easily developed for the discrete case.

Camin and Sokal (1965) studied the reconstruction of phylogenetic trees by creating artificial animals endowed with discrete morphological characters that could change according to certain simple rules (the most restrictive of which was the assumption that evolution was irreversible). Camin and Sokal (1965:311) then observed that

> those trees which most closely resembled the true cladistics invariably required for their construction the least number of postulated evolutionary steps for the characters studied. Subsequently we exam-

ined the possibility of reconstructing cladistics by the principle of evolutionary parsimony.

This seems to have been the first use of the word *parsimony* to describe the method of minimum evolution, but Camin and Sokal did not refer to Ockham's razor. Rather, having arrived at the method empirically, they advanced the justification that "the correctness of our approach depends on the assumption that nature is indeed parsimonious" (Camin and Sokal, 1965:323). Although they referred to Edwards and Cavalli-Sforza (1964) as one of the background papers, they did not remark on the justification of the method there offered (of expected similarity with the maximum-likelihood solution). We criticized Camin and Sokal's justification: the method of minimum evolution's "success with discrete data must ... be attributed to its closeness to the solution which a proper probability model would give" (Cavalli-Sforza and Edwards, 1967:555). By not arguing in a probabilistic framework, Camin and Sokal had not been in a position to distinguish between logical features of the method and logical features of the model and were therefore beguiled into predicating the success of the minimum method on a supposed minimum feature of the model.

In a useful review, Kluge (1984:26) made the above distinction by referring to methodological parsimony versus evolutionary parsimony, but unfortunately he attributed the wrong one to me: "According to Estabrook (1972), A. W. F. Edwards first suggested the idea of evolutionary parsimony to R. R. Sokal in 1963, and Edwards and Cavalli-Sforza described a method of minimum evolution in that same year." It was, of course, methodological parsimony that I had discussed with Sokal, following Cavalli-Sforza's delivery of our paper (Cavalli-Sforza and Edwards, 1964). Estabrook (1972) had not himself made any logical distinction between the two kinds but was clearly referring to methodological parsimony. His source of information was my reply, dated 22 September 1971, to a letter from him asking for offprints, in which I wrote

> At the International Congress of Genetics I discussed it [the method of minimum evolution] with Sokal (who probably also heard the paper read—I forget), and I was consequently surprised to find it stated as a new method by Camin and Sokal in 1965. Indeed, they give a justification which I could not accept (see Cavalli-Sforza & Edwards, 1967), and at this time I changed the name from the *Principle* to the *Method* to avoid any further misunderstanding of its logical basis.

Estabrook replied, "I fully agree that minimum evolution is a method and not a principle." The change in name had little effect, and at scientific meetings I have often had to argue against the supposed justification. Years later my frustration even surfaced in a book review (Edwards, 1981: 12):

> Some years ago, in collaboration with L. L. Cavalli-Sforza, I introduced the "method of minimum evolution" as a procedure for estimating the structure of evolutionary trees, and it very rapidly found its way into the literature as a method justified by an appeal to the principle that evolution attains its goals by proceeding along some minimum path, like a sentient being going shopping. It is nonsense, of course, but it has proved compelling nonsense.

I elaborated the point less intemperately in my vote of thanks to Felsenstein (1983: 263):

> One of the difficulties which all of us who have worked in this field have had to face is the enthusiasm with which, in certain quarters, the so-called "parsimony" methods are embraced. It has repeatedly been argued that such methods are justified because evolution proceeds parsimoniously. Beneath this assertion lurks a philosophy that has tainted much of modern evolutionary biology, namely the false analogy with physics or with game theory according to which biological systems are striving to achieve certain ends, to win certain contests, to minimize this or to maximize that. It is the fascination of the potential function, the lure of the directed force, the mirage of the goal, that misleads. In truth, however closely the mathematics of these biological systems mimics the mathematics of physics or game theory, the systems are not *trying* to achieve anything: they are responding to pressures. The proper justification of parsimony as a method of estimation lies, therefore, not in supposing evolution to have a will, but in demonstrating that the most parsimonious solution is an approximation to the maximum-likelihood solution on some acceptable model. Dr Felsenstein, and Dr Elizabeth Thompson, have shown that, on the Brownian-motion model, the approximation is not very good in general; but in my view it is good

enough to explain the popularity of the method of minimum evolution.

Minimum Mutation Distance

When the discrete characters under consideration are the amino acid sequences of proteins, a count of the number of residue differences between sequences can be expected to be related to the time back to their common ancestor (Zuckerkandl and Pauling, 1962), but the accumulating knowledge about the genetic code led Margoliash (1963:676) to suggest that "an estimate of the minimal number of mutational steps that can lead from one residue to another may be possible when the nucleotide base sequences of all such 'codes' will have been determined." Margoliash and Smith (1965:232) went on to say that "this code will indeed provide a means of estimating the minimal number of mutational events and the most probable genetic pathway. . . ."

The assumption that the minimum number of mutations corresponds to the most probable genetic pathway may be justified by a simple probability argument. The mutational transitions between different DNA base pairs may all be supposed to occur with certain probabilities, which need not be known exactly for the argument to succeed. The transition from any one amino acid residue to any other may occur as a result of any number of mutations in the DNA down to a minimum number that will be specific to the two residues in question. For the simple model of independent events, the probability of any particular sequence of mutations will be the product of the individual probabilities, and if these probabilities are not too diverse this product will decrease monotonically with the number of steps in the sequence, the minimum number of steps thus almost certainly having the highest probability (if the probabilities are all equal, the uncertainty is removed). The most probable number of steps is not the expected number because the distribution is skewed (see Holmquist, 1972). Once again, however, there is no need to invoke Ockham's razor.

In 1965, Zuckerkandl (1965:117) wrote in a popular article, "In postulating a possible ancestral amino acid residue one should prefer the residue that can be reached by invoking the fewest number of mutations in the genetic message." The Poisson probability model for changes in amino acid sequences was proposed by Zuckerkandl and Pauling (1965a) and Margoliash and Smith (1965). Full mathematical analyses soon followed, both for amino acid changes and changes in the underlying DNA, enabling minimum mutation distances to be replaced with improved estimates. These estimates then became the basic data for phylogenetic analysis (for accounts, see Nei, 1975, 1987).

Discrete Probability Models for Evolutionary Trees

The first probability models for complete trees using discrete characters, with suggestions for their analysis, were proposed in 1966.

Eck and Dayhoff (1966)

Eck and Dayhoff (1966:165), in their *Atlas of Protein Sequence and Structure*, proposed a model of random changes in amino acids and the reconstruction of the resulting evolutionary tree by the method of minimum evolution: "In principle, for a family of closely related proteins, we seek the topology that has the minimum number of mutations" (amino acid residue changes). As we now know, a probability argument like the one given in the previous section does not carry over to a complete tree, and Eck and Dayhoff (1966:165) offered no detailed justification for their adoption of the principle: "This minimal tree is the best approximation we can presently make within the stated simple assumptions. The discrepancy from the true tree will be small and it should decrease in any region where many sequences have been observed." This question has been the subject of later controversy (see Felsenstein, 1978; Sober, 1988).

Eck and Dayhoff were evidently not familiar with the 1963 proposal of the method of minimum evolution or the use of it in the years 1963–1965. As with the original proposal, their suggestion of it also stemmed from a desire to develop specific rules suitable for computers. In the earlier papers by Pauling and Zuckerkandl (1963), Zuckerkandl (1964, 1965), and Zuckerkandl and Pauling (1965a, 1965b), these authors got as far as considering the minimum mutation distance but did not extend this consideration to a global principle covering the whole tree and all the data simultaneously, as did Eck and Dayhoff. Another paper quoted by Eck and Dayhoff, Doolittle and Blombäk (1964), contains a strong "minimizing" thread without, however, being explicit.

Edwards (1966)

In an article in the same year (Edwards, 1966), I similarly introduced a stochastic model for a complete tree, and I proposed likelihood as a criterion for estimation and described the method of minimum evolution. To explain how random inherited changes provide information that enables phylogenetic inferences to be made, I imagined a custom in which, for example, each species possesses a pole on which are threaded 100 discs, either black or white. Every year, a disc is chosen at random and its color is changed. Species do this independently, and when a species splits into two daughter species each inherits the current parental pattern of black and white discs. This was as simple a model of random evolution as possible, but it was a stochastic model and therefore appropriate for statistical treatment. The first full analysis of the likelihood approach for discrete characters was by Felsenstein (1973a). But if we want to use the method of minimum evolution instead, then

> for discontinuous characters, such as the discs on the pole, the method selects the tree with the minimum total number of changes. (Edwards, 1966: 440)

> Nature, in endowing us with genes arranged on chromosomes rather as the discs are arranged on the pole, has unwittingly presented us with an excellent tool for reconstructing the past. (Edwards, 1966:439)

> [But] its success with discrete [as well as with continuous] data must also be attributed to its closeness to the solution which a proper probability model would give. (Cavalli-Sforza and Edwards, 1967:555)

Essentially the same idea of a simple stochastic model for a discrete-character evolutionary tree may be found *de novo* in Felsenstein's (1968) Ph.D. dissertation, accompanied by the fullest mathematical analysis extant at that time.

Hennig's Phylogenetic Systematics

According to Stewart (1993:604), "Hennig's (1966) ideas spawned the school of thought called cladistics from which modern parsimony analysis developed." This and similar statements from other authors are examples of what Rackham (1990:23) has wittily called a *factoid*: "A factoid looks like a fact, is respected as a fact, and has all the properties of a fact except that it is not true." Factoids have a life of their own, and the mere refusal of well-informed writers to repeat them is insufficient to ensure their demise. It is necessary to present the counterevidence, and in that spirit I offer the following observations.

By "modern parsimony analysis," Stewart meant the modern application of the method of minimum evolution to discrete data through a variety of computer programs (but especially PAUP; Swofford, 1992). The first of such programs were reported in 1965 (Camin and Sokal) and 1966 (Eck and Dayhoff), and Camin and Sokal, although probably not Eck and Dayhoff, were familiar with the principle of minimum evolution enunciated in 1963.

The Hennigian claim can be countered at two levels. First, even if Hennig had proposed a parsimony solution in his 1966 book, modern parsimony analysis had already been initiated by then. Second, Hennig did not in fact propose such a solution anyway.

The first point is a matter of simple timing, which, with respect to molecular data especially, is obvious from the above dates. The principal route by which parsimony methods for discrete data entered nonmolecular cladistics was through papers by Kluge and Farris (1969) and Farris (1970). These authors did not refer to Hennig for the idea of the most-parsimonious tree but to Camin and Sokal (1965), who were familiar with Edwards and Cavalli-Sforza (1964). Kluge and Farris (1969:6) also wrote that "the method of Wagner (1961) . . . is a simplified procedure for producing a most parsimonious tree in the sense of Camin and Sokal," and Farris (1970:83) wrote that "Wagner Trees are one possible generalization of the most parsimonious trees of Camin and Sokal (1965)." However, these authors seem to be reading into Wagner's (1961) method a principle that, on examination, it does not contain and that Wagner (1980) did not claim. The same type of hindsight has affected the interpretation of Hennig's contribution.

Edwards and Cavalli-Sforza (1964) did not refer to either Hennig or Wagner, and I can confidently assert that we had heard of neither. Kluge and Farris (1969) and Farris (1970) also made no reference to Hennig; the sole reference to him is by Camin and Sokal (1965) and concerns criteria for making judgments about character coding rather than parsimony. In contrast, Farris (1969, 1970) and Camin and Sokal (1965) referred to Edwards and Cavalli-Sforza (1964), although without remarking explicitly on the method of minimum evolution.

The second point, that Hennig (1965, 1966) did not in fact propose the parsimony solution then or at an earlier time, is, in the nature of a negative, more difficult to establish. Duncan and Stuessy (1985:3) stated that "Hennig's original method was modified to include a criterion of parsimony in Farris, Kluge, and Eckardt (1970)." According to Farris et al. (1970), the key statement is to be found on page 121 of the English version of Hennig's book (1966). Because there had been some criticism of the translation of Hennig's original German manuscript, Farris et al. (1970:174) gave their own translation as well as Hennig's German sentence: "The more characters certainly interpretable as apomorphous (not characters in general) that there are present in a number of species, the better founded is the assumption that these species form a monophyletic group." This unexceptionable statement seems to be devoid of operational interpretation. It says that derived (apomorphous) characters common to a group of species are evidence of monophyly, with which all will agree, and of course implicit in this statement is the corollary that the greater the number of derived characters, the stronger the evidence. But how is one to use this idea as a principle for phylogenetic reconstruction? Hennig might have gone on to suggest that a tree with the minimum number of homoplasies (shared derived characters of independent origin) would perhaps possess "well-founded" monophyletic groupings, just as Cavalli-Sforza and I thought that a minimum-evolution tree might have high likelihood, but apparently he did not; Hennig's "parsimony" is local, not global.

Farris et al. (1970:176) correctly observed,

> Unfortunately [Hennig's statement] is not sufficiently detailed to allow us to select a unique criterion for choosing a most preferable tree. . . . [It] deals only with single monophyletic groups and does not tell us how to evaluate a tree consisting of several monophyletic groups. One widely known criterion—parsimony—could be used to select trees.

They attributed "the method of most parsimonious trees" to Camin and Sokal (1965) and not to Hennig at all.

Felsenstein (1982) described what he called Hennig's dilemma, by which he meant the absence of any procedure in Hennig's proposals for resolving incompatibilities in the data: "Hennig's method works only as long as there is no internal conflict in the data" (Felsenstein, 1982:381). He then reviewed the methods of parsimony and compatibility for resolving the dilemma: "Parsimony or minimum-evolution methods were first proposed by Edwards and Cavalli-Sforza (1963, 1964)" (Felsenstein, 1982:382).

Sober (1988:114), who thought that Hennig should be "properly viewed as the source of cladistic parsimony," could only state weakly (1988:115) that "we find in the works of Willi Hennig a positive endorsement of ideas that lead naturally to the principle of parsimony, even though

Hennig, to my knowledge, never called his idea by that name."

Finding arguments for global parsimony in the works of Hennig is an example of that phenomenon well known to the history of science: reading into early writers more than they actually wrote. Hindsight is a dangerous viewpoint, and famous scientists in particular often have later principles attributed to them. Thus, Fisher's principle of maximum likelihood has variously been attributed to Gauss, Karl Pearson, and Edgeworth (see Edwards, 1974). One can often discern an ulterior motive in such attributions. Hull (1984:18) put the matter in perspective:

> If the essence of cladism is strictly methodological, the precise nature of this methodology was not immediately clear to everyone concerned. I find the desire to claim that the essence of cladism lay buried in the writings of Hennig ... paradoxical. If Hennig set out the essence of cladistic analysis in the beginning, then what credit can later cladists claim for themselves? As I look back over the past decade [1971–1981], its events do not look like a gradual uncovering of a truth that lay buried in Hennig's early writings. On the contrary, I think that later cladists vastly improved on Hennig's system. Perhaps they did not totally transform it, but transform it they did, and they deserve considerable credit for the improvements they made. If someone were to ask me where to go to learn the principles of cladistic analysis, I would not recommend Hennig's *Phylogenetic Systematics*.
>
> ... Although there is a strong tendency to claim that all truth resides in the works of some great scientist (was Darwin really a cladist?), I do not think this belief gives proper credit to later workers. Later cladists did not simply uncover the basic principles of cladistic analysis by a careful reading of Hennig. To some extent, they created those principles.

Conclusion

The method of minimum evolution for reconstructing phylogenetic trees, enunciated in 1963, did not stem from Ockham's razor (the principle of parsimony), or from the mistaken notion that evolution proceeds parsimoniously, or from Hennig's phylogenetic systematics. As the original papers make clear, it arose as an approximation of the maximum-likelihood solution obtained on the basis of a model for random evolution (whether continuous or discrete), itself an approach justified by modern concepts of statistical inference. Any approach that attempts to grapple with inference under uncertainty without using ideas from the theory of probability is unlikely to command scientific respect.

Acknowledgments

I am grateful to two referees for their careful and perceptive comments, most of which I have gladly acted on.

References

Camin, J. H., and R. R. Sokal. 1965. A method for deducing branching sequences in phylogeny. Evolution 19:311–326.

Cannings, C., and E. A. Thompson. 1981. Genealogical and genetic structure. Cambridge Univ. Press, Cambridge, England.

Cavalli-Sforza, L. L., I. Barrai, and A. W. F. Edwards. 1964. Analysis of human evolution under random genetic drift. Cold Spring Harbor Symp. Quant. Biol. 29:9–20.

Cavalli-Sforza, L. L., and A. W. F. Edwards. 1964. Analysis of human evolution. Genet. Today 3:923–933.

Cavalli-Sforza, L. L., and A. W. F. Edwards. 1966. Estimation procedures for evolutionary branching processes. Bull. Int. Stat. Inst. Proc. 41:803–808.

Cavalli-Sforza, L. L., and A. W. F. Edwards. 1967. Phylogenetic analysis: Models and estimation procedures. Evolution 21:550–570. [Am. J. Hum. Genet. Suppl. 19:233–257.]

Cotterman, C. W. 1941. Relatives and human genetic analysis. Sci. Monthly 53:227–234.

Cotterman, C. W. 1975. A calculus for statistico-genetics. Pages 157–272 *in* Genetics and social structure: Mathematical structuralism in population genetics and social theory (P. Ballonoff, ed.). Dowden, Hutchinson and Ross, Stroudsburg, Pennsylvania.

Courant, R., and H. Robbins. 1961. What is mathematics? Oxford Univ. Press, Oxford, England.

Cronin, H. 1991. The ant and the peacock. Cambridge Univ. Press, Cambridge, England.

Darwin, C. 1859. On the origin of species by means of natural selection. Murray, London.

Doolittle, R. F., and B. Blombäck. 1964. Amino-acid sequence investigations of fibrinopeptides from various mammals: Evolutionary implications. Nature 202:147–152.

Duncan, T., and T. F. Stuessy (eds.). 1985. Cladistic theory and methodology. Van Nostrand Reinhold, New York.

Eck, R. V., and M. O. Dayhoff. 1966. Atlas of protein sequence and structure. National Biomedical Research Foundation, Silver Spring, Maryland.

Edwards, A. W. F. 1966. Studying human evolution by computer. New Sci. 30:438–440.

Edwards, A. W. F. 1969. Genetic taxonomy. Pages 140–142 *in* Computer applications in genetics (N. E. Morton, ed.). Univ. Hawaii Press, Honolulu.

Edwards, A. W. F. 1970. Estimation of the branch

points of a branching diffusion process. J. R. Stat. Soc. B 32:155–174.
EDWARDS, A. W. F. 1972. Likelihood. Cambridge Univ. Press, Cambridge, England.
EDWARDS, A. W. F. 1974. The history of likelihood. Int. Stat. Rev. 42:9–15.
EDWARDS, A. W. F. 1981. Hereditary genius: Review of *Statistics in Britain*, by D. MacKenzie. Lond. Rev. Books 6–19 Aug:12–13.
EDWARDS, A. W. F. 1987. Pascal's arithmetical triangle. Oxford Univ. Press, New York.
EDWARDS, A. W. F. 1992. Likelihood. Johns Hopkins Univ. Press, Baltimore, Maryland.
EDWARDS, A. W. F., AND L. L. CAVALLI-SFORZA. 1963a. A method for cluster analysis. Fifth International Biometrics Conference, Cambridge, England.
EDWARDS, A. W. F., AND L. L. CAVALLI-SFORZA. 1963b. The reconstruction of evolution. (Abstr.) Heredity 18:553; Ann. Hum. Genet. 27:104–105.
EDWARDS, A. W. F., AND L. L. CAVALLI-SFORZA. 1964. Reconstruction of evolutionary trees. Pages 67–76 *in* Phenetic and phylogenetic classification. Systematics Association, London.
ESTABROOK, G. F. 1972. Cladistic methodology: A discussion of the theoretical basis for the induction of evolutionary history. Annu. Rev. Ecol. Syst. 3:427–456.
FARRIS, J. S. 1969. A successive approximations approach to character weighting. Syst. Zool. 18:374–385.
FARRIS, J. S. 1970. Methods for computing Wagner trees. Syst. Zool. 19:83–92.
FARRIS, J. S., A. G. KLUGE, AND M. J. ECKARDT. 1970. A numerical approach to phylogenetic systematics. Syst. Zool. 19:172–189.
FELSENSTEIN, J. 1968. Statistical inference and the estimation of phylogenies. Ph.D. Dissertation, Univ. Chicago, Chicago.
FELSENSTEIN, J. 1973a. Maximum likelihood and minimum-steps methods for estimating evolutionary trees from data on discrete characters. Syst. Zool. 22:240–249.
FELSENSTEIN, J. 1973b. Maximum-likelihood estimation of evolutionary trees from continuous characters. Am. J. Hum. Genet. 25:471–492.
FELSENSTEIN, J. 1978. Cases in which parsimony or compatibility methods will be positively misleading. Syst. Zool. 27:401–410.
FELSENSTEIN, J. 1981a. Evolutionary trees from gene frequencies and quantitative characters: Finding maximum likelihood estimates. Evolution 35:1229–1242.
FELSENSTEIN, J. 1981b. A likelihood approach to character weighting and what it tells us about parsimony and compatibility. Biol. J. Linn. Soc. 16:183–196.
FELSENSTEIN, J. 1982. Numerical methods for inferring evolutionary trees. Q. Rev. Biol. 57:379–404.
FELSENSTEIN, J. 1983. Statistical inference of phylogenies (with discussion). J. R. Stat. Soc. A 146:246–272.
FELSENSTEIN, J. 1988. The detection of phylogeny. Pages 112–127 *in* Prospects in systematics (D. L. Hawksworth, ed.). Systematics Association and Clarendon Press, Oxford, England.
FISHER, R. A. 1921. On the "probable error" of a coefficient of correlation deduced from a small sample. Metron 1:3–32.
FITCH, W. M., AND E. MARGOLIASH. 1967. Construction of phylogenetic trees. Science 155:279–284.
FITCH, W. M., AND J. V. NEEL. 1969. The phylogenic relationship of some Indian tribes of Central and South America. Am. J. Hum. Genet. 21:384–397.
GOLDMAN, N. 1990. Maximum likelihood inference of phylogenetic trees, with special reference to a Poisson process model of DNA substitution and to parsimony analyses. Syst. Zool. 39:345–361.
HARVEY, P. H., AND M. D. PAGEL. 1991. The comparative method in evolutionary biology. Oxford Univ. Press, Oxford, England.
HENNIG, W. 1965. Phylogenetic systematics. Annu. Rev. Entomol. 10:97–116.
HENNIG, W. 1966. Phylogenetic systematics. Univ. Illinois Press, Urbana.
HOLMQUIST, R. 1972. Theoretical foundations for a quantitative approach to paleogenetics. J. Mol. Evol. 1:115–133.
HULL, D. L. 1984. Cladistic theory: Hypotheses that blur and grow. Pages 5–23 *in* Cladistics: Perspectives on the reconstruction of evolutionary history (T. Duncan and T. F. Stuessy, eds.). Columbia Univ. Press, New York.
JEFFERYS, W. H., AND J. O. BERGER. 1992. Ockham's razor and Bayesian analysis. Am. Sci. 80:64–72.
KIDD, K. K., AND L. A. SGARAMELLA-ZONTA. 1971. Phylogenetic analysis: Concepts and methods. Am. J. Hum. Genet. 23:235–252.
KLUGE, A. G. 1984. The relevance of parsimony to phylogenetic inference. Pages 24–38 *in* Cladistics: Perspectives on the reconstruction of evolutionary history (T. Duncan and T. F. Stuessy, eds.). Columbia Univ. Press, New York.
KLUGE, A. G., AND J. S. FARRIS. 1969. Quantitative phyletics and the evolution of anurans. Syst. Zool. 18:1–32.
MARGOLIASH, E. 1963. Primary structure and evolution of cytochrome *c*. Proc. Natl. Acad. Sci. USA 50: 672–679.
MARGOLIASH, E., AND E. L. SMITH. 1965. Structural and functional aspects of cytochrome *c* in relation to evolution. Pages 221–242 *in* Evolving genes and proteins (V. Bryson and H. J. Vogel, eds.). Academic Press, New York.
NEI, M. 1975. Molecular population genetics and evolution. North-Holland, Amsterdam.
NEI, M. 1987. Molecular evolutionary genetics. Columbia Univ. Press, New York.
PAULING, L., AND E. ZUCKERKANDL. 1963. Chemical paleogenetics: Molecular "restoration studies" of extinct forms of life. Acta Chem. Scand. 17(Suppl. 1):S9–S16.
RACKHAM, O. 1990. Trees and woodland in the British landscape. Dent, London.
SOBER, E. 1988. Reconstructing the past: Parsimony, evolution, and inference. MIT Press, Cambridge, Massachusetts.

STEWART, C.-B. 1993. The powers and pitfalls of parsimony. Nature 361:603–607.

SWOFFORD, D. L. 1992. PAUP: Phylogenetic analysis using parsimony. Illinois Natural History Survey, Champaign.

THOMPSON, E. A. 1973. The method of minimum evolution. Ann. Hum. Genet. 36:333–340.

THOMPSON, E. A. 1975. Human evolutionary trees. Cambridge Univ. Press, Cambridge, England.

WAGNER, W. H., JR. 1961. Problems in the classification of ferns. Recent Adv. Bot. 1:841–844.

WAGNER, W. H. 1980. Origin and philosophy of the groundplan-divergence method of cladistics. Syst. Bot. 5:173–193.

ZUCKERKANDL, E. 1964. Perspectives in molecular anthropology. Pages 243–272 *in* Classification and human evolution (S. L. Washburn, ed.). Methuen, London.

ZUCKERKANDL, E. 1965. The evolution of hemoglobin. Sci. Am. 212:110–118.

ZUCKERKANDL, E., AND L. PAULING. 1962. Molecular disease, evolution, and genic heterogeneity. Pages 189–225 *in* Horizons in biochemistry (M. Kasha and B. Pullman, eds.). Academic Press, New York.

ZUCKERKANDL, E., AND L. PAULING. 1965a. Evolutionary divergence and convergence in proteins. Pages 97–166 *in* Evolving genes and proteins (V. Bryson and H. J. Vogel, eds.). Academic Press, New York.

ZUCKERKANDL, E., AND L. PAULING. 1965b. Molecules as documents of evolutionary history. J. Theor. Biol. 8:357–366.

Received 6 January 1995; accepted 2 August 1995
Associate Editor: Allan Larson

Paper 173 (2001)

Darwin and Mendel united: the contributions of Fisher, Haldane and Wright up to 1932
In *Encyclopedia of Genetics*, ed. E. C. R. Reeve, London: Fitzroy Dearborn, 77–83.

DARWIN AND MENDEL UNITED: THE CONTRIBUTIONS OF FISHER, HALDANE AND WRIGHT UP TO 1932

A.W.F. Edwards
Gonville and Caius College, Cambridge University, Cambridge, UK

Since the 1930s, it has been widely recognized and oft repeated that R.A. Fisher (1890–1962), J.B.S. Haldane (1892–1964) and S. Wright (1889–1988) successfully developed evolutionary Mendelian genetics and applied it to Darwin's theory of natural selection, thereby creating "neo-Darwinism" by supplying the mechanism which that theory had perforce lacked. Their achievement was all the more important because, before the rediscovery of Mendel's work in 1900, and for some years afterwards, Darwinism had been in the doldrums and many biologists had continued to doubt the possibility of accounting for evolution in terms of natural selection or of natural selection alone. In showing that the Mendelian mechanism could indeed account for the main features of evolution, Fisher, Haldane and Wright thus revivified Darwin's theory and ensured it a place not only in biology but also in the development of human thought as a whole. Modern evolutionary biology rests on their achievements, but their influence is also felt as far afield as game theory, sociology, economics and in specialist areas such as genetic algorithms in computer science.

R.A. Fisher (Sir Ronald Fisher) was born in London and educated at Cambridge, where he studied mathematics and mathematical physics. He was successively statistician to Rothamsted Experimental Station, Galton Professor of Eugenics at University College London and Arthur Balfour Professor of Genetics at Cambridge. He died in Adelaide, South Australia. J.B.S. Haldane was born and educated in Oxford, also studying mathematics, but then transferring to "Greats", that is, classics and philosophy. Initially a don at Oxford, Haldane became Reader in Biochemistry at Cambridge and subsequently a professor at University College London, finally holding the Weldon Professorship of Biometry. S. Wright, who normally published under his full name Sewall Wright, was born in Massachusetts and educated in Illinois, undertaking graduate study in biology at the University of Illinois followed by 3 years at Harvard. He worked at the Animal Husbandry Division of the US Department of Agriculture in Washington, DC, and subsequently at the University of Chicago, and enjoyed a productive retirement at the University of Wisconsin, Madison. Each man also made contributions outside evolutionary biology: Fisher notably creating much of modern statistical theory, Haldane ranging widely over most of genetics as well as being a gifted popular writer and Wright contributing the inbreeding coefficient, path analysis and many studies in physiological genetics. Each has been the subject of a full biography (Clark, 1968; Box, 1978; Provine, 1986).

The first of the three to show an interest in the application of Mendelism was Fisher who, as an undergraduate in 1911, gave a lecture (a copy of which has survived) entitled "Mendelism and biometry" which foreshadows his subsequent work in many respects. Fisher's early reading was an ideal preparation for the synthesis of Darwinism and Mendelism, for he had chosen the 13-volume edition of the collected works of Charles Darwin as a school prize in 1906 and had bought William Bateson's *Mendel's Principles of Heredity*, with its translation of Mendel's paper, in his first undergraduate year (1909). Studying for the mathematical tripos while relaxing with Darwin and Mendel could not have been a better preparation for the work that was to follow. Fisher published his first paper in 1912 and by 1916 had completed the paper which was to appear 2 years later as *The correlation between relatives on the supposition of Mendelian inheritance*, which established his early reputation.

In a 1914 paper (originally read in 1912), Fisher pointed out how the genes of a childless man could nevertheless prosper if he had enough nephews, thus formulating the idea of inclusive fitness or "kin selection". Another early paper of evolutionary significance was *The evolution of sexual preference* published in 1915 in which, starting from Darwin's theory of sexual selection, he took the fundamental further step of "showing how mate choice itself would evolve as a consequence of the very process of sexual selection it produced" (O'Donald, 1990). His next work of evolutionary significance was the seminal 1922 paper which we consider below.

Haldane similarly showed an early interest in genetics, helping his sister Naomi to breed guinea pigs in 1908. In 1915, they and a fellow-student, A.D. Sprunt, were able to publish a paper reporting linkage from experiments with mice. Haldane's first paper, like Fisher's, had appeared in 1912. In 1919, he published two papers on the statistical estimation of linkage, but not until 1924 did the first paper in his series *A mathematical theory of natural and artificial selection* appear. In 1927, he wrote in one of his popular scientific essays about how the concept of inclusive fitness would explain the evolution of altruism.

Wright also published his first paper in 1912 and, a few months before the Haldane–Sprunt paper announced a linkage in the mouse, Wright and W.E. Castle published a similar observation in the rat, the first linkage to be observed in mammals. Throughout the 1920s, Wright worked on the statistical analysis of pedigrees, inventing path coefficients, the inbreeding coefficient and the concept of effective population size in the process. An important result in this work was the calculation, in 1921, of the rate of decrease in heterozygosis in regular systems of mating (without selection).

In 1929, Wright's first papers on evolution appeared, although it is known that in 1925 he had prepared a long paper on the subject, which was to be the basis of his influential 1931 paper and to which we will return.

As early as 1931, Lancelot Hogben (Hogben, 1931) linked together the names of Fisher, Haldane and Wright as the leaders of the neo-Darwinian revolution and, in 1932, Haldane, with characteristic immodesty, repeated the claim. The key period requiring detailed study is the decade 1922–32, starting with Fisher's 1922 paper and ending with the *Proceedings of the Sixth International Congress of Genetics* and Haldane's book *The Causes of Evolution*, both of which appeared in 1932. The decade embraces the publication of Fisher's *The Genetical Theory of Natural Selection* (1930) and Wright's influential 1931 and 1932 papers. We shall explore it chronologically.

In his 1922 paper, Fisher introduced stochastic models into evolutionary genetics. He was concerned "to discuss the distribution of the frequency ratio of the allelomorphs of dimorphic factors, and the conditions under which the variance of the population may be maintained". The second point is a kind of lodestar in Fisher's work of this period, for he had been the first to perceive clearly that one major distinction between Mendel's theory and pre-Mendelian theories of blending inheritance was that, under the former, the rate of loss of genetic variability would be infinitely less and indeed due only to finite population size and the consequential stochastic losses. How would the rate of loss compare with the gains through mutation under differing assumptions about selection? Fisher's concentration on the variance (a word he had coined in his 1918 paper) and its analysis ran right through to his enunciation of the "fundamental theorem of natural selection" in 1930.

First Fisher breaks new ground by analysing the simple model of selection at a single diallelic locus with constant fitnesses, pointing out that "if the selection favours the heterozygote, there is a condition of stable equilibrium". Given that there had been calculations on the effects of selection on Mendelian models over the previous 7 years by workers such as H.T.J. Norton and H.C. Warren, it is remarkable that this theory of the balanced polymorphism had to wait until 1922. Even when Haldane inaugurated his long series of papers in 1924, he did not notice Fisher's result (all the more remarkable given that the address on Haldane's papers was Trinity College and on Fisher's was Gonville and Caius College, neighbouring colleges of the University of Cambridge).

Next in the paper, Fisher introduces stochastic considerations, first treating "the survival of individual genes" by means of a branching process analysed by functional iteration and then setting up the "chain-binomial" model and analysing it by a diffusion approximation. The latter model is the keystone of stochastic population genetics (and, incidentally, stochastic diffusion theory itself), yet it often goes by the name of "the Fisher–Wright model" or even the "Wright–Fisher model". Not only is it due to Fisher alone, but Wright's main paper on it did not appear until a further 9 years had elapsed. We consider it below. Fisher goes on to consider the distributions of the gene frequency (actually in the form of the logarithm of the frequency ratios) for various models of selection, and for no selection, finding that in the latter case a population of n individuals would have its variance reduced by a factor e in $4n$ generations (the "relaxation time"). He concludes: "The rate of mutation necessary to maintain the variance of the species may be calculated from these distributions."

Wright first met Fisher in 1924 in Washington, DC, at which time Wright had not yet seen Fisher's paper, but Fisher promised him a copy and sent it on his return to England. "Until reading Fisher's 1922 paper, it had not occurred to Wright to extend his own quantitative analysis to the statistical distribution of genes in populations" (Provine, 1986). Fisher's 1922 paper may indeed be said to have inaugurated neo-Darwinism.

Between 1924 and 1934, Haldane published a series of ten papers under the general title *A mathematical theory of natural and artificial selection* in which he worked out the consequences of selection for a large variety of Mendelian models, frequently solving the resulting difference equations by deriving and solving the corresponding differential equation in continuous time. For the most part, these papers report the necessary, but somewhat ponderous, mathematical analysis of particular models with infinite populations and discrete generations, and they have naturally tended to be neglected because of the existence of Haldane's 1932 book *The Causes of Evolution*, whose appendix *Outline of the mathematical theory on natural selection* summarizes them. However, certain historical points may be noted, as follows.

In Part I (1924), Haldane makes the first attempt to estimate the "intensity of natural selection", from data on the spread of melanism in the peppered moth in Manchester. In Part IV (1927) he gives the first account of the theory of natural selection in an age-structured population, making a generous acknowledgement of H.T.J. Norton's priority in unpublished work that Haldane had seen in 1922 and which eventually appeared in 1928. In Part V (also 1927), Haldane takes a rare excursion into the stochastic theory, developing Fisher's 1922 branching-process treatment to make the first calculation of the probability of fixation of an advantageous mutant gene; in the Appendix to *The Causes of Evolution*, Haldane wrote with uncharacteristic modesty "The investigation of the case where the total population is finite has been wholly due to Fisher (1930) and Wright (1931)". In Part VIII (1931), Haldane inaugurates the mathematical study of two linked loci in a paper whose single diagram may have had some influence on Wright because it depicted population trajectories in the two-dimensional space of gene frequencies.

Continuing chronologically, we should next notice in passing two 1928 papers of Fisher's on the evolution of dominance, because these drew a comment from Wright the following year and inaugurated a topic which eventually involved many people, including Haldane. Important though the ensuing controversy is for the light it throws on the differing emphases of Fisher and Wright on the mechanisms of evolution, it is too much of a side issue to be considered in detail here. Also to be recorded is the 1929 abstract of a paper in which Wright gave the first published intimation of his "shifting balance" theory of evolution.

In 1929, Wright sent Fisher a draft of his forthcoming 1931 paper *Evolution in Mendelian populations* in which, as we shall see, he used a different method of determining the "relaxation time" required for a population to have its variance reduced by a factor e and found it to be $2n$ generations rather than Fisher's $4n$. Fisher acknowledged his mistake in 1930 in a paper in which he also gave "a more rigorous and comprehensive treatment of the whole subject" of his 1922 paper. Of this paper Motoo Kimura later wrote "In my opinion this is one of the most beautiful papers ever written on the mathematical theory of population genetics". However, the principal event of 1930 was the publication of Fisher's *magnum opus* in this field, *The Genetical Theory of Natural Selection*.

The depth, difficulty and originality of *The Genetical Theory* ensured that its initial impact was limited, just as Leonard Darwin, Charles's fourth son, to whom the book is dedicated, had predicted: "it will be slowly recognized as a very important contribution to the subject. But I am afraid it will be slow, because so few will really grasp all that it means" (Bennett, 1983). Haldane and Wright were perhaps the only two people really capable of appreciating it at the time. In the introduction to *The Causes of Evolution*, Haldane said that he was "much indebted" to Fisher's "brilliant book" and, in a 1931 essay (Haldane, 1931), he remarked:

> The book before us is of the latter class [books which, though often inaccurate in detail, state a new point of view, and lay the foundations of new branches of science] . . . No serious future discussion either of evolution or eugenics can possibly ignore it.

and

> during the next generation any discussions of the problem of gradual evolution which are likely to be of permanent value will take the form of a development, discussion, and perhaps in some cases a refutation, of the arguments stated in the book before us.

Wright reviewed *The Genetical Theory* at length for the *Journal of Heredity*, stating: "It is a book which is certain to take rank as one of the major contributions to the theory of evolution".

Only a brief description of *The Genetical Theory* is possible here. Chapter 1 contrasts the conservation of variance under Mendelian inheritance with its decay under the blending theory of inheritance assumed by Darwin. In Chapter 2, the Malthusian parameter of population increase is introduced (which Fisher had previously acknowledged to be equivalent to A.J. Lotka's "intrinsic rate of natural increase"), along with the novel concept of "reproductive value" for each age of life. The genic element in the overall variance is then discussed, leading to the "fundamental theorem of natural selection", which relates the partial rate of increase in the mean fitness of a population ascribable to natural selection acting through changes in gene frequencies to this genic variance in fitness. Long misunderstood, this theorem is now leading to new insights in population genetics and evolutionary biology. Chapter 3 is devoted to the evolution of dominance. Chapters 4 and 5 *Variation as determined by mutation and selection* contain Fisher's mature treatment of the topics he first examined in 1922. Once again, his concentration on the variance of a population is evident. Some section headings are: *The chances of survival of an individual gene*; *Low mutation rates of beneficial mutations*; *Single origins not improbable*; *Distribution of gene ratios in factors contributing to the variance*; and *The observed connection between variability and abundance*. Then, in *Stable gene ratios*, Fisher reproduces his 1922 result about an equilibrium under heterozygotic advantage, but this time demonstrating convergence to the equilibrium, and, in *Equilibrium involving two factors*, he opens the debate on selection at two linked loci (soon to be discussed by Haldane) and, in *Simple metrical characters*, the debate on selection for continuous characters. Chapter 6 contrasts sexual and asexual reproduction and the "runaway" process of sexual selection which Fisher had discussed in 1915, and ends with the highly influential discussion of *Natural selection and the sex-ratio* which, though not original to Fisher, has had ramifications in many areas of evolutionary biology. In Chapter 7, *Mimicry*, kin selection, already mentioned in 1914, is invoked to explain the evolution of distastefulness in insects. The last five chapters are on the implications of the entire theory for man and society, and include an emphasis on the influential concept of parental expenditure, an idea which Fisher attributes to Leonard Darwin.

The Genetical Theory of Natural Selection was a watershed in the development of neo-Darwinism because within the compass of a single book it provided, for the first time, a comprehensive account of the synthesis of Mendelian genetics and Darwinian evolutionary theory. As Fisher later remarked, one of the main purposes of his book was to demonstrate how little basis there was for the opinion that "the discovery of Mendel's laws of inheritance was unfavourable, or even fatal, to the theory of natural selection". However, it had the added importance that its argument was based on selection acting on the individual rather than on the group, and it was consequently to be highly influential in disposing not only of objections to Darwinism, but also of the group selection version of Darwinism then in the ascendant. Moreover, where Haldane's papers treated mainly the immediate effects of selection acting in various models, Fisher's book already discussed the second-order effects, such as the evolutionary modification of dominance, the tightening of linkage between loci, natural selection and the sex ratio, and selection for fertility, which occupy such an important place in modern evolutionary biology.

Wright's sole publication of any kind in 1930 was his review of *The Genetical Theory*, but this in itself was to have lasting influence in an unexpected way. In it, Wright commented on the importance Fisher evidently attached to his fundamental theorem of natural selection, which in turn prompted a comment from Fisher in the letter to Wright in which he expressed his delight at the review. Wright's letter in reply, dated 3 February 1931, included the following statement:

> Some aspects of the ideas which I tried to express in pages 353 to 355 of my review might be visualised as follows: Think of the field of visible joint frequencies of all genes as spread out in a multidimensional space. Add another dimension measuring degree of fitness. The field would be very humpy in relation to the latter because of epistatic relations, groups of mutations which were deleterious individually producing a harmonious result in combination.

The letter contains a drawing of a continuous line with three humps of different height (fitness) followed by a description of how a species might be seen to evolve in this representation. It would "tend to move steadily" up the slope towards the nearest maximum "under the influence of selection", but, if this maximum was not the absolute maximum, the species would only be able to escape from it and make further progress in fitness through "something other than the steady pressure of selection". Wright suggested four factors: (1) a changing environment would continually change the system of humps, (2) new mutations would add further dimensions to the field, allowing new paths of advance, and (3) random genetic drift in a small species would allow stochastic jumps from one hump to another, as would (4) a subdivided large species. Wright's image was that of a potential function expressing the selective effect in a multidimensional space, accompanied by stochastic variation. It was an image that Fisher was quick to dismiss, but one which was to achieve great popularity and influence.

In January 1931, Haldane delivered lectures at the National University of Wales, Aberystwyth, which the following year were to become his book *The Causes of Evolution*, but before describing the book the sequence of events first requires an examination of Wright's major 1931 paper *Evolution in Mendelian populations*, the final form of the manuscript which he had been polishing since 1925.

The paper occupies 63 pages of the journal *Genetics*. After a brief introduction *Theories of evolution*, it continues with a discussion of simple Mendelian selection models in *Variation of gene frequency*, with mentions of mutation, migration and multiple alleles, and a section *Random variation of gene frequency*. This opens with a brief discussion of Wright's own 1921 result on the rate of decrease of heterozygosis under inbreeding and continues: "Another phase of this question was opened by Fisher (1922) who attempted to discover the distribution of gene frequencies ultimately reached in a population as a result of the above process." The remainder of Wright's long paper is largely devoted to the elucidation of such distributions under various conditions, in *The distribution of gene frequencies and its immediate consequences*, and to the conclusions to be drawn from them, in the final section *The evolution of Mendelian systems*. Notable among these conclusions is his "shifting balance" theory of evolution according to which selection will be most effective in populations of intermediate size (defined, of course, in relation to the mutation rate) or in large populations which are subdivided into "partially isolated local races of small size".

As we noted earlier, this famous paper of Wright's was based on a typescript which he wrote in 1925 (which has not survived), the year after he had received from Fisher the offprint of the latter's 1922 paper. By the time it was published in 1931, *The Genetical Theory of Natural Selection* had already appeared and was indeed the object of four of his footnotes. Just as Wright had reviewed *The Genetical Theory*, Fisher published a notice of *Evolution in Mendelian populations*. Short and friendly, it is of interest for including Fisher's first criticism of the "shifting balance" theory of evolution. It is important to note that Wright's 1931 paper did not include any reference to the "adaptive surface" metaphor, whose origin is thus almost certainly no earlier than the February letter to Fisher. However, by August 1932, when he spoke at the 6th International Congress of Genetics at Cornell, Ithaca, US, he had seen how to modify this adaptive surface metaphor so as to give pictorial representation to his shifting balance theory, by replacing the space of *gene frequencies*, in which each point represents a *population* as in Haldane's 1931 diagram for two linked loci, by a space in which each point represents a *genotype*, and a population is thus represented by a sort of bounded cloud of points depicting its individual members. Fitness was now an attribute of the genotype, not of the population. Fisher and Haldane both heard Wright's paper at the Congress, which his biographer, W.B. Provine, called "probably the most influential paper he ever published", for it contained the first drawings of this "adaptive (or fitness) surface" diagram which had had its origin in Wright's contemplation of Fisher's fundamental theorem. More importantly for the future popularity of the adaptive surface, the lecture was heard (and seen) by Th. Dobzhansky, who was enthralled by it.

The metaphor captivated many others besides Dobzhansky and dominated evolutionary biology for a generation. Its appearance in 1932 essentially completed the theoretical basis of the neo-Darwinian revolution, not so much because it contributed to the mathematical theory itself, but because it used a graphical metaphor which, whether or not it was an accurate reflection of nature, was simple enough to persuade evolutionary biologists that the mathematical geneticists really had succeeded in understanding the Mendelian mechanism of evolutionary change.

Fisher's and Haldane's papers at the Congress did not have the same impact as Wright's. Haldane's was a general discussion entitled *Can evolution be explained in terms of known genetical facts?* (to which question he answered "probably") while in *The evolutionary modification of genetic phenomena* Fisher stood Haldane's "analytic and deductive" approach on its head by adopting the "inductive and statistical" approach involved in enquiring "Can genetical phenomena be explained in terms of known evolutionary causes?" under which Fisher then discussed the evolution of dominance.

1932 also saw the publication of Haldane's *The Causes of Evolution*, which is now summarized. It opens in Chapter 1 with a discussion of "what is meant by evolution" and of Darwinism and its critics. Chapter 2 discusses variation within a species and Chapter 3 the genetical analysis of interspecific differences, both chapters with many examples taken from nature. Chapter 4 describes natural selection at work and in Chapter 5 the question *What is fitness?* is answered with emphasis on "the fallacy . . . that natural selection will always make an organism fitter in its struggle with the environment". Here Haldane is echoing *The Genetical Theory* in stressing the importance of selection within populations as opposed to group selection, and there is a further similarity as Haldane turns to man and, like Fisher, invokes the example of social insects as a simile for the evolution of social behaviour. However, Haldane is more specific in discussing altruism: "For in so far as it makes for the survival of one's descendants and near relations, altruistic behaviour is a kind of Darwinian fitness, and may be expected to spread as the result of natural selection." Chapter 6 is a wide-ranging *Conclusion*.

The main interest of the book, however, is the Appendix, *Outline of the mathematical theory on natural selection*, which was not part of the lectures. Naturally, it summarizes many of the author's results from *A mathematical theory of natural and artificial selection* and also from Fisher's *Genetical Theory*. In a brief section, *Wright's theory*, Haldane mentions that Wright's 1931 paper was only published after his book was written. The most original section of the Appendix is *Socially valuable but individually disadvantageous characters* in which Haldane initiates the mathematical discussion of the selective effects of altruism.

Fisher, in an unpublished review of *The Causes of Evolution* (Bennett, 1983), reciprocated Haldane's epithet of 'brilliant' for *The Genetical Theory*, but complained about the loose and egocentric argument which the lecture format had encouraged. Privately, in a letter to Leonard Darwin, Fisher wrote that the book was "an utter disappointment" to him (Bennett, 1983). Wright, in an assessment of *Haldane's contribution to population and evolutionary genetics* written in 1969, mentioned the series of papers *A mathematical theory of natural and artificial selection*, but not *The Causes of Evolution*.

The publication of *The Genetical Theory of Natural Selection* in 1930, of *Evolution in Mendelian populations* in 1931, of *The Causes of Evolution* and the *Proceedings of the 6th International Congress of Genetics* in 1932, brings to a close the decade which opened with Fisher's 1922 paper. The theory of neo-Darwinism had been established, but its propagation through the biological community was yet to come. As with all scientific revolutions, its long-term impact is likely to be rather different from its immediate reception and this article therefore closes with an attempt at a prediction of the former rather than a description of the latter. For the history of the spread of the ideas in the period 1930 to 1950, the reader is referred to W.B. Provine's article *The role of mathematical population geneticists in the evolutionary synthesis of the 1930s and 1940s* (Provine, 1978).

The achievement of the decade 1922–32 was a revolution in the understanding of the role of natural selection in evolution. Mendelian inheritance was shown to be an adequate mechanism for the operation of Darwin's theory, so that the theory, shorn of its Lamarckian tendencies, could itself be accepted as the explanation of evolutionary change. Dismissive references to the theory as being incapable of accounting for evolution, such as Sir John Herschel's description of it in 1861 as "the law of the higgledy-piggledy", were replaced by the realization that "natural selection is a mechanism for generating an exceedingly high degree of improbability" (reported by Julian Huxley (Huxley *et al.*, 1954) to be one of Fisher's favourite remarks). The mathematical theory developed by Fisher, Haldane and Wright contained within it sufficient flexibility to accommodate the enormous variety of observations from nature and from selection experiments.

In the years following 1932, the different emphases of Fisher and Wright, Fisher thinking that "evolution would proceed most surely and rapidly in a large, panmictic, population" and Wright arguing that "the most favourable condition for evolution lies in a structured population" (the descriptions are those of Crow, 1992), were the subject of much controversy, but nature is so varied that there is surely a need for both points of view, and the same can be said in connection with the evolution of dominance. The one contribution which, though it was extremely influential, has not stood the test of time, is Wright's "adaptive surface" metaphor which, as we have seen, started life as a corollary to Fisher's fundamental theorem with gene frequency as the variate, and then underwent a metamorphosis into a picture in which there was no variate as such but the fitness of individual genotypes was "mapped". In 1935, the picture reverted to a surface in which the mean fitness of the population was once again graphed against gene frequencies and, in this form, was used as the basis for a great deal of theorizing. Unfortunately, the loose identification of this adaptive surface with the potential functions of physics then led to misapprehensions about the role of Fisher's fundamental theorem, and the net result was a model of the operation of natural selection which was untenable in detail and historically extremely confusing, Fisher being supposed to have championed a "potential function" view about which his resolute denials were of no avail. It is unlikely that this controversy will be long remembered.

The notable features of the neo-Darwinian theory are, first, that the Mendelian mechanism of inheritance (in conjunction, of course, with other known influences) accounts satisfactorily for the similarities between relatives and for the changes observed in selection experiments and inbred populations. Second, although changes under natural selection are difficult to observe, the Mendelian mechanism coupled with reasonable values for such parameters as mutation rates and effective population sizes (the latter determined by population structure as well as size) accounts satisfactorily for the kind of intrapopulation selection which appears to be the raw material of evolution. Third, detailed mathematical analysis of the stochastic properties of populations demonstrates that the survival probabilities of genes of varying degrees of advantage or disadvantage are compatible with the requirements of the evolutionary model. Fourth, the theory has been successfully applied to the evolution of genetic systems themselves, such as linkage, dominance, altruism and the sex ratio.

So successful has the neo-Darwinian revolution been in incorporating Mendelian genetics into the theory of evolution that the modern tendency in evolutionary biology is to take the genetics for granted by formulating models in terms of game theory and other similar concepts. This has its own dangers, both conceptual and technical. On the one hand, it introduces inappropriate words with a teleological flavour such as "strategy" and, on the other hand, it presumes that the underlying Mendelian mechanism will always be able to deliver the predicted result even though it has been ignored in the formulation of the problem. A study of the original papers and books by Fisher, Haldane and Wright is a sensible antidote to this tendency.

R.A. Fisher, J.B.S. Haldane and Sewall Wright jointly engineered a revolution. Though they often disagreed over its details, Fisher's remark to Wright in 1931 could be held to apply to all three equally: "However, there is a substantial body of theory on which I think we do agree and that after all is of infinitely more interest to the world at large than the very obscure points still in dispute."

Primary literature

All the papers of R.A. Fisher referred to in this article will be found in *Collected Papers of R.A. Fisher*, vols 1–5, edited by J.H. Bennett, University of Adelaide (1971–1974). A bibliography appears in vol. 1, and also in the biography of Fisher (Box, 1978). Fisher's 1911 lecture *Mendelism and biometry* is reprinted in *Natural Selection, Heredity, and Eugenics: Including Selected Correspondence of R.A. Fisher with Leonard Darwin and Others*, edited by J.H. Bennett, Oxford: Clarendon Press and New York: Oxford University Press (1983), which also includes correspondence with Haldane and Wright and a very valuable *Introduction* by Bennett which covers the period of the present article in detail. Fisher's 1922 paper is: On the dominance ratio, *Proceedings of the Royal Society of Edinburgh* 42: 321–41. *The Genetical Theory of Natural Selection* was published by Clarendon Press, Oxford, in 1930, and reprinted in a revised edition by Dover, New York, in 1958. A variorum edition prepared by J.H. Bennett was published in 1999 by Oxford University Press.

There is no collected edition of J.B.S. Haldane's papers, but a bibliography appears in his biography (Clark, 1968), and shortened references to the series of papers *A mathematical theory of natural and artificial selection* are given in his book *The Causes of Evolution*, originally published in London and New York by Longmans Green in 1932. There is a facsimile reprint of *The Causes of Evolution* published by Cornell University Press in 1966 and a reprint with a new *Introduction* and *Afterword* by E.G. Leigh Jr was published by Princeton University Press in 1990.

Most of the papers of Sewall Wright referred to in this article will be found in *Evolution: Selected Papers*, edited by W.B. Provine, Chicago University Press (1986), which also includes a complete bibliography, as does Provine's biography of Wright (see below). Wright's 1931 paper is: Evolution in Mendelian populations, *Genetics* 16: 97–159, and his 1932 paper is: The roles of mutation, inbreeding, crossbreeding and selection in evolution, *Proceedings of the Sixth International Congress of Genetics* 1: 356–66. Many years after the period described in this article Wright published *Evolution and the Genetics of Populations* in four volumes, University of Chicago Press (1968–78).

The important papers of Fisher, Haldane and Wright have been frequently reprinted. Two valuable collections are in the *Benchmark Papers in Genetics* series published by Dowden, Hutchinson and Ross, Stroudsburg, Pennsylvania: no. 7, *Stochastic Models in Population Genetics*, edited by W-H. Li (1977) and no. 8, *Evolutionary Genetics*, edited by D.L. Jameson (1977).

Secondary literature referred to in the text

Bennett, J.H. (1983) *Natural Selection, Heredity, and Eugenics. Including Selected Correspondence of R.A. Fisher with Leonard Darwin and Others*, Oxford: Clarendon Press, and New York: Oxford University Press

Box, J.F. (1978) *R.A. Fisher: The Life of a Scientist*, New York: Wiley

Clark, R. (1968) *J.B.S: The Life and Work of J.B.S. Haldane*, London: Hodder and Stoughton

Crow, J.F. (1992) Haldane, Fisher and Wright. In *J.B.S. Haldane: A Tribute*, Calcutta: Indian Statistical Institute

Haldane, J.B.S. (1931) Mathematical Darwinism. *Eugenics Review* 23: 115–17

Hogben, L. (1931) *Genetic Principles in Medicine and Social Science*, London: Williams and Norgate

Huxley, J.S., Hardy, A.C. & Ford, E.B. (eds) (1954) *Evolution as a Process*, London: Allen and Unwin; 2nd edition, 1958

O'Donald, P. (1990) Fisher's contributions to the theory of sexual selection as the basis of recent research. *Theoretical Population Biology* 38: 285–300

Provine, W.B. (1978) The role of mathematical population geneticists in the evolutionary synthesis of the 1930s and 1940s. *Studies in History of Biology* 2: 167–92

Provine, W.B. (1986) *Sewall Wright and Evolutionary Biology*, Chicago: Chicago University Press

Other secondary literature of interest

The secondary literature about Fisher, Haldane and Wright is now quite extensive. Many important contributions are to be found in the editorial material of the sources listed under **Primary literature** and in obituaries published by learned societies. The occasion of the centenary of Fisher's birth resulted in a number of publications in 1990 which are listed in Edwards (1992) and the publication of the variorum edition of *The Genetical Theory of Natural Selection* in 1999 prompted a number of articles including Edwards (2000).

Edwards, A.W.F. (1992) Celebration of the centenary of the birth of Sir Ronald Aylmer Fisher, 1990. *Historia Mathematica* 19: 81–82

Edwards, A.W.F. (2000) The genetical theory of natural selection. In *Perspectives*, edited by J.F. Crow & W.F. Dove, *Genetics* 154: 1419–26

See also:

Provine, W.B. (1971) *The Origins of Theoretical Population Genetics*, Chicago: University of Chicago Press

Sarkar, S. (ed.) (1992) *The Founders of Evolutionary Genetics*, Dordrecht: Kluwer

and articles in the series *Oxford Surveys in Evolutionary Biology* published by Oxford University Press.

GLOSSARY

allelomorph former term for allele

diallelic locus specific site on a chromosome where two different alleles are located

game theory mathematical theory to determine the most favourable strategy in situations of competition or conflict

heterozygosis percentage of heterozygotes for a given locus in a population

inbreeding coefficient of an individual, the probability that the pair of alleles carried by the male and female gametes that produced it is identical by descent from a common ancestor, as a result of inbreeding

path coefficient a standardized regression coefficient measuring the influence of one of a number of variables on another variable

---

See also **Gregor Mendel** (p.62); **Charles Darwin and the Theory of Evolution** (p.68)

Paper 192 (2003)

Human genetic diversity: Lewontin's fallacy
BioEssays 25, 798–801.

Note from A. W. F. Edwards: A paper such as this does not come out of the blue. It owes its existence to the prompting of my brother, Professor J. H. Edwards, FRS, Professor of Genetics at Oxford, and Dr Adam Wilkins, editor of *BioEssays*, at the time in Cambridge, both of whom it ought to have acknowledged. When my brother alerted me to the problem, he called it 'the death of phylogeny' and I immediately opened a file of that name and started to collect information, though it was a long time before I tracked down Lewontin's calculations and conclusions. My brother, Dr Wilkins, and I fell to discussing these informally at a meeting of the Genetical Society, and Dr Wilkins encouraged me to write a paper, since I seemed to have the measure of the fallacy.

Human genetic diversity: Lewontin's fallacy

A.W.F. Edwards

Summary
In popular articles that play down the genetical differences among human populations, it is often stated that about 85% of the total genetical variation is due to individual differences within populations and only 15% to differences between populations or ethnic groups. It has therefore been proposed that the division of *Homo sapiens* into these groups is not justified by the genetic data. This conclusion, due to R.C. Lewontin in 1972, is unwarranted because the argument ignores the fact that most of the information that distinguishes populations is hidden in the correlation structure of the data and not simply in the variation of the individual factors. The underlying logic, which was discussed in the early years of the last century, is here discussed using a simple genetical example. *BioEssays* 25:798–801, 2003.

"When a large number of individuals [of any kind of organism] are measured in respect of physical dimensions, weight, colour, density, etc., it is possible to describe with some accuracy the population of which our experience may be regarded as a sample. By this means it may be possible to distinguish it from other populations differing in their genetic origin, or in environmental circumstances. Thus local races may be very different as populations, although individuals may overlap in all characters;..." R.A. Fisher (1925). See Ref. 1.

"It is clear that our perception of relatively large differences between human races and subgroups, as compared to the variation within these groups, is indeed a biased perception and that, based on randomly chosen genetic differences, human races and populations are remarkably similar to each other, with the largest part by far of human variation being accounted for by the differences between individuals. Human racial classification is of no social value and is positively destructive of social and human relations. Since such racial classification is now seen to be of virtually no genetic or taxonomic significance either, no justification can be offered for its continuance". R.C. Lewontin (1972). See Ref. 2.

Gonville and Caius College, Cambridge, CB2 1TA, UK.
E-mail: awfe@cam.ac.uk
DOI 10.1002/bies.10315
Published online in Wiley InterScience (www.interscience.wiley.com).

"The study of genetic variations in *Homo sapiens* shows that there is more genetic variation within populations than between populations. This means that two random individuals from any one group are almost as different as any two random individuals from the entire world. Although it may be easy to observe distinct external differences between groups of people, it is more difficult to distinguish such groups genetically, since most genetic variation is found within all groups." *Nature* (2001). See Ref. 3.

Introduction

In popular articles that play down the genetical differences among human populations it is often stated, usually without any reference, that about 85% of the total genetical variation is due to individual differences within populations and only 15% to differences between populations or ethnic groups. It has therefore been suggested that the division of *Homo sapiens* into these groups is not justified by the genetic data. People the world over are much more similar genetically than appearances might suggest.

Thus an article in *New Scientist*[4] reported that in 1972 Richard Lewontin of Harvard University "found that nearly 85 per cent of humanity's genetic diversity occurs among individuals within a single population." "In other words, two individuals are different because they are individuals, not because they belong to different races." In 2001, the *Human Genome* edition of *Nature*[3] came with a compact disc containing a similar statement, quoted above.

Such statements seem all to trace back to a 1972 paper by Lewontin in the annual review *Evolutionary Biology.*[2] Lewontin analysed data from 17 polymorphic loci, including the major blood-groups, and 7 'races' (Caucasian, African, Mongoloid, S. Asian Aborigines, Amerinds, Oceanians, Australian Aborigines). The gene frequencies were given for the 7 races but not for the individual populations comprising them, although the final analysis did quote the within population variability. "The results are quite remarkable. The mean proportion of the total species diversity that is contained within populations is 85.4%.... Less than 15% of all human genetic diversity is accounted for by differences between human groups! Moreover, the difference between populations within a race accounts for an additional 8.3%, so that only 6.3% is accounted for by racial classification."

Lewontin concluded "Since... racial classification is now seen to be of virtually no genetic or taxonomic significance..., no justification can be offered for its continuance" (full quotation given above).

Lewontin included similar remarks in his 1974 book *The Genetic Basis of Evolutionary Change*[5] "The taxonomic division of the human species into races places a completely disproportionate emphasis on a very small fraction of the total of human diversity. That scientists as well as nonscientists nevertheless continue to emphasize these genetically minor differences and find new 'scientific' justifications for doing so is an indication of the power of socioeconomically based ideology over the supposed objectivity of knowledge."

The fallacy

These conclusions are based on the old statistical fallacy of analysing data on the assumption that it contains no information beyond that revealed on a locus-by-locus analysis, and then drawing conclusions solely on the results of such an analysis. The 'taxonomic significance' of genetic data in fact often arises from correlations amongst the different loci, for it is these that may contain the information which enables a stable classification to be uncovered.

Cavalli-Sforza and Piazza[6] coined the word 'treeness' to describe the extent to which a tree-like structure was hidden amongst the correlations in gene-frequency data. Lewontin's superficial analysis ignores this aspect of the structure of the data and leads inevitably to the conclusion that the data do not possess such structure. The argument is circular. A contrasting analysis to Lewontin's, using very similar data, was presented by Cavalli-Sforza and Edwards at the 1963 International Congress of Genetics.[7] Making no prior assumptions about the form of the tree, they derived a convincing evolutionary tree for the 15 populations that they studied. Lewontin,[2,5] though he participated in the Congress, did not refer to this analysis.

The statistical problem has been understood at least since the discussions surrounding Pearson's 'coefficient of racial likeness'[8] in the 1920s. It is mentioned in all editions of Fisher's *Statistical Methods for Research Workers*[1] from 1925 (quoted above). A useful review is that by Gower[9] in a 1972 conference volume *The Assessment of Population Affinities in Man*. As he pointed out, "...the human mind distinguishes between different groups *because* there are correlated characters within the postulated groups."

The original discussions involved anthropometric data, but the fallacy may equally be exposed using modern genetic terminology. Consider two haploid populations each of size n. In population 1 the frequency of a gene, say '+' as opposed to '−', at a single diallelic locus is p and in population 2 it is q, where $p+q=1$. (The symmetry is deliberate.) Each population manifests simple binomial variability, and the overall variability is augmented by the difference in the means. The natural way to analyse this variability is the analysis of variance, from which it will be found that the ratio of the within-population sum of squares to the total sum of squares is simply $4pq$. Taking $p=0.3$ and $q=0.7$, this ratio is 0.84; 84% of the variability is within-population, corresponding closely to Lewontin's figure. The probability of misclassifying an individual based on his gene is p, in this case 0.3. The genes at a single locus are hardly informative about the population to which their bearer belongs.

Now suppose there are k similar loci, all with gene frequency p in population 1 and q in population 2. The ratio of the within-to-total variability is still 84% at each locus. The total number of '+' genes in an individual will be binomial with mean kp in population 1 and kq in population 2, with variance kpq in both cases. Continuing with the former gene frequencies and taking $k=100$ loci (say), the mean numbers are 30 and 70 respectively, with variances 21 and thus standard deviations of 4.58. With a difference between the means of 40 and a common standard deviation of less than 4.6, there is virtually no overlap between the distributions, and the probability of misclassification is infinitesimal, simply on the basis of counting the number of '+' genes. Fig. 1 shows how the probability falls off for up to 20 loci.

One way of looking at this result is to appreciate that the total number of '+' genes is like the first principal component in a principal component analysis (Box 1). For this component the between-population sum of squares is very much greater than the within-population sum of squares. For the other components the reverse will hold, so that overall the between-population sum of squares is only a small proportion (in this example 16%) of the total. But this must not beguile one into thinking that the two populations are not separable, which they clearly are.

Each additional locus contributes equally to the within-population and between-population sums of squares, whose

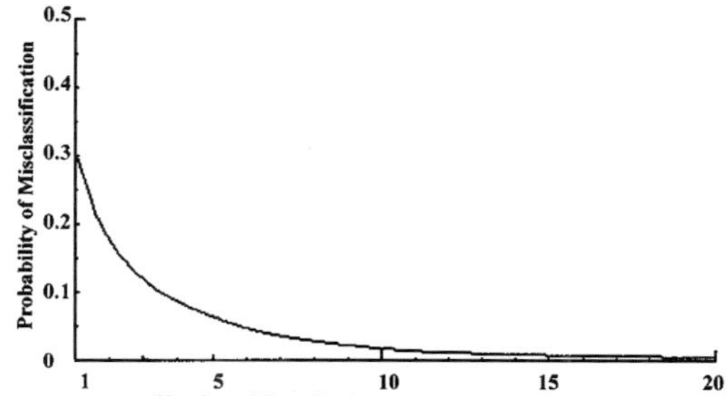

Figure 1. Graph showing how the probability of misclassification falls off as the number of gene loci increases, for the first example given in the text. The proportion of the variability within groups remains at 84% as in Lewontin's data, but the probability of misclassification rapidly becomes negligible.

Box 1. Principal component analysis

Principal Component Analysis (PCA) is a way of teasing out the more important information in multivariate data, where the high dimensionality renders simple graphical presentation impossible. The procedure can easily be understood even with just two variates, though its use might then be unnecessary. Taking an example from anthropometry where PCA originated, we might have data on the lengths and breadths of a number of human skulls. Each skull can be represented by a point in a diagram whose two axes are length and breadth. Since length and breadth will almost certainly be associated to some extent, the points will tend to be spread out preferentially in a certain direction, stretching from short length and breadth (a small skull) to long length and breadth (a large skull).

PCA defines this direction precisely as that of the line for which the sum of the squares of the perpendicular distances from the points to the line is a minimum. This line passes through the centre of gravity of the points, and a simple application of Pythagoras's Theorem shows that the one-dimensional array of the points defined by the feet of the perpendiculars from the points to the line then has the maximum possible sum of squares. In other words, the variability of the data has been partitioned into two components, one of which, along this line, is known as the (First) Principal Component because it encapsulates as much of the variability as can be represented in one dimension. The Second Component, at right angles to the First, encapsulates the remainder, which is, of course, a minimum.

These two components can be used as replacement axes on the graph. Sometimes the First Component will have an obvious meaning, as would be the case with the skulls, where it is clear that it corresponds in a general way to 'size'. Similarly the Second Component corresponds in some sense to 'shape', because a skull whose data-point is far from the line of the First Component will either be longer and narrower than the norm, or shorter and broader.

The procedure generalises to any number of variates, and the successive First, Second, Third, . . . Compo-Components are then mutually-orthogonal directions partitioning the total variability into ever-decreasing amounts. A graph of the first two components will represent as much of the information as is possible using only two dimensions.

proportions therefore remain unchanged but, at the same time, it contributes information about classification which is cumulative over loci because their gene frequencies are correlated.

Classification

It might be supposed, though it would be wrong, that this example is prejudiced by the assumptions that membership of the two populations is known in advance and that, at each locus, it is the same population that has the higher frequency of the '+' gene. In fact the only advantage of the latter simplifying assumption was that it made it obvious that the total number of '+' genes is the best discriminant between the two populations.

To dispel these concerns, consider the same example but with '+' and '−' interchanged at each locus with probability ½, and suppose that there is no prior information as to which population each individual belongs. Clearly, the total number of '+' genes an individual contains is no longer a discriminant, for the expected number is now the same in each group. A cluster analysis will be necessary in order to uncover the groups, and a convenient criterion is again based on the analysis of variance as in the method introduced by Edwards and Cavalli-Sforza.[10] Here the preferred division into two clusters maximises the between-clusters sum of squares or, what is the same thing, minimises the sum of the within-clusters sums of squares.

As pointed out by these authors, it is extremely easy to compute these sums for binary data, for all the information is contained in the half-matrix of pairwise distances between the individuals, and at each locus this distance is simply 0 for a match and 1 for a mismatch of the genes. Since interchanging '+' and '−' makes no difference to the numbers of matches and mismatches, it is clear that the random changes introduced above are irrelevant. Continuing the symmetrical example, the probability of a match is $p^2 + q^2$ if the two individuals are from the same population and $2pq$ if they are from different populations. With k loci, therefore, the distance between two individuals from the same population will be binomial with mean $k(p^2 + q^2)$ and variance $k(p^2 + q^2)(1-p^2-q^2)$ and if from different populations binomial with mean $2kpq$ and variance $2kpq(1-2pq)$. These variances are, of course, the same.

Taking $p = 0.3$, $q = 0.7$ and $k = 100$ as before, the means are 58 and 42 respectively, a difference of 16, the variances are 24.36 and the standard deviations both 4.936. The means are thus more than 3 standard deviations apart (3.2415). The entries of the half-matrix of pairwise distances will therefore divide into two groups with very little overlap, and it will be possible to identify the two clusters with a risk of misclassification which tends to zero as the number of loci increases.

By analogy with the above example, it is likely that a count of the four DNA base frequencies in homologous tracts of a genome would prove quite a powerful statistical discriminant for classifying people into population groups.

Conclusion

There is nothing wrong with Lewontin's statistical analysis of variation, only with the belief that it is relevant to classification

It is not true that "racial classification is...of virtually no genetic or taxonomic significance". It is not true, as *Nature* claimed, that "two random individuals from any one group are almost as different as any two random individuals from the entire world", and it is not true, as the *New Scientist* claimed, that "two individuals are different because they are individuals, not because they belong to different races" and that "you can't predict someone's race by their genes". Such statements might only be true if all the characters studied were independent, which they are not.

Lewontin used his analysis of variation to mount an unjustified assault on classification, which he deplored for social reasons. It was he who wrote "Indeed the whole history of the problem of genetic variation is a vivid illustration of the role that deeply embedded ideological assumptions play in determining scientific 'truth' and the direction of scientific inquiry".[5] In a 1970 article *Race and intelligence*[11] he had earlier written "I shall try, in this article, to display Professor Jensen's argument, to show how the structure of his argument is designed to make his point and to reveal what appear to be deeply embedded assumptions derived from a particular world view, leading him to erroneous conclusions."

A proper analysis of human data reveals a substantial amount of information about genetic differences. What use, if any, one makes of it is quite another matter. But it is a dangerous mistake to premise the moral equality of human beings on biological similarity because dissimilarity, once revealed, then becomes an argument for moral inequality. One is reminded of Fisher's remark in *Statistical Methods and Scientific Inference*[12] "that the best causes tend to attract to their support the worst arguments, which seems to be equally true in the intellectual and in the moral sense."

Epilogue

This article could, and perhaps should, have been written soon after 1974. Since then many advances have been made in both gene technology and statistical computing that have facilitated the study of population differences from genetic data. The magisterial book of Cavalli-Sforza, Menozzi and Piazza[13] took the human story up to 1994, and since then many studies have amply confirmed the validity of the approach. Very recent studies[14,15] have treated *individuals* in the same way that Cavalli-Sforza and Edwards treated *populations* in 1963, namely by subjecting their genetic information to a cluster analysis thus revealing genetic affinities that have unsurprising geographic, linguistic and cultural parallels. As the authors of the most extensive of these[15] comment, "it was only in the accumulation of small allele-frequency differences across many loci that population structure was identified."

References

1. Fisher RA. Statistical Methods for Research Workers. Edinburgh: Oliver and Boyd. 1925.
2. Lewontin RC. The apportionment of human diversity. In: Dobzhansky T, Hecht MK, Steere WC, editors. Evolutionary Biology 6. New York: Appleton-Century-Crofts. 1972. p 381–398.
3. The Human Genome. Nature 2001;409:following p 812.
4. Ananthaswamy A. Under the skin. New Scientist 2002;174:34–37.
5. Lewontin RC. The Genetic Basis of Evolutionary Change. New York: Columbia University Press. 1974.
6. Cavalli-Sforza LL, Piazza A. Analysis of evolution: evolutionary rates, independence and treeness. Theor Pop Biol 1975;8:127–165.
7. Cavalli-Sforza LL, Edwards AWF. Analysis of human evolution. Proc. 11th Internat. Congr. Genetics, The Hague 1963, Genetics Today 3. Oxford: Pergamon. 1965. p 923–933.
8. Pearson K. On the coefficient of racial likeness. Biometrika 1926;18: 105–117.
9. Gower JC. Measures of taxonomic distance and their analysis. In: Weiner JS, Huizinga J, editors. The Assessment of Population Affinities in Man. Oxford: Clarendon. 1972. p 1–24.
10. Edwards AWF, Cavalli-Sforza LL. A method for cluster analysis. Biometrics 1965;21:362–375.
11. Lewontin RC. Race and intelligence. Bulletin of the Atomic Scientists. March 1970;2–8.
12. Fisher RA. Statistical Methods and Scientific Inference. Edinburgh: Oliver and Boyd. 1956.
13. Cavalli-Sforza LL, Menozzi P, Piazza A. The History and Geography of Human Genes. Princeton University Press. 1994.
14. Pritchard JK, Stephens M, Donnelly P. Inference of population structure using multilocus genotype data. Genetics 2000;155:945–959.
15. Rosenberg NA, Pritchard JK, Weber JL, Cann HM, Kidd KK, Zhivotovsky LA, Feldman MW. Genetic structure of human populations. Science 2002;298:2381–2385.

Paper 197 (2004)

Parsimony and computers
In *Milestones in Systematics*, eds. D. M. Williams and P. L. Forey, The Systematics Association Special Volume Series 67, Boca Raton, FL: CRC Press, 181–190.

Chapter 8

Parsimony and Computers

A.W.F. Edwards

Abstract

The introduction and use of parsimony methods in phylogenetic estimation is described and discussed against the background of Ockham's razor and *extremum* principles in science generally. It is argued that the method of maximum likelihood applied to realistic models of evolutionary divergence is the optimum procedure, to which parsimony methods are often an acceptable approximation.

Historical Introduction

Parsimony or *maximum parsimony* are terms now in regular use to describe methods of phylogenetic estimation that involve seeking the tree that minimizes some function of evolutionary change such as the total number of mutations or the total length of the branches in a continuous character space (see, for example, the recent books by Weir [1996], Page and Holmes [1998] and Durbin et al. [1998]). The adoption of the word was perhaps unfortunate because it already had a well-established use in mathematics to describe the classical notion (Pappus, fourth century) that geometrical proofs should rely only on ruler-and-compass constructions, and that in general the minimum number of postulates should be used in mathematical proofs. It was also unfortunate because it had a well-established use in philosophy as "multiplicity ought not to be posited unnecessarily" (*"pluralitas non est ponenda sine necessitate"* [William of Ockham, fourteenth century]).

However, these were not the considerations that led Camin and Sokal (1965) to coin the phrase "principle of evolutionary parsimony" when they noticed that on their simulated data "those trees which most closely resembled the true cladistics invariably required for their construction the least number of postulated evolutionary steps for the character studied," since their justification was "the assumption that nature is indeed parsimonious." In a *Scientific American* article Sokal (1966:115) explained what they meant by this: "that nature is fundamentally parsimonious, so that the diversity in character states within a given group was achieved at or close to the minimum number of evolutionary steps." This teleological justification, embodying the idea that nature is trying to achieve something, quickly attracted criticism as being unacceptable (Edwards, 1966; Cavalli-Sforza and Edwards, 1967).

In fact there was already a publication and justification of maximum parsimony in the systematics literature, for at the Systematics Association meeting, Phenetic and Phylogenetic Classification, in Liverpool in 1964 (Heywood and McNeill, 1964:Figure 8.1) Cavalli-Sforza and I had referred to the use of our method of minimum evolution and had commented, "It is probable that this method gives a tree which is approximately the same as the projection

0-415-28032-X/04/$0.00+$1.50

SYSTEMATICS ASSOCIATION

SYMPOSIUM ON

PHENETIC
AND PHYLOGENETIC
CLASSIFICATION

TO BE HELD IN

THE UNIVERSITY OF LIVERPOOL

ON THE

8th and 9th April 1964

Speakers will include

R. E. BLACKWELDER	Carbondale, Illinois, U.S.A.
B. L. BURTT	Edinburgh, Scotland
L. L. CAVALLI-SFORZA	Pavia, Italy
Y. DELEVORYAS	Yale, Connecticut, U.S.A.
G. A. HARRISON	Oxford, England
A. D. J. MEEUSE	Amsterdam, Netherlands
E. C. OLSON	Chicago, U.S.A.
L. G. SILVESTRI	Milano, Italy

Topics to be discussed include

Recent advances in numerical taxonomy including character weighting and evolutionary interpretations. Developments in phenetic classification of hominids and other groups. Nature of characters and biological significance of correlation 'pleiades'. Fact and fiction in phylogeny.

A detailed programme will be available in January 1964. Enquiries should be addressed to
Dr. J. McNEILL, THE HARTLEY BOTANICAL LABORATORIES, THE UNIVERSITY, LIVERPOOL, 3.

FIGURE 8.1 Poster advertising the Systmatics Association meeting, Phenetic and Phylogenetic Classification, in Liverpool in 1964. (Heywood, V.H. and McNeill, J., Eds. With permission.)

of the maximum-likelihood tree onto the 'now' character space" (Edwards and Cavalli-Sforza, 1964:75). Maximum-likelihood trees are statistical estimates of trees based on the criterion of likelihood applied to a specific probability model, and likelihood itself is a well-known concept in statistics (see, for example, Edwards, 1972, 1992).

Our method of minimum evolution, originally called a principle until we realized that the word was misleading, was presented in 1963 (Edwards and Cavalli-Sforza 1963a: 104–105), and it is worth recalling its precise formulation:

The Reconstruction of Evolution

> Darwin's theory of evolution invites us to believe that closely similar species are closely related. In cases in which there is no fossil record, or other means of following evolutionary history, this concept provides the only means of drawing inferences about the course of evolution, and it is important to see whether it can be defined sufficiently precisely to be used for estimating the most likely form of an evolutionary tree solely from data on the similarities and differences between species. A detailed consideration of this problem leads to the conclusion that the following principle may be employed:

The Principle of Minimum Evolution

> The most plausible estimate of the evolutionary tree is that which invokes the minimum nett amount of evolution.

It is not hindsight to note the careful wording. We wanted to draw an inference by defining *sufficiently precisely* what was meant by similar species being closely related so that we could *estimate* the *most likely* form of an evolutionary tree. The precision was important because we needed an algorithm as the basis for our computer programs, while *estimate* and *most likely* are standard statistical parlance. The computer revolution, in full swing in the early 1960s, had two distinct aspects (Edwards, 1966; Sokal, 1966). Most obviously, it facilitated far larger calculations of all kinds, and in phylogenetic reconstruction this meant calculations that embodied searches through multitudinous tree forms. Less obviously, but just as importantly, the discipline of writing a computer program imposed an absolute requirement for clear thinking and thus for the consideration of basic principles.

It happened that Cavalli-Sforza and I were computer-literate geneticists, but exactly the same phenomenon had been appreciated by the computer-literate authors Sokal and Sneath (1963) in their contemporary book *Principles of Numerical Taxonomy,* to which it is a pleasure to pay tribute. Indeed, the session at the 1964 meeting in which Cavalli-Sforza and I referred to our method of minimum evolution was entitled "Numerical Taxonomy" and introduced by Sneath with some wise words about phylogenies (Sneath, 1964). Sneath (1995) has recently written a historical review.

Ever since Darwin's time phylogeneticists had been implicitly using minimum-evolution or parsimony arguments, but their explicit use had to wait until the requirements of computers compelled algorithmic thinking. When we presented some results of our phylogenetic analysis at the XIth International Congress of Genetics at The Hague in 1963 (Cavalli-Sforza and Edwards, 1964) Professor Silvestri of Milan queried the method of minimum evolution and in reply I said, "It seems to us that our principle has been implicitly used in all evolutionary studies, including cases in which fossil evidence has been available, and all we have done is to admit it explicitly and develop it as the basis for statistical estimation."

Silvestri was also at the Systematics Association meeting, and in his paper with Hill he enlarged on his doubts:

> In more general terms, going beyond the limits of microbiology and also those of biology itself, it does not seem possible for us to determine the stages through which a particular system has evolved simply by studying the final state of that system, if historical documentation is not available, just as an analysis of the position of chess pieces at a particular stage of the game does not allow us to reconstruct the moves made to reach that stage. At the most, all that is possible is to determine, for example, the most economical sequence of moves, or any other sequence which conforms with other principles. In this case, however, the hypothetical reconstruction can only be accepted if the postulated principle, e.g., of economy, can be verified. This is precisely the objection that could be raised against the presentation of the problem by Edwards and Cavalli-Sforza (1963[a]). Using data relative to the present horizon, they reconstruct a possible evolution of races of *Homo sapiens* according to the principle of minimum evolution, which is postulated but not verified. (Silvestri and Hill, 1964:91)

I do not know how good chess enthusiasts are at inferring recent moves from a glance at a mid-game position, but the biological problem was precisely the one that Cavalli-Sforza and I claimed to be able to solve by statistical methods, contrary to these authors' opinion. Their strictures (with which we were already familiar through our contact with Silvestri in Milan) were valuable, however, in forcing us to think about the basis of our methods and to realize that our principle of economy was justified by an appeal to probability models and maximum likelihood.

In order to implement a computer program "which invokes the minimum nett amount of evolution" it is necessary to define *amount of evolution,* and to us geneticists that presented no problem. We happened to be primarily studying human populations characterized by their gene frequencies at the common blood-group loci, and an amount of evolution was obviously a distance measure in some suitable space of gene frequencies, the details of which need not now concern us but which were dictated by the population geneticists' standard model of gene-frequency change through random genetic drift.

With discrete data, to which our principle was also designed to apply, amount of evolution was simply measured by the number of changes, whether of characters or "evolutionary steps" (as in Camin and Sokal, 1965) or mutations in the DNA sequences (as with modern molecular phylogenetics). In a recent paper (Edwards, 1996) I examined, with all the care that I could summon, the literature of the 1960s to see if I could find other explicit statements of the method of minimum evolution, and the earliest rival I unearthed was Eck and Dayhoff (1966), who without offering a justification, sought "the [tree] topology that has the minimum number of mutations [amino acid residue changes to proteins]" (Sneath, 1995 has a reference to the same *Atlas of Protein Sequence and Structure* of the previous year, which I have been unable to consult).[1] Further references can be found in Edwards (1996), to which I have nothing to add except that Dr. E. Zuckerkandl, in commenting on that paper, assures me that he had the principle in mind in late 1964 and regrets not having then stated it explicitly.

It is hardly surprising that geneticists, particularly population geneticists interested in Darwinian evolution, applied their newly won computing skills to the problem of phylogenetic reconstruction. Cavalli-Sforza mentioned in 1991 that he had "started thinking about ... the reconstruction of where human populations originated and the paths by which they spread throughout the world" more than 40 years previously while he was in R.A. Fisher's department of genetics at Cambridge. Fisher himself (1958) remarked in a BBC broadcast, the recording of which unfortunately no longer exists, "The reconstruction of racial history, though a distant aim, is not an unreasonable one." We were all thoroughly familiar with the work of A.E. Mourant, which had culminated in his book, *The Distribution of the Human Blood Groups* (1954).

On reflection, if one asks what kind of training and background would be valuable if one wanted to initiate research on how to construct phylogenies based on data from contemporary populations, races, species, etc., whose variability is accepted as being due to splitting, genetic isolation and evolutionary change, the answer would have to be evolutionary biology and statistical inference. Cavalli-Sforza and I had both been in Fisher's department, and Fisher was the author of both *The Genetical Theory of Natural Selection* (1930) and *Statistical Methods for Research Workers* (1925), two of the most famous and influential scientific books of the twentieth century.

[1] According to the Margaret Oakley Dayhoff Memorial website (http://www.dayhoff.cc/modbooks.html), the 1965 atlas is listed as "Atlas of Protein Sequence and Structure, 1965 (Black Cover) by Dayhoff, Eck, Chang, and Sochard, 93 printed one-sided pages."

The Influence of Traditional Procedures

But what about traditional systematics and taxonomy? The 1964 Systematics Association meeting was my introduction to these fields, and much of the discussion at that meeting was about philosophical issues. There were many amusing exchanges. At one point I am recorded as having replied to Sneath, "What we are searching around for is for some taxonomists to say why they are doing taxonomy — as opposed to phylogeny for instance. Mostly they are not engaged in pure information retrieval taxonomy but are undertaking applied taxonomy — but for what purpose? Until this purpose is declared then the required answers cannot be provided."

It all seems less of a mystery nearly 40 years on. Of course a taxonomic classification is, or should be, an efficient information-retrieval system, but it will also possess predictive value through the simple circumstance that characters are correlated (or else no tree structure would have emerged). The predictive value arises because the correlations are not confined to the data actually used. As to why these correlations exist, an underlying tree structure is the obvious answer, which will usually be justified by evolutionary considerations. In this connection, I should like to draw attention to the work of Cavalli-Sforza and Piazza (1975) on what they called "treeness." Although it has not been of much practical importance, their treeness test is an interesting application of the idea that a distance matrix contains within it evidence about any underlying tree structure that is amenable to statistical testing. An accessible account is contained in Cavalli-Sforza et al. (1994).

These remarks on taxonomy bring me to the vexing question of the influence of taxonomic procedures on modern phylogenetic analysis. Here I have little to add to Edwards (1996), where I reported on my efforts to trace the origins of various methods, notably parsimony. Farris and Kluge (1997) disputed my conclusions, preferring to emphasize the contributions of W. Hennig and W.H. Wagner (for references see Edwards, 1996). It is quite difficult to see what the argument is about. I have already quoted my 1964 remark about minimum evolution having been implicitly used in all evolutionary studies "and all we have done is to admit it explicitly and develop it as a basis for statistical estimation." Wagner (1984:115) was of the same opinion with respect to his own groundplan-divergence (GPD) method: "Students of the history of systematic biology will recognize that there is nothing wholly new or original in the elements of the GPD method. Most of its components are based on time-honoured concepts that are as old or nearly as old as evolutionary systematics itself." Perceiving in retrospect that algorithmic methods such as those attributed to Hennig and to Wagner make use of some form of parsimony argument is interesting, but it is not the same thing as starting with an explicit statement of a quantity to be minimized and then deducing the consequences in the manner of Edwards and Cavalli-Sforza (1963a), Camin and Sokal (1965) and Eck and Dayhoff (1966), leading eventually to the perfecting of computer algorithms.

The only reference to Hennig in the 1964 Systematics Association volume was by Sneath (1964:45), who referred to him as having pointed out "that the relation by ancestry of adults and larvae is necessarily the same, while their phenetic relationships need not be so." In the dominant field of modern molecular phylogeny, the three recent books I mentioned at the outset (Weir [1996], Page and Holmes [1998], and Durbin et. al. [1998]) contain hundreds and hundreds of references and indexed names, but not Hennig's.

In this account I have tended to use *parsimony* and *minimum evolution* as synonyms, but it should be mentioned that in recent writing (for example, Nei et al., 1998) there has been a tendency to polarize the terms so they apply to the discrete and the continuous, respectively. There seems to be no good reason for this, though it is a harmless enough convention. A further use of *parsimony* which has recently been noted by Steel and Penny (2000) refers to the preference for assuming as little as possible in the way of an underlying evolutionary model.

Maximum Likelihood

The method of maximum likelihood is the standard estimation procedure in statistical inference. Whether one looks at the inference problem from the point of view of classical repeated-sampling theory or Bayesian theory or straightforward likelihood theory, maximizing the likelihood emerges as the preferred procedure. There really is no dispute about this in regular estimation problems, and phylogenetic inference does seem to be unexceptional from a statistical point of view, even though it took a little while for the initial difficulties in the application of maximum likelihood to be sorted out. This was mainly done by Felsenstein (1968) and Thompson (1974) in their Ph.D. dissertations and subsequent publications.

It became clear quite early that there was no question of minimum evolution providing the maximum-likelihood solution exactly in the case of continuous characters (Cavalli-Sforza and Edwards, 1967; Edwards, 1969), and since there is no difference in principle between the continuous and discrete cases, it came as no surprise that a similar situation was found for discrete models (Felsenstein, 1973). It was also clear on intuitive grounds that the two methods (and, for that matter, almost any sensible algorithmic method that could be devised) were nevertheless going to give quite similar phylogenetic trees on any data with high treeness. The important point is that, as already mentioned in 1964, minimum-evolution algorithms are justified by likelihood considerations. Such considerations may also explain the appeal of Ockham's razor, as Sober (1988) concluded in his book *Reconstructing the Past: Parsimony, Evolution and Inference* and later in "Let's Razor Ockham's Razor" (1990). Edwards 1996 contains a full discussion. For a detailed discussion on the relations between minimum evolution and maximum likelihood the papers by Goldman (1990) and Steel and Penny (2000) should be consulted.

Sometimes attempts have been made to rank methods according to the probability of them giving the right result from simulated data. This is an old canard in statistical inference, and in connection with phylogenetic trees I tried to shoot it down in a letter to *Science* (Edwards, 1995). The approach confuses before-trial evaluation methods that are usually right and after-trial evaluation methods that are credible on each occasion of use. I can do no better than quote the last paragraph of my letter:

> The message from statistical theory is simple: You cannot simulate evolutionary trees without a probability model, and if you possess a probability model you should be performing efficient after-trial evaluation using the method of maximum likelihood. Simulation is valuable in assessing the robustness of a model, but not the suitability of an estimation procedure: modern statistical theory has solved that problem already. (Edwards, 1995)

However, the point has not yet been taken. For example Nei et al. (1998) used computer simulation to compare various methods, one of which, parsimony, they still claim has as its "theoretical basis" William of Ockham's "philosophical idea."

Extremum Principles in Science

Extremum principles, minimizing this or maximizing that, exert a peculiar fascination in science, and biologists have been eager to emulate the success of their physics colleagues in harnessing them. Alas, the situations are quite different. The classic physics example is from optics, where Fermat in the seventeenth century established that when a ray of light passes from a point in one medium through a plane interface to a point in another medium with a different refractive index, its route is the one that minimizes the time it takes for the light to travel. That is, on various assumptions about the speed of light in the different media, he showed that this solution corresponds to Snel's sine law of refraction. Modern books on optics find it necessary to frame Fermat's principle rather differently, not using the notion of a minimum at all, but the point for us is that in its original form it solves at least some optical path problems by minimization. The mathematics works out just right, but that is not to agree with Fermat that "nature is economical."

In the eighteenth and nineteenth centuries mathematical physics unearthed a whole range of *extremum* principles of great practical value because they provided a unifying point of view that gave results that are mathematically identical to the older approaches. Ideas such as potential energy and its minimization led to the notions of potentials and fields of force in gravitational and electromagnetic theory. The great eighteenth century developments in calculus not only provided the necessary mathematical tools but also encouraged thinking in terms of maxima and minima. No one supposed any longer that nature was economical, or quoted Ockham's razor. Rather, there was an implicit appeal to the classical concept of parsimony, that the simplest mathematical approach to any problem should be preferred, and this often involved an *extremum* principle. Ockham's razor cut away the unnecessary mathematics but offered no explanation of nature. The success of mathematical physics has nothing to teach phylogeneticists trying to infer an evolutionary tree, though it did unfortunately beguile evolutionary biologists into thinking they could emulate the physicists by constructing theories of fitness maximization — the survival of the fittest — so that they could use the metaphors of potential theory without troubling themselves with the mathematics. The whole saga of "adaptive surfaces" (or "topographies" or "landscapes") is a warning against such analogical seduction (Edwards, 1994).

A quite different kind of *extremum* principle grew up more or less simultaneously in statistics in the eighteenth and nineteenth centuries, the most famous example being the method of least squares originating in the work of Legendre and Gauss. Far from being a mathematically economical way of solving some problem as is Fermat's principle, it is a method for selecting among probability models the one that most closely fits some observed data (in Gauss's case, the observations of cometary positions). There is an element of arbitrariness in the choice of the criterion to be minimized (here the sum of the squares of the residual errors unexplained by the model), but subsequent research in statistical inference has produced plenty of justifications, and Fisher's method of maximum likelihood is one of them.

These outstandingly successful developments involving *extremum* principles contributed to the idea that one can elevate into a principle the minimization of some quantity or other — in our case a quantity of evolution — in order to infer an evolutionary tree. But the situations are quite different, and an amount of evolution is analogous neither to the potentials of physics nor the errors of statistics.

As we have seen, the true justification of any method of minimum evolution lies in an appeal to a probability model and the primitive notion of likelihood, and to that extent is closer to the method of least squares than to the methods of mathematical physics. A matter of minor historical interest is that the first attempt at a statistical solution of the phylogeny problem did in fact invoke least-squares:

> An additive model may be appropriate, in which the lengths of the segments are so estimated that the observed distance between two points is closely approximated by the expected distance along the tree, formed by the sum of the connecting segments. Least squares provides an elegant estimation procedure, and we are currently engaged in applying this approach to taxonomic data. Our results will be published elsewhere. (Edwards and Cavalli-Sforza, 1963b)

Epilogue

It is exactly 40 years since I joined Professor L.L. Cavalli-Sforza at the University of Pavia to work in his program of statistical human genetics. He introduced me to the phylogenetic trees problem and made the above suggestion about using least-squares on distance data. By the time I left for the United States in September 1964 we had working computer programs for least-squares and for minimum-evolution on distance data, and one for maximum-likelihood estimation that was at the prototype stage, handling only four populations. We were fully aware of contemporary work both in numerical taxonomy and in cluster analysis (through personal contact and from studying Sokal and Sneath, 1963). These programs were, of course, primitive compared to those available today. Personal computers did not exist, and suitable mainframe machines were often not available locally. For my later work in Pavia I had to travel to Milan to use an IBM machine, and even when at the University of Aberdeen from 1965 onwards I had initially to travel to WHO in Geneva to find adequate facilities. Nor could the primitive programs be easily made available to other workers, first because to do so required either the repunching of the program punch cards by the recipient or the mailing of heavy parcels of cards and also because each computer had its own idiosyncrasies that the program had to meet. Nevertheless, the programs, especially EVOMIN, our minimum-evolution program, did travel and were used by others. It was eventually written up as a student project by Thompson (1973).

My overwhelming impression of those years is that our small group, assisted by colleagues such as Professor L. Sylvestri of Milan and Dr. P.A.H. Sneath of London, argued over most of the methodological issues that have since occupied the ensuing four decades. We quickly came to the conclusion that although any reasonable method of cluster analysis would come up with a reasonable tree (and inventing reasonable methods was not difficult), a proper scientific approach required a probability model and maximum-likelihood estimation. Anything less, including the method of minimum evolution, could only be justified as an approximation. I can only repeat once more what we reported at the 1964 Systematics Association meeting: "It is probable that this method gives a tree which is approximately the same as the projection of the maximum-likelihood tree onto the 'now' character space."

At the 1987 Systematics Association meeting Felsenstein (1988:113) remarked that our 1964 paper, "the first to propose a statistical approach to inferring phylogenies," though "extremely influential," was "little known and even less read." He concluded his paper, which is a fine summary of the history, by saying,

> With a bit of luck and much hard work, we will find ourselves in an era in which there is a much more fruitful interaction between systematists and population geneticists. Ancient prejudices of each group against the other will have to be modified. We will need, on the way, to set aside the hypothetico-deductive model of phylogenetic inference, and adopt a statistical one. Whether we can treat statistical models in phylogenetic inference with the appropriate mixture of reverence and scepticism remains to be seen.' (Felsenstein, 1988:124)

This paper is intended as a contribution to the realization of Felsenstein's hopes.

References

Camin, J.H. and Sokal, R.R., A method for deducing branching sequences in phylogeny, *Evolution*, 19, 311–326, 1965 (reprinted in Duncan and Stuessy, 1985).

Cavalli-Sforza, L.L., Genes, peoples and languages, *Sci. Am.*, 265, 104–110, 1991.

Cavalli-Sforza, L.L. and Edwards, A.W.F., Analysis of human evolution, in *Genetics Today, Proceedings of the 11th International Congress of Genetics,* 1963, 3, 923–933, 1964.

Cavalli-Sforza, L.L. and Edwards, A.W.F., Phylogenetic analysis: models and estimation procedures, *Evolution,* 21, 550–570; *Am. J. Hum. Genet.* (Supplement), 19, 233–257, 1967.

Cavalli-Sforza, L.L., Menozzi, P., and Piazza, A., *The History and Geography of Human Genes,* Princeton University Press, Princeton, NJ, 1994 (abridged paperback ed. 1996).

Cavalli-Sforza, L.L. and Piazza, A., Analysis of evolution: evolutionary rates, independence and treeness, *Theoret. Popul. Biol.,* 8, 127–165, 1975.

Duncan, T. and Stuessy, T.F., Eds., *Cladistic Theory and Methodology,* Van Nostrand Reinhold, New York, 1985.

Durbin, R., Eddy, S., Krogh, A., and Mitchison, G., *Biological Sequence Analysis: Probalistic Models of Proteins and Nucleic Acids,* Cambridge University Press, Cambridge, 1998.

Eck, R.V. and Dayhoff, M.O., *Atlas of Protein Sequence and Structure,* National Biomedical Research Foundation, Silver Spring, MD, 1966.

Edwards, A.W.F., Studying human evolution by computer, *New Sci.,* 30, 438–440, 1966.

Edwards, A.W.F., Genetic taxonomy, in *Computer Applications in Genetics,* Morton, N.E., Ed., University of Hawaii Press, Honolulu, 1969, pp. 140–142.

Edwards, A.W.F., *Likelihood,* Cambridge University Press, Cambridge, 1972.

Edwards, A.W.F., *Likelihood,* expanded edition, Johns Hopkins University Press, Baltimore, 1992.

Edwards, A.W.F., The fundamental theorem of natural selection, *Biol. Rev.,* 69, 443–474, 1994.

Edwards, A.W.F., Assessing molecular phylogenies, *Science,* 267, 253, 1995.

Edwards, A.W.F., The origin and early development of the method of minimum evolution for the reconstruction of evolutionary trees, *Syst. Biol.,* 45, 79–91, 1996.

Edwards, A.W.F. and Cavalli-Sforza, L.L., The reconstruction of evolution: abstract, *Heredity,* 18, 553; *Ann. Hum. Genet.,* 27, 104–105, 1963a.

Edwards, A.W.F. and Cavalli-Sforza, L.L., A method for cluster analysis, Preprints 5th International Biometrics Conference, Cambridge, 1963b.

Edwards, A.W.F. and Cavalli-Sforza, L.L., Reconstruction of evolutionary trees, in *Phenetic and Phylogenetic Classification,* Publication 6, Heywood, V.H. and McNeil, J., Eds., Systematics Association, London, 1964, pp. 67–76 (reprinted in Duncan and Stuessy, 1985).

Farris, J.S. and Kluge, A.G., Parsimony and history, *Syst. Biol.,* 46, 215–218, 1997.

Felsenstein, J., Statistical Inference and the Estimation of Phylogenies, PhD dissertation, University of Chicago, 1968.

Felsenstein, J., Maximum-likelihood estimation of evolutionary trees from continuous characters, *Am. J. Hum. Genet.,* 25, 471–492, 1973.

Felsenstein, J., The detection of phylogeny, in *Prospects in Systematics,* Systematics Association Special Vol. 36, Hawksworth, D.L., Ed., Clarendon Press, Oxford, 1988, pp. 112–127.

Fisher, R.A., *Statistical Methods for Research Workers,* Oliver & Boyd, Edinburgh, 1925.

Fisher, R.A., *The Genetical Theory of Natural Selection,* Clarendon Press, Oxford, 1930 (2nd ed., Dover, New York, 1958; variorum ed., Oxford University Press, 1999).

Fisher, R.A., The discontinuous inheritance, *Listener,* 60, 85–87, 1958.

Goldman, N., Maximum likelihood inference of phylogenetic trees, with special reference to a Poisson process model of DNA substitution and to parsimony analyses, *Syst. Zool.,* 39, 345–361, 1990.

Heywood, V.H. and McNeill, J., Eds., *Phenetic and Phylogenetic Classification,* Publication 6, Systematics Association, London, 1964.

Mourant, A.E., *The Distribution of the Human Blood Groups,* Blackwell, Oxford, 1954.

Nei, M., Kumar, S., and Takahashi, K., The optimization principle in phylogenetic analysis tends to give incorrect topologies when the number of nucleotides or amino acids used is small, *Proc. Natl. Acad. Sci. Washington,* 95, 12390–12397, 1998.

Page, R.D.M. and Holmes, E.C., *Molecular Evolution: A Phylogenetic Approach,* Blackwell, Oxford, 1998.

Silvestri, L.G. and Hill, L.R., Some problems of the taxometric approach, in *Phenetic and Phylogenetic Classification*, Publication 6, Heywood, V.H. and McNeil, J., Eds., Systematics Association, London, 1964, pp. 87–103.

Sneath, P.H.A., Introduction, in *Phenetic and Phylogenetic Classification*, Publication 6, Heywood, V.H. and McNeil, J., Eds., Systematics Association, London, 1964, pp. 43–45.

Sneath, P.H.A., Thirty years of numerical taxonomy, *Syst. Biol.*, 44, 281–298, 1995.

Sober, E., *Reconstructing the Past: Parsimony, Evolution, and Inference*, MIT Press, Cambridge MA, 1988.

Sober, E., Let's razor Ockham's razor, in *Explanation and its Limits*, Knowles, D., Ed., Cambridge University Press, Cambridge, 1990, pp. 73–93.

Sokal, R.R., Numerical taxonomy, *Sci. Am.*, 215(6), 106–116, 1966.

Sokal, R.R. and Sneath, P.H.A., *Principles of Numerical Taxonomy*, Freeman, San Francisco, 1963.

Steel, M. and Penny, D., Parsimony, likelihood and the role of models in molecular phylogenetics, *Mol. Biol. Evol.*, 17, 839–850, 2000.

Thompson, E.A., The method of minimum evolution, *Ann. Hum. Genet.*, 36, 333–340, 1973.

Thompson, E.A., Mathematical Analysis of Human Evolution and Population Structure, PhD dissertation, University of Cambridge, 1974.

Wagner, W.H., Jr., Applications of the concepts of groundplan-divergence, in *Cladistics: Perspectives on the Reconstruction of Evolutionary History*, Duncan, T. and Stuessy, T.F., Eds., Columbia University Press, New York, 1984, pp. 95–118.

Weir, B.S., *Genetic Data Analysis II*, Sinauer, Sunderland, MA, 1996.

Paper 225 (2009)

Statistical methods for evolutionary trees
Genetics 183, 5–12.

Perspectives

Anecdotal, Historical and Critical Commentaries on Genetics

Statistical Methods for Evolutionary Trees

A. W. F. Edwards[1]

Gonville and Caius College, Cambridge CB2 1TA, United Kingdom

ABSTRACT

In 1963 and 1964, L. L. Cavalli-Sforza and A. W. F. Edwards introduced novel methods for computing evolutionary trees from genetical data, initially for human populations from blood-group gene frequencies. The most important development was their introduction of statistical methods of estimation applied to stochastic models of evolution.

> The comparative ease with which certain aspects of population analysis are made possible by blood-group studies has already revealed ethnographic distinctions with a detailed precision quite unattainable without their aid. Although inevitably some ethnographers have at first felt ill at case with the new facts, and what is indeed much more important with the new *potentialities* of serological studies, yet from a wider point of view it must be evident that all that ethnography strives to do will be facilitated beyond precedent as our knowledge of the blood-groups is extended. The traces left on existing populations by such past events as race mixture, by the relative expansion or attenuation of numbers, or by evolutionary modification under environmental influences, have certainly not by existing means been successfully unravelled. In considering the use of blood-group evidence in clarifying the ethnic history of Man, the peculiar advantage should be emphasized that independent evidence is within reach from different parts of the human germ plasm, so that independent confirmation should often be obtainable from the evidence of different factors independently inherited.
>
> R. A. Fisher (1956)

ANY historian of science who chronicles events in which he himself has taken part should issue a disclaimer. First, his own memory is fallible and may even be innocently selective. Thus, John Venn, the leading exponent of the genre of university and college biographical history (and incidentally the inventor of the Venn diagram), gave the date of his degree of doctor of science as 1883 instead of 1884 in his own entry in a volume that he himself compiled (Venn 1898). Second, in the development of scientific methodology there is no new thing under the sun. Every "new" idea is like another node in a spreading network. When placed in its historical setting, not only will an idea be seen to be the intersection of many advancing lines, but more often than not the very idea will be found earlier in the network, born before its time and recognizable only with the advantage of hindsight. As the historian of science K. O. May (1975) warned:

> On closer examination, however, the hope of finding a "first" comes to grief because of the historically dynamic character of ideas. If we describe a result with sufficient vagueness, there seems to be an endless sequence of those who had something within the vague specifications. Even plagiarists usually introduce innovations! If we specify the idea or result precisely, it turns out that exact duplications seldom occur, so that every mathematical event is a "first," and the priority question becomes trivial.

In this account of a development in which I was personally involved, I describe it from the point of view of a historian of science, stressing the influence of social connections and the scientific atmosphere as much as the contributions of particular individuals.

In the 1960s, I collaborated with L. L. Cavalli-Sforza in his project to develop methods for constructing phylogenetic trees from genetical data using modern statistical methods and the then-new computers. In a relatively short period of intense activity at the University of Pavia Italy, we developed three methods: (1) least-squares based on an "additive" tree; (2) minimum evolution also called maximum parsimony; and (3) maximum likelihood. Although this was before DNA sequences

[1]*Author e-mail:* awfe@cam.ac.uk

were available, these methods underlie modern work on phylogenetic analysis.[1]

I preface the account with a description of how the statistician and geneticist R. A. Fisher (1890–1962) bequeathed an "intellectual milieu" (to use his own expression; FISHER 1959) that might almost have been created specifically for the task at hand. A detailed account of the development of the methods follows.

THE FISHERIAN BACKGROUND

As early as 1925, in his pioneering book *Statistical Methods for Research Workers*, FISHER had clearly stated what many have not understood (see EDWARDS 2003):

> When a large number of individuals [of any kind of organism] are measured in respect of physical dimensions, weight, colour, density, etc., it is possible to describe with some accuracy the population of which our experience may be regarded as a sample. By this means it may be possible to distinguish it from other populations differing in their genetic origin, or in environmental circumstances. Thus local races may be very different as populations, although individuals may overlap in all characters . . .

Fisher's essential point was that the ability to investigate meaningful population differences from data such as human blood-group frequencies depends on the accumulation of information from a variety of blood-group systems, no one of which will reveal the phylogenetic structure by itself. The same is true today when the wealth of genetic material available for analysis in all species is effectively boundless.

At the Galton Laboratory, University College London, Fisher established a Serological Unit in 1935, which was funded by the Rockefeller Foundation to study human blood groups and other genetic markers with a view to initiating a linkage map of humans (EDWARDS 2004a). One of those who joined the unit was R. R. Race, who, after the Second World War, published with Ruth Sanger the pioneering book *Blood Groups in Man* (RACE and SANGER 1950) to which Fisher contributed the foreword. Race and Sanger became among Fisher's closest friends and kept him abreast of the latest developments in serology. At the same time (1951), the Nuffield Foundation made a grant to the Royal Anthropological Institute, London, "to aid the study of blood groups as a clue to the understanding of human diversity," and this enabled A. E. Mourant to write *The Distribution of the Human Blood Groups* (MOURANT 1954; foreword by H. J. Fleure, from which the above quotation about the purpose of the grant is taken).

[1]I gave an earlier and more detailed account of the Pavia developments in a talk entitled, "History and philosophy of phylogeny methods" at the EC Summer School "Methods for Molecular Phylogenies," Newton Institute, Cambridge, United Kingdom, on August 10, 1998, on which parts of the present article are based. The typescript of the talk was distributed to participants and is available from the author.

Fisher's Galton Unit was evacuated to Cambridge during the war, and Fisher was reunited with it on becoming Professor of Genetics in 1943. When, in 1945, Race was promoted to be the Director of the Galton Unit following the death of the first director, G. L. Taylor, his position was filled by Mourant (MOURANT 1995). Therefore, for a brief period, Fisher, Race, Sanger, and Mourant were in regular contact in Cambridge. Fisher made great efforts to persuade the University of Cambridge to offer both Race and Mourant positions in his department after the war ended in 1945, but without success. In 1946, both took up posts at the Lister Institute in London as Directors of Medical Research Council Units (BOX 1978) but continued to keep in close touch with Fisher.

In 1946 on a visit to Milan, Race had been introduced to Cavalli (*b.* 1922; later Cavalli-Sforza), then a young bacteriologist with a medical degree and an interest in genetics who "mentioned that he was then grappling with the study of Fisher's statistical books" (BOX 1978). Race wrote to Fisher about how impressed he had been with him. Cavalli was hoping to spend some time in England, and in March 1948 he arrived at the John Innes Horticultural Institution on a scholarship to spend 6 months working in the laboratory of K. Mather, a pre-war colleague of Fisher's at University College. In July he traveled to Stockholm for the 8th International Congress of Genetics and met Fisher for the first time (he had already seen and heard him at meetings of the Genetical Society) "and after five minutes of conversation, he offered me a job at Cambridge in his department for setting up a laboratory to work on crossing over in bacteria" (CAVALLI-SFORZA 1990; CAVALLI-SFORZA and CAVALLI-SFORZA 2005). Cavalli accepted and started in Cambridge in October. Both Race and Mather also attended the Stockholm Congress and perhaps acted as midwives to Fisher's offer. Cavalli spent 2 years in Cambridge and has written (CAVALLI-SFORZA 1991) about the genesis during his time at Cambridge of his idea of reconstructing evolutionary trees:

> More than 40 years ago, when I was studying bacterial genetics in the laboratory of Sir Ronald A. Fisher of the University of Cambridge, the place was saturated with mathematical theorizing. Thus, it is not surprising that I started thinking about a project so ambitious it seemed almost crazy: the reconstruction of where human populations originated and the paths by which they spread throughout the world. I reasoned that the task could be accomplished by measuring how closely living populations are related to one another and by deducing from this information a comprehensive family tree.

Cavalli returned to Milan in October 1950 as a bacteriologist, taking with him the lasting friendship, personal and scientific, of Race, Sanger, Mourant, and Fisher. These were the world's leaders in blood-group genetics, in the application of genetics to anthropology, and in statistics, especially as applied to problems of genetical

FIGURE 1.—Italy, 1963: standing—Luca and Alba Cavalli-Sforza and three of their children; seated—Hiroko Kimura, Anthony Edwards, Akio Kimura, and Thomas, Catharina, and Ann Edwards. Photograph by Motoo Kimura; reproduced by courtesy of Hiroko Kimura.

estimation. All that was missing from the realization of Cavalli's project was a suitable academic environment and adequate computing facilities.

The decade 1950–1960 saw astonishing advances in electronic computers for scientific use. In Cambridge, one of the leading centers, the pioneer machine EDSAC 1 performed its first calculation on May 6, 1949, and its last on July 11, 1958, by which time its successor EDSAC 2 had taken over (WILKES 1985). Appropriately, the first use of an electronic computer in biology was to tabulate the solution to a differential equation for Fisher (FISHER 1950).

By 1961, Cavalli-Sforza (as he became in 1950) had left bacteriological work and moved into the field of human population genetics with a professorship at the University of Pavia. The university had a modern computer on order, an Olivetti Elea 6001. Research funds were available to use it for human genetics and in particular to make associated appointments. Early in the year Cavalli-Sforza came back to his old Cambridge department and sought me out. As an undergraduate I had initially read physics and allied subjects but in 1956 applied to do genetics under Fisher in his last year as professor (see EDWARDS 1990). By 1961, I had completed a Ph.D. on the statistics of the human sex ratio, out of which had already come a number of publications. As part of my course I had been sent to Race and Sanger's laboratory to learn elementary blood grouping and had met Mourant as well. I had also been sent to Mather, by then in Birmingham, to study chromosomes. I had spent a month in Uppsala working with M. Fraccaro, an Italian colleague of Cavalli-Sforza's. Any of these researchers might have mentioned my name, as Fisher might have, for I had been the last undergraduate to be admitted to his department before he retired in 1957. Cavalli-Sforza offered me an appointment in Pavia and, after a short visit there, I accepted.

Like Cavalli-Sforza, I was not trained as a statistician and had similarly learned most of my statistics from Fisher's books (EDWARDS 2005). As a research student I had taught myself to use the Cambridge computer EDSAC 2 to tackle problems in genetics (EDWARDS 1961). It would be hard to imagine a better training for the problem of constructing phylogenetic trees from genetical data using statistical methods and electronic computers.

During a visit to Pavia in 1963 by Motoo Kimura and his wife Hiroko and son Akio, the Cavalli-Sforzas invited the Kimura and Edwards families to accompany them on an outing to the Alps, during which Motoo Kimura took a group photograph (Figure 1). This appears to be the only photograph showing Cavalli-Sforza and me together at the time of our collaboration.

PHYLOGENETIC ANALYSIS BEFORE THE COMPUTER

Acceptance of Darwin's view that *Homo sapiens* formed a single species whose races descended from a common stock led rapidly to the construction of phylogenies, most memorably that by HAECKEL (1876; reproduced in EDWARDS 2001), which placed the origin in Lemuria, a mythical continent in the Indian Ocean that delivered branches of humankind to Africa, India, the Far East, and Australasia before conveniently disappearing beneath the waves. Haeckel recognized 12 races some of which he subdivided. All such phylogenies were of course, based on classification by observable or measurable anthropological characters. Population differences in the *ABO* blood-group gene frequencies were first noticed during World War I and, as more blood groups were discovered and samples were collected from populations worldwide, "contour" maps depicting the frequencies of particular genes were derived (see especially MOURANT 1954, *The Distribution of the Human Blood Groups*). BOYD (1952), in *Genetics and the Races of Man*, reviewed attempts to use blood-group information to define races, coming down in favor of six genetically defined races, although with no provision for a Pacific race because the blood-group evidence there was too discordant. But he stopped short of suggesting a phylogenetic tree, inclining in any case toward a polyphyletic origin. The use of the blood groups to construct a phylogenetic tree of human populations is probably the first example of the use of genetic data in phylogeny because it is unlikely that a comparable amount of data would have accrued for any other species.

At first sight it might seem natural to describe the development of the new methods of phylogenetic analysis against the background of taxonomic practice in the middle of the 20th century, but in fact it took place almost independently of the field of taxonomy. Rather, it was an attempt by geneticists and statisticians to build from scratch methods designed primarily to handle human blood-group gene-frequency data of the kind already mentioned. Unlike contemporary practice in taxonomy, the new methods of phylogenetic analysis belonged to the quite separate tradition of genetical model building and statistical estimation theory used, for example, in genetic linkage (MATHER 1951; BAILEY 1961).

The closest connection to taxonomic methods was with the "numerical taxonomy" promoted by SNEATH and SOKAL in a long article in Nature in March 1962 and subsequently in their book, *Principles of Numerical Taxonomy* (SOKAL and SNEATH 1963). In Pavia, we were familiar with this development through personal contact with Sneath, but it hardly overlapped with our thoughts because of the absence from it of any approaches using statistical estimation procedures. Of course, it was encouraging to find that others were also grappling with the problem of coaxing the early electronic computers into producing useful taxonomic output, and the discussion of clustering methods by Sokal and Sneath will have been helpful. But the main point to note is that, insofar as Sokal and Sneath were building on contemporary taxonomy, their book is a convenient reflection of that field as it stood in 1963. SNEATH (1995) reviewed the development in "Thirty years of numerical taxonomy," although to describe numerical phylogenetics then as one of the "four areas of numerical taxonomy that were only faintly foreshadowed in 1963" is perhaps more monophyletic than the evidence can bear. FELSENSTEIN (2004) noted that MICHENER and SOKAL (1957), in "A quantitative approach to a problem in classification," one of the founding articles of numerical taxonomy, derived a tree whose "interpretation as a phylogeny was made by Michener; Sokal saw it as a classification that did not necessarily have any validity as a phylogeny."

Another useful guide to taxonomic thinking at the time is the book, *Phenetic and Phylogenetic Classification*, from the Systematics Association meeting held in Liverpool (United Kingdom) in April 1964 (HEYWOOD and MCNEILL 1964b). The meeting had also featured prominently in Nature: "The main aim of the symposium was to consider the principles which should be followed in the construction and arrangement of taxonomic groups and how these principles differ if the purpose of the classification is evolutionary or not" (HEYWOOD and MCNEILL 1964a). Apart from our own article at the meeting (EDWARDS and CAVALLI-SFORZA 1964b), to which I return below, again there is no evidence of any overlap with our statistical genetics approach.

To maintain that contemporary taxonomic thinking made little contribution to the development of our methods of phylogenetic analysis is not, of course, to deny that with the benefit of hindsight certain similarities could be recognized. This is particularly true of the "method of minimum evolution," called elsewhere the "principle of parsimony." I return to this point after describing our developments.

THE DEVELOPMENT OF THE METHODS

Least-squares estimation on an additive tree: When I took up my post in Pavia in October 1961, the promised computer, an Olivetti Elea 6001, was late in delivery, so that it was not possible to start work on the computer analysis of the human population data as planned. Instead, in many discussions over the next few months, Cavalli-Sforza introduced me to his idea of developing methods for a computer-based construction of evolutionary trees from contemporary data, his particular interest being, as Fisher's had been, the analysis of human populations using information from the blood groups. Initially, I was skeptical. I had studied linkage estimation theory under Fisher himself and I knew how difficult it was. I had visions of evolutionary tree estimation being much the same but with the addition of the need to estimate the form of the tree itself, surely a fatal complexity: my intuition was that there would be insufficient data for the task.

Cavalli-Sforza was very persuasive, however. Let us choose, he said, a number of distinct human populations, say15, and from their gene frequencies at various blood-group loci let us compute the pairwise genetic distances between them. Now imagine a phylogenetic tree uniting the 15 populations, with a particular topology but unknown branch lengths. To each pairwise distance there corresponds a distance along the branches of the tree, being the sum of the relevant branch segments. These can now be estimated by the method of least squares. Different topologies can be tried and the one with the least residual sum of squares adopted. Because the data for the method are naturally presented as a square matrix of pairwise distances between the populations, this has often been called the "distance matrix" method. It was wholly Cavalli-Sforza's idea, from its inception right up to the specification of the estimation procedure. As a matter of record, the method was first mentioned in an article on cluster analysis presented at the International Biometrics Congress in September 1963 (EDWARDS and CAVALLI-SFORZA 1963b):

> Alternatively, an additive model may be appropriate, in which the lengths of the segments are so estimated that the observed distance between two points is closely approximated by the expected distance along the tree, formed by the sum of the connecting segments. Least squares provides an elegant estimation procedure...

My first task was to write the computer program to implement Cavalli-Sforza's proposal. The Olivetti computer was operational by May 1962, but as yet without any high-level language such as FORTRAN so that progress was quite slow. A FORTRAN II compiler became available early in 1963 and probably by June the additive tree least-squares program was working for 15 populations. I had made a special visit to Mourant in London and with his help had selected 15 human populations from around the world for which there was adequate information about gene frequencies for the commoner blood groups. The method and the results were reported by Cavalli-Sforza in an invited paper at the 11th International Genetics Congress at The Hague in September 1963 (Figure 2; CAVALLI-SFORZA and EDWARDS 1965).

The method of minimum evolution: The article also referred to an alternative method in which the criterion to be minimized was not the residual sum of squares but simply the total branch length of the tree in the multidimensional character space. This came about through my realizing that, although the points representing the 15 populations were displayed in a Euclidean space, the tree generated by Cavalli-Sforza's least-squares procedure was not. Of course, one could simply add the condition that the nodes of the tree must be points in the same space, but, thinking about this early in the spring of 1963, it suddenly occurred to me that of all the possible trees in the space, one must be the shortest overall. Given that the metric of the character space was an "amount" of evolution, this shortest tree was the one that joined all the data points invoking the minimum amount of evolution.

This intuitive algorithmic solution was obviously quite sensible, for points close to each other in the character space would tend to be on the same branch of the tree. But it was also very easy to write the associated computer program EVOMIN to search for this minimum Steiner tree in a multi-dimensional space. A copy of EVOMIN still exists. I presented the idea at the summer meeting of the Genetical Society of Great Britain in London in July 1963 (Figure 3; EDWARDS and CAVALLI-SFORZA 1963a) and its use on the blood-group data was then reported at the Genetics Congress in September.

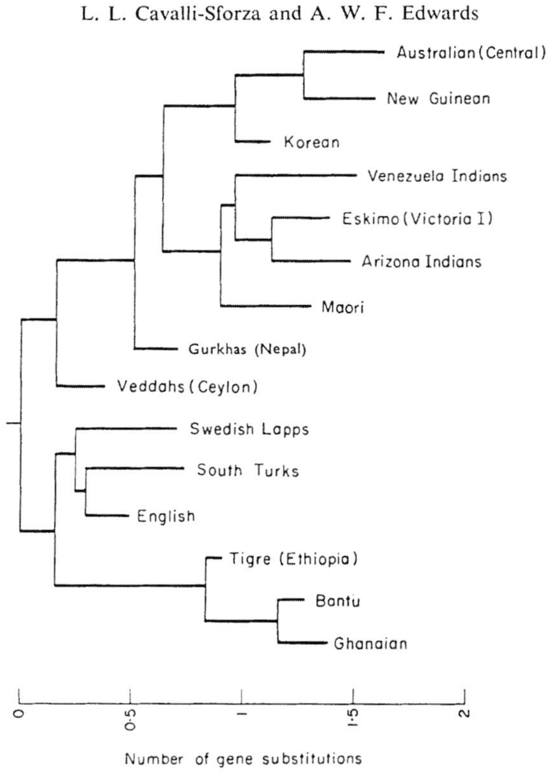

FIGURE 2.—The first numerical phylogeny produced by a combination of least-squares and parsimony methods (11th International Congress of Genetics, 1963, CAVALLI-SFORZA and EDWARDS 1965). The blood-group gene frequencies of the 15 populations used to produce Figure 2 have never been published but are available from the author.

The method of maximum likelihood: Just as I had not been entirely satisfied with Cavalli-Sforza's additive-tree method, so he was not satisfied with the logic of my method of minimum evolution, and in truth neither was I. It just seemed a very neat computable idea that would give a sensible result. We both strove for something better, something more statistical, and during many discussions came to see that the basis of any statistical method comparable to, say, the estimation of recombination fractions in genetic linkage, would be a stochastic model for the evolutionary process whose parameters (including the form of the tree) we would then estimate. With this clarification, we chose a branching diffusion process as our model for evolution. We added a time dimension to the character space, and the challenge became one of estimating not only the form (topology) of the tree but the position in this space-time of each of its nodes. Naturally, the chosen estimation procedure was Fisher's method of maximum likelihood. The development was first reported at the Systematics Association meeting in April 1964 (EDWARDS and CAVALLI-SFORZA 1964b).

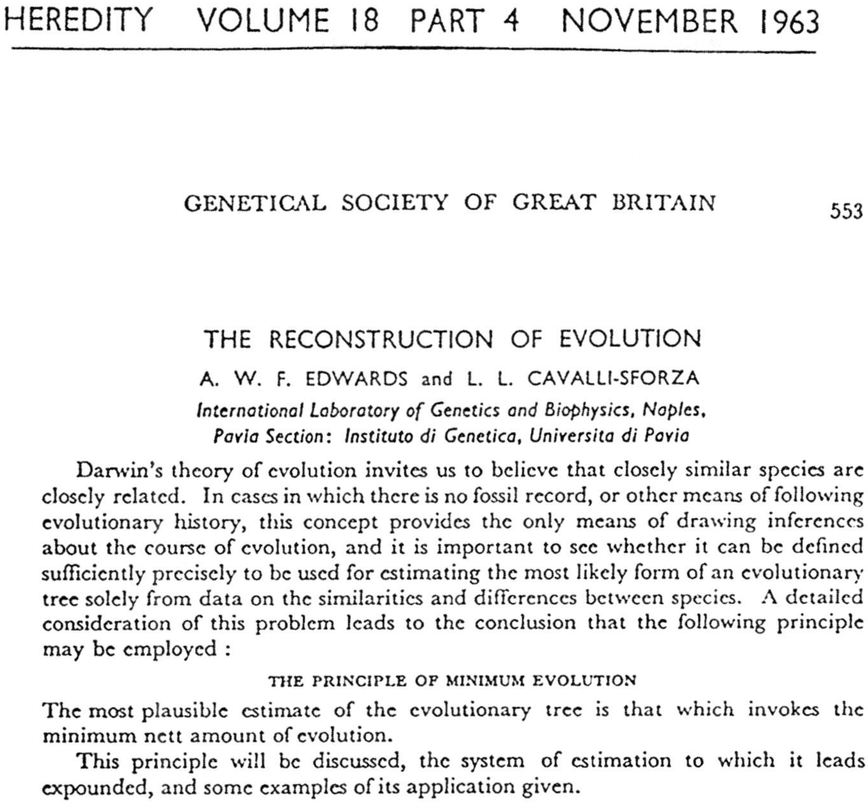

HEREDITY VOLUME 18 PART 4 NOVEMBER 1963

GENETICAL SOCIETY OF GREAT BRITAIN 553

THE RECONSTRUCTION OF EVOLUTION

A. W. F. EDWARDS and L. L. CAVALLI-SFORZA

International Laboratory of Genetics and Biophysics, Naples, Pavia Section: Instituto di Genetica, Universita di Pavia

Darwin's theory of evolution invites us to believe that closely similar species are closely related. In cases in which there is no fossil record, or other means of following evolutionary history, this concept provides the only means of drawing inferences about the course of evolution, and it is important to see whether it can be defined sufficiently precisely to be used for estimating the most likely form of an evolutionary tree solely from data on the similarities and differences between species. A detailed consideration of this problem leads to the conclusion that the following principle may be employed :

THE PRINCIPLE OF MINIMUM EVOLUTION

The most plausible estimate of the evolutionary tree is that which invokes the minimum nett amount of evolution.

This principle will be discussed, the system of estimation to which it leads expounded, and some examples of its application given.

FIGURE 3.—The first announcement of the method of minimum evolution or parsimony (EDWARDS and CAVALLI-SFORZA 1963a).

Thus, toward the end of 1963 was born the "likelihood approach" to phylogenetic analysis, although it was to be a long time before all the logical and computational difficulties were ironed out. I spent much of my 1964–1965 year at Stanford University struggling to extend the four-population prototype program to 15 populations, reporting on it at the November 1965 meeting of the Genetical Society in London. Eventually, the program EVOTREE was distributed, but it was soon overtaken by packages such as Felsenstein's PHYLIP. A copy of EVOTREE exists, having recently been found by Felsenstein. The subsequent development of the approach is considered briefly below.

In retrospect, the real contribution from Pavia in 1962–1964 was the embedding of modern phylogenetic analysis in the Fisherian tradition of stochastic model building and efficient statistical estimation by maximum likelihood and the use of genetical data. The detailed articles on this development are CAVALLI-SFORZA and EDWARDS (1966, 1967) and EDWARDS (1970); a number of other relevant contemporary publications are included in the list of references. Perversely, however, it was the earlier distance-matrix and minimum-evolution methods that caught the popular scientific imagination through being promoted by others after Cavalli-Sforza and I had introduced the maximum-likelihood approach. The next section summarizes this.

CONTEMPORARY INTRODUCTIONS OF DISTANCE-MATRIX AND MINIMUM-EVOLUTION METHODS

As indicated in the quotation from K. O. May given in the Introduction, there is little point in priority-hunting when it comes to ideas. One can report only what happened and leave it at that. In his book *Inferring Phylogenies*, FELSENSTEIN (2004) devotes a chapter to history and philosophy that leaves no stone unturned. Distance-matrix methods using least squares became popular through the article by FITCH and MARGOLIASH (1967), who used a weighted variant of Cavalli-Sforza's method. I had discussed this method with Fitch during a visit to Madison, Wisconsin, in the summer of 1965.

The case of the method of minimum evolution is more interesting because it was clearly an idea whose time had come. It became popular under the name "parsimony" through the article by CAMIN and SOKAL (1965). Sokal in fact had heard Cavalli-Sforza deliver our paper in September 1963 at the 11th International Genetics Congress at which the method of minimum evolution was mentioned, and I discussed it with him afterward. Their use of parsimony, however, was independent, having originated with Camin (SOKAL 1966). Others who are known to have had the idea include ECK and DAYHOFF (1966); for further information, see EDWARDS (1996, 2004b).

It has occasionally been represented that the taxonomic methods of W. Hennig (1913–1976) involved an implicit use of the idea of minimum evolution on a tree. My own reading of his work failed to find this (EDWARDS 1996); for further views, see SOBER (1988), SNEATH (1995), and FELSENSTEIN (2004).

Two things should be said about the 1963 announcement of the method of minimum evolution (originally called a "principle"). First, it was so explicitly stated that one could—and did—immediately start thinking about computer programs to implement it, for both discrete and continuous data: "The most plausible estimate of the evolutionary tree is that which invokes the minimum net amount of evolution." Second, in addition to its being presented at a meeting of Britain's Genetical Society, the abstract containing the principle in full appeared in 1963 in two of the leading genetics journals, Heredity (Figure 3) and the Annals of Human Genetics (EDWARDS and CAVALLI-SFORZA 1963a). No earlier publication of such a principle has been noticed.

EPILOGUE: THE SUBSEQUENT DEVELOPMENT OF LIKELIHOOD METHODS

Distance-matrix and minimum-evolution methods enjoyed considerable popularity partly because they were essentially algorithmic and thus relatively easy to program for computers and perhaps also because they bore some resemblance to traditional taxonomic practice and were therefore more accessible to taxonomists. Even so, they were attempts to "find" a phylogenetic tree and were soon seen to be no substitute for the application of the theory of statistical estimation to the supposed stochastic model that had generated the tree in the first place.

The concept of likelihood was one of R. A. Fisher's great contributions to statistics [for further information, see my book *Likelihood* (EDWARDS 1992)]. The method of maximum likelihood rapidly became, and has remained, the standard method for the estimation of statistical parameters, so it was natural for EDWARDS and CAVALLI-SFORZA (1964b) to propose its adoption in the case of phylogenetic trees. The initial difficulties of implementation (see especially CAVALLI-SFORZA and EDWARDS 1967) were surmounted only once the mistake of treating all the unknowns in the tree as parameters in the true sense was recognized (FELSENSTEIN 1968; EDWARDS 1970).

Progress, largely by Felsenstein himself starting with his Ph.D. thesis (FELSENSTEIN 1968), is best traced through his book *Inferring Phylogenies* (FELSENSTEIN 2004). Computationally feasible methods were developed for both continuous data such as gene frequencies and discrete data such as DNA sequences [these of course were not available in 1964, but EDWARDS and CAVALLI-SFORZA (1964b) had already observed that in a discrete case the character space would be a lattice of points]. Computer packages incorporating likelihood methods were written and made widely available.

Some indication of the dominance that the statistical approach via maximum-likelihood and likelihood-based Bayesian methods has now achieved is in the book *Computational Molecular Evolution* by YANG (2006), where the author states in his preface:

> The time appears ripe to summarize the methodological advancements in the field, and this book is such an attempt. I make no effort to be comprehensive in the coverage. There is hardly such a need now, thanks to recent publication of Joseph Felsenstein's (2004) treatise, which has discussed almost everything relevant to phylogenies. Instead I take the view that molecular evolutionary analysis, including reconstruction of phylogenies and inference of the evolutionary process, is a problem of statistical inference (CAVALLI-SFORZA and EDWARDS 1967). Thus well-established methods such as likelihood and Bayesian are described as standard.

(For the avoidance of misunderstanding, this should not be taken to mean that Felsenstein's view is any different: it was, after all, the main argument of his Ph.D. thesis.)

On the occasion of the presentation of 12 Darwin–Wallace Medals in 2008 by the Linnean Society (Professor Felsenstein being one of the recipients), Professor Mohamed Noor responded on behalf of all the medalists. He described his choice of "three major innovations [which] stand out as having played a major role in driving progress in understanding evolution these past 50 years": (1) "the rise of the field of phylogenetic systematics"; (2) "the explosion of computational power and availability;" and (3) "the amazing growth of the field of genetics." Of the first, he said:

> We now know that physical or DNA-sequence similarity does not necessarily demonstrate closer relationship, and we apply quantitative formulae to sort relationships using shared-derived characters in particular. This field also heralded the application of model-based statistical methods for inferring species relationships, including both Bayesian and maximum likelihood based approaches. Overall, phylogenetic systematics is thriving and providing an irreplaceable framework that essentially didn't exist for understanding how and when evolutionary processes operated in the past. (NOOR 2009)

I am indebted to Professor J. Felsenstein for his help, to Professor Cavalli-Sforza for answering several queries about times long ago, and, as ever, for a collaboration of consuming interest.

LITERATURE CITED

For the sake of historical completeness, this list includes some early references not mentioned in the text.

BAILEY, N. T. J., 1961 *Introduction to the Mathematical Theory of Genetic Linkage*. Clarendon Press, Oxford.

BOX, J. F., 1978 *R. A. Fisher, The Life of a Scientist*. John Wiley & Sons New York.

Boyd, W. C., 1952 *Genetics and the Races of Man.* Little, Brown, Boston.

Camin, J. H., and R. R. Sokal, 1965 A method for deducing branching sequences in phylogeny. Evolution **19:** 311–326.

Cavalli-Sforza, L. L., 1966 Population structure and human evolution. Proc. R. Soc. Lond. B Biol. Sci. **164:** 362–379.

Cavalli-Sforza, L. L., 1990 Recollections of Whittingehame Lodge. Theor. Popul. Biol. **38:** 301–305.

Cavalli-Sforza, L. L., 1991 Genes, peoples and languages. Sci. Am. **265**(5): 72–78.

Cavalli-Sforza, L. L., and F. Cavalli-Sforza, 2005 *Perché la scienza.* Mondadori, Milan.

Cavalli-Sforza, L. L., and A. W. F. Edwards, 1965 Analysis of human evolution. Proceedings of the 11th International Congress of Genetics, The Hague 1963, Genetics Today **3:** 923–933. (*N.B. The year of this paper is often given as 1964, which is indeed the date on the offprints.*)

Cavalli-Sforza, L. L., and A. W. F. Edwards, 1966 Estimation procedures for evolutionary branching processes. Bulletin of the International Statistical Institute (Proceedings of the 35th session) **41:** 803–808.

Cavalli-Sforza, L. L., and A. W. F. Edwards, 1967 Phylogenetic analysis: models and estimation procedures. Evolution **21:** 550–570 and Am. J. Hum. Genet. **19:** 233–257.

Cavalli-Sforza, L. L., I. Barrai and A. W. F. Edwards, 1964 Analysis of human evolution under random genetic drift. Cold Spring Harb. Symp. Quant. Biol. **29:** 9–20.

Eck, R. V., and M. O. Dayhoff, 1966 *Atlas of Protein Sequence and Structure 1966.* National Biomedical Research Foundation, Silver Spring, MD.

Edwards, A. W. F., 1961 The population genetics of sex-ratio in *Drosophila pseudoobscura.* Heredity **16:** 291–304.

Edwards, A. W. F., 1965 The reconstruction of evolution: estimation by maximum likelihood. (Abstract). Genetical Society programme, November. 1965.

Edwards, A. W. F., 1966 Studying human evolution by computer. New Sci. **30:** 438–440. Reprinted in *Evolution,* edited by M. Ridley, 1997. Oxford University Press, Oxford.

Edwards, A. W. F., 1969a Genetic taxonomy, pp. 140–142 in *Computer Applications in Genetics,* edited by N. E. Morton. University of Hawaii Press, Honolulu.

Edwards, A. W. F., 1969b Genetic taxonomy. (Abstr.) Biometrics **25:** 610.

Edwards, A. W. F., 1970 Estimation of the branch points of a branching diffusion process. J. R. Stat. Soc. Series B **32:** 155–174.

Edwards, A. W. F., 1990 Fisher, *W,* and the fundamental theorem. Theor. Popul. Biol. **38:** 276–284.

Edwards, A. W. F., 1992 *Likelihood,* Expanded Edition. Johns Hopkins University Press, Baltimore.

Edwards, A. W. F., 1996 The origin and early development of the method of minimum evolution for the reconstruction of evolutionary trees. Syst. Biol. **45:** 79–91.

Edwards, A. W. F., 2001 La ricostruzione degli alberi evolutivi, pp. 23–30 in *Le radici prime dell'Europa,* edited by G. Bocchi and M. Ceruti. Mondadori, Milan.

Edwards, A. W. F., 2003 Human genetic diversity: Lewontin's fallacy. BioEssays **25:** 798–801.

Edwards, A. W. F., 2004a Mendelism and man 1918–1939, pp. 33–46 in *A Century of Mendelism in Human Genetics,* edited by M. Keynes, A. W. F. Edwards and R. Peel. CRC Press, Boca Raton, FL.

Edwards, A. W. F., 2004b Parsimony and computers, pp. 181–190 in *Milestones in Systematics,* Special Volume Series 67, edited by D. M. Williams and P. L. Forey. Systematics Association, London.

Edwards, A. W. F., 2005 R. A. Fisher, Statistical Methods for Research Workers, pp. 856–870 in *Landmark Writings in Western Mathematics, 1640–1940,* Ed. 1, edited by I. Grattan-Guinness. Elsevier Science Publishers, Amsterdam.

Edwards, A. W. F., and L. L. Cavalli-Sforza, 1963a The reconstruction of evolution. (Abstr.) Heredity **18:** 553, and Annals of Human Genetics **27**, 104–105.

Edwards, A. W. F., and L. L. Cavalli-Sforza, 1963b A method for cluster analysis. Preprints of the 5th International Biometrics Conference, Cambridge, UK, September, 1963.

Edwards, A. W. F., and L. L. Cavalli-Sforza, 1964a A method for cluster analysis. (Abstr.) Biometrics **20:** 383.

Edwards, A. W. F., and L. L. Cavalli-Sforza, 1964b Reconstruction of evolutionary trees, pp. 67–76 in *Phenetic and Phylogenetic Classification,* Systematics Association Publication No. 6, London. Reprinted in *Cladistic Theory and Methodology,* edited by T. Duncan and T. F. Stuessy, 1985. Van Nostrand Reinhold, New York.

Edwards, A. W. F., and L. L. Cavalli-Sforza, 1965 A method for cluster analysis. Biometrics **21:** 362–375.

Felsenstein, J., 1968 Statistical inference and the estimation of phylogenies. Ph.D. Thesis, University of Chicago, Chicago.

Felsenstein, J., 2004 *Inferring Phylogenies.* Sinauer, Sunderland, MA.

Fisher, R. A., 1925 *Statistical Methods for Research Workers.* Oliver & Boyd, Edinburgh.

Fisher, R. A., 1950 Gene frequencies in a cline determined by selection and diffusion. Biometrics **6:** 353–361.

Fisher R. A., 1956 Blood-groups and population genetics. Proceedings 1st International Congress of Human Genetics. Acta Genet. Stat. Med. **6:** 507–509.

Fisher, R. A., 1959 Natural selection from the genetical standpoint. Aust. J. Sci. **22:** 16–17.

Fitch, W. M., and E. Margoliash, 1967 Construction of phylogenetic trees. Science **155:** 279–284.

Gomberg, D., 1968 "Bayesian" postdiction in an evolution process. Istituto di Genetica, Pavia. (*Unpublished.*)

Haeckel, E., 1876 Frontispiece, *History of Creation,* Vol. II. King, London.

Heywood, V. H., and J. McNeill, 1964a Phenetic and phylogenetic classification. Nature **203:** 1220–1224.

Heywood, V. H., and J. McNeill (Editors), 1964b *Phenetic and Phylogenetic Classification,* Systematics Association Publication No. 6, London.

Mather, K., 1951 *The Measurement of Linkage in Heredity,* Ed. 2. Methuen, London.

May, K. O., 1975 Historiographic vices. II. Priority chasing. Historia Mathematica **2:** 315–317.

Michener, C. D., and R. R. Sokal, 1957 A quantitative approach to a problem in classification. Evolution **11:** 130–162.

Mourant, A. E., 1954 *The Distribution of the Human Blood Groups.* Blackwell, Oxford.

Mourant, A. E., 1995 *Blood and Stones: An Autobiography.* La Haule, Jersey, Channel Islands.

Noor, M., 2009 (In) Presentation of the Darwin–Wallace Medals 2008. Linnean **25**(2): 3–9.

Race, R. R., and R. Sanger, 1950 *Blood Groups in Man.* Blackwell, Oxford.

Sneath, P. H. A., 1995 Thirty years of numerical taxonomy. Syst. Biol. **44:** 281–298.

Sneath, P. H. A., and R. R. Sokal, 1962 Numerical taxonomy. Nature **193:** 855–860.

Sober, E., 1988 *Reconstructing the Past.* MIT Press, Cambridge, MA.

Sokal, R. R., 1966 Numerical taxonomy. Sci. Am. **215**(6): 106–116.

Sokal, R. R., and P. H. A. Sneath, 1963 *Principles of Numerical Taxonomy.* Freeman, San Francisco.

Venn, J. (Compiler), 1898 *Biographical History of Gonville and Caius College,* Volume II, Cambridge University Press, Cambridge, UK.

Wilkes, M. V., 1985 *Memoirs of a Computer Pioneer.* MIT Press, Cambridge, MA.

Yang, Z., 2006 *Computational Molecular Evolution.* Oxford University Press, Oxford.

Paper 227 (2011)

Mathematizing Darwin

Behavioral Ecology and Sociobiology 65, 421–430.

Note from A. W. F. Edwards: This paper was a response to a request to stand in for an invited 'keynote' speaker who had had to withdraw through family illness from the meeting 'Mathematical Models in Ecology and Evolution: Darwin 2009' in Bristol. It was possible to accept at short notice because in Cambridge not only were we similarly celebrating the Darwin bicentenary, but Gonville and Caius College and the Department of Genetics were celebrating the centenary of the admission of Fisher to the college. I thus had much of the material ready to hand.

Notice may be taken of the emphasis on the distinction in Fisher's work between his inductive or synthetic approach and the analytic approach of Wright and Haldane. I take this opportunity to revise my interpretation of the word 'statistical' in the quotation from Fisher on page 276 (this volume) in which he wrote 'inductive and statistical'. I now think Fisher was referring not to estimation theory and significance testing but to linear regression and the analysis of variance. This point of view is further developed in [248].

Mathematizing Darwin

A. W. F. Edwards

Received: 17 July 2010 / Revised: 17 November 2010 / Accepted: 17 November 2010 / Published online: 5 January 2011

Abstract Ernst Mayr called the first part of the evolutionary synthesis the 'Fisherian synthesis' on account of the dominant role played by R.A. Fisher in forging a mathematical theory of natural selection together with J.B. S. Haldane and Sewall Wright in the decade 1922–1932. It is here argued that Fisher's contribution relied on a close reading of Darwin's work to a much greater extent than did the contributions of Haldane and Wright, that it was synthetic in contrast to their analytic approach and that it was greatly influenced by his friendship with the Darwin family, particularly with Charles's son Leonard.

Keywords Darwin · Fisher · Haldane · Wright · Evolution · Mathematical theory of natural selection

Introduction

The celebration of the sesquicentenary of the publication of *The Origin of Species* and the bicentenary of Darwin's birth is a good moment to look at the initial mathematical development of Darwin's theory of evolution by natural selection, especially at a meeting on Mathematical Models in Ecology and Evolution. About ecology I shall say nothing, and I shall also make a distinction between the study of evolution and that of mathematical models in Mendelian genetics. The two are closely related, of course, but whereas mathematical genetics is an analytic approach, mathematical evolution is synthetic, attempting to draw out general principles rather than pursuing the reductionist approach that Ernst Mayr so memorably called 'beanbag genetics'.

No better example of the synthetic approach can be found than Düsing's mathematical treatment of Darwin's argument for natural selection favouring an equality of the sexes, forgotten until I drew attention to it in an article in *The American Naturalist* in 1998. This famous argument, though it used to be attributed to R.A. Fisher in *The Genetical Theory of Natural Selection* published in 1930, we now know was given by Darwin in the first edition of *The Descent of Man and Selection in Relation to Sex* of 1871 but omitted from the second edition of 1874. In that edition, Darwin remarked 'but I now see that the whole problem is so intricate that it is safer to leave its solution to the future', as Fisher noted.

In 1883 and 1884, Carl Düsing of Jena took Darwin's argument from the first edition and expounded it mathematically. I published an English translation of the relevant pages of Düsing's 1884 book in 2000. You will of course already have observed that in 1883 Mendel's paper had not yet received the attention it had from 1900 onwards, so that Düsing's treatment is pre-Mendelian, synthetic. There is no genetical model, merely the Darwinian assumption that there exists heritable variation in the tendency to produce one sex in excess.

In the initial years of the twentieth century following the rediscovery of Mendel's paper, two major reconciliations had to take place. The first was the synthesis of the findings of Karl Pearson's biometrical school and the Mendelian model of inheritance. Initiated by Udny Yule in 1902, this culminated in Fisher's first major publication *The correlation between relatives on the supposition of Mendelian*

Communicated by Guest Editor A. Houston

This contribution is part of the Special Issue 'Mathematical Models in Ecology and Evolution: Darwin 200' (see Marshall et al. 2010).

A. W. F. Edwards (✉)
Gonville and Caius College, University of Cambridge, Cambridge, UK
e-mail: awfe@cam.ac.uk

inheritance of 1918. The details of this famous paper, the foundation of biometrical genetics, do not now concern us because evolution is not its topic. The second reconciliation was, of course, the synthesis of Darwinian evolution by natural selection and Mendelian genetics. This proved to be more drawn out because, as Peter Bowler in his 1983 book *The Eclipse of Darwinism* and Jean Gayon in *Darwinism's Struggle for Survival* (English edition 1998) have chronicled, in the early years of the century the followers of the Mendelian school of William Bateson, and many others, were strongly opposed to the suggestion that evolution could proceed by the accumulation of the small changes that Mendelism seemed to offer and Darwinism require.

I shall try to persuade you that once again the youthful Fisher was the principal contributor, and that his 1930 book *The Genetical Theory of Natural Selection* (and some papers that preceded it) was the major event of the first part of what Julian Huxley was to call 'the evolutionary synthesis', that complete reconciliation of Darwin's theory not only with Mendelian genetics but also with the natural world described by Huxley, Dobzhansky, Mayr, Simpson and other natural historians. Mayr went so far as to call this first part 'the Fisherian synthesis'. The important decade is 1922–1932, and I have described it in detail in a 2001 encyclopaedia article 'Darwin and Mendel united: The contributions of Fisher, Haldane and Wright up to 1932'. I shall argue that Haldane and Wright were not only later than Fisher, but that their work was primarily analytic, starting from simple Mendelian models and deducing the consequences, whereas Fisher achieved a synthesis based to a great extent on his reading of Darwin. He made the distinction himself in his paper at the Sixth International Congress of Genetics in 1932, when he contrasted his approach with Haldane's:

> The one approach [Haldane's] … is analytic and deductive. Genetic studies are regarded as revealing the mechanism connecting cause and effect, from a knowledge of which the workings of the machine can be deduced and the course of evolutionary change inferred. The other approach is inductive and statistical; genetics supplies the facts as to living things as they now are, facts which … have an evolutionary history and may be capable of an evolutionary explanation … .

In using the word 'statistical', Fisher is here referring to the great strides that he and others were making at the time in the inductive subject of statistical inference through the introduction of maximum likelihood and significance testing. The title of Fisher's paper was 'The evolutionary modification of genetic phenomena', and in this and in an address to the Royal Society of Dublin earlier in 1932 'The bearing of genetics on theories of evolution', one can see how far Fisher was already concerned not just with deducing the evolutionary consequences of genetic systems but, like Darwin, with the subtler problem of drawing inductive conclusions about the evolution of those systems themselves from the mass of biological observations often requiring statistical interpretation. With good reason Richard Dawkins remarked in *The Blind Watchmaker* that Fisher was the greatest of Darwin's successors.

Fisher the Darwinist

(I call a 'Darwinist' anyone who closely studies and develops Darwin's own arguments, in contrast to a 'Darwinian', who simply subscribes to Darwin's theory of evolution by natural selection.)

Fisher came up to Gonville and Caius College in the University of Cambridge to read mathematics in October 1909, shortly after the University had celebrated Darwin's first centenary. He has left us with some reminiscences: 'I first came to Cambridge in 1909, the year in which the centenary of Darwin's birth and the jubilee of the publication of *The Origin of Species* were being celebrated. The new school of geneticists using Mendel's laws of inheritance was full of activity and confidence, and the shops were full of books good and bad from which one could see how completely many writers of this movement believed that Darwin's position had been discredited'. 'The fiftieth anniversary of the publication of *The Origin of Species* was being celebrated, apart from other things, by the publication of Bateson's book *Mendel's Principles of Inheritance*' (actually *Mendel's Principles of Heredity*). Fisher bought the book: 'It includes a translation of Mendel's paper on "Pisum"'. Caius and the Cambridge Department of Genetics are mounting a small exhibition to commemorate Fisher's admission to the college a hundred years ago.

In his last year at Harrow School, Fisher had chosen the collected works of Darwin for a school prize, so that, in the words of his biographer Joan Box, 'he went up to Cambridge in possession of volumes he was to read and reread with loving care throughout his life'. These 13 volumes of the John Murray edition, bound in green cloth, are preserved in Adelaide, where Fisher died in 1962. In Cambridge, he received as a College Prize the book of essays *Darwin and Modern Science* published by Cambridge University Press as part of the Darwin Celebrations ('a remarkable collection of able essays' Fisher was later to remark) and as a gift from his undergraduate friend C.S. Stock, *The Foundations of the Origin of Species: Two Essays Written in 1842 and 1844 by C. Darwin.*

So great was the impact of *The Origin of Species* on Fisher and Stock that a few years later, in 1915, they were to write in response to one of Darwin's critics:

> So melancholy a neglect of Darwin's work suggests reflections upon the use of those rare and precious possessions of man—great books. It was, we believe, the custom of the late Professor Freeman to warn his students that mastery of one great book was worth any amount of knowledge of many lesser ones. The tendency of modern scientific teaching is to neglect the great books, to lay far too much stress upon relatively unimportant modern work and to present masses of detail of doubtful truth and questionable weight in such a way as to obscure principles. … How many biological students of today have read *The Origin*? The majority know it only from extracts, a singularly ineffective means, for a work of genius does not easily lend itself to the scissors; its unity is too marked. Nothing can really take the place of a first-hand study of the work itself.
> (Professor E.A. Freeman was Regius Professor of Modern History at Oxford, 1884–1892.)

Fisher was from the outset influenced by Darwin's writings to a much greater extent than any of his contemporaries; he was a *successor* in the truest sense. In 1948, he received the Darwin Medal of the Royal Society and in replying to D.J. Finney's congratulations had this to say:

> Of course it was an immense satisfaction to me to have the Darwin Medal awarded, as I have worked for a good many years, and indeed saw the need nearly 40 years ago, to reverse the trend then prevalent of misrepresenting and minimizing the importance of Darwin's achievement. The books and articles to be bought in Cambridge in 1909, the year in which the centenary of Darwin's birth was celebrated, make very strange reading today, and it is relevant to anyone really interested in the way science makes progress that the writers of the first 10 years of the century, which began with the rediscovery of Mendel's work, were so biased against Darwin and natural selection by the controversies preceding this rediscovery that much that Mendel himself said in his 1865 paper was completely overlooked.

Fisher's 1915 paper with Stock, from which I have already quoted, is a vigorous defence of Darwin against criticism of his theory of evolution by L. Cuénot, and in the same year Fisher published his first evolutionary paper, 'The evolution of sexual preference', with the opening sentence: 'Of the branches of biological science to which Charles Darwin's life-work has given us the key, few, if any, are as attractive as the subject of Sexual Selection'. As Peter O'Donald remarked, 'Fisher took the theory a fundamental step further than Darwin by showing how mate choice itself would evolve as a consequence of the very process of sexual selection it produced'. Fifteen years later, Fisher devoted a section of *The Genetical Theory of Natural Selection* to this topic.

A further reason for Fisher's emergence as the best-informed Darwinist of his generation was that his knowledge of Darwin's writings was reinforced by the friendship of two of Darwin's sons, Horace and Leonard, and the acquaintance of two more, Francis and George. In 1911, Fisher and some fellow-undergraduates canvassed support for the formation of a Cambridge University Eugenics Society, and succeeded in forming a Council which included Dampier Whetham, R.C. Punnett, John Maynard Keynes (not a bad choice as Treasurer), A.C. Seward and Horace Darwin, Charles's youngest son, as well as Fisher himself. In February 1912, Major Leonard Darwin, Charles's fourth surviving son, who had become President of the national Eugenics Education Society in 1911, came to address the Cambridge Society. As Joan Box tells us 'The link was strengthened that summer when the First International Eugenics Congress took place in London under Major Darwin's presidency, when Fisher, with other Cambridge members, served as stewards at the meetings'. Two more Darwin brothers, Francis and George, were also involved in the Cambridge Society. Francis was to give the first Galton Lecture of the Eugenics Education Society in 1914 and George the Lecture in 1939. In 1930, Charles Galton Darwin, George's son, was to review *The Genetical Theory* for the *Eugenics Review.*

Leonard Darwin in particular soon became a major influence on Fisher. From 1915 onwards, the two men conducted an extensive correspondence, much of it reproduced by J.H. Bennett in his 1983 book *Natural Selection, Heredity and Eugenics*, which also gives the fullest account of their interaction. It suffices to say here that Leonard was at once a friend, a father-figure, a supporter and a wise adviser, always ready to discuss his father's views. Fisher dedicated *The Genetical Theory of Natural Selection*

TO
MAJOR LEONARD DARWIN
In gratitude for the encouragement,
given to the author, during the last
fifteen years, by discussing many
of the problems dealt with
in this book

We may also note in passing that Fisher's 1918 paper 'The correlation between relatives on the supposition of Mendelian inheritance' ends with the words 'Finally, it is a pleasure to acknowledge my indebtedness to Major

LEONARD DARWIN, at whose suggestion this inquiry was first undertaken, and to whose kindness and advice it owes its completion'. Indeed, Darwin saw to its publication following its withdrawal from the Royal Society after unfavourable reports by Karl Pearson and R.C. Punnett ('both of whom I later succeeded', Fisher used to say).

The Genetical Theory of Natural Selection

Whilst many of Fisher's papers refer explicitly to Charles Darwin (see the name index in Volume 5 of his *Collected Papers*), it is mainly to *The Genetical Theory* that we must turn to substantiate the view that he was the leading mathematizer of Darwin's theory. It would be facile simply to repeat the opinions of distinguished scientists about this book, but many may be found in three dedicated publications, the Introduction in Bennett's book, his Foreword to the Variorum Edition of *The Genetical Theory* and my own 'Perspectives' article devoted to the book in *Genetics* in April 2000.

A word must be said about Fisher's style before embarking on a discussion of the Darwinian content of *The Genetical Theory*. Fisher was a consummate mathematician trained in the hard school of the Cambridge Mathematical Tripos, as those of you who have studied his statistical work will know, but when writing for a more general audience he often suppressed the mathematical basis of his arguments in an attempt (usually unsuccessful) to make them more accessible, especially to biologists. Indeed, many of us have had to struggle to construct (or reconstruct) the mathematics, as for example Walter Bodmer and I did in 1960 with Fisher's discussion of natural selection and the sex ratio (thus covering the same ground as Düsing of course, unknown to us or anyone else at the time), and O'Donald has done with sexual selection and the evolution of dominance.

In his Foreword to the variorum edition, Bennett devotes 11 pages to an account of *The Genetical Theory*. Attention should also be drawn to the Summaries with which Fisher ended each chapter. Here, I shall only discuss the topics that have a clear connection to Darwin's writings. That connection could hardly be stronger than in Chapter I where Fisher contrasts the tendency of a blending theory of inheritance to reduce the variance of a character in a population (described by some simple mathematics) with the conservation of the variance implicit in Mendelian inheritance, a comparison he had first made in 1924. Throughout the chapter, Fisher shows great familiarity not only with Darwin's books but also with his correspondence and with his essays of 1842 and 1844. Michael Bulmer has given an extended account of the problem in his book *Francis Galton: Pioneer of Heredity and Biometry*, pointing out that Darwin's theory of pangenesis actually only involved partial blending 'since the patent elements fuse but the latent elements do not'.

In Chapter II of *The Genetical Theory*, in a section entitled 'The genetic element in variance', Fisher introduces his 'fundamental theorem of natural selection' showing that the instantaneous increase in the mean fitness of a population ascribable to natural selection acting through changes in gene frequencies is exactly equal to its additive genetic, or genic, variance at that time. Much has been written about the fundamental theorem, a lot of it misleading, and this is not the moment to review how a full understanding of Fisher's meaning has been arrived at.

Rather, it is important to see this theorem as part of the central Darwinian theme that variation is the meat and drink of evolution by natural selection. The first chapter of *The Origin of Species* is entitled 'Variation under Domestication' and is preliminary to Chapter II 'Variation under Nature'. Following Darwin's lead, in 1922 Fisher had set out to investigate how the variance of a population is maintained or increased, and how it decays. After showing (for the first time) that heterozygote advantage leads to balanced polymorphism, he studied the survival of individual mutant genes by introducing the Galton–Watson branching process and analysing it by functional iteration. He then set up the chain-binomial model for a simple population (often now incorrectly called the Wright–Fisher model), treating it by a diffusion approximation (thereby inventing stochastic diffusion theory in the process) and deriving partial differential equations for the study of gene-frequency distributions in populations of finite size. As we shall see, these topics are developed in Chapters IV and V of *The Genetical Theory*. For the moment I just want you to note the date, 1922, and the Darwinian context. Indeed Fisher quotes what he calls 'the dictum of Charles Darwin' that 'wide ranging, much diffused and common species vary most' from Chapter II of *The Origin*.

Chapter III of *The Genetical Theory* is devoted to the evolution of dominance, a matter which had been engaging Fisher's attention in several papers at the time of writing the book, involving exchanges with both Haldane and Wright. Bennett writes 'This work contributed in an important wa to the growth of the concept of the gene-complex' but it is of course, essentially post-Darwin in its subject matter. It i an example of how Fisher's thinking was already directe at the evolution of genetic systems, as I mentioned earlier.

Chapters IV and V essentially constitute one long chapte entitled 'Variation as determined by mutation and selec tion'. The Summary itself covers more than three page This is the heart of the book, and the most mathematica The 1922 results are corrected and extended as Fishe studies the effects of mutation rate, population size an selection on the variance, and on the survival of individu genes, concluding, as is well known, that natural selection operating in large populations is likely to be of much greater evolutionary significance than chance variation in small ones. His analysis leads him to conclude, in accordance with Darwin's dictum, that the more numerous species will tend to be the more variable, and will make the most rapid evolutionary progress, supplanting less abundant groups, as Darwin foresaw. Fisher's stance in the long-running controversy with Wright and his 'shifting-balance' theory of evolution has a very definite Darwinian origin. Looking again at stable polymorphism, this time Fisher proves convergence to and not just the stability of the equilibrium in the case of heterozygote advantage. He also discusses the population genetics of two linked loci, once again putting into English what might have been easier to follow in mathematics, as others (such as Haldane) have since done.

Chapter VI 'Sexual reproduction and sexual selection' is more Darwinian than even Fisher realized, for it contains the section 'Natural selection and the sex-ratio' which, as we have seen, he did not trace to its origin in the first edition of *The Descent of Man*. In 'Fission of species' he describes sympatric speciation—'the case of a species subject to different conditions of survival and reproduction at opposite ends of its geographical range'—and shows how 'The constant elimination in each extreme region of the genes which diffuse from it to the other, must involve incidentally the elimination of those types of individuals which are most apt to diffuse' leading to an increasing contrast in the genetic composition of the two parts and their ultimate independence. This is a typically Fisherian addition to a Darwinian theory, the theory of divergence in Chapter IV of *The Origin*. By adding the notion that the aptitude to diffuse is also heritable, Fisher has supplied a powerful dynamic.

Here I might mention that many years ago, in 1962, I wrote a short paper called 'Migrational selection' which I had (quite properly) forgotten. Now that I recall it, I find that I had been reading the section 'Fission of species', which is heavily marked up in my original paperback copy of *The Genetical Theory*. In a footnote I mention 'Fisher (1950) has since [1930] given a mathematical formulation of this problem', and this is a rare example of him mathematizing a *Genetical Theory* discussion. But his paper is also memorable for quite a different reason. The differential equation Fisher obtains required numerical solution, so he appealed to his friend Maurice Wilkes, the constructor of the first Cambridge EDSAC computer, who asked a student, David Wheeler, to tabulate it. I believe this to be the first use of a computer to tackle a problem in biology, and I have been stating this for so long without a challenge that it may even be true. A personal reason for now remembering my paper is that I asked the editor of *Heredity*, Kenneth Mather, if I could dedicate it to Fisher, who had just died. He said no.

The long section on 'Sexual selection' in Chapter VI derives directly from *The Descent of Man* and, as we have already seen, adds substantially to the theory by introducing the 'runaway' process Fisher first described in 1915. This is another example where Fisher has suppressed the mathematics, for Bennett has written 'Fisher never published anything on the quantitative basis for his runaway process although he had, in fact, set this out in correspondence with C.G. Darwin in 1932'. O'Donald's 1980 book *Genetic Models of Sexual Selection* gives a detailed historical account as well as many mathematical models. Last in Fisher's chapter comes the section 'Natural selection and the sex-ratio' already mentioned.

In Chapter VII 'Mimicry', Fisher remarks that the theory of mimicry is 'the greatest post-Darwinian application of Natural Selection'. The chapter is notable particularly for the section 'The evolution of distastefulness' [in insects] which he explains by what is now known as 'kin selection', often attributed to Haldane but in fact suggested by Fisher already in a student talk in 1912 published in 1914 when he considered how a childless man killed in war could be replaced genetically speaking by his nephews.

The last five chapters, on Man, which Fisher regarded as 'strictly inseparable' from the rest of the book, are more Galtonian than Darwinian, but they do reflect the influence of Leonard Darwin, whose *The Need for Eugenic Reform* had appeared in 1926. Dedicated to the memory of his father, in the preface Leonard thanks Fisher for statistical advice and proof-reading. In Chapter VIII of *The Genetical Theory*, 'Man and Society', Fisher discusses the evolution of fertility, recalling how Leonard Darwin, taking the parasitic cuckoo as his example, had in his book introduced the concept of parental expenditure on the young in this connection (also used by Fisher when discussing natural selection and the sex-ratio in Chapter VI). David Lack much later wrote an influential paper 'The evolution of reproductive rates' in ignorance of Fisher's discussion. Reviewing the problem in 1992, David Burbridge observed that Leonard Darwin had thought his own contribution original, but in fact 'his father had discussed the problem in the first edition of *The Descent of Man* and given essentially the same solution'. But he removed it from the second edition and 'As later biologists seldom consulted the scarce first edition, Darwin's contribution seems to have been forgotten even by his own son'. This is exactly what had happened with natural selection and the sex-ratio, so once again we find that a Fisherian discussion actually derives from Charles Darwin without Fisher realizing it. But in this case Darwin was himself following an insight of Herbert Spencer's.

In my 'Perspectives' article I remarked, in connection with Fisher's literary style, that 'it is illuminating to think of

The Genetical Theory as a kind of mathematical–Mendelian appendix to *The Origin of Species*'; I should have added 'and *The Descent of Man* (first edition)'. My article should be consulted for further information about *The Genetical Theory*, such as its writing and its reception. In addition, let us not forget that whilst writing it Fisher was the full-time Head of the Statistics Department at Rothamsted Experimental Station and had, in 1925, published his book *Statistical Methods for Research Workers*, which was to become so renowned that it was the only statistical writing to be granted a chapter in the volume *Landmark Writings in Western Mathematics 1640–1940* edited by Ivor Grattan-Guinness.

Sewall Wright and J.B.S. Haldane

Fisher's description of Haldane's approach as 'analytic and deductive' can also be applied to Wright's. To a much greater extent than Fisher they both established genetical models and pursued their consequences, Haldane from 1924 in a series of papers and Wright from 1931 starting with his influential paper 'Evolution in Mendelian populations', the final version of a manuscript that he had been polishing since 1925. I have discussed their contributions at length in my 'Darwin and Mendel united' article of 2001, where summary biographical details will also be found. But my present purpose is to study the mathematization of Darwin and therefore to enquire only into the extent of Darwin's influence on Wright and Haldane.

Wright was certainly well-read in Darwin, whose *Origin of Species* he had studied as a student as well as other books about Darwinism, but what the historian of science W.B. Provine called his 'two most seminal early papers on evolutionary theory' contain no references to Darwin's writings. These two papers in 1931 and 1932 set the scene for much of Wright's extensive subsequent publications on evolution, but he hardly ever mentioned Darwin.

It is interesting to note that Wright first learnt of Mendelian heredity through the 1911 article by Punnett in the 11th edition of the *Encyclopedia Britannica* which he had been set to read by Mrs Key, his teacher at Lombard College, Galesburg, Illinois. Punnett was a young Fellow of Caius College in 1909 when Fisher was admitted, and the author of the little book *Mendelism* (1905, second edition 1907). Provine also mentions that later in graduate school Wright read Lock's (1906) *Recent Progress in the Study of Variation, Heredity and Evolution*, which was known in the United States through being used as a text book by E.B. Wilson at Columbia University. R.H. Lock, an associate of William Bateson, was also a young Fellow of Caius until 1910, though he left for Ceylon in 1908. Wilson had been one of the recipients of an Honorary Doctorate at the Cambridge Darwin Celebrations and would probably have met Lock then. Thus, Wright initially shared Fisher's intellectual background to some extent, but without such a strong emphasis on Darwin. Provine writes 'Until reading Fisher's 1922 paper [in 1924], it had not occurred to Wright to extend his own quantitative analysis to the statistical distribution of genes in populations'.

Haldane's intellectual background has been described by his biographer Ronald Clark and is not noted for any special influence of Darwin's writings. His own *The Causes of Evolution* of 1932 arose from a series of lectures he gave at the University of Wales, Aberystwyth, in 1931, entitled 'A Re-examination of Darwinism'. They show a wide knowledge of Darwinism and its development but, as is entirely understandable in lectures to a general audience, do not suggest any deep study of Darwin's work. Indeed, rather amusingly, in five pages of bibliography the only Darwin reference is 'Darwin and Wallace (1858), Journ. Linn. Soc. July 1' which he mentioned in the first sentence of the book! In a long Appendix 'Outline of the mathematical theory on natural selection' Haldane summarized much of his own work in the series of papers starting in 1924.

Wright and Haldane were indeed Darwinians, but they were not Darwinists in the sense in which I have used the word, and, with their analytic model-building approach, cannot be considered to have been 'mathematizing Darwin' to the extent that Fisher was. Mayr's description 'the Fisherian synthesis' is understandable.

The second part of the evolutionary synthesis

Long before he coined the phrase 'Fisherian synthesis', Mayr (1980) divided the evolutionary synthesis into two 'conclusions':

> The term "evolutionary synthesis" was introduced by Julian Huxley in *Evolution: The Modern Synthesis* (1942) to designate the general acceptance of two conclusions: gradual evolution can be explained in terms of small genetic changes ("mutations") and recombination, and the ordering of this genetic variation by natural selection; and the observed evolutionary phenomena, particularly macroevolutionary processes and speciation, can be explained in a manner that is consistent with the known genetic mechanisms.

Julian Huxley, grandson of T.H. Huxley, 'Darwin's bulldog', is now remembered more as a writer about the evolutionary synthesis than as a contributor to it, especially through *Evolution: The Modern Synthesis*. In the first chapter Huxley did of course summarize Darwin's achievement (and coin the phrase 'The eclipse of Darwinism', later to be used by Bowler as a book title), but in the preface he acknowledges his primary influences: 'I owe a great debt to J.B.S. Haldane's *The Causes of Evolution'* and 'My debt to R.A. Fisher's work is obvious'. Interestingly, he attended the Cambridge Darwin celebrations in 1909: 'As a Huxley and a budding biologist, I was invited, and was deeply impressed by the stream of addresses stressing the importance of Darwin's many-sided work'. He added:

> I resolved that all my scientific studies would be undertaken in a Darwinian spirit and that my major work would be concerned with evolution, in nature and man. This was not so much a turning point in my career as a crystallization of my ideas, a clear vision and inspiration which I can truly say remained with me all through my life.

Huxley was a true Darwinian, subscribing to and promoting Darwin's theory and its subsequent development, but not a Darwinist.

In Mayr's second 'conclusion' of the evolutionary synthesis, that 'the observed evolutionary phenomena … can be explained in a manner that is consistent with the known genetic mechanisms', the three most influential contributions were the books by Theodosius Dobzhansky, G.G. Simpson and Mayr himself. All emerged from Columbia University Press under the editorship of L.C. Dunn; the first two paid homage to Darwin in their titles: *Genetics and the Origin of Species* (Dobzhansky 1937) and *Systematics and the Origin of Species* (Mayr 1942). Simpson's (1944) could also have done so as 'Paleontology and the Origin of Species' but was in fact called *Tempo and Mode in Evolution*.

Neither Dobzhansky nor Simpson gave any of Darwin's works in their copious bibliographies. Mayr gave just *The Origin of Species* and offered a number of incidental comments on Darwin's contributions. All three were one stage removed from Darwin and were able to rely on the supposition that the work of Fisher, Haldane and Wright had achieved the synthesis of Mendelism and Darwinian evolution by natural selection. Indeed, for the most part they did not follow this work closely, mainly because of its substantial mathematical content. Dobzhansky relied heavily on Wright's work but the details defeated him. In his long opening address to the 1955 Cold Spring Harbor Symposium on Quantitative Biology 'A review of some fundamental concepts and problems of population genetics' he ventured to describe what happens at a diallelic locus with heterozygotic advantage. He says that the significant fact is that the mean fitness reaches a maximum at equilibrium. Thus 'Selection maximises the adaptive value of the population, even at the expense of making the latter contain some handicapped individuals [the homozygotes]. The fitness of some individuals is sacrificed for the fitness of the population as a whole'. I don't think we would let an undergraduate get away with that.

Mayr himself became a historian of the evolutionary synthesis, and together with Provine edited the results of a 1974 conference held to discuss it (*The Evolutionary Synthesis*, Mayr and Provine 1980), a starting point for any student of the topic today. He became fascinated by Darwin even if his early work had not been particularly influenced by him. In the Preface to his *One Long Argument—Charles Darwin and the Genesis of Modern Evolutionary Thought*, Mayr wrote 'My interest in Darwin's thought arose in my university years, but my more active preoccupation began in 1959 with the centenary of the publication of *On the Origin of Species*. I studied Darwin's writings even more intensely when I prepared an introduction to a facsimile of the first edition of the *Origin*, published in 1964'.

In my 2000 'Perspectives' article on *The Genetical Theory*, I criticized Mayr since he 'never seems to have come to terms with [*The Genetical Theory*'s] importance'. 'In *The Evolutionary Synthesis* (Mayr 1980) he contributed a "Prologue: Some thoughts on the history of the evolutionary synthesis" without once mentioning *The Genetical Theory* (unless a remark about "the supposedly evolutionary writings of … R.A. Fisher" counts as a mention)'. Mayr and I had been corresponding intermittently for years, starting in 1964 or earlier, and in 2001 we were discussing Fisher's view of sympatric speciation in the course of which I sent him a copy of my 'Perspectives' article and drew attention to the criticisms of him it contained. This led to much further friendly correspondence in which Mayr objected to Fisher's 'gene-centred' view of evolution (which I maintained was a misunderstanding on Mayr's part) but most interestingly he wrote 'Actually I don't think I read Fisher before I came to Harvard in 1953, and even then perhaps not right away'. The correspondence continued through 2003.

Conclusion

Fisher's reputation as having an oversimplified 'gene-centred' view of evolution is often promoted by those who, like Mayr, have not studied *The Genetical Theory* properly and have fallen into the trap of assuming that he was ploughing the same furrow as Haldane and Wright. Indeed, it is customary to follow Lancelot Hogben's 1931 initiative and mention Fisher, Haldane and Wright in the same breath. But as we have seen, Fisher was a Darwinist, and, like his hero, adopted a synthetic approach to evolution in contrast to Haldane and Wright, whose approaches were much more analytic, exploring the evolutionary consequences of the genetical models they set up. Ironically for one who established the fundamental model of balanced polymorphism in 1922, Fisher rarely constructed specific models. Indeed when, in 1959, O'Donald asked him why he had written so little on [mathematical genetics] after 1930, he replied that it had seemed to him that the mathematics had already by then gone beyond the available data'.

Failure to distinguish the synthetic and analytic approaches is the main reason why for so many years the fundamental theorem of natural selection was itself misunderstood, to the detriment of both Fisher's reputation and the advance of understanding. Mathematicians (mainly) viewed it analytically and became excited when they discovered population-genetic models in which the mean fitness was not an increasing function. The rest of us were more cautious and finally came to see that it was central to Fisher's mathematizing of Darwin. Wright's analytical version, in which he attempted to saddle a population's mean fitness with the properties of a potential function, was one of the principle butts of Fisher's invective.

The myth that Fisher's whole approach was analytic and 'single-locus' dies hard. As recently as last year, even the respected author of *The Eclipse of Darwinism*, Bowler, could write:

> '… we arrive at what Ronald Aylmer Fisher called in the title of his 1930 book *The Genetical Theory of Natural Selection*. Fisher's theory ignored the possibility that genes might interact with one another and worked with a simple model in which each gene coded for a particular variant form of a character. … it was then a straightforward procedure to assign a degree of "fitness" to each gene … .'

Countering such widespread misapprehension requires the same dedication to reading Fisher, especially *The Genetical Theory*, as Fisher applied to reading Darwin, especially *The Origin of Species* and *The Descent of Man*.

I shall leave you with some advice of Fisher's. Exactly 50 years after he came up to Cambridge, there was another Darwin anniversary to celebrate, this time the centenary of the publication of *The Origin*. Fisher had this to say:

> A centenary celebration is an occasion for retrospect, yet I submit, though the view is an old-fashioned one, that the purpose of retrospect is to prepare ourselves for the future, by avoiding the unnecessary repetition of the errors of the past.
>
> More attention to the History of Science is needed, as much by scientists as by historians, and especially by biologists, and this should mean a deliberate attempt to understand the thoughts of the great masters of the past, to see in what circumstances or intellectual *milieu* their ideas were formed, where they took the wrong turning or stopped short on the right track.

I hope I have demonstrated to you that the intellectual *milieu* in which Fisher's contributions were formed was the writings of Darwin, and that this placed him in a unique position to lead the *synthesis* of Darwinism with Mendelism. A century after the immersion of Fisher in Darwinian Cambridge, we can see even more clearly why Dawkins called him the greatest of Darwin's successors.

Acknowledgements My understanding of all matters relating to Darwin and Fisher has, for more than 50 years, benefitted from conversations with Dr Peter O'Donald, Fellow of Emmanuel College, to whom I am most grateful. In particular, he emphasized Fisher's distinction between the inductive and deductive approaches to the study of evolution in the introductory chapter to his 1980 book *Genetic Models of Sexual Selection*, which I had the pleasure of reading in draft.

Only on recently re-reading Alan Grafen's 2003 lecture 'Fisher the evolutionary biologist', which I had heard him deliver, did I discover that he described Fisher as 'Darwin's mathematician', thus no doubt sowing the seed of the title of this lecture in my subconscious mind.

Bibliography

Bateson W (1909) Mendel's principles of heredity. Cambridge University Press, Cambridge

Bennett JH (ed) (1983) Natural selection, heredity, and eugenics. Clarendon, Oxford

Bodmer WF, Edwards AWF (1960) Natural selection and the sex ratio. Ann Hum Genet 24:239–244

Bowler PJ (1983) The eclipse of Darwinism. Johns Hopkins University Press, Baltimore

Bowler PJ (2008) Survival of the "fittest". Regal 1:116–138

Box JF (1978) R.A. Fisher: the life of a scientist. Wiley, New York

Bulmer M (2003) Francis Galton: pioneer of heredity and biometry. Johns Hopkins University Press, Baltimore

Burbridge D (1992) Lacking a solution? Nature 355:118

Clark R (1968) J.B.S.: the life and work of J.B.S. Haldane. Hodder and Stoughton, London

Crew FAE (1969) Recollections of the early days of the Genetical Society In: Jinks J (ed) Fifty years of genetics. Oliver & Boyd, Edinburgh

Crow JF (1990a) R.A. Fisher, a centennial view. Genetics 124:207–211

Crow JF (1990b) Fisher's contributions to genetics and evolution Theor Popul Biol 38:263–275

Darwin C (1859) On the origin of species by means of natura selection. Murray, London (facsimile reprint with an introductio by E. Mayr, Harvard University Press, 1964)

Darwin C (1871) The descent of man, and selection in relation to sex Murray, London (2nd edn., 1874, Murray, London)

Darwin C (1909) The foundations of the origin of species, two essay written in 1842 and 1844 by C. Darwin. In: Darwin F (ed Cambridge University Press, Cambridge

Darwin CG (1930) Natural selection. Eugen Rev 22:127–130

Darwin L (1926) The need for eugenic reform. Murray, London

Dawkins R (1986) The blind watchmaker. Longman, London

Dobzhansky T (1937) Genetics and the origin of species. Columbi University Press, New York

Dobzhansky T (1955) A review of some fundamental concepts an problems of population genetics. Cold Spring Harbor Syn Quant Biol 20:1–15

Edwards AWF (1963) Migrational selection. Heredity 18:101–106
Edwards AWF (1990) R.A. Fisher: twice professor of genetics: London and Cambridge *or* "A fairly well-known geneticist". Biometrics 46:897–904
Edwards AWF (1992) Likelihood (expanded edition). Johns Hopkins University Press, Baltimore
Edwards AWF (1993) Mendel, Galton, Fisher. Aust J Stat 35:129–140
Edwards AWF (1994) The fundamental theorem of natural selection. Biol Rev 69:443–474
Edwards AWF (1995) Fiducial inference and the fundamental theorem of natural selection. Biometrics 51:799–809
Edwards AWF (1998) Natural selection and the sex ratio: Fisher's sources. Am Nat 151:564–569
Edwards AWF (2000a) Carl Düsing (1884) on the regulation of the sex ratio. Theor Popul Biol 58:255–257
Edwards AWF (2000) The genetical theory of natural selection. In: Crow JF, Dove WF (ed) Perspectives. Genetics 154:1419–1426
Edwards AWF (2001) Darwin and Mendel united: the contributions of Fisher, Haldane and Wright up to 1932. In: Reeve ECR (ed) Encyclopedia of genetics. Fitzroy Dearborn, London, pp 77–83
Edwards AWF (2005) R.A. Fisher, statistical methods for research workers, First Edition (1925). In: Grattan-Guinness I (ed) Landmark writings in Western mathematics, 1640–1940. Elsevier, Amsterdam, pp 856–870
Fisher RA (1914) Some hopes of a eugenist. Eugen Rev 5:309–315
Fisher RA (1915) The evolution of sexual preference. Eugen Rev 7:184–192
Fisher RA (1918) The correlation between relatives on the supposition of Mendelian inheritance. Trans R Soc Edinb 52:399–433
Fisher RA (1922) On the dominance ratio. Proc R Soc Edinb 42:321–341
Fisher RA (1924) The biometrical study of heredity. Eugen Rev 16:189–210
Fisher RA (1925) Statistical methods for research workers. Oliver & Boyd, Edinburgh
Fisher RA (1930a) The distribution of gene ratios for rare mutations. Proc Roy Soc Edinb 50:205–220
Fisher RA (1930) The genetical theory of natural selection. Clarendon Press, Oxford (2nd edn., 1958, New York, Dover. Variorum edition, Bennett JH (ed), 1999, Oxford University Press)
Fisher RA (1932a) The bearing of genetics on theories of evolution. Sci Prog 27:273–287
Fisher RA (1932b) The evolutionary modification of genetic phenomena. Proc 6th Int Congr Genetics 1:165–172
Fisher RA (1947) The renaissance of Darwinism. Listener 37:1001
Fisher RA (1950a) Gene frequencies in a cline determined by selection and diffusion. Biometrics 6:353–361
Fisher RA (1950b) Contributions to mathematical statistics. Wiley, New York
Fisher RA (1952) Statistical methods in genetics. Heredity 6:1–12
Fisher RA (1954) Retrospect of criticisms of the theory of natural selection. In: Huxley J, Hardy AC, Ford EB (eds) Evolution as a process, 84–98. Allen & Unwin, London
Fisher RA (1959) Natural selection from the genetical standpoint. Austr J Sci 22:16–17
Fisher RA (1971–74) Collected papers of R.A. Fisher, volumes I–V. In: Bennett JH (ed). University of Adelaide, Australia
Fisher RA, Stock CS (1915) Cuénot on preadaptation. A criticism. Eugen Rev 7:46–61
Gayon J (1998) Darwinism's struggle for survival. Cambridge University Press, Cambridge
Grafen A (2003) Fisher the evolutionary biologist. The Statistician 52:319–329
Haldane JBS (1931) A mathematical theory of natural selection. Pt. VIII. Metastable populations. Proc Camb Phil Soc 27:137–142
Haldane JBS (1932) The causes of evolution. Longman Green, London
Hamilton WD (1964) The genetical evolution of social behaviour, II. J Theor Biol 7:17–51
Hogben L (1931) Genetic principles in medicine and social science. Williams and Norgate, London
Hunt BJ (1991) The Maxwellians. Cornell University Press, New York
Huxley JS (1936) Natural selection and evolutionary progress. British Association Annual Meeting Report, 81–100
Huxley JS (1942) Evolution, the modern synthesis. Allen & Unwin, London
Huxley JS (1954) The evolutionary process. In: Huxley J, Hardy AC, Ford EB (eds) Evolution as a process. Allen & Unwin, London, pp 1–23
Huxley JS (1970) Memories I. Allen & Unwin, London
Jones S, Keynes M (eds) (2007) Twelve Galton lectures. Galton Institute, London
Kendall DG (1990) Obituary, Andrei Nikolaevich Kolmogorov (1903–1987). Bull Lond Math Soc 22:31–100
Kimura M (1964) Diffusion models in population genetics. Methuen's monographs on applied probability and statistics. Supplementary Review Series in Applied Probability 2
Lack D (1954) The evolution of reproductive rates. In: Huxley J, Hardy AC, Ford EB (eds) Evolution as a process, 143–156. Allen & Unwin, London
Lock RH (1906) Recent progress in the study of variation, heredity and evolution. Murray, London
Marshall J, McNamara J, Houston A (2010). The state of Darwinian theory. Behav Ecol Sociobiol. doi:10.1007/s00265-010-1121-y
Mayr E (1942) Systematics and the origin of species. Columbia University Press, New York (reprinted 1999 with a new Introduction, Harvard University Press, Cambridge)
Mayr E (1982) The growth of biological thought. Belknap, Cambridge
Mayr E (1980) Prologue, some thoughts on the history of the evolutionary synthesis. In: Mayr E, Provine WB (eds) The evolutionary synthesis, 1–48. Harvard University Press, Cambridge
Mayr E (1991) One long argument: Charles Darwin and the genesis of modern evolutionary thought. Harvard University Press, Cambridge
Mayr E (1999) see 1942
Mazumdar PMH (1992) Eugenics, human genetics and human failings. Routledge, London
O'Donald P (1986) Models for the evolution of dominance. Proc R Soc B 171:127–143
O'Donald P (1980) Genetic models of sexual selection. Cambridge University Press, Cambridge
O'Donald P (1990) Fisher's contributions to the theory of sexual selection as the basis of recent research. Theor Popul Biol 38:285–300
Provine WB (1978) The role of mathematical population geneticists in the evolutionary synthesis of the 1930s and 1940s. Stud Hist Biol 2:167–192
Provine WB (1986) Sewall Wright and evolutionary biology. University of Chicago Press, Chicago
Provine WB (1990) Discussion: population genetics. Bull Math Biol 52:201–207
Punnett RC (1905) Mendelism. Bowes and Bowes, Cambridge
Punnett RC (1911) Mendelism. In: The Encyclopædia Britannica, 11th edn., 18:115–120. New York
Regal B (ed) (2008) Icons of evolution. Greenwood, Westport
Seward AC (ed) (1909) Darwin and modern science. Cambridge University Press, Cambridge
Simpson GG (1944) Tempo and mode in evolution. Columbia University Press, New York
Shaw RF, Mohler JD (1953) The selective significance of the sex ratio. Am Nat 87:337–342
Wilkins A (2008) Neo-Darwinism. Regal 2:491–515

Williams GC (1996) Adaptation and natural selection. Princeton University Press, Princeton
Wright S (1930) The genetical theory of natural selection, a review. J Hered 21:340–356
Wright S (1931) Evolution in Mendelian populations. Genetics 16:97–159
Wright S (1932) The roles of mutation, inbreeding, crossbreeding and selection in evolution. Proc Sixth Int Congr Genet 1:356–366
Wright S (1968–78). Evolution and the genetics of populations (I Genetic and biometric foundations; II The theory of gene frequencies; III Experimental results and evolutionary deductions; IV Variability within and among natural populations). University of Chicago Press, Chicago
Yule GU (1902) Mendel's laws and their probable relations to intra-racial heredity. New Phytol 1(193-207):222–238

Paper 235 (2013)

Robert Heath Lock and his textbook of genetics, 1906

Genetics 194, 529–537.

Robert Heath Lock and His Textbook of Genetics, 1906

A. W. F. Edwards[1]
Gonville and Caius College, Cambridge, CB2 1TA, United Kingdom

ABSTRACT Robert Heath Lock (1879–1915), a Cambridge botanist associated with William Bateson and R. C. Punnett, published his book *Recent Progress in the Study of Variation, Heredity, and Evolution* in 1906. This was a remarkable textbook of genetics for one appearing so early in the Mendelian era. It covered not only Mendelism but evolution, natural selection, biometry, mutation, and cytology. It ran to five editions but was, despite its success, largely forgotten following Lock's early death in 1915. Nevertheless it was the book that inspired H. J. Muller to do genetics and was remembered by A. H. Sturtevant as the source of the earliest suggestion that linkage might be related to the exchange of parts between homologous chromosomes. Here we also put forward evidence that it had a major influence on the statistician and geneticist R. A. Fisher at the time he was a mathematics student at Cambridge.

ROBERT HEATH LOCK (1879–1915; Figure 1) is a largely forgotten figure today but was highly influential in the nascent field of genetics in the early 1900s. As an undergraduate, he had counted William Bateson as one of his lecturers and had graduated in Botany in 1902, soon afterward becoming a Fellow of Gonville and Caius College, University of Cambridge. Caught up in the excitement of the development of Mendelism in Cambridge, as championed in particular by Bateson, he nevertheless kept his distance from Bateson and wrote his own book-length treatment of genetics, *Recent Progress in the Study of Variation, Heredity, and Evolution*. This book appeared at the end of November 1906 (Lock 1906) and in subsequent editions in 1909, 1911, 1916, and 1920. Although it was not the first book to embrace the new science of genetics, following the rediscovery of Mendel's paper, it was the first in English that can be described as a textbook. As such it was widely used and admired in Britain and the United States. Its influence, both acknowledged and inferred, was substantial, but it was essentially forgotten after the First World War (1914–1918) and the death of its still young author, in 1915, from heart failure following influenza.

An indication of the book's importance is that it was used as a text by E. B. Wilson in his lectures at Columbia University, inspiring H. J. Muller to take up the study of genetics (Crew 1969; Carlson 1981, p.27) and leaving its mark on T. H. Morgan and A. H. Sturtevant as we shall see. Another famous American geneticist Sewall Wright read it in graduate school (Provine 1971, p.25).

There is also the possibility that the second edition was the primary source of information about genetics and evolution for the undergraduate R. A. Fisher, who entered Gonville and Caius College in October 1909. Fisher, famous as a statistician, held the Arthur Balfour Professorship of Genetics at Cambridge from 1943 until 1957 and in addition to his genetical work he had (and continues to have) a great influence on evolutionary biology through his book *The Genetical Theory of Natural Selection* (Fisher 1930; see the *Perspectives* by Edwards 2000, in *GENETICS*, Vol. 154, No. 4). In July 1909 Lock had presented a copy of the second edition of his book to the college library, and I shall offer evidence that this actual copy was used by Fisher. If this is true then the role of Lock's book on the development of genetics through its influence on Fisher will equal that through its influence on the Columbia school of *Drosophila* geneticists in the United States.

In asserting that *Variation, Heredity, and Evolution* was the first textbook in English of Mendelian genetics, I should mention two other books for the avoidance of misunderstanding. Bateson's *Mendel's Principles of Heredity* of 1902, unlike its namesake of 1909, was no textbook, but principally "A Defense of Mendel's Principles of Heredity" against the criticism of W. F. R. Weldon, a prominent member of K. Pearson's group of biometricians who were opposed to

doi: 10.1534/genetics.113.151266
[1]Address for correspondence: Gonville and Caius College, Cambridge, CB2 1TA, United Kingdom. Email: awfe@cam.ac.uk

Figure 1 R. H. Lock, from the 4th (1916) edition of *Recent Progress in the Study of Variation, Heredity, and Evolution.*

Bateson's Mendelian line of thought. Second, R. C. Punnett's *Mendelism* (1905) was, in its first edition, a tiny octavo volume of vii + 63 pages with neither contents nor index: essentially just a single chapter. It was subsequently expanded into a substantial textbook in its third edition of 1911, and by the time of its seventh and last edition in 1927 it had acquired a considerable reputation.

Mention should also be made of J. P. Lotsy's *Vorlesungen über Deszendenztheorien* of 1906. Lock wrote in his Preface (p. viii):

> Whilst I was still engaged upon my task [of writing], the first volume of Dr. Lotsy's admirable "Vorlesungen über Descendenztheorien" made its appearance. But for the fact that most of the following pages had then already been written, I might have hesitated to pursue my project, since a book not altogether unlike the present might be produced by the comparatively simple process of making a series of judicious extracts from Dr. Lotsy's work.

Bateson reviewed Lotsy's first volume in *Nature* of June 14, 1906 under the heading *A TEXTBOOK OF GENETICS*, this being the first time his word "genetics" appeared in print:

> As the moment is favourable, may it be suggested that the branch of science the rapid growth of which forms the occasion of Prof. Lotsy's book should now receive a distinctive name? ... the physiology of heredity and variation is a definite branch of science, and if we knew nothing of evolution that science would still exist. To avoid further periphrasis, then, let us say genetics.

(On July 31 Bateson also proposed "genetics" in his presidential address to the Third International Congress on Hybridization and Plant Breeding 1906, with such success that when the *Report* was published, the title of the congress was changed to "Genetics" retrospectively.) Lock intimated in the preface of his book (p. vii), dated October 23, 1906, that he would have liked to have used "genetics" in its title. "Since, however, the meaning of the word 'genetics' is not yet clearly understood by everybody, it seemed better to adopt in the present instance a somewhat more descriptive title."

Lotsy's book seems to have had little impact in the English-speaking world. Thompson (1908, p. 564) records it as "a valuable treatise," but it was rarely referenced, not even by Waddington (1939) in his extensive bibliography with many entries in German. Bateson (1909) lists it, but apparently without reference to it, as did Doncaster (1910), while Sturtevant (1965) does not include it.

Although Lock has been almost completely forgotten, many readers will in fact have come across his name in those tables often reproduced showing the results of experiments with peas in the first decade of the 20th century when early Mendelians sought to confirm Mendel's results. Such tables, which usually give the name "Lock" without initials and without a reference, originate from the one he published in "The present state of knowledge of heredity in Pisum" (Lock 1908). He there mentioned: "Many of the original crosses upon which the most recent of these results are based were made by Mr. Bateson. I desire to record here my indebtedness to him for handing over to me, in 1905, the greater part of his material relating to Pisum."

In this *Perspectives* I first give a biography of Lock, necessarily scanty given his neglect for nearly a century, and then a description of his book. There follow two sections discussing its influence, first on the Columbia school and second on R. A. Fisher.

R. H. Lock, 1879–1915

The posthumous fourth edition of *Variation, Heredity, and Evolution* (1916) carries a biographical note by "B. L.," Lock's widow Bella, formerly Bella Sidney Woolf (1877–1960), eldest daughter of Sidney Woolf, QC. Her brother Leonard married Virginia Stephen, the second daughter of Sir Leslie Stephen and better known today under her married name of Virginia Woolf. The details of Lock's early life, given below, are based largely on this biographical note.

Robert Heath Lock was born on January 19, 1879 at Eton College where his father, the Reverend John Bascombe Lock, was a master. His mother Emily was the daughter of Edwin Baily of Cirencester. He was educated at Charterhouse School and entered Gonville and Caius College with an entrance exhibition on October 1 1898, by which time his father, a mathematician, had moved to Cambridge and then become bursar of the college. The son prospered in the Natural Sciences Tripos, attending Bateson's lectures "On the practical study of evolution" and being placed in the first class in both part I (1900) and part II (Botany 1902). While still an undergraduate Lock accompanied Bateson abroad, who wrote to his wife Beatrice from Tours, France, in April 1902, commenting on Lock, "My little companion ... [who] knows no French of any kind, and when I say no French I mean it. 'What is French for 'to eat?' ... German, he reads with fair ease. He has a prodigious memory for verse: knows most of Hamlet, Macbeth and Gilbert and Sullivan by heart, and can recite more ephemeral verses than ever I heard of" (Cock and Fordyce 2008, p. 290).

His Tripos results gained Lock the Frank Smart studentship of the college for two years, which he spent at the Royal Botanic Gardens at Peradeniya in Ceylon (Sri Lanka), where another Caian (as members of the college are called) who had also been a Frank Smart student, J. C. Willis, was director. During this time, Lock engaged in an extensive correspondence with Bateson. Described by Cock and Forsdyke (2008, p. 290), these authors remark "In principle, the correspondence was all that a free and open scientific correspondence should be."

Returning to England, Lock was elected a Drosier fellow of the college on January 11, 1905, finding himself next in seniority to R. C. Punnett, who had been elected in 1901 (for Punnett, see the *Perspectives* by Edwards 2012, *GENETICS*, Vol. 192, No. 9). The interval of three years reflected not only their difference in age but the timing of their graduation, Lock entering Caius in the term after Punnett graduated. Though they did not overlap then, they did in the academic year 1901–1902 when Punnett was junior fellow and Lock a fourth-year undergraduate. Very probably their friendship started at that time. On p. 43 of *Variation, Heredity, and Evolution* (1906) Lock refers to "my friend Mr. R. C. Punnett," and there is much evidence of friendly banter between them later on in the Fellows' Betting Book. Punnett had joined Bateson as a collaborator at the start of 1904, so when Lock returned at the end of the year he would soon witness their famous discovery of "partial coupling" in the sweet pea (later "linkage"; for details see Edwards 2012). Lock was also appointed curator of the University Herbarium in the Botany School.

As bachelor fellows, Punnett and Lock both lived in college, and will have done much of their writing in their college rooms. These were in the 16th century buildings of Caius Court (M2 and K3, respectively). The preface of Punnett's book *Mendelism* is dated May 1905, so that for the first months of 1905 he was writing *Mendelism* and Lock was writing *Variation, Heredity, and Evolution* within a few yards of each other and in daily contact as resident college fellows. There was surely an understanding between them that they were engaged in quite different tasks, Punnett writing a short essay on just Mendelism, published in Cambridge, and Lock a survey of recent progress in genetics and evolution published in London by John Murray, Darwin's publisher. "The idea of writing this little book," he wrote in his preface (p. vii), "occurred to me while reading Mr. W. C. D. Whetham's volume on *The Recent Development of Physical Science*." "I found the story of the modern progress of physics so interesting as to encourage the belief that a similar account of the subjects with which I was myself more particularly familiar might prove of a like interest to other people."

Lock held his fellowship for six years, though not resident for the last two-and-a-half. In August 1906 he attended the Third International Conference on Hybridization and Plant Breeding (later styled "Genetics") in London with Bateson, Punnett, R. H. Biffen (another former Frank Smart student) and T. B. Wood (also a Caian). The "Newnham College Mendelians" (Richmond 2001) were also well represented, so Bateson, who was president of the congress, was supported by a substantial Cambridge contingent of his colleagues, helpers, and former students, mostly from Caius and Newnham. Further information about the "Cambridge Mendelians" is given by Richmond (2006a) and also in her account of the 1909 Darwin Celebration by the University (Richmond 2006b).

In the spring of 1908 Lock returned to Peradeniya as assistant director of the gardens under Willis. "We understand," says the Caius College magazine, "that he is to have complete charge of the Economic Development of the Gardens." There he met his Trinity College contemporary Leonard Woolf and, on the voyage out, Leonard's sister Bella who was visiting him. At the end of June 1909, Punnett also traveled to Ceylon (to study mimicry) on a three-month trip, visiting Lock. No doubt their conversation naturally turned to the appointment of a first director of the newly founded John Innes Horticultural Institute then under consideration. Lock was one of three who had been short listed from among 30 applicants (Cock and Forsdyke 2008, p. 382) but the council deemed none satisfactory and ultimately, in November, Bateson, who had not applied, was offered the directorship. He accepted. There are further letters from Lock to Bateson during this period.

In 1910 Lock returned to England on short leave and married Bella on July 7. Two days previously he had dined in college and settled a bet with Mr. Duncan Jones, whom he had bet on March 11, 1906 "will be engaged to be married before him." He had won, and Mr. Duncan Jones paid (with the customary wine after dinner). Returning to Ceylon with Bella, Lock conducted experiments on rubber tapping and rice breeding, but was also heavily engaged in administrative work, especially as acting director when Willis was on leave.

In 1911 Willis moved to Rio de Janiero as director of the botanic gardens there, but Lock, not wishing to succeed him in Ceylon, returned to England in 1912, spending the first few months after his return writing a second book *Rubber and Rubber Planting* (1913). In June 1912 Bateson was invited to return to Cambridge for the newly created Arthur Balfour Professorship of Genetics, but he declined it and suggested Punnett (see Edwards 2012). Had the professorship been filled by a board of electors rather than Arthur Balfour and the prime minister, Lock might have stood a chance, but Punnett was effectively the sitting tenant and the possibility might never have crossed Lock's mind. Only with hindsight can we see what a good choice he might have been.

He then took up a post of inspector at the Board of Agriculture and Fisheries. When the First World War broke out in 1914 he sought to join the forces, but was asked instead to add the chairmanship of a Vegetable Drying and

Fruit Preserving Committee to his existing duties, a government body charged with furthering work on food preservation as a wartime necessity.

Struck down by a severe bout of influenza in February 1915 from which he never fully recovered, and overworked with his wartime duties, Lock died from heart failure at Eastbourne on June 26, 1915 at the age of 36. He is buried in the Ascension Burial Ground, Cambridge, where also lie his parents John and Emily. His widow, Bella, went on, as Bella Woolf, to be a successful author.

Robert and Bella Lock had no children, but Robert's three brothers were also educated at Caius College, and two of them started family lines that have ensured four and five generations of Locks at Caius, a tradition that had started with John Bascombe in 1867.

Recent Progress in the Study of Variation, Heredity, and Evolution

In the preface to *Variation, Heredity, and Evolution* (p. ix) Lock is at pains courteously to distance himself from his teacher Bateson. After thanking a number of his friends he continues:

> Adequately to acknowledge Mr. Bateson's influence upon these pages is a more difficult matter, and not the less so because I have deliberately refrained as far as possible from consulting him while the book was in course of preparation, in order that it might retain if possible some traces of individuality. It is therefore clear that he is in no way responsible for its deficiencies. But, apart from the fact that I am conscious of having quoted his ideas at more points than could possibly be acknowledged *seriatim*, I owe to Mr. Bateson both my first introduction to the science of genetics, and a continual fund of encouragement in the prosecution of studies connected with it.

Lock's book is indeed characterized by a transparent willingness to give both sides of every controversial point, and this is especially true of the saltationist view of evolution championed by Bateson as opposed to the gradualist Darwinian mechanism. Cock and Forsdyke (2008, p. 382) mention in connection with the John Innes directorship that "Bateson's fair-mindedness had been sorely tried by [Lock's] support of de Vries and even, on occasions, of Pearson." "Fair-mindedness," however, is a quality that one comes to associate more with Lock than with Bateson. Bella's biographical note reveals a character that is sensitive, thoughtful, and considerate. She quotes "an old college friend" as writing about the photograph of Lock (Figure 1) "I think, in a very charming way, it reflects the outstanding feature of his moral and mental personality. He liked cleverness, but wouldn't put up with it for a moment if it were superficial. His love of truth was something more than the habit of a trained scientist." In contrast to Punnett, who worked happily with Bateson (see Edwards 2012), Lock kept his distance. After his death Bateson reread Lock's book with a view to writing an obituary notice for him: "I have made a start on my notice of Lock. I don't like the book better on closer acquaintance" (Cock and Forsdyke 2008, p. 459).

Variation, Heredity, and Evolution is an astonishing work for its time, partly because of the encyclopedic knowledge of its 27-year-old author and partly because of its original and influential musings on the relationship between the phenomenon of partial coupling discovered by Bateson and Punnett and the chromosomal theory of heredity (see the next section). The 10 chapters of the first edition bring together discussion of evolution by natural selection, biometry (at quite an advanced level) including the law of ancestral heredity, the theory of mutation, Mendelism (including the contributions of Yule and Pearson), cytology, and the implications of genetics to human affairs. Chapter topics are listed in the contents, and reproduced below in an appendix. The historical background is amply covered: Darwin, Wallace, and Mendel of course, but Malthus and Lamarck and Galton and other still more distant figures. Of recent writers Lock is thoroughly familiar with de Vries, Weismann, Johannsen, Tschermak, Correns, Wilson, Boveri, Yule, and Pearson, as well as with his Cambridge contemporaries Bateson, E. R. Saunders, Punnett, and Biffen. Some indication of the extent of Lock's reading can be gauged from the list of 80 references at the end of a paper he published just before returning to England in the autumn of 1904 "Introductory: The work of Mendel and an account of recent progress on the same lines; with some new illustrations. Part I of Studies in Plant Breeding in the Tropics" (Lock 1904).

The second edition (1909) adds bibliographies to each chapter. An important addition is made to Chapter VIII: where the first edition briefly alludes to "some coupling" between two Mendelian factors, the second discusses Bateson and Punnett's case of *partial gametic coupling* in the sweet pea, the first observation of what came to be known later as *linkage* (see Edwards 2012). Chapter X "Eugenics" is entirely new, amplifying some remarks contained in the concluding chapter of the first edition. Lock's views are typical of the period. He summarizes Galton's arguments and adds that "The point of view which has been adopted in the present chapter is very well summed up in the following paragraph by Prof. Karl Pearson," which starts "As we have found conscientiousness is inherited, so I have little doubt that the criminal tendency descends in stocks" and continues in like vein. No reference is given, but the paragraph is from *The Scope and Importance to the State of the Science of National Eugenics* (Pearson 1909), the Robert Boyle Lecture in Oxford in 1907. The views expressed in the lecture are more extreme than those customarily attributed to Pearson.

The third edition (1911) differs little from the second. The fourth, posthumous, edition (1916) was a revision by Lock's friend Leonard Doncaster, who inserted substantial alterations and additions at three points. We note the first two, to do with linkage, in the next section. The third concerned sex determination and sex linkage, particular interests of Doncaster (1914). Doncaster had himself written a small book *Heredity in the Light of Recent Research* (1910) in which he recorded his debt to Bateson (1909) but added "I have not hesitated to make use of Lock's *Recent*

Progress in the Study of Variation, Heredity and Evolution, Thomson's *Heredity* [see below], and some other books dealing with general aspects of the subject." His own third edition (1921) lists Lock's book in a fifth edition, 1920, which he had revised but little further.

Lock wrote to Bateson on March 8, 1909 that his first edition had sold so well that the publisher was proposing a second (Cock and Forsdyke 2008, p. 382). The first edition had already been reprinted twice, in April 1907 and March 1909. Copies of all editions are not hard to find.

There was a sympathetic full-page review of the first edition in *Nature* by "F.A.D." (1907; probably Dixey), and a hostile one in *Biometrika* by W. P. Elderton (1907), a statistical colleague of Pearson's. J. Arthur Thompson, in his *Heredity* (1908, p. 563), thought the book "a useful introduction especially to biometric methods and Mendelian experiments," while Alfred Russel Wallace's (1908) opinion merits a full quotation:

> In conclusion, I would suggest to those of my readers who are interested in the great questions associated with the name of Darwin, but who have not had the means of studying the facts either in the field or the library, that in order to obtain some real comprehension of the issue involved in the controversy now going on they should read at least one book on each side. The first I would recommend is a volume by Mr. R. H. Lock on "Variation, Heredity and Evolution" (1906) as the only recent book giving an account of the whole subject from the point of view of the Mendelians and Mutationists. ... .

This quotation is printed on the fly leaf of the second (1909) and subsequent editions of *Variation, Heredity, and Evolution*.

The influence of *Variation, Heredity, and Evolution* on the Columbia school

The 2000 reprint of A. H. Sturtevant's *A History of Genetics* (1965) says, on the back cover:

> In the small "Fly Room" at Columbia University, T. H. Morgan and his students, A. H. Sturtevant, C. B. Bridges, and H. J. Muller, carried out the work that laid the foundations of modern, chromosomal genetics. The excitement of those times, when the whole field of genetics was being created, is captured by this book, written by one of those present at the beginning. ... it is worth remembering that the world's first genetic map was created in 1913 by A. H. Sturtevant .. .

Appendix A in the book gives a chronology, recording for 1906 "Lock: suggested the relation between linkage and the exchange of parts between homologous chromosomes" (2000, p. 137). In the text, Sturtevant wrote (2000, p. 43) "The possibility that linkage might result from genes lying on the same chromosome had been suggested by Lock in 1906, in his elaboration of de Vries' idea that exchange of material between homologous chromosomes could account for independent segregation." The reference is to the American edition of *Variation, Heredity, and Evolution* (Dutton, New York 1906; identical to the British one) and Chapter IX Recent Cytology. Lock was, of course, familiar with the recent discovery of partial coupling of characters in the sweet pea found by Bateson and Punnett in 1904 and 1905, later called linkage, and on p. 187 he briefly alludes to it. Then on p. 252, after describing de Vries's (erroneous) hypothesis for independent segregation, Lock writes:

> In cases where the phenomenon of correlation or coupling has been observed we must suppose that there is some mechanism which causes the representative particles of the respective characters concerned to remain in company during the process by which the other allelomorphs are being reassorted between the chromosomes. Of this process of coupling the cytologists have not yet been able to observe any visible indication in the behaviour of the chromosomes, any more than they can really see the redistribution of the supposed factors carried by the chromosomes. But apart from this it must be allowed that the facts of the experiment and of microscopic observation fit in with one another in a remarkable way, and that the Mendelian theory throws considerable light on the minute features of cell anatomy.

E. B. Wilson, professor of zoology at Columbia, used Lock's book for his course "Heredity, variation and the chromosomes." The minds of the geneticists of the Columbia school, for whom *Variation, Heredity, and Evolution* had been their textbook, were thus exposed to the germ of an idea about a chromosomal explanation for linkage.

H. J. Muller had taken general biology in his freshman year, much of it taught by J. H. MacGregor, and

> After his first successful year, Muller spent the summer as a hotel clerk in New Jersey. He knew that in the fall he would be taking Wilson's course on the cell, and he knew from MacGregor that the text for it would be R. H. Lock's *Recent Progress in the Study of Variation, Heredity, and Evolution* (1906). Muller read the book while he hid, out of sight, under the stairwell of the hotel when the bell wasn't ringing, and he again experienced the electrifying thrill he had when he first read Mendeleev's periodic tables. Here was a system that brought evolution, cell biology, and plant and animal breeding into one common scheme (Carlson 1981 p. 27).

In the words of Crew (1969), it was Lock's book "that did so much to attract Muller to genetics." Carlson (2013) attended Muller's course "Mutation and the gene" in 1955 and prints a page of his lecture notes recording: "1906. R. H. Lock. Recent progress in the study of variation and evolution (*sic*)—The completely modern viewpoint for its time." Muller (1943) himself wrote that in 1908 he took E. B. Wilson's "thrilling one semester course on heredity and the chromosomes, variation and evolution. In this course, the text chosen by Wilson was Lock's extraordinary book of 1906—too far "ahead of its time" to be sufficiently remembered now—which, with less caution and fewer qualifications than employed by Wilson himself, advocated the sufficiency of Mendelism, multiple factors, the chromosome theory (including exchange of linearly arranged genes during parasynapsis, after de Vries), and the natural selection of mutations, as the basis of all heredity and evolution."

Muller (1916) did not mention Lock in his Ph.D. thesis "The mechanism of crossing-over," and neither had Morgan (1911) when he advanced his hypothesis that "partial coupling—linkage—is a reflection of 'associations of factors' that are located near together in the chromosomes." By then Morgan was able to rely on the cytological observations of Janssens (in 1909; see the *Perspectives* by Koszul *et al.* 2012, in *GENETICS*, Vol. 191, Num. 2). But when all four of the Columbia school combined to publish *The Mechanism of Mendelian Heredity* (Morgan *et al.* 1915) in Chapter I they wrote "In the same year [1906] ... Lock called attention to the possible relation between the chromosome hypothesis and linkage" (p. 5) and did not there mention Janssens. Consciously or subconsciously, they had absorbed Lock's point. But memory of Lock was fading: they gave him no initials, either there or in the index, and his book is not in the bibliography. By the time Whitehouse (1965) wrote his *Toward an Understanding of the Mechanism of Heredity* with its substantial historical content, Lock had disappeared almost completely, squeezed out between de Vries and Morgan rather than perceptively linking the two. Besides Sturtevant (1965) perhaps only Cock (1983), in his study of Bateson's reaction to the chromosome theory, remembered: "Lock, in particular, gave vigorous support to the chromosome theory as early as 1906." Indeed, in his second edition (1909, p. 201) when he added a report of the segregation ratios proposed by Bateson and Punnett on their "reduplication hypothesis" (see Edwards 2012), Lock did not mention the hypothesis by name. His account of their idea evinces skepticism and concludes "There is good reason for hoping that we shall soon be able to describe the facts of correlation [*i.e.* partial coupling] in terms as definite as those which can now be applied to the phenomena of segregation." Janssens was just round the corner.

Wilson attended the 1909 Darwin Centenary Celebrations in Cambridge, staying with the Batesons (Cock and Forsdyke 2008, p. 321) and receiving an honorary degree from the university. It seems probable that Lock was in Ceylon at the time and missed the opportunity to meet the most influential user of his book.

From a historical point of view, it was unfortunate but understandable that when Doncaster inserted new material about linkage in Lock's fourth edition (pp. 200–202, 269–271) he removed Lock's suggestion for the phenomenon of coupling, quoted above, and replaced it with a description of Morgan's work.

R. A. Fisher's Sources of Information About Genetics, 1909–1911

> I first came to Cambridge in 1909, the year in which the centenary of Darwin's birth and the jubilee of the publication of *The Origin of Species* were being celebrated. The new school of geneticists using Mendel's laws of inheritance was full of activity and confidence, and the shops were full of books good and bad from which one could see how completely many writers of this movement believed that Darwin's position had been discredited (Fisher 1947).

"The fiftieth anniversary of the publication of *The Origin of Species* was being celebrated, apart from other things, by the publication of Bateson's book *Mendel's Principles of Inheritance*" (Fisher 1959; actually *Mendel's Principles of Heredity*). Fisher bought the book. He also received as a college prize the book of essays *Darwin and Modern Science* (Seward 1909) published by Cambridge University Press as part of the Darwin Celebrations ("a remarkable collection of able essays," Fisher was to say in 1959).

This is all the definite information there is about Fisher's reading in genetics, and includes Mendel's paper, which was given in translation in *Mendel's Principles of Heredity.* But among the "books good and bad" that he encountered then, or soon after, we must surely include Punnett's *Mendelism* in one of its first two editions (1905, 1907). Punnett, as we have seen, was a resident fellow of Caius, and Fisher certainly came to know him during his undergraduate days (Mazumdar 1992, p. 99). However, Fisher may not have met Lock. The only time when they might both have been in college was when Lock dined there on July 5, 1910 just before his wedding (see above). But Lock had presented his second edition (1909) to the college library in July, and even if Punnett had not drawn attention to it Fisher would surely have come across it. An undergraduate who notes "books good and bad" on a subject of consuming interest to him in the Cambridge bookshops is hardly likely to ignore his own college library. Be that as it may, Fisher never referred to Lock's book in his writings, but then he never referred to Punnett's either.

There is some evidence that Fisher used this particular copy of Lock's second edition. The book has been moderately well used, and of all the potential Caius undergraduate readers of it, Fisher seems the most likely. But more direct evidence appears on p. 87. Lock there reproduces a diagram from Wallace's *Darwinism* (1889, p. 58). Neither axis is labeled, and in the Caius copy someone has added "No of birds" for the *x*-axis and someone else (the hand is different) an indistinct "length of wing" or some such for the *y*-axis. As one who is familiar with Fisher's spidery handwriting, often very small because of his extreme short sightedness, I believe "No of birds" (Figure 2) is in his hand. I can cite other instances of Fisher having annotated library books in Cambridge, including one other in the Caius library. There is also of course the possibility that Fisher remembered the diagram from Wallace, whose *Darwinism* he had read by 1915 (Fisher 1915).

By the time Fisher gave his talk "Mendelism and Biometry" on November 10, 1911 (printed in Bennett 1983, p. 51) he was to varying degrees familiar with the work of Weismann, de Vries, and Johannsen. His first example of Mendelism is the Andalusian fowl, which he could have got from any of the three books of Bateson, Punnett, and Lock. He is familiar with the work of Pearson, and in particular with the problem of parent–offspring correlation and the differing views of Pearson and of Yule (whom he does not mention by name). The only source among the books so far mentioned seems to be Lock (1909), Chapter VIII

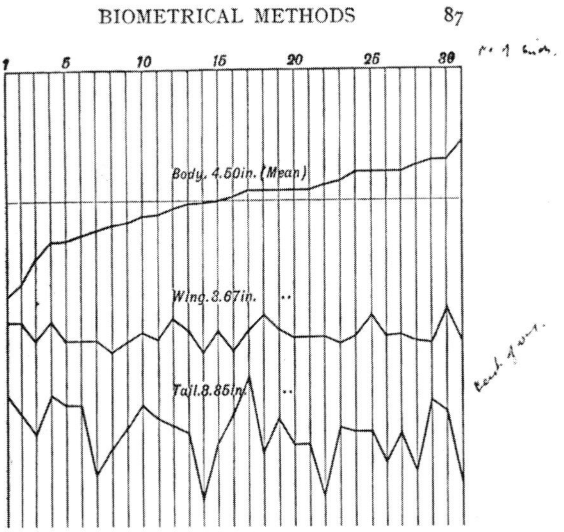

Figure 2 From the 2nd (1909) edition of *Recent Progress in the Study of Variation, Heredity, and Evolution*, annotated "No of birds" in a hand probably R. A. Fisher's.

Mendelism (*continued*) with a section "Mendelism and Biometry"—the very title of Fisher's talk—"researches of Yule and Pearson" (p. 227). Fisher mentions the good fit of Mendel's data to his hypothesis, revealing an apparent familiarity with Weldon (1902), a paper in *Biometrika* (see Edwards 1986, Magnello 1998). This seems to be a direct result of perusing *Biometrika* itself. He has read Galton's *Hereditary Genius* (1869). Lock (1909) devoted a whole new Chapter X Eugenics (p. 282) to Galton and his work.

In addition to this evidence for Lock's book in its second (1909) edition having been an important source for Fisher, it contains a number of quite suggestive remarks bearing on some of Fisher's later work. We have already noted the opening of the topic of the correlation between relatives, which matured into Fisher (1918). On the subject of classification Lock wrote in his introduction (p. 10)

> ... though at first sight it may appear almost paradoxical, ... it is quite possible for groups to be perfectly distinct, although individual members of them may have deviated so far, each from its proper type, as to render impossible the task of deciding from their appearance which group any of these individuals belong to.

In Fisher's (1925, p. 37) *Statistical Methods for Research Workers* we find the same thought:

> When a large number of individuals are measured in respect of physical dimensions, weight, colour, density, etc., it is possible to describe with some accuracy the *population* of which our experience may be regarded as a sample. By this means it may be possible to distinguish it from other populations differing in their genetic origin, or in environmental circumstances. Thus local races may be very different as populations, although individuals may overlap in all characters;

Furthermore, this follows two pages after a figure (his figure 3) constructed on the same strange principle as Wallace's diagram referred to above, which was reproduced by Lock (p. 87). As Fisher was still a fellow of Caius when writing *Statistical Methods* he would have had ready access to Lock in the library.

A remarkably prescient remark is made by Lock on p. 99, to the effect that the standard deviation "happens to be determinable with greater accuracy from an actual series of variates" than the mean quartile, and in a footnote he gives the root-mean-square derivation. Fisher (1920) explored the whole question in intricate detail and proved that *The whole of the information respecting* σ [the standard deviation of a normal distribution], *which a sample provides, is summed up in the value of* σ_2 [the sample standard deviation calculated from the mean-square deviation from the mean]. This is the first appearance of the notion of statistical *information* and led to the concept of *sufficient* statistics. Fisher (1950) notes that the paper arose from an examination of a statement by A. S. Eddington, but that does not preclude an initial stimulus from his reading of Lock.

Finally, when "quasi-linkage," or association between Mendelian factors on different chromosomes, was first discussed (Michie 1953, Wallace 1953) Fisher suggested the word "affinity" to describe it (Wallace 1958). Lock (1906, p. 185, 1909 p. 197) had written of "the existence of some kind of affinity occurring in the same individual between allelomorphs which belong to distinct pairs."

But perhaps equal to all these pieces of evidence is the unique feature of Lock's book that it brought together (to quote from its chapter headings) evolution, the theory of natural selection, biometry, the theory of mutation, Mendelism, cytology, and eugenics, all in a single volume. Nowhere else could the young Fisher have found such a guide to the subjects that fascinated him over and above his student work for the Mathematical Tripos. He had to look no further than his college library for the book, which would lay out the ground that he was to till so fruitfully in *The Genetical Theory of Natural Selection* (1930). Lock's *Recent Progress in the Study of Variation, Heredity, and Evolution* was, I suggest, as much an inspiration for R. A. Fisher as it had been for H. J. Muller.

Acknowledgments

I am grateful to Michael Buttolph, Elof Carlson, Donald Forsdyke, Stephen Stigler, and Adam Wilkins, who all helped to draw my attention to sources and facts about Lock and suggested improvements to this account.

Literature Cited

Bateson, W., 1902 *Mendel's Principles of Heredity: A Defence*. Cambridge University Press, Cambridge.

Bateson, W., 1906 A text-book of genetics. A review of Lotsy (1906). Nature 74: 146–147.

Bateson, W., 1909 *Mendel's Principles of Heredity*. Cambridge University Press, Cambridge.

Bennett, J. H. (Editor), 1983 *Natural Selection, Heredity, and Eugenics*. Clarendon Press, Oxford.
Carlson, E. A., 1981 *Genes, Radiation, and Society: The Life and Work of H. J. Muller*. Cornell University Press, Ithaca, NY.
Carlson, E. A., 2013 H. J. Muller's contributions to mutation research. Mutat. Res. 752: 1–5.
Cock, A. G., 1983 William Bateson's rejection and eventual acceptance of chromosome theory. Ann. Sci. 40: 19–59.
Cock, A. G., and D. R. Forsdyke, 2008 *Treasure Your Exceptions: The Science and Life of William Bateson*. Springer-Verlag, New York.
Crew, F. A. E., 1969 Recollections of the early days of the Genetical Society, *Fifty Years of Genetics*, edited by J. . Jinks Oliver and Boyd, Edinburgh.
D, F. A. 1907 The new evolution. Review of Lock (1906). Nature: 75, 578–579.
Doncaster, L., 1910 *Heredity in the Light of Recent Research*. Cambridge University Press, Cambridge.
Doncaster, L., 1914 *The Determination of Sex*. Cambridge University Press, Cambridge.
Edwards, A. W. F., 1986 Are Mendel's results really too close? Biol. Rev. Camb. Philos. Soc. 61: 295–312.
Edwards, A. W. F., 2000 The genetical theory of natural selection. Genetics 154: 1419–1426.
Edwards, A. W. F., 2012 Reginald Crundall Punnett: First Arthur Balfour Professor of Genetics, Cambridge, 1912. Genetics 192: 3–13.
Elderton, W. P., 1907 Review of Lock (1906). Biometrika 5: 478–480.
Fisher, R. A., 1918 The correlation between relatives on the supposition of Mendelian inheritance. Trans. R. Soc. Edinb. 52: 399–433.
Fisher, R. A., 1920 A mathematical examination of the methods of determining the accuracy of an observation by the mean error, and by the mean square error. Mon. Not. R. Astron. Soc. 80: 758–770.
Fisher, R. A., 1925 *Statistical Methods for Research Workers*. Oliver and Boyd, Edinburgh.
Fisher, R. A., 1930 *The Genetical Theory of Natural Selection*. Clarendon Press, Oxford. Ed 2, Dover, New York (1958); variorum edition, Oxford University Press, Oxford (1999).
Fisher, R. A., 1947 The renaissance of Darwinism. Listener 37: 1001.
Fisher, R. A., 1950 *Contributions to Mathematical Statistics*. Wiley, New York.
Fisher, R. A., 1959 Natural selection from the genetical standpoint. Aust. J. Sci. 22: 16–17.
Galton, F., 1869 *Hereditary Genius*. Macmillan, London.
Genetics, *Report of the Third International Conference*, 1906. Roy. Horticultural Soc., London.
Koszul, R., M. Meselson, K. Van Doninck, J. Vandenhaute, and D. Zickler, 2012 The centenary of Janssens's chiasmatype cheory. Genetics 191: 309–317.
Lock, R. H., 1904 Studies in plant breeding in the tropics. Ann. Roy. Botanic Gdns Peradeniya 2: 299–356.
Lock, R. H., 1906 Recent Progress in the Study of Variation, Heredity and Evolution. Murray, London.
Lock, R. H., 1908 The present state of knowledge of heredity in Pisum. Ann. Roy. Botanic Gdns Peradeniya 4: 93–111.
Lock, R. H., 1913 *Rubber and Rubber Planting*. Cambridge University Press, Cambridge.
Lotsy, J. P., 1906 *Vorlesungen über Deszendenztheorien*, Vol. I. Fischer, Jena.
Magnello, M. E., 1998 Karl Pearson's mathematization of inheritance: from ancestral heredity to Mendelian genetics. Ann. Sci. 55: 35–94.
Mazumdar, P. M. H., 1992 *Eugenics, Human Genetics and Human Failings*. Routledge, London.
Michie, D., 1953 Affinity: A new genetic phenomenon in the house mouse. Evidence from distant crosses. Nature 171: 26–27.
Morgan, T.H., 1911 Random segregation *vs.* coupling in Mendelian inheritance. Science (n.s.) 34: 384.
Morgan, T. H., A. H. Sturtevant, H. J. Muller, and C. B. Bridges, 1915 *The Mechanism of Mendelian Heredity*. Holt, New York.
Muller, H. J. 1916 The mechanism of crossing-over. Amer. Natur. 50: 193–221, 284–305, 350–366, 421–434.
Muller, H. J. 1943 Edmund B. Wilson: an appreciation. Amer. Natur. 77: 5–37, 142–172.
Pearson, K., 1909 *The Scope and Importance to the State of the Science of National Eugenics*. Dulau, London.
Provine, W. B., 1971 *The Origins of Theoretical Population Genetics*. University of Chicago Press, Chicago.
Provine, W. B., 1986 *Sewall Wright and Evolutionary Biology*. University of Chicago Press, Chicago.
Punnett, R. C., 1905 *Mendelism*. Macmillan and Bowes, Cambridge; 1907 Bowes and Bowes, Cambridge; 1911, 1912, 1919, 1922, 1927 Macmillan, London.
Richmond, M. L., 2001 Women in the early history of genetics. Isis 92: 55–90.
Richmond, M. L., 2006a The domestication of heredity: the familial organisation of geneticists at Cambridge University, 1895–1910. J. Hist. Biol. 39: 565–605.
Richmond, M. L., 2006b The 1909 Darwin Celebration. Isis 97: 447–484.
Seward, A. C. (Editor), 1909 *Darwin and Modern Science*. Cambridge University Press, Cambridge.
Sturtevant, A. H., 1965 *A History of Genetics*. Cold Spring Harbor Laboratory Press, Cold Spring Harbor, NY.
Thompson, J. A., 1908 *Heredity*. Murray, London.
Waddington, C. H., 1939 *An Introduction to Modern Genetics*. Allen and Unwin, London.
Wallace, A. R., 1889 *Darwinism*. Macmillan, London.
Wallace, A. R., 1908 The present position of Darwinism. Contemp. Rev. (Aug): 140.
Wallace, M. E., 1953 Affinity: a new genetic phenomenon in the house mouse. Evidence from within laboratory stocks. Nature 171: 27–28.
Wallace, M. E., 1958 Experimental evidence for a new genetic phenomenon. Philos. Trans. R. Soc. Lond., B 241: 211–254.
Weldon, W. F. R., 1902 Mendel's laws of alternative inheritance in peas. Biometrika 1: 228–254.
Whitehouse, H. L. K., 1965 *Towards an Understanding of the Mechanism of Heredity*. Arnold, London.

Communicating editor: A. S. Wilkins

Appendix

The CONTENTS of the first edition of

Recent Progress in the Study of Variation, Heredity, and Evolution

Paper 238 (2014)

R. A. Fisher's gene-centred view of evolution and the Fundamental Theorem of Natural Selection

Biological Reviews 89, 135–147.

R.A. Fisher's gene-centred view of evolution and the Fundamental Theorem of Natural Selection

A. W. F. Edwards*

Gonville and Caius College, Cambridge, CB2 1TA, UK

ABSTRACT

The background to R.A. Fisher's enunciation of his Fundamental Theorem of Natural Selection in 1930 is traced and the Theorem in its original form explained. It can now be seen as the centrepiece of Fisher's introduction of the gene-centred approach to evolutionary biology. Although this paper is a sequel to Edwards (1994) it is not a review of the recent literature on the Theorem, to which, however, reference is made at the end.

Key words: R.A. Fisher, Fundamental Theorem of Natural Selection, gene's-eye view.

CONTENTS

'Many authors have maintained that the theorem holds only under very special conditions, while only a few (e.g. Edwards, 1967*a*) have thought that Fisher may have been correct—if only we could understand what he meant! It will be shown here that this latter view is correct. Fisher's theorem does indeed hold with the generality that he claimed for it.'

G.R. Price (1972*b*).

'I believe that no one who is familiar, either with mathematical advances in other fields, or with the range of special biological conditions to be considered, would ever conceive that everything could be summed up in a single mathematical formula, however complex.'

R.A. Fisher (1932).

I. INTRODUCTION

The central tenet of R.A. Fisher's development of Darwin's theory of evolution by natural selection is the gradual replacement of one type of allele by another through the stochastic processes of Mendelian segregation and the

* Address for correspondence (E-mail: awfe@cam.ac.uk).

selective forces arising from differential rates of reproduction and death. 'Natural Selection is not Evolution' is the opening sentence of his *magnum opus* in evolutionary biology *The Genetical Theory of Natural Selection* (Fisher, 1930*b*), and the very title of the book defines the subject-matter. Fisher viewed this aspect of evolution through the twin lenses of statistical theory and the mathematical theory of gases. The genes were the molecules and the approach was through the study of the means and variances of the distributions of their frequencies. The variance was the measure of the variability on which selection acts; the mean the result of the action. The rates of change in each were the heart of the matter. How rapid was the loss of variance through the stochastic effects of Mendelian segregation in finite populations? How much mutation was required to maintain a steady state? What were the chances of a mutation surviving, and how were they altered by it having a selective advantage? How rapidly would such forces of selection change the mean?

Fisher's researches started in 1915 under the encouragement of Leonard Darwin, Charles Darwin's fourth surviving son, and culminated in 1930 in the publication of *The Genetical Theory*, dedicated to Leonard. The book initiated the gene-centred point of view canvassed by Richard Dawkins in 1976 in *The Selfish Gene*. Dawkins there acknowledged that 'the dominant figures' among the originators of the new ideas were 'G.C. Williams, J. Maynard Smith, W.D. Hamilton, and R.L. Trivers' (Dawkins, 1976, p. x; 1989, p. vi). But in the Preface to the 1989 edition he acknowledged that 'The gene's-eye view of Darwinism is implicit in the writings of R.A. Fisher and the other great pioneers of neo-Darwinism in the early thirties, but was made explicit by W.D. Hamilton and G.C. Williams in the sixties' (Dawkins, 1989, p. ix). In fact the other great pioneers, J.B.S. Haldane and Sewall Wright, started their theoretical treatments of evolution later than Fisher, and though one can say that a gene's-eye view is implicit in any genetical model of the analytical kind favoured by Haldane and Wright, in Fisher's work it was explicit (Fisher, 1930*b*, p. 95):

> 'To put the matter in another way, each gene is constantly tending to create genetic situations favourable to its own survival, so that an increase in numbers due to any cause will in its turn react favourably upon the selective advantage which it enjoys.'

He described his approach as 'inductive and statistical' in contrast to theirs, which was 'analytic and deductive' (Fisher, 1932).

The expression 'gene-centred' was not used by Fisher, and in associating it with his approach it is necessary to avoid a common misunderstanding. He was not simply attaching fitnesses to genes and ignoring all questions of interaction and other non-linear components. He was not advancing a dynamic model of the evolutionary process but creating an explanatory one. As Kempthorne remarked to Robertson (1955) 'It is a descriptive and not a mechanistic model'. In evaluating a population by its gene frequencies Fisher was not ignoring its other features but concentrating on the one defining feature capable of rigorous analysis. Genotype frequencies were by comparison transient, affected by mating systems or, in plants, incompatibility systems. The point of view is indeed that of the 'gene's eye', but the scenery is forever changing. 'There is no place for "levels" of selection in such a representation of evolution. There is only one kind of selection, the selection of genes' (Gayon, 1998, p. 326).

The centre-piece of Fisher's gene-centred view of selection is Chapter 2 of *The Genetical Theory*, 'The Fundamental Theorem of Natural Selection', in which he indicates how the rate of change of any heritable character is proportional to the additive genetic variance in the character and, if that character is fitness itself, how the rate of change in its mean value in the population is equal to the genetic variance in fitness. To be precise, as we shall be below, he is writing about the rate of change *ascribable to changes in the gene frequencies*. As Okasha (2008) concluded at the end of his long 'philosophical analysis' of the Theorem, 'That there exists this close conceptual link between the FTNS and the gene's eye viewpoint is the main conclusion of the foregoing analysis, and one which highlights the enduring relevance of Fisher's ideas for contemporary evolutionary theory'.

Our account here is chronological. In the next section we look at how Fisher's education and early interests prepared him for the work ahead. Sections III and IV are devoted to non-technical descriptions of the first stages of Fisher's argument in the papers of 1918 and 1922. Section V shows how the ideas there developed culminated in the Fundamental Theorem of Natural Selection in 1930, while in sections VI and VII we look at its history up to 1958. In section VIII we remark on the long period before evolutionary biologists noticed *The Genetical Theory*, just as Leonard Darwin had predicted '... it will be slowly recognized as a very important contribution to the subject. But I am afraid it will be slow, because so few will really grasp all that it means' (Darwin to Fisher, 9 June 1930; Bennett, 1983, p. 121). In section IX we list without discussion some of the more important papers on the Fundamental Theorem since 1993.

II. FISHER'S EDUCATION

Fisher was born in 1890. His precocious brilliance, especially in mathematics and biology, has been well documented by his daughter and biographer Joan Fisher Box (Box, 1978). At Harrow School he won the Neeld Medal for mathematics in 1906 and in his last school year he chose the 13-volume collected works of Charles Darwin for one of his prizes. Even earlier, at his previous school, his mathematics prizes were books in natural history (Bennett, 1983, p. 7). Then in October 1909 he entered Gonville and Caius College in the University of Cambridge with a Scholarship to read for the Mathematical Tripos.

As an undergraduate he threw himself into the study of Mendelism and its implications for evolution and the human species in particular. A recent account is Edwards (2011). Fisher's talk 'Mendelism and biometry' to the Cambridge

University Eugenics Society in November 1911 (printed in Bennett, 1983, p. 51; an uncorrected copy was incorporated in Norton & Pearson, 1976) shows just how far he had advanced in his reading and understanding of the new science.

After graduating with 'firsts' in both parts of the Mathematical Tripos Fisher spent the academic year 1912–1913 as Wollaston Student of Gonville and Caius College in Physics studying statistical mechanics and quantum theory under J.H. Jeans (later Sir James) and the theory of errors under F.J.M. Stratton (Box, 1978, p. 33). There has to be some doubt about the extent of his contact with Jeans since there are no lectures listed by Jeans in 1912–1913 and he resigned his Stokes Lectureship 'in 1912' (which would ordinarily mean at the end of the academic year in September), retiring to Guildford in Surrey. However, in the Lent Term 1912 Jeans had lectured on 'Statistical mechanics in its application to the kinetic theory of gases and radiation', which Fisher could have attended in his final undergraduate year. Even earlier, on 10 November 1911, in a talk 'Mendelism and biometry' Fisher refers to a fact that 'recently' impressed him, that 'if you put a kettle over a fire it will probably boil, but it is not a certainty; it may freeze; it is true that the odds against such an event are very large; but it remains a possibility, or so my 'theory of gases' tells me' (Bennett, 1983, p. 56). Jeans had published a book *The Dynamical Theory of Gases* in 1904 which he no doubt used as a text for his lectures, and in which this example appears (p.166).

Fisher could hardly have had a better training for the task ahead. A first-class mathematician well read in Darwin's works, encouraged by the Darwin family, particularly Leonard, and deeply versed in the two fields he was to apply to the problems of evolution by natural selection. On the one hand statistical mechanics and the theory of gases gave him the background to attack the stochastic theory of the gene, and on the other the 'theory of errors' enabled him to launch into the astonishing advances in statistics by which he became famous. Of these, as we shall see, the analysis of variance and the theory of regression were to be central to the discussion of selection.

In 1915 the advance began. In that year Leonard Darwin's earliest surviving letters to Fisher 'set out various problems which he hoped Fisher would solve; these are concerned mostly with biological variation and inheritance, Galton's law of ancestral heredity, parental correlation and regression, as well as natural selection and mutation' (Bennett, 1983, p. 13). Jointly with his friend from undergraduate days, C.S. Stock, Fisher wrote (Fisher & Stock, 1915):

> 'Changes in the constitution of a mixed population depend primarily upon selection. ... We may borrow an illustration from the kinetic theory of gases. The several molecules are conceived to move freely in all directions with greatly varying velocities, but the statistical result is a perfectly definite measurable pressure. Controversy may rage round the nature and properties of the atom, yet our knowledge of general principles enables us to calculate gas pressures with accuracy.'

Stock had read history, and we may reasonably conclude that this analogy was Fisher's.

III. THE REGRESSION MODEL FOR GENIC EFFECT

By 1916 Fisher had completed a path-breaking but extremely difficult paper 'The correlation between relatives on the supposition of Mendelian inheritance', but it was not favourably viewed by the two referees for the Royal Society, K. Pearson and R.C. Punnett (Norton & Pearson, 1976) and was eventually published by the Royal Society of Edinburgh in 1918 (Fisher, 1918). It concludes 'Finally, it is a pleasure to acknowledge my indebtedness to Major Leonard Darwin, at whose suggestion this enquiry was first undertaken, and to whose kindness and advice it owes its completion'. A major commentary on the paper was written in 1966 by the distinguished mathematical geneticists P.A.P. Moran and C.A.B. Smith and published alongside the full text (Moran & Smith, 1966), but for present purposes it is sufficient to note Fisher's introduction of the statistical model of regressing genotypic values on the number of genes of a particular kind in the genotype (0, 1 or 2) and the associated concept of variance and its partitioning. Indeed the paper contains the coining of the word *variance* for the square of the standard deviation. Moran & Smith write in their introduction:

> 'Sir Ronald Fisher's 1918 paper on the correlation between relatives is one of the classical papers of scientific literature. A few papers had previously appeared giving the expected value of the correlations on very simple Mendelian assumptions. Fisher succeeded in dealing with all the more obvious complications such as complete or partial recessivity, multiple allelism, epistacy, linkage, and assortative mating, and indeed with combinations of these, in one single paper....
>
> It is also of interest that we can see in this paper the beginning of some of Fisher's most important statistical ideas. Thus he sets out the idea of partitioning the variance into components. This presumably led to the Analysis of Variance. Fisher uses in this paper a technique which is very closely related to the analysis of variance applied to linear regression.'

In fact the 'technique' that is central to the present discussion is the very one introduced by Fisher at the outset, regressing genotypic value on gene number as follows. We use the notation that Fisher was later to adopt (Fisher, 1941).

Let the genotypes for a single Mendelian factor be *GG*, *Gg* and *gg* with associated genotypic values i, j and k and frequencies P, $2Q$ and R, respectively. Fisher and other early writers in Mendelian genetics were careful to distinguish between *factor*, *locus*, *allele*, and *gene* (see Fisher, 1922*a*); *gene* is now often used where *factor* would be more appropriate. The regression coefficient of genotypic value on the number of G (say) genes in the genotype, 0, 1 or 2, may now be found by the standard least-squares procedure, which leads to the regression coefficient, say α,

$$\alpha = \frac{P(Q+R)(i-j) + R(P+Q)(j-k)}{P(Q+R) + R(P+Q)}. \quad (1)$$

Since this is the slope of the regression line it corresponds to the effect on the 'expected' genotypic values—the regression values – of substituting a gene of one type with a gene of the other type, that is, of changing *gg* to *Gg* or *Gg* to *GG*. This was later given as the definition of the *average effect of the gene substitution* by Fisher (1941). The variance of the linear regression is known in modern terminology as the additive genetic variance or the *genic variance* and is seen to be

$$(P+R)\alpha^2 - (P\alpha - R\alpha)^2. \quad (2)$$

This, according to Fisher (1918), is 'the additive part which reflects the genetic nature without distortion':

> 'The contributions of imperfectly additive genetic factors divide themselves for statistical purposes into two parts: an additive part which reflects the genetic nature without distortion, ...; and a residue which acts in much the same way as an arbitrary error introduced into the measurements.'

Here a word of caution is necessary. If the differences $i-j$ and $j-k$ are unequal then the average effect defined in this way will only be a true average of these differences if

$$P(Q+R) = R(P+Q), \text{which is } P = R, \quad (3)$$

a very special case indeed. In the general case α is a weighted average as can be seen from the expression for it above, as noted by Fisher in 1941. Nevertheless, Fisher thought of the average effect as the proper measure for gauging the rate of change in the mean of the character value with respect to a change in the gene frequency in the population. We shall look at this in more detail later, and mention it here only to record the birth of his idea that variation not included in the variance of the linear regression is not to be considered as genetic but to be counted as 'environmental'. From a gene's point of view, the nature of the allele it is paired with in a zygote is part of its environment, as indeed is the whole of the rest of the genome.

IV. THE DISTRIBUTION OF GENE FREQUENCIES

We now turn aside for a moment from the concept of regression, so important for the development of the Fundamental Theorem, to Fisher's second path-breaking paper in mathematical genetics, 'On the dominance ratio' (Fisher, 1922*b*), like the '1918' paper also published by the Royal Society of Edinburgh (and likewise communicated by Professor J. Arthur Thompson). The title will only have conveyed meaning to those familiar with the earlier paper, whose findings Fisher starts by summarising. He turns from considering correlations between relatives in a population of infinite size in which sampling effects are not present, to the consequences of finite population size on the survival of genes. The central theme is the distribution of gene frequencies. Again it is the gene's-eye view, and Fisher takes up the analogy with the theory of gases once more:

> '... the whole investigation may be compared to the analytical treatment of the Theory of Gases, in which it is possible to make the most varied assumptions as to the accidental circumstances, and even the essential nature of the individual molecules, and yet to develop the general laws as to the behaviour of gases, leaving but a few fundamental constants to be determined by experiment.'

This time, however, Fisher has moved on from his 1915 analogy with Stock, and the molecules of gas are no longer the individuals of a population but the frequencies of the genes themselves.

Fisher first treats a non-stochastic matter 'Equilibrium under selection', showing that if selection favours the heterozygote in a random-mating diallelic population there will be a stable equilibrium. This is the original demonstration of the consequences of heterozygote advantage. If 'selection favours the heterozygote, there is a condition of stable equilibrium, and the factor will continue in the stock'. In *The Genetical Theory* (Fisher, 1930*b*, p. 100) Fisher additionally proved convergence to such an equilibrium.

This section sits rather oddly in a paper devoted to the stochastic theory but the discovery is likely to have had a profound effect on Fisher's interpretation of his own regression model of 4 years earlier. As we have seen, he had there analysed the total genotypic variance into 'an additive part which reflects the genetic nature' and 'a residue which acts in much the same way as an arbitrary error'. Thinking about the simple case of a random-mating population in equilibrium through heterozygote advantage he will surely have noticed (as did Bennett, 1956) that although there is then a non-zero genotypic variance, the gene frequency is no longer changing *and the genic variance is zero*. The regression is a horizontal line. Away from equilibrium, might not the rate of change of the gene frequency depend in some way on the genic variance? This, I suggest, is the thought that matured into the Fundamental Theorem of Natural Selection, though, as we shall see, since rates of increase of each genotype in continuous time are used in the proof and these are all equal at equilibrium ironically the genotypic variance is then zero too.

Fisher now turns his attention to the stochastic treatment in a section 'The Survival of Individual Genes' in which he uses a branching process analysed by functional iteration. In the next section 'Factors not acted on by selection' he sets up the 'chain-binomial' model, sometimes called the Wright–Fisher model though Wright's paper on it did not appear until a further nine years had elapsed (Wright, 1931). This Fisher analyses using a diffusion approximation, thereby introducing stochastic diffusion theory itself (Kendall, 1990), finding the distribution of the logarithm of the 'frequency ratio' $\log(p/q)$ for two alleles with frequencies p and q: 'The distribution of the frequency ratio for different factors may be calculated from the condition that this distribution is stable, as is that of velocities in the Theory of Gases'. He finds that the variance in a population of n individuals would decay by a factor e in $4n$ generations. 'The rate of mutation necessary to maintain the variance of the species may be calculated from these distributions'.

When selection is introduced, the first case considered is 'uniform genetic selection' in which the fitness of the heterozygote is taken to be the geometric mean of the two homozygotes so that the problem resolves itself into genetic (genic) fitnesses, and genotypes still need not be considered. A second case is 'genotypic selection' in which dominance is complete, but this is only treated approximately by showing it to be equivalent to genic selection when the fitness differences are small.

In Fisher (1930*a*) he corrects an error in the 1922 paper, acknowledging that Wright's result of $2n$ generations is correct for the decay time rather than $4n$, and extends his treatment by the use of functional equations. As he notes in a 1950 introduction to this paper 'The evolutionary consequences are developed in The Genetical Theory of Natural Selection, which is based for these questions on the present paper' (Fisher, 1950). A detailed account of the contemporary approaches by Wright and by Haldane may be found in Edwards (2001).

The influence of the theory of gases on Fisher's genetical thinking has run its course, leaving behind the gene-centred viewpoint which permeates *The Genetical Theory*. But not without a final reference: 'The particulate theory of inheritance resembles the kinetic theory of gases with its perfectly elastic collisions, whereas the blending theory resembles a theory of gases with inelastic collisions, and in which some outside agency is required to be continually at work to keep the particles astir' (Fisher, 1930*b*, p. 11).

We can now return to the 1918 regression model for gene effect and watch Fisher use it to obtain the Fundamental Theorem of Natural Selection.

V. THE FUNDAMENTAL THEOREM OF NATURAL SELECTION

The object of the Fundamental Theorem is to find the effect on the mean fitness of a population when gene frequencies are changed by natural selection operating through differential genotypic fitnesses. It establishes that at any moment the rate of increase of the mean fitness on this account *as defined by the regression model* is exactly equal to the additive genetic variance, or 'genic variance', in fitness. Introducing it in *The Genetical Theory* Fisher writes (p. 22) 'The object of the present chapter [II] is to combine certain ideas derivable from a consideration of the rates of death and reproduction of a population of organisms, with the concepts of the factorial scheme of inheritance, so as to state the principle of Natural Selection in the form of a rigorous mathematical theorem, by which the rate of improvement of any species of organisms in relation to its environment is determined by its present condition'.

In fact it is neither necessary nor helpful to introduce the parameters by which Fisher deals with 'the rates of death and reproduction' when explaining the Fundamental Theorem. Most readers of Chapter II of *The Genetical Theory* were not equipped to follow Fisher's mathematical derivation of the *intrinsic* rate of population increase which 'measures the relative rate of increase or decrease of a population when in the steady state appropriate to [any given system of rates of death and reproduction]' (p. 26). Fisher called this intrinsic rate the 'Malthusian parameter'. His next section 'Reproductive value' is an extension enquiring into the contributions of persons of different ages to future generations, and is similarly not needed for an explanation of the Theorem, though it is noteworthy for his explicit recognition that in developing the concept he is ignoring the 'indirect effects in cases in which an animal favours or impedes the survival or reproduction of its relatives' (p. 27), that is, inclusive fitness, another indication of his recognition of a gene-centred view.

This is certainly not to suggest that Malthusian parameters and reproductive values are irrelevant to Fisher's total argument. Indeed he is most careful to stress that 'the rigour of the demonstration requires that the terms employed should be used strictly as defined' (p. 35). Were all the necessary facts available enabling their calculation in a particular case they should be used, but for explaining the essence of the Theorem it is sufficient to attach simple growth rates to the genotypes as the measure of their 'fitnesses', just as in leading up to the Theorem Fisher initially uses human stature as the measure associated with genotypes.

Evaluating Malthusian parameters for genotypes is in fact a complicated and advanced question. Moran (1962, p. 60) pointed out that 'changes in the relative frequencies of genotypes with differing Malthusian parameters will result in the age composition changing' so that an exact treatment would be impossible without the use of integral equations. Further discussion with references may be found in Charlesworth's (1980) *Evolution in age-structured populations*. Whether one uses intrinsic or direct rates of increase or decrease, in a population maintained at equilibrium through heterozygote advantage the rates of increase *using the chosen rates* will be zero for all genotypes, and both the genotypic and genic variances will have converged on zero.

We now simply postulate that the genotypic values already indicated in Section III by i, j and k are exponential growth rates or 'fitnesses' such that, for example, a growth rate of i means an increase by a factor of e^{idt} in time dt, which to first order is $1+idt$.

The two parts to which Fisher referred in stating his object (quoted above) are neatly summarized by the equation

Genic rate of change of the mean fitness m is

$$\frac{dm}{dt} = \frac{dp}{dt} \times \frac{dm}{dp}, \qquad (4$$

in which dp/dt is to be found from the growth rates of the genotypes and dm/dp by an appeal to the 1918 regression model, Fisher's 'factorial scheme of inheritance'. Here p is the gene frequency, say of allele G, and $q = 1-p$ is the frequency of allele g. Avoiding the notation of partial differentiation, dm/dp and hence dm/dt are to be understood in the '1918' regression sense. The proportionate changes dp and dq in the gene frequencies are $dp/p = d(\ln p)$ and $dq/q = d(\ln q)$ with $dq = -dp$. Call the difference between the rates of change $d(\ln p)/dt$ and $d(\ln q)/dt$ the *average excess* a in fitness of G over g. Then $adt = (1/p + 1/q)dp$ and $dp/dt = pqa$.

To find a consider the genes of type G. P of them have growth rate i and Q (that is, half the heterozygotes Gg) growth rate j. But the combined growth rate will not be exactly $(Pi+Qj)/(P+Q)$ because a population whose two parts are growing exponentially at different rates is not itself growing exponentially. However, the theorem is in continuous-time so that we need only consider the first-order changes. Growth rates are thus additive to first order, and we may indeed set

$$a = \frac{Pi + Qj}{P + Q} - \frac{Qj + Rk}{Q + R}. \tag{5}$$

Turning now to dm/dp and replacing the actual values i, j and k by their regression values on Fisher's (1918) model, $dm/dp = 2\alpha$, since α is the slope of the regression per gene substitution and there are two genes at each locus. Thus

$$\frac{dm}{dt} = \frac{dp}{dt} \times \frac{dm}{dp} = pqa \times 2\alpha = 2pq\,a\alpha. \tag{6}$$

Now the genic variance was found in Section III to be

$$(P + R)\,\alpha^2 - (P\alpha - R\alpha)^2 \tag{7}$$

which, on recalling the expression for the average excess a and the relations $P+Q = p$, $Q+R = q$, $p+q = 1$, reduces to $2pqa\alpha$. Hence the genic rate of increase in mean fitness at any time, that is, the rate of change attributable solely to changes in gene frequency as computed from the additive genetic regression model, is equal to the genic variance in fitness at that time and the Fundamental Theorem is proved. In 1941 Fisher's phrase was 'ascribable to a change in gene frequency'.

Note that the theorem arrives at the expression $2pqa\alpha$ from two directions, as a genic rate of change in the mean fitness and as the genic variance in fitness. All its component parts are themselves changing and the contemplation of the consequences of this must not be limited to the first interpretation. The variance is changing in exactly the same way too, as Price (1972*b*) noted.

In retrospect the Fundamental Theorem of Natural Selection is not at all a surprising result. If one adopts the regression model for gene effect and argues that the change in the mean for a change in the gene frequency is relevantly measured by the slope of the regression it follows naturally. In 1926 Fisher had already written 'it is easily demonstrable that in a species in which a higher proportion of the total variance is ascribable to genetic causes, the effective selection will be more intense than in species in which the variance is to a larger extent ascribable to environmental variations' (Fisher & Ford, 1926, 1928). As I have previously reported (Edwards, 1994), Lady Fisher told me that Fisher dictated Chapter II to her. The Fundamental Theorem 'came to him as he dictated', and we need not be surprised.

Fisher was repeatedly at pains to stress that the genic variance is not, except in special circumstances, the total genotypic variance. 'The genetic [genic] variance as here defined is only a portion of the variance defined genotypically, and this will differ from, and usually be somewhat less than, the total variance to be observed' (1930, p. 34). As a side issue Fisher (1941) later derived the conditions under which the rate of increase 'ascribable to a change in gene frequency' is equal to the total rate of increase. In addition to the case of no dominance he showed that a particular constraint on P, $2Q$ and R would render them equal. Since $m = iP + 2jQ + kR$ the total change in the mean fitness is $dm = idP + 2jdQ + kdR$, so that for $2\alpha dp$ to be exactly equal to this whatever i, j and k, their respective coefficients in the two expressions must be equal (Edwards, 1967*b*):

$$dP = 2P\left(Q + R\right) dp/S,\ \ 2dQ = 2Q\,(R-P)\,dp/S,$$
$$dR = -2R\left(P + Q\right) dp/S, \tag{8}$$
$$\text{where}\ \ S = P\left(Q + R\right) + R\left(P + Q\right). \tag{9}$$

This implies the restriction $dP/P + dR/R = 2dQ/Q$ which in the limit has solution $Q^2 = \lambda PR$, where λ is a constant.

This, however, has no bearing on the validity of the Fundamental Theorem. Fisher points out that any changes in the genotypic frequencies ascribable only to changes in the gene frequencies are those that would occur if λ were constant. He regards any change in λ itself as a change in the 'genic environment' to be counted with other environmental changes, and insists that his 'exact specification of the meaning of the *average effect* of a gene substitution is necessary for clarity of thought, and obviously also necessary for the development of its properties' (Fisher, 1941). The adoption of the 1918 regression model is, in other words, essential to his argument. As he was to remark in 1935 'What [Hogben and Haldane] do not see is that we ordinarily count as genetic only such part of the genetic effect as may be included in a linear formula and that we make a present to the environmentalists of such variation due to the combined action of genetic and environmental causes as is not expressible in such a formula' (Bennett, 1983, p. 260).

There is no doubt that Fisher's (1930) description of the regression coefficient as the 'average effect of a gene substitution' hindered the acceptance of the Fundamental Theorem, because it is not in general the unweighted average of such effects (see Section III). Falconer (1985) closely examined what Fisher's meaning might be and concluded that neither of the interpretations he could contemplate corresponded to the regression coefficient. Yet in retrospect one has some sympathy with Fisher's terminology in the context of his 1918 model. When he states (1930*b*, p. 31) 'By whatever rules mating, and consequently the frequency of different gene combinations, may be governed, the substitution of a small proportion of the genes of one kind by the genes of another will produce a definite proportional effect upon the average stature' he is already using that model, for otherwise he would have had to add the condition among the frequencies of different gene combinations [genotypes] he later showed to be constant λ. As the person who coined the phrase 'degree of freedom' in its statistical sense, Fisher would not have been blind to the existence of the degree of freedom represented by λ even if he did not identify it specifically until 1941. On the 1918 model the regression coefficient is indeed the average of the effects of each type of gene substitution because these are the same when the linear regression values of the genotypic fitnesses are used.

VI. THE RECEPTION OF THE FUNDAMENTAL THEOREM: BENNETT (1956) AND KEMPTHORNE (1957)

I provided an extensive history of the Fundamental Theorem in Edwards (1994) up to and a little beyond the publication of Ewens's (1989) influential clarification and proof. Some comments and revisions are in order. The most obvious is that the present account of the theorem is in continuous time whereas my discussion in 1994 was based on the discrete-time version introduced by Ewens. Fisher's development was in continuous time, however, and I now believe that the account above is an accurate rendering of it. Though I possess written notes that I made in 1964 or 1965 with the same mathematical structure, I was not then sufficiently familiar with the context of the theorem to understand it fully or confident enough to expound it. I no longer think, as in 1994, that Fisher 'discovered the exact growth-rate theorem' prior to applying it, but rather that it came out of his genetical analysis.

Waddington (1939) should be added to the list of references, for in his book *An Introduction to Modern Genetics*, p. 291, he has a section 'Fisher's Fundamental Theorem of Natural Selection' in which he attempts to summarise Fisher's argument, but he hardly comments on the result and leaves the impression that he may not have fully understood it himself.

Of much greater importance is Bennett (1956), which in 1994 I cited only as a 'Reference added in proof' with the comment that it seemed to have been 'entirely overlooked since its publication'. It expounds Fisher's (1930/1941) argument in mathematical form with the greatest clarity, differing only in the account given above in that Bennett justifies the equation $dm/dp = 2\alpha$ by proving it true for constant λ whereas I, following Fisher (1941), accept $dm/dp = 2\alpha$ as a consequence of adopting the additive regression model and prove that constant λ is then a condition for the genic rate of increase to equal the total rate of increase. However, Bennett then argues, like Fisher, that the difference due to a change in λ 'cannot be considered as an evolutionary change in the specific constitution' since it belongs to 'causes other than gene substitution'. He then proceeds to give the Theorem without explicitly stating that the rate of increase in it does not include any contribution from a changing λ, though it is of course implied. It is much to be regretted that Bennett's paper was overlooked by subsequent writers interested in understanding Fisher's derivation of his Theorem. My similar version in the present paper was reached independently, but that is not to my credit.

It is of considerable relevance that Bennett held the post of Assistant in Research in Fisher's Cambridge Department of Genetics at the time he wrote his paper. It had followed on from his review of Li's (1955) book *Population Genetics* (see Edwards, 1994). I have not yet been able to ascertain from Professor Bennett whether Fisher passed him Li's book to review (as he once passed to me a book he had been asked to review), but I suppose that Fisher, as Head of Department, is likely to have seen and approved Bennett's paper.

Kempthorne (1957), in *An Introduction to Genetic Statistics*, Chapter 16, p. 343, considered the whole question in great mathematical detail, finding the expression for the rate of change in the mean genotypic fitness to be given by the genic variance plus a term involving Fisher's λ. However, he 'found Fisher (1930, 1941) entirely obscure with regard to his derivation of the result' of the condition that the genic rate of increase equals the total rate, and regarded the condition as necessary for the validity of the Fundamental Theorem, writing it into his statement of the Theorem (p. 360). Of his own proof he wrote (p. 355) it 'is possibly tedious to the reader. ... Dr M. Kimura (private communication) has developed a simpler proof based on quantities he has devised called "coefficients of departure from random mating" ... . we look forward to publications by Dr Kimura'. We outline Kimura's approach below, but we note in advance that Kempthorne, unlike Kimura, did not at all understand that the condition was not necessary for the validity of the Theorem. In reply to a letter from Kempthorne enclosing a draft of a paper, Fisher wrote (31 January 1955; Bennett, 1983, p. 227) 'I do not at all agree with the last sentence of the opening paragraph of your introduction, "Later in 1941 Fisher showed that this is true only if the quantity Q^2/PR remains constant" ... '. He explained his position at length, starting his last paragraph 'Due to all this I am completely puzzled by the statement in your letter that the rate of evolutionary change may be equated to the total variance rather than to the genetic [genic] component of variance as I had done'.

Fisher wrote again to Kempthorne on 18 February (p. 229), including the memorable statement '... the only evolutionary effect, either in increased fitness or in anything else, that I can recognize as such, is constituted by the changes in gene ratio ... '. Kempthorne made no attempt to meet Fisher's point in his book, and in 1983 wrote:

> 'It seemed never to occur to Fisher that he could be wrong. He made algebraic mistakes, and these he could acknowledge. But the possibility that the two ideas that I surmise he valued above all else, fiducial inference and the fundamental

theorem of natural selection, were both wrong simply did not enter his mind.'

I took this as the text for my Fisher Memorial Lecture in 1994 (Edwards, 1995).

Further letters in the correspondence between Fisher and Kempthorne have now been published on the Adelaide University Library website (http://digital.library.adelaide.edu.au/dspace/handle/2440/3860) and show how Kempthorne's manuscript (assuming there was only one such) made the rounds of *Heredity*, *Proceedings of the Royal Society*, and *Annals of Human Genetics* as a result of Fisher's interventions, no doubt so exasperating Kempthorne that he gave up separate publication and simply published his results in his book. Kempthorne's letters do, however, show his concerted efforts to understand Fisher's point of view.

VII. THE RECEPTION OF THE FUNDAMENTAL THEOREM: KIMURA (1958)

In the case of Kimura (1958), my remarks in Edwards (1994) were unjustified and what here follows may be considered as a replacement for them. It was not correct to say 'Mathematically Kimura's result was indeed true, but since it treated the total change in mean fitness, and not the partial change, it was not really Fisher's theorem'. On a more careful reading of Kimura's paper and his correspondence with Fisher it becomes clear that, unlike Kempthorne, he had taken Fisher's point.

Kimura's interest in the fundamental theorem stemmed from his association with J.F. Crow at the University of Wisconsin, Madison, in 1953–1956. Indeed, the sole acknowledgment in the paper is the expression of his appreciation to Crow 'for reading the manuscript and making valuable suggestions'. Crow (1955) had initiated the topic, which was continued in Crow & Kimura (1956).

In a letter to Fisher dated 26 April 1957 Kimura wrote to Fisher 'Last year while I was still in the United States, I wrote letters to you hoping that I could hear your opinion on my thoughts concerning to the "fundamental theorem of natural selection". For my letters, you were most kind in answering my questions'. Fisher had replied to these questions on 3 May and 14 July 1956, the salient points of his letters being reproduced in Bennett (1983, p. 229). In the first he wrote:

> 'In considering the original statement of what I ventured to call 'the fundamental theorem of natural selection', I had, of course, considered the relation between such a situation and that in which a potential function existed, for my mathematical education lay in the field of mathematical physics. As you realize, I preferred to develop the theory without this assumption, which of course in another aspect is a restriction. Of course, I do not question that the selective intensities acting instantaneously may well be equivalent to those derivable from such a function, but I think it should be emphasized that both changes in time, that is in the environmental *milieu* and in the gene ratios themselves, that is the heritable constitution of the organism, will change this virtual function in a way that cannot be specified in terms of the quantities used in formulating the fundamental theorem.'

Kimura replied on 8 July 1956, and on 14 July Fisher wrote elaborating his point of view of what constituted the environment 'in which the advantage, or disadvantage, of any particular gene is determined'. The letter is similar in content to an addition Fisher made to the Dover edition of *The Genetical Theory of Natural Selection* (Fisher, 1930*b*, p. 40; see Fisher, 1930*b*, p. 278). It goes back to his correspondence with Oscar Kempthorne the previous year (see above).

As a result of these exchanges Kimura was able to incorporate Fisher's view into his 1958 paper. Kimura's letter to Fisher of 26 April 1957 continued (on from the opening already quoted):

> 'Your letters were really enlightening and it was through those correspondences that I could correct my wrong notion that existance of some kind of a potential function is required for the fundamental theorem to hold.
>
> Since then I have done some works on this line and recently I have written a short paper entitled "On the change of population fitness by natural selection". Under a separate cover, I am going to send a copy of this paper to you with the hope that you would go over the manuscript more or less in detail to consider its acceptability for "HEREDITY".
>
> Though I am rather happy of various results I have obtained, still I am afraid that you have known them already or you might consider that they do not add very much to what you have published already. Even though I venture to think that my paper may serve to clear up some of the confusion on the implication of the fundamental theorem which especially in the United States is quite widespread among statistical geneticists.
>
> Meanwhile I am looking forward for the publication of your book "The Genetical Theory of Natural Selection" in Dover Edition.'
>
> (This letter is quoted from a personal copy sent to me by Dr Kimura in 1967).

Fisher was one of the Editors of *Heredity*, with his colleague K. Mather as the Senior Editor. We now know, from the correspondence in Adelaide, that Fisher replied almost immediately, on 3 May:

> 'I have looked through your typescript with interest, and believe it should be acceptable for Heredity. I should like, however, to add to the summary a short explication in accordance with your last two letters in case it should be thought, as it might be from your wording as it now stands, that your conclusions differed from mine, and were not an explicit reformulation of effectively identical conclusions.'

He suggests adding the two sentences that conclude section 1 of the Summary of Kimura (1958), and these are exactly Fisher's own words:

> 'It is the reformulation of the results published by Fisher in 1930 under the title *The Fundamental Theorem of Natural Selection*, and in no way differs from it in substance. θ corresponds with λ of Fisher's (1941) treatment.'

Sir Walter Bodmer tells me that he recalls Fisher saying when he received Kimura's typescript that he supposed Kimura had to prove things for himself in order to be convinced.

Instead of defining the slope of the regression to be the most meaningful causal agent of the change in gene frequency, and deducing the consequences, Kimura had adopted the same linear model without this intuitive assumption of Fisher's and had worked out its full consequences for the change in the mean fitness. His complete mathematical analysis showed precisely how the change in the mean fitness separated into the part that concerned Fisher which depended on the genic variance as Fisher had found, and the part that Fisher considered irrelevant to his argument which depended on changes to the genotypic frequencies that were not caused by changes in gene frequencies.

He showed that 'the rate of increase of population fitness ... is expressed as the sum of three terms: A term due to genic variance, one due to change in environment, and one due to the joint effect of genetic interactions (dominance and epistasis) and change in mating system'. By 'genic variance' Kimura meant the additive genetic variance; by 'change in environment' he meant any changes that alter genotypic fitnesses; and by 'change in mating system' he meant any changes that alter the way the genes are distributed amongst the genotypes (as measured by changes in coefficients of departure from random mating).

Kimura continued:

> 'From the standpoint of the genetical theory of evolution, the first term V_g [the genic variance] is by far the most important, since it represents the part of increase in fitness due to the improvement of the *genic* constitution of the species. The effects due to the remaining terms are generally speaking unpredictable from the *genic* constitution of the species and may conveniently be treated as environmental factors.' [Italics original]

No wonder that Fisher approved.

There the matter might have been expected to rest, and we shall not pursue the subsequent history here except to add another rider to Edwards (1994). In that account I took the by-then accepted line that it was the paper by Price (1972) that finally turned the tide in favour of Fisher's theorem as he intended it. But Price did not notice Bennett (1956), and in his Introduction he quoted Kempthorne (1957) only as saying that he had found Fisher's derivation 'entirely obscure' and Crow & Kimura (1970, p. 210) only as saying in *An Introduction to Population Genetics Theory* that they had found it 'recondite'. This last was particularly reprehensible because in the same three-line paragraph they had added 'For a straightforward, general derivation, see Kimura (1958)' which they then gave for the simpler two-allele case. And on introducing the Theorem Crow & Kimura (1970, p. 209) had said:

> 'One interpretation of the theorem is to say that it measures the rate of increase in fitness that would occur if the gene frequency changes took place, but nothing else changed. The theorem thus gives the effect of gene frequency changes alone, isolated from the other things that are happening.'

In fact Price referenced Kimura (1958) but only to say parenthetically 'Kimura treats these non-additive effects in an interesting way in a paper based on the usual interpretation of Fisher's theorem as an equation for dM/dt', that is, on the usual interpretation that Price regarded as erroneous. He is similarly dismissive of Fisher's (1941) 'brief summary of the main part of the derivation of the theorem. Unfortunately this is just as obscure as its 1930 counterpart'. Yet it is in 1941 that Fisher explicitly defines the average effect to be the regression of genotypic value on gene number as a replacement for the awkward verbal 1930 definition.

Finally, in his Introduction, Price damns a short note of mine with faint praise: 'only a few (e.g. Edwards, 1967*a*) have thought that Fisher may have been correct—if only we could understand what he meant!' This letter to *Nature* was simply a protest at the name 'Fisher's fundamental theorem' being attached to calculations based on random-mating discrete-generation models when Fisher had developed it for continuous time and without the assumption of random mating. Though it erred in following Kempthorne's (1957) view that constant Q^2/PR was a 'further condition' for the validity of the Theorem, it also said 'Kimura (1958) has given a very general formulation'. The letter would not be worth recalling were it not for the fact that it added 'Several points remain to be cleared up and I shall examine them elsewhere[8]', and reference 8 said '*Ann. Hum. Genet.* (submitted for publication)'.

The *Annals of Human Genetics* was, of course, the journal in which Price's paper was published five years later, of which C.A.B. Smith was the Joint Editor who dealt with the mathematical and statistical material. I had submitted my paper to Smith in 1967 (Edwards, 1967*b*) and we had corresponded about it. I still possess the copy on which he made notes in pencil. The paper fails to stress Fisher's essential viewpoint that in his Theorem he is concerned only with the influence of the change in gene frequency on the change in the mean fitness, but it is in many respects a useful review, lacking only a reference to Bennett (1956). In particular it says 'Of his work Kimura modestly writes "It is the reformulation of the results published by Fisher in 1930 under the title The Fundamental Theorem of Natural Selection, and in no way differs from it in substance"'. Kimura was one of those to whom I sent a duplicated copy of the paper. He replied on 19 May 1967 'I went through it rather rapidly and think that it is very nice, though it might still produe [*sic*] some heated discussion with Sir Ronald had he alive'.

Price had been made welcome in the Galton Laboratory, University College London, by Smith on 24 September 1968 after having shown him the covariance part of his famous 'Price Equation' (Harman, 2010, p. 209). He submitted his paper on the fundamental theorem in September 1971, thanking Smith 'for much help during the long period of time during which work on the problem of understanding Fisher was intermittently pursued'. Since Price referred to Kimura (1958), and I had also done so in my 1967 submission to Smith, it is perplexing that he, as editor, did not require him to consider Kimura's paper further when it explicitly claimed to be a reformulation of Fisher's theorem. Smith will also have known that Fisher was one of the editors of *Heredity*. Thus the view that Price 'was the first to crack the most obscure and misunderstood of R.A.Fisher's many oracles: the fundamental theorem of natural selection' (Frank, 1995) has become embedded in the literature, and Kimura (1958) overlooked.

VIII. THE ADVANCE OF GENE-CENTRED THINKING

Returning to the remark by Dawkins quoted in Section I, 'The gene's-eye view of Darwinism is implicit in the writings of R.A. Fisher and the other great pioneers of neo-Darwinism in the early thirties, but was made explicit by W.D. Hamilton and G.C. Williams in the sixties' (Dawkins, 1989, p. ix), we can see how the Fundamental Theorem of Natural Selection is in fact the explicit realisation of Fisher's gene-centred thinking about selection. In an article celebrating the seventieth anniversary of *The Genetical Theory* (Edwards, 2000) I discussed the neglect it had suffered during much of this period, and then explained in some detail how both Hamilton and Williams were greatly influenced by it. In Hamilton's case the influence was direct. He studied genetics in the Cambridge Department from which Fisher had only just retired, and he celebrated Bennett's (1999) variorum edition of the *Genetical Theory* in these words:

> 'This is a book which, as a student, I weighed as of equal importance to the entire rest of my undergraduate BA course and, through the time I spent on it, I think it notched down my degree. Most chapters took me weeks, some months; .... And little modified by molecular genetics, Fisher's logic and ideas still underpin most of the ever broadening paths by which Darwinism continues its invasion of human thought.
>
> For a book that I rate only second in importance in evolution theory to Darwin's "Origin" (this as joined with its supplement, "of Man"), and also rate as undoubtedly one of the greatest books of the present century, the appearance of a variorum version is a major event. ...
>
> By the time of my ultimate graduation, will I have understood all that is true in this book and will I get a First? I doubt it. In some ways some of us have overtaken Fisher; in many, however, this brilliant, daring man is still far in front.' (Hamilton; dust-jacket of Fisher, 1930*b*).

When Ridley (1996) reviewed the first volume of Hamilton's collected papers (1996) he claimed so many of Fisher's ideas for Hamilton that I was moved to dissent in *The Times Literary Supplement* (Edwards, 1996). Hamilton wrote to me agreeing with my 'general comment that Matt Ridley's review was over the top regarding my contributions to evolution theory and it is good that someone should give a better historical perspective' (17 December 1996).

Williams (1966) quoted *The Genetical Theory* at a number of points in his influential *Adaptation and Natural Selection*, but in the Preface to the 1996 reprint he singled out the work of Shaw & Mohler (1953), Lack (1954) and Hamilton (1964) as of particular influence on him. But as I pointed out in Edwards (2000), Shaw & Mohler's starting point was Fisher's discussion on natural selection and the sex ratio in *The Genetical Theory* (as Williams of course knew) and, as Burbridge (1992) noted, Lack's topic 'The evolution of reproductive rates' had already been discussed by Fisher in *The Genetical Theory* unbeknown to him.

Historians of science of considerable distinction have retarded the understanding of Fisher's contribution to evolutionary biology by a distracting emphasis on his involvement in the eugenics movement. They seem to have been unable to put themselves in the position of a first-rate young mathematician well read in Darwin and natural history who, just nine years after the start of the Mendelian revolution in genetics, finds himself an undergraduate at the very place where the liveliest discussion of it is taking place, and turns his mind, as Galton did when he read *The Origin of Species*, to its profound implications for mankind. Without a proper study of Fisher's scientific thinking they too readily drew the unwarranted inference that it was driven by social considerations rather than that the unfolding new science was generating the social concern. Mackenzie (1981, p. 189) thought Fisher in 1918 'sought not to *reconcile* Mendelism and biometry, but to *use* Mendelism to vindicate biometric eugenics' (for a critical review of Mackenzie's book see Edwards, 1981).

Mackenzie also quoted Norton (1978) who wrote that Fisher (1918) should be seen 'predominantly as a contribution to the hereditarian social ideology of eugenics' (also quoted by Bennett, 1983, p. 17). Norton thought that 'To understand Fisher's 1918 paper historically, we must surely attempt to uncover the problems that generated it' and 'Until recently, Fisher's decision to become involved in this sort of work has remained somewhat mysterious'. Unfortunately Norton did not have available the essential evidence contained in Fisher's biography (Box, 1978) and his correspondence with Leonard Darwin together with Bennett's commentary (Bennett, 1983). But he had recently been able to consult the records of the Cambridge University Eugenics Society and in particular had found a typescript of Fisher's (1911) lecture mentioned in Section II above (Norton & Pearson, 1976). He filled the void with what he had found: 'Recently uncovered documents show that Fisher's problems were ideological rather than biological'. 'From this one might suppose that the motivation for Fisher's (1918) paper was an eugenic one'. Finally, 'We should see it ... as a stunning contribution to eugenics'. It would not have been necessary to dwell on this but for the fact that it has been so influential.

IX. RECENT LITERATURE

Since 1958 quite a large literature has grown up about the interpretation of the Fundamental Theorem. The list of references in Edwards (1994) includes the main papers and books up to 1993 or so, many of them discussed in that paper. For the sake of completeness we may now add in particular Lessard (1997), Frank (1997), Gayon (1998), Edwards (2002), Plutynski (2006), Okasha (2008), and Ewens (2011). Of these, none mentions Bennett (1956) and only Lessard examines Kimura (1958) closely (though Gayon, 1998, p. 392, has as an aside that he was 'the only theoretician to have thoroughly proved Fisher's theorem', which he rather spoils by adding in parentheses 'as a good approximation'). The last three papers, by Plutynski, Okasha and Ewens, constitute an essential reading list for a modern understanding of the Theorem based on the discrete-time version developed by Ewens (1989) rather than Fisher's original development considered here. Finally, Grafen's (2003) 'Fisher the evolutionary biologist' is valuable for placing the Theorem in the context of modern evolutionary biology and its origins.

In a forthcoming paper 'The causal meaning of Fisher's average effect' J.J. Lee and C.C. Chow (*Genetics Research*, in press) offer a very careful discussion of the concept of the average effect in the light of the distinction between causation and mere correlation so often emphasized in Fisher's other work.

X. EPILOGUE

Finally, we may speculate on the whole train of thought that led Fisher to the Fundamental Theorem by filling in the gaps in the above account. Starting with a schoolboy's love of natural history and of Darwin's books, excited by the implications of the recently uncovered Mendelian laws, possessed of first-class mathematical ability, trained in statistical mechanics and the theory of gases, befriended by the sons of Charles Darwin and in particular Leonard, Fisher completed 'The correlation between relatives on the supposition of Mendelian inheritance' in 1916. In 1919 he started to write a book about natural selection (some parts of the typescript, which is in the Adelaide archive, eventually appeared in *The Genetical Theory*) in which he stated how both variable rates of increase, and a correlation between the differing rates and the heritable character in question, were necessary for evolutionary change to take place.

In 1922 Fisher considered the stochastic effects of finite population size on the survival of genes in the absence of selection, continuing with some extensions incorporating selection. By 1926 he was clear that selection would be more intense 'in a species in which a higher proportion of the total variance is ascribable to genetic causes', and when writing *The Genetical Theory* he had clearly in mind the definition of the genetic variance (from the 1918 paper—now 'genic' variance) and that the effect of selection was measured by the regression of fitness (rates of increase) on gene number (again 1918) and thus by the genic variance. Obviously (to Fisher) the other necessary link, between fitness and the character, was measured by the covariance between the two, leading to the theorem that the rate of change in the mean value of a heritable character is given by the covariance of the character with fitness.

As an aside we may note that, just as Fisher defined the technical meaning of the word *variance* in the 1918 paper, so the first recorded occurrence of the word *covariance* is in *The Genetical Theory* (1930*b*, p. 195). Surprisingly, the above theorem did not appear explicitly in the literature until Robertson (1966). A more general form making allowance for changes in the character values themselves was given by Price (1970, 1972*a*).

It seems that only during the writing of *The Genetical Theory* did Fisher then realise that fitness itself could be considered as the character under selection, so that, since fitness is perfectly correlated with itself, the Fundamental Theorem emerges: the rate of change in fitness ascribable to gene-frequency change is equal to the genic variance.

Grafen (2003), already mentioned above, is the best account of the place of the Fundamental Theorem in modern evolutionary biology. Yet a word of caution is in order. Fisher was opposed to any dynamical mathematical theory (see, for example, his letter to Kimura dated 3 May 1956 quoted in Section VII above). In the Fundamental Theorem he isolated just one aspect of evolutionary change, a measure of the rate of change of the mean fitness on account of changes in gene frequencies alone, and that was as far as he would go. I do not share Grafen's view that some developments in modern evolutionary biology 'are parts of a Fisherian tradition of treating individual organisms as optimising agents'. I do not recognize that tradition as Fisherian. Grafen (2008) goes further and states that Fisher certainly thought that the Fundamental Theorem has an 'optimization interpretation'. My point is not to argue now against such an interpretation, but only to question whether it was Fisher's. In his writings I see many criticisms of this view, and such firm qualifications when the issue does arise (as in his discussion of the Fundamental Theorem itself). I conclude, therefore, with a particularly delightful reflection of his point of view from his Presidential Address to the Royal Statistical Society in 1952 (Fisher, 1953):

> 'In truth, when I try to explain the Theory of Natural Selection to students, the phrases I find least helpful are those rather journalistic slogans, "The Struggle for Existence" and "the Survival of the Fittest". Each, of course, has a defensible meaning, but it is, I fear, only their dramatic violence that makes them effective.
>
> Now it is very natural for the human race, and very unphilosophical, to imagine that results of towering and majestic importance, such as the evolution of living creatures, must be brought about by equally powerful or violent causes. If we can imagine my winkles, or my primroses, *tense with struggle*, red in their non-existent teeth and claws, it becomes easier to believe that great things are happening to them, that they are being hammered into shape by cosmic forces. Whereas all we know is that each is living its own quiet life with the life Table and the rates of reproduction characteristic of its genotype, and that for this reason, and for no other these species are probably changing, imperceptibly slowly but with a speed sufficient to make them perceptibly differen

if they keep it up for some millions, or tens of millions, of years.'

XI. CONCLUSIONS

(1) The 'gene-centred' view of Darwinian evolution by natural selection has its origin in R.A. Fisher's (1918) paper 'The correlation between relatives on the supposition of Mendelian inheritance' where he introduces the linear regression of gene number on genotypic effect, thus identifying the 'additive part which reflects the genetic nature without distortion'.

(2) Fisher fully develops the point of view a dozen years later in his book *The Genetical Theory of Natural Selection*, the very title of which speaks volumes. Its centrepiece is 'The Fundamental Theorem of Natural Selection', analysed and expounded in the present account with the aid of the writings of Bennett, Kempthorne and Kimura in 1956–1958 which were not fully appreciated at the time, and with the benefit of access to Fisher's correspondence, not published until 1983 and even then not in its entirety.

(3) *The Genetical Theory* extends the gene's-eye view beyond the immediate limitations of the Fundamental Theorem in important ways. 'Each gene is constantly tending to create genetic situations favourable to its own survival'. (The concepts of kin selection and parental investment are also introduced.) Although the rise in gene-centred thinking has been correctly attributed to G.C. Williams and W.D. Hamilton it is easy to trace the influence of *The Genetical Theory* on them, as they were the first to recognize.

(4) The common suggestion that Fisher developed his advances in evolutionary biology as a 'contribution to eugenics' rather than out of concern for the evolutionary consequences of the Mendelian theory of inheritance for man is refuted by a study of his education in mathematics and mathematical physics and his schoolboy enthusiasm for natural history and the writings of Charles Darwin.

(5) Fisher did not support any theory that natural selection was a maximising 'force' in some sense. It was a major factor in his criticism of Sewall Wright's theory of adaptive surfaces, and is made particularly clear in his correspondence with Kimura. No-one was better qualified than he to bring in to evolutionary biology the concepts of potential functions, of entropy, and of information, from mathematical physics, statistical mechanics, and statistics, but he eschewed them, understanding with Darwinian insight that evolution was too complex a process. The Fundamental Theorem was the most he would distil by way of general theory, and even there an analogy with entropy was tempered with such caution as to warn off prospective enthusiasts.

XII. ACKNOWLEDGEMENTS

I owe my revived interest in expounding Fisher's original Fundamental Theorem to the penetrating questions of a French student visiting Oxford, Jean-Baptiste Grodwohl. Brian Charlesworth introduced me to the literature on Malthusian parameters and shaped my view on what I could leave out. Warren Ewens persuaded me that in a recent publication I had not done justice to Fisher's contribution to the gene's-eye view. I am grateful to all of them.

XIII. REFERENCES

BENNETT, J. H. (1956). Population genetics and natural selection. *Genetica* **28**, 297–307.

BENNETT, J. H. (1983). *Natural Selection, Heredity, and Eugenics*. Clarendon Press, Oxford.

BOX, J. F. (1978). *R.A. Fisher: The Life of a Scientist*. Wiley, New York.

BURBRIDGE, D. (1992). Lacking a solution? *Nature* **355**, 118.

CHARLESWORTH, B. (1980). *Evolution in Age-structured Populations*. Cambridge University Press.

CROW, J. F. (1955). General theory of population genetics: synthesis. *Cold Spring Harbor Symposia on Quantitative Biology XX: Population Genetics: The Nature and Causes of Genetic Variability in Populations*, 54–59.

CROW, J. F. & KIMURA, M. (1956). Some genetic problems in natural populations. In *Proceedings Third Berkeley Symposium on Mathematical Statistics and Probability*, Volume **4**, pp. 1–22.

CROW, J. F. & KIMURA, M. (1970). *An Introduction to Population Genetics Theory*. Harper & Row, New York.

DAWKINS, R. (1976). *The Selfish Gene*. Oxford University Press (New edition 1989).

EDWARDS, A. W. F. (1967*a*). On the fundamental theorem of natural selection. Circulated typescript.

EDWARDS, A. W. F. (1967*b*). Fundamental theorem of natural selection. *Nature* **215**, 537–538.

EDWARDS, A. W. F. (1981). Hereditary genius. Review of *Statistics in Britain, 1865–1930: The Social Construction of Scientific Knowledge* by D. MacKenzie. *London Review of Books* **3**(14), 12–13.

EDWARDS, A. W. F. (1994). The fundamental theorem of natural selection. *Biological Reviews* **69**, 443–474.

EDWARDS, A. W. F. (1995). Fiducial inference and the fundamental theorem of natural selection. XVIIIth Fisher Memorial Lecture, London, 20th October 1994. *Biometrics* **51**, 799–809.

EDWARDS, A. W. F. (1996). W.D. Hamilton's Darwinian precursors. *Times Literary Supplement*, December 6th, No. 4888, p. 17.

EDWARDS, A. W. F. (2000). The genetical theory of natural selection. In Perspectives, ed. J.F. Crow & W.F. Dove. *Genetics* **154**, 1419–1426.

EDWARDS, A. W. F. (2001). Darwin and Mendel united: the contributions of Fisher, Haldane and Wright up to 1932. In *Encyclopedia of Genetics* (ed. E. C. R. REEVE), pp. 77–83. Fitzroy Dearborn, London.

EDWARDS, A. W. F. (2002). Fundamental theorem of natural selection. *Theoretical Population Biology* **61**, 335–337.

EDWARDS, A. W. F. (2011). Mathematizing Darwin. *Behavioral Ecology and Sociobiology* **65**, 421–430.

EWENS, W. J. (1989). An interpretation and proof of the fundamental theorem of natural selection. *Theoretical Population Biology* **36**, 167–180.

EWENS, W. J. (2011). What is the gene trying to do? *The British Journal for the Philosophy of Science* **62**, 155–176.

FALCONER, D. S. (1985). A note on Fisher's 'average effect' and 'average excess'. *Genetical Research* **46**, 337–347.

FISHER, R. A. (1918). The correlation between relatives on the supposition of Mendelian inheritance. *Transactions of the Royal Society of Edinburgh* **52**, 399–433.

FISHER, R. A. (1922*a*). Darwinian evolution by mutations. *The Eugenics Review* **14**, 31–34.

FISHER, R. A. (1922*b*). On the dominance ratio. *Proceedings of the Royal Society of Edinburgh* **42**, 321–341.

FISHER, R. A. (1930*a*). The distribution of gene ratios for rare mutations. *Proceedings of the Royal Society of Edinburgh* **50**, 205–220.

FISHER, R. A. (1930*b*). *The Genetical Theory of Natural Selection*. Clarendon Press, Oxford Second Edition: Dover, New York, 1958; variorum edition: Oxford University Press, 1999.

FISHER, R. A. (1932). The evolutionary modification of genetic phenomena. In *Proceedings 6th International Congress of Genetics*, Volume **1**, pp. 165–172.

FISHER, R. A. (1941). Average effect and average excess of a gene substitution. *Annals of Eugenics* **11**, 53–63.

FISHER, R. A. (1950). *Contributions to Mathematical Statistics*. Wiley, New York.

FISHER, R. A. (1953). The expansion of statistics. Address of the President, Royal Statistical Society, 19th November 1953. *Journal of the Royal Statistical Society A* **116**, 1–6.

FISHER, R. A. & FORD, E. B. (1926). Variability of species. *Nature* **118**, 515–516.
FISHER, R. A. & FORD, E. B. (1928). The variability of species in the Lepidoptera, with reference to abundance and sex. *Transactions of the Royal Entomological Society, London* **76**, 367–379.
FISHER, R. A. & STOCK, C. S. (1915). Cuénot on preadaptation. A criticism. *The Eugenics Review* **7**, 46–61.
FRANK, S. A. (1995). George Price's contributions to evolutionary genetics. *Journal of Theoretical Biology* **175**, 373–388.
FRANK, S. A. (1997). The Price equation, Fisher's Fundamental Theorem, kin selection, and causal analysis. *Evolution* **51**, 1712–1729.
GAYON, J. (1998). *Darwinism's Struggle for Survival*. Cambridge University Press.
GRAFEN, A. (2003). Fisher the evolutionary biologist. *The Statistician* **52**, 319–329.
GRAFEN, A. (2008). The simplest formal argument for fitness optimization. *Journal of Genetics* **87**, 421–433.
HAMILTON, W. D. (1964). The genetical evolution of social behaviour, I. *Journal of Theoretical Biology* **7**, 1–16 Reprinted in Hamilton (1996).
HAMILTON, W. D. (1996). *Narrow Roads of Gene Land* (Volume **I**). Spektrum, Freeman, Oxford.
HARMAN, O. (2010). *The Price of Altruism*. Bodley Head, London.
JEANS, J. H. (1904). *The Dynamical Theory of Gases*. Cambridge University Press.
KEMPTHORNE, O. (1957). *An Introduction to Genetic Statistics*. Wiley, New York.
KEMPTHORNE, O. (1983). A review of "R.A. Fisher: an appreciation" by Fienberg & Hinkley. *Journal of the American Statistical Association* **78**, 482–490.
KENDALL, D. G. (1990). Obituary, Andrei Nikolaevich Kolmogorov (1903–1987). *Bulletin of the London Mathematical Society* **22**, 31–100.
KIMURA, M. (1958). On the change of population fitness by natural selection. *Heredity* **12**, 145–167.
LACK, D. (1954). The evolution of reproductive rates. In *Evolution as a Process* (eds J. HUXLEY, A. C. HARDY and E. B. FORD), pp. 143–156. Allen & Unwin, London.
LESSARD, S. (1997). Fisher's fundamental theorem of natural selection revisited. *Theoretical Population Biology* **52**, 119–136.
LI, C. C. (1955). *Population Genetics*. University of Chicago Press.
MACKENZIE, D. A. (1981). *Statistics in Britain, 1865–1930*. Edinburgh University Press.
MORAN, P. A. P. (1962). *The Statistical Processes of Evolutionary Theory*. Clarendon Press, Oxford.
MORAN, P. A. P. & SMITH, C. A. B. (1966). *Commentary on R.A. Fisher's Paper on the Correlation between Relatives on the Supposition of Mendelian Inheritance, Eugenics Laboratory Memoirs, Volume XLI*. Cambridge University Press, for the Galton Laboratory, University College London.
NORTON, B. (1978). Fisher and the Neo-Darwinian synthesis. In *Human Implications of Scientific Advance, Proceedings XV International Congress of the History of Science* (ed. E. G. FORBES), pp. 481–494. Edinburgh University Press.
NORTON, B. & PEARSON, E. S. (1976). A note on the background to, and refereeing of, R.A. Fisher's 1918 paper 'On the correlation between relatives on the supposition of Mendelian inheritance'. *Notes and Records of the Royal Society* **31**, 151–162.
OKASHA, S. (2008). Fisher's fundamental theorem of natural selection—a philosophical analysis. *The British Journal for the Philosophy of Science* **59**, 319–351.
PLUTYNSKI, A. (2006). What was Fisher's fundamental theorem of natural selection and what was it for? *Studies in History and Philosophy of Biological and Biomedical Sciences* **37**, 59–82.
PRICE, G. R. (1970). Selection and covariance. *Nature* **227**, 520–521.
PRICE, G. R. (1972*a*). Extension of covariance selection mathematics. *Annals of Human Genetics* **35**, 485–490.
PRICE, G. R. (1972b). Fisher's 'fundamental theorem' made clear. *Annals of Human Genetics* **36**, 129–140.
RIDLEY, M. (1996). The honorary animal. In *Review of Narrow Roads of Gene Land* (Volume **1**, ed. W. D. HAMILTON *Times Literary Supplement*, 14 June), pp. 3–4.
ROBERTSON, A. (1955). Selection in animals: synthesis. In *Cold Spring Harbor Symposia on Quantitative Biology XX: Population Genetics: The Nature and Causes of Genetic Variability in Populations*, pp. 225–229.
ROBERTSON, A. (1966). A mathematical model of the culling process in dairy cattle. *Animal Production* **8**, 95–108.
SHAW, R. F. & MOHLER, J. D. (1953). The selective significance of the sex ratio. *American Naturalist* **87**, 337–342.
WADDINGTON, C. H. (1939). *An Introduction to Modern Genetics*. Allen & Unwin, London.
WILLIAMS, G. C. (1966). *Adaptation and Natural Selection*. Second Edition (1996). Princeton University Press.
WRIGHT, S. (1931). Evolution in Mendelian populations. *Genetics* **16**, 97–159.

(*Received 7 January 2013; revised 30 May 2013; accepted 31 May 2013; published online 20 August 2013*)

Paper 248 (2016)

Analysing nature's experiment: Fisher's inductive theorem of natural selection

Theoretical Population Biology 109, 1–5.

Note from A. W. F. Edwards: The substance of this paper is already in [238], but here I show how a comparison with Fisher's statistical work at Rothamsted Experimental Station further illuminates the development and structure of his fundamental theorem.

Analysing nature's experiment: Fisher's inductive theorem of natural selection

A.W.F. Edwards

Gonville and Caius College, Cambridge, CB2 1TA, UK

ARTICLE INFO

Article history:
Received 1 September 2015
Available online 12 November 2015

Keywords:
Fisher's fundamental theorem
Inductive analysis
Ewens and Lessard (2015)

ABSTRACT

The paper by Ewens and Lessard (2015) adds to the progress that has been made in exploring the discrete-generation analytical version of Fisher's Fundamental Theorem of Natural Selection introduced by Ewens (1989). Fisher's continuous-time theorem differs from the version described by Ewens and Lessard by using a different concept of fitness. Ewens and Lessard use the conventional 'viability' concept whereas for Fisher the fitness of a genotype was its relative rate of increase or decrease in the population. The sole purpose of the present paper is to emphasize the alternative inductive nature of Fisher's theorem, as presented by him in 1930, by placing it in the context of his contemporary development of the analysis of variance in agricultural experiments. It is not a general discussion of the theorem itself.

1. Introduction

Ewens and Lessard (2015) 'On the interpretation and relevance of the Fundamental Theorem of Natural Selection' is a rigorous exploration of one way of approaching Fisher's Fundamental Theorem ('FTNS' Fisher, 1930). Here I argue that their approach is not a true reflection of Fisher's purpose and therefore that the 'problems' that they see in the theorem are particular to their own development, as they partly recognize.

Rather than as an attempt to model natural selection, Fisher's FTNS may be viewed as an analysis of the effect of selection on a character at a single moment in time, comparable to the analysis of an agricultural experiment. The major differences are (1) that nature's experiment of Darwinian evolution is not a designed one and (2) that the error terms which occur in an agricultural analysis are in the genetical case high-order effects, and the population is considered infinite. Tests of significance do not arise. These differences are irrelevant to the force of the analogy with agricultural experiments since the basic idea of Fisher's analysis of variance carries over into FTNS. Moreover, in one general respect the analogy is particularly appropriate: unlike the development in Ewens and Lessard the purpose of FTNS is inductive, drawing out information about the present effect of selection in just the same way as the analysis of a field experiment draws out information about the effect of, say, different types of fertilizer. Nature's experiment is a factorial experiment and the genes (or gene-complexes) are the factors.

I present three types of evidence in support of this view. First, the evidence of what Fisher wrote, both in *The Genetical Theory of Natural Selection* (1930) and in subsequent publications and letters. Secondly, the background of his path-breaking developments in statistics at Rothamsted which were contemporary with the writing of *The Genetical Theory*. Thirdly, Fisher's later insistence that FTNS was not a potential theory to be compared with Wright's adaptive surfaces.

I disclaim any originality for this viewpoint. Naturally I believe it was Fisher's. Of the commentaries on FTNS prior to Price (1972), Bennett (1956) and Kimura (1958) understood it, while Kempthorne (1957) would have done had he not fallen at the last fence by failing to interpret his mathematics in Fisher's sense that FTNS referred to the increase in mean fitness attributable solely to the change in gene frequencies.

Price (1972), who was not familiar with Bennett (1956), was rather dismissive of 'The geneticists who have published derivations of what they thought was Fisher's theorem' because they 'have in most cases shown that the relation [total rate of change = genic variance] holds under special conditions'. Presumably this is a reference to Kempthorne (1957) and Kimura (1958), and while this may be fair to Kempthorne it is not fair to Kimura, whose paper, wrote Price, was 'based on the usual interpretation of Fisher's theorem as an equation for the [total rate of change]'. On the contrary, Kimura partitioned the rate of change into three components and came to the same conclusion as Price fourteen years later, that the additive genetic variance only contributed 'the part

E-mail address: awfe@cam.ac.uk.

http://dx.doi.org/10.1016/j.tpb.2015.11.002
0040-5809/© 2015 Elsevier Inc. All rights reserved.

of increase in fitness due to the improvement of the *genic* constitution of the species'. Ewens (1989) called this 'the partial change'. Price's is a valuable commentary with many supporting quotations from Fisher (1930, 1941), but it is not an ideal introduction for a first study of FTNS, and I myself have been more influenced by rereading Kimura (1958). For further information on these historical points and an interpretation of Fisher' continuous-time original see Edwards (2014).

2. The analysis-of-variance structure

FTNS is a development starting from Fisher (1918) 'The correlation between relatives on the supposition of Mendelian inheritance' in which he wrote 'The contributions of imperfectly additive genetic factors divide themselves for statistical purposes into two parts: an additive part which reflects the genetic nature without distortion, ..., and a residue which acts in much the same way as an arbitrary error introduced into the measurements'. Here in embryo is the regression of genotypic value on gene number and the associated analysis of variance. In 1935 Fisher remarked in a letter (Bennett, 1983, p.260): '...we ordinarily count as genetic only such part of the genetic effect as may be included in a linear formula and ...we make a present to the environmentalists of such variation due to the combined action of genetic and environmental causes as is not expressible in such a formula' (an explicit recognition of his idiosyncratic use of 'environment' to include all the non-linear effects such as dominance and epistasis). We should note that Fisher's 1918 view that the 'additive part ...reflects the genetic nature without distortion' needs careful interpretation because the effects of the genes inherited by the next generation will depend on how they are distributed among the genotypes (see Falconer, 1985). A gene does not have a unique effect: in the presence of dominance its effect depends on the allele it is paired with, so one may only speak of its average effect in a particular population.

Ewens and Lessard observe that 'The relevance of the additive genetic variance in plant and animal breeding programs via the concept of heritability is central'. It is the same in FTNS. The analysis of variance identifies the 'main effects' of the genes with the additive genetic variance, Fisher treating the higher-order terms as environmental.

3. The inductive purpose of FTNS

At the 6th International Congress of Genetics at Cornell Fisher (1932) contrasted his approach to the study of evolution in the light of Mendelian genetics to that of J.B.S. Haldane. In 1924 Haldane had started publishing a long series of papers 'A mathematical theory of natural and artificial selection' (Haldane, 1924) in which he worked out the consequences of selection for a large variety of Mendelian models, usually with discrete generations and infinite population size. Fisher called this approach 'analytic and deductive': 'Genetic studies are regarded as revealing the mechanism connecting cause and effect, from a knowledge of which the workings of the machine can be deduced and the course of evolutionary change inferred' (p.165). His own approach, by contrast, was 'inductive and statistical'. That is, he was – by implication in *The Genetical Theory* and in particular FTNS – using statistical methods such as the analysis of variance and fixed-x regression to elucidate inductively the fundamental feature of the operation of natural selection. There was no intention to predict the course of evolutionary change by adopting a specific Mendelian model and none was implied by his analysis beyond the basic laws of Mendelian inheritance.

A more recent recognition of the inductive nature of FTNS is in The inductive theory of natural selection' (Frank, 2014). According to Frank 'Fisher's fundamental theorem is a simple invariance. ...[It] shows that the change in mean fitness by selection is invariant to all details of variability in the population except the variance associated with the transmissible predictors'.

4. The influence of Rothamsted

From his appointment as statistician to Rothamsted Experimental Station in 1919 and throughout the 1920s Fisher developed the analysis of variance, randomization and experimental design at an astonishing rate, writing *The Genetical Theory* in the evenings. Already in 1925 the first edition of *Statistical Methods for Research Workers* (Fisher, 1925) had contained two sections 'Technique of Plot Experimentation' and 'The Latin Square', but the most convenient summary of the background to these developments is his later *The Design of Experiments* (Fisher, 1935) which grew out of the material in *Statistical Methods*.

Introducing FTNS in *The Genetical Theory* in 1930 Fisher explained (p.22) 'The object of the present chapter [II] is to combine certain ideas derivable from a consideration of the rates of death and reproduction of a population of organisms, with the concepts of the factorial scheme of inheritance, so as to state the principle of Natural Selection in the form of a rigorous mathematical theorem, by which the rate of improvement of any species of organisms in relation to its environment is determined by its present condition'. To interpret Fisher's word 'factorial' reference may be made to *The Design of Experiments* Chapter VI 'The factorial design in experimentation'. As Fisher remarked in his Bateson Lecture in 1951 (Fisher, 1952) '...the "factorial" method of experimentation ...derives its structure, and its name, from the simultaneous inheritance of Mendelian factors' (the usual word employed by the early Mendelians).

In FTNS the 'factors' are the genes themselves and their additive effects are the 'main effects' in a 'factorial scheme'. The treatment is that of the analysis of variance applied to the regression of genotypic measure on gene number. Concentrating on a single diallelic locus, Fisher called the change caused by substituting one allele by another as represented by the slope of the regression line 'the average effect of a gene substitution' on the measure (his example was human stature). With this definition the effect of changing from one homozygote to the heterozygote is the same as changing from the heterozygote to the other homozygote, so the average effect is this quantity too. Fisher called the variance accounted for by the regression the 'genetic variance' (using the word 'genetic' as the adjective from 'gene') as opposed to the overall 'genotypic variance'. His contrast in meaning between 'genetic' and 'genotypic' escaped some commentators and the genetic variance soon became the 'additive genetic variance' and latterly the 'genic variance', a phrase which incidentally had Fisher's approval.

By contrast, the 'average excess (in stature) associated with the gene substitution in question' was arrived at by imagining half the heterozygotes to be grouped with each of the homozygotes and taking the difference between the average statures of these imaginary groups. Unlike the average effect, this was an exact concept not dependent on regression. Of course if the genotypic values were themselves linear (the case of no dominance) then the two measures were the same, but Fisher did not assume this.

As to 'certain ideas derivable from a consideration of the rates of death and reproduction of a population of organisms', this is a reference to his development of Malthusian parameters of population increase, but it is a complexity not necessary for an understanding of FTNS, where, as we shall see, ordinary rates of increase can be used in the first instance (Edwards, 2014).

5. FTNS and potential theory

The final reason for adopting the view of FTNS as an inductive analysis of the genetical operation of natural selection is the negative one that Fisher explicitly rejected any 'mechanistic' approach:

> In considering the original statement of what I ventured to call 'the fundamental theorem of natural selection', I had, of course, considered the relation between such a situation and that in which a potential function existed, for my mathematical education lay in the field of mathematical physics. As you realize, I preferred to develop the theory without this assumption, which of course in another aspect is a restriction. Of course, I do not question that the selective intensities acting instantaneously may well be equivalent to those derivable from such a function, but I think it should be emphasized that both changes in time, that is in the environmental *milieu* and in the gene ratios themselves, that is the heritable constitution of the organism, will change this virtual function in a way that cannot be specified in terms of the quantities used in formulating the fundamental theorem. Fisher, 1956, in a letter to Kimura (Bennett, 1983, p.229).

He was caustic about Sewall Wright's 'adaptive landscapes': 'It has been proposed... that the environment, picturesquely renamed the landscape, governs the course of evolutionary change, much as the field of force determines the trajectory of a comet' (1936), and 'Wright's conception... that selective intensities are derivable, like forces in a conservative system, from a simple potential function dependent on the gene ratios as a whole, has led him into extensive but untenable speculation' (1941). Finally:

> I have never indeed written about *W*bar and its relationships, and now that the alleged relationship has been brought to my attention, I must point out that the existence of such a "potential function" as that which Wright designates by *W*bar, is not a general property of natural populations, but arises only from the special and restricted cases which Wright has chosen to consider. Selective tendencies are not, in general, analogous to what mechanicians describe as a conservative system of forces. (Fisher, 1958).

We may note, however, that Wright (1955) accepted that 'Fisher's form of statement indicates that it is intended to apply only momentarily rather than to define "fitness" as a quantity that governs the course of evolution, analogously to a potential function', thus distancing his own approach from FTNS.

6. The genesis and interpretation of FTNS

As early as 1919 in an unpublished manuscript now in the Adelaide archive Fisher had moved from considering the correlation between relatives (Fisher, 1918) to the question of evolutionary change: 'To produce such a change groups which differ in their rates of increase must also differ on the average in their heritable qualities. Moreover, the association of rate of increase with heritable quality must continue from generation to generation' (Fisher, 1919). In 1924 he was more specific: 'The regression coefficient shows just how effective such selection is' and in 1926 and 1928 even more specific, writing with E.B. Ford: 'it is easily demonstrable that in a species in which a higher proportion of the total variance is ascribable to genetic causes, the effective selection will be more intense than in species in which the variance is to a larger extent due to environmental variation'. To use a word he had not yet coined (but did in 1930), the rate of change in a character under selection is proportional to the covariance of the character with fitness. We may note in passing that this is the basis of the 'Breeder's Equation', Robertson's Equation (1966) and Price's Equation (1970).

Continuing this line of thought we arrive at the fundamental theorem. Taking the character to be fitness itself instead of human stature, the covariance becomes a variance and 'The rate of increase in the mean fitness of any organism at any time ascribable to natural selection acting through changes in gene frequencies is exactly equal to its genic variance in fitness' (my rewording Edwards, 1994). Fisher demonstrated the equality in the diallelic case by showing that both sides of the equation were equal to the product of the average effect, the average excess, and $2pq$, where p and q are the frequencies of the two genes (see Edwards, 2014).

7. From 'genotypic value' to 'relative rate of increase'

Here we must pause to consider Fisher's meaning of genotypic 'fitness' because it is different from the usual fitness attributed to genotypes in population genetics. Rather than the 'viability' of the genotype as is customary in discrete-generation models it is simply the current rate of growth of the frequency of the genotype in the continuous-time treatment of a population of genotypes. Denniston (1978) discussed this definition of fitness with great care, particularly in relation to FTNS and Kimura (1958). Fisher (1930, pp.33–34) slides easily from 'genotypic value' into 'relative rate of increase or decrease'. The genotypic value to which he refers is human stature (p.30). But then 'The definitions given above may be applied to any characteristic whatever; it is of special interest to apply them to the special characteristic m which measures the relative rate of increase or decrease'. (Interestingly, he removed this sentence in the 1958 edition.)

For an understanding of FTNS, Fisher's introduction here of the Malthusian parameter of population increase (or decrease) m which 'measures the relative rate of increase or decrease of a population when in the steady state appropriate to such a system' is unfortunate. It applies when the distributions of ages at birth and death have reached equilibrium, but it is not obvious to me that this is more appropriate for FTNS than the actual rates obtaining at the present moment, which I use in this account. The matter has been explored most recently by Grafen (2015). What seems clear is that it is an unnecessary preliminary to an understanding of FTNS (Edwards, 2014).

The 'fundamental' in FTNS was possibly intended to convey the idea that in studying the change in the mean of any heritable character through selection one could first find the change in the mean fitness and then that of the character through the correlation of the two, leading to the replacement of the variance in the theorem by the covariance as in Robertson's and Price's Equations. Thus the fundamental quantity is the change in the mean fitness which can then be applied to any heritable character correlated with fitness.

8. The analogy with entropy

It is appropriate to remark on Fisher's cautious and qualified comparison of FTNS with entropy in *The Genetical Theory*. Because a variance is necessarily non-negative, FTNS implies that the genic contribution to the change in the mean fitness is positive (or at least not negative) assuming an additive analysis. But as Fisher points out '(3) Fitness may be increased or decreased by changes in the environment', 'environment' here being interpreted in his particular sense. To understand Fisher's point of view we have to recall that he regarded a population as an assemblage of genes (or, more generally, gene-combinations) and that the measurable effect of natural selection, though acting through the genotypes is to change gene frequencies. How the genes are assorted into genotypes he considered a transient question of no great interest

'... the only evolutionary effect, either in increased fitness or in anything else, that I can recognize as such, is constituted by the change in gene ratio [frequency]' (Fisher, 1955, in a letter to Kempthorne (Bennett, 1983, p.229)).

The growth-rate theorem may be helpful to a consideration of the entropy analogy. It simply states that 'in a subdivided population the rate of change in the overall growth-rate is proportional to the variance in growth-rates'. 'This elementary theorem is applicable wherever there are differential growth-rates, in economics for example (Edwards, 1987) as well as in biology. It opens a window on the Fundamental Theorem, yet never seems to have been published in its own right, by Fisher or anyone else' (Edwards, 1994). It clearly lies behind the increasing property in FTNS, for since a variance is necessarily positive when the growth-rates vary, the mean growth-rate must be increasing. It is indeed true that entropy in physics is also an increasing quantity but it is hardly unique in that respect. For the analogy to have any force there would need to be further similarities, but the ones Fisher offers (1930, p.36) are not convincing.

9. Discussion of Ewens and Lessard (2015)

The modern discussion of FTNS started with Ewens (1989) who credited Price (1972) with the insight that FTNS referred to the partial rate of increase in the mean fitness. However, Ewens used a discrete-time model with distinct generations and conventional genotypic fitnesses, though Fisher had used continuous time, growth-rate fitnesses, and no mention of generations at all. Ewens cleverly showed how to prove a version of FTNS that then applied but (to return to the distinction between analytical methods and inductive ones) his model was analytical, firmly in the tradition of the mathematics of population genetics. It is this approach that he and Lessard have now fully developed which has led them to some unfavourable conclusions.

The introduction of an analytical model with discrete time and generations clearly draws attention to the fact that the average effects are, in general, changing, which in turn makes it obvious that the model cannot say 'anything about long-term gene frequency changes'. 'It cannot be used to calculate gene frequency changes even two generations in advance'. '*On the other hand, the theorem does not claim to calculate long-term gene frequency changes, so that this comment is not a criticism of the theorem*' (my italics). The sense of disappointment in Ewens and Lessard arises from an expectation that an analytical development of FTNS such as they have essayed might have led to a predictive theory, but it did not.

10. Conclusion

Thus Fisher's original theorem is not a mechanism 'from a knowledge of which the workings of the machine can be deduced and the course of evolutionary change inferred'. Rather, it is like the analysis of an agricultural or other experiment in which the main effect of the factors at one particular moment in time is teased out by regression and the analysis of variance. It captures the precise relation that Fisher developed in his mind between 1918 and 1928. The growth-rate theorem shows that the variance is the quantity relevant to the rate of change of a mean, and partitioning the variance as an explanatory procedure is one of Fisher's great statistical contributions. He separates out the variance associated with the regression of genotypic measure on gene number as 'the natural measure of the average effect of the gene substitution' and argues that the other effects 'are irrelevant to the effect of the gene substitution considered, in the actual circumstances in which we wish to evaluate it' (Fisher, 1941).

I agree with the analysis of Kimura (1958). Rather more to the point, so did Fisher. He asked Kimura to add to his paper 'It is the reformulation of the results published by Fisher in 1930 ... and in no way differs from it in substance ...', and Kimura agreed to. Fisher wrote that Kimura's analysis was 'an explicit reformulation of effectively identical conclusions' (quoted by Edwards, 2014).

In his Croonian Lecture to the Royal Society Fisher (1953) summed up FTNS: 'In this way the principle of natural selection is put upon an exact quantitative basis, and the course of spontaneous population change is related to its state at any instant. ... Natural selection does not imply that every species is progressively better off, but that all species tend by all possible changes of gene ratio to improve their average condition'.

Acknowledgements

I am indebted to Dr. Peter O'Donald for our continuing discussion of FTNS over many years. In O'Donald (1970) he noted the inductive nature of Darwin's argument and in *Genetic Models of Sexual Selection* (O'Donald, 1980) he drew attention to Fisher's contrast between deductive and inductive arguments by quoting from Fisher (1932) at greater length than I have done.

References

Bennett, J.H., 1956. Population genetics and natural selection. Genetica 28, 297–307.
Bennett, J.H., 1983. Natural Selection, Heredity, and Eugenics. Clarendon Press, Oxford.
Denniston, C., 1978. An incorrect definition of fitness revisited. Ann. Hum. Genet. 42, 77–85.
Edwards, A.W.F., 1987. Fisher, Ronald Aylmer. In: The New Palgrave: A Dictionary of Economics, Volume II. Macmillan, London, pp. 376–377;
Reprinted in 1990. Time Series and Statistics. Macmillan, London, pp. 95–97.
Edwards, A.W.F., 1994. The fundamental theorem of natural selection. Biol. Rev. 69, 443–474.
Edwards, A.W.F., 2014. R.A. Fisher's gene-centred view of evolution and the Fundamental Theorem of Natural Selection. Biol. Rev. 89, 135–147.
Ewens, W.J., 1989. An interpretation and proof of the fundamental theorem of natural selection. Theor. Popul. Biol. 36, 167–180.
Ewens, W.J., Lessard, S., 2015. On the interpretation and relevance of the Fundamental Theorem of Natural Selection. Theor. Popul. Biol. 104, 59–67. http://dx.doi.org/10.1016/j.tpb.2015.07.002.
Falconer, D.S., 1985. A note on Fisher's 'average effect' and 'average excess'. Genet. Res. 46, 337–347.
Fisher, R.A., 1918. The correlation between relatives on the supposition of Mendelian inheritance. Trans. Roy. Soc. Edinburgh 52, 399–433.
Fisher, R.A., 1919. Unpublished manuscript in the Adelaide University Library Special Collections.
Fisher, R.A., 1925. Statistical Methods for Research Workers. Oliver & Boyd, Edinburgh.
Fisher, R.A., 1930. The Genetical Theory of Natural Selection. Clarendon Press, Oxford; second ed: Dover, New York, 1958; variorum edition: Oxford University Press, 1999.
Fisher, R.A., 1932. The evolutionary modification of genetic phenomena. In: Proceedings 6th International Congress of Genetics, Vol. 1, pp. 165–172.
Fisher, R.A., 1935. The Design of Experiments. Oliver & Boyd, Edinburgh.
Fisher, R.A., 1941. Average effect and average excess of a gene substitution. Ann. Eugen. 11, 53–63.
Fisher, R.A., 1952. Statistical methods in genetics. Heredity 6, 1–12.
Fisher, R.A., 1953. Population genetics. Proc. R. Soc. Lond. Ser. B 141, 510–523.
Fisher, R.A., 1958. Polymorphism and natural selection. Bull. Int. Stat. Inst. 36, 284–289; 1958. J. Ecol. 46, 289–293.
Frank, S.A., 2014. The inductive theory of natural selection. arXiv:1412.1285v1.
Grafen, A., 2015. Biological fitness and the fundamental theorem of natural selection. Amer. Nat. 186, 1–14.
Haldane, J.B.S., 1924. A mathematical theory of natural and artificial selection. I. Trans. Camb. Philos. Soc. 23, 19–41.
Kempthorne, O., 1957. An Introduction to Genetic Statistics. Wiley, New York.
Kimura, M., 1958. On the change of population fitness by natural selection. Heredity 12, 145–167.
O'Donald, P., 1970. Natural selection for quantitative characters. Heredity 27, 137–153.
O'Donald, P., 1980. Genetic Models of Sexual Selection. Cambridge University Press.
Price, G.R., 1970. Selection and covariance. Nature 227, 520–521.
Price, G.R., 1972. Fisher's 'fundamental theorem' made clear. Ann. Hum. Genet. 36, 129–140.
Robertson, A., 1966. A mathematical model of the culling process in dairy cattle. Anim. Prod. 8, 95–108.
Wright, S., 1955. Classification of the factors of evolution. Cold Spring Harb. Symp. Quant. Biol. 20, 16–24.

Part 2

Commentaries

Anthony Edwards's Seminal Contributions to Phylogenetics, Likelihood, and Understanding R. A. Fisher and the History of Genetics

WALTER BODMER

It is a pleasure to take part in this tribute to Anthony Edwards's pioneering research and insightful writings on problems of phylogenetics, statistical inference, and the interpretation of Fisher's work, especially the Fundamental Theorem of Natural Selection, and on the history of genetics.

Anthony Edwards started his third undergraduate year in the genetics department at the University of Cambridge in 1956, the same time that I began my PhD under R. A. Fisher. We remained together in Cambridge until 1961, two years after I had completed my PhD and just as Anthony finished his. For both of us, these five years under the influence of that extraordinary genius, R. A. Fisher, were important, formative years. One result of this shared experience is that for the last 40 years or so we have enjoyed the joint responsibility of looking after the R. A. Fisher Memorial Trust, he as treasurer, and I as chairman. This activity has kept us in close touch, affording the opportunity for many Fisherian discussions, ranging across the whole field of population genetics and its history as well as issues concerning statistical inference. Anthony and I share another close academic bond; we have both worked closely with the noted human geneticist, Luca Cavalli-Sforza, who was also mentored by Fisher as he developed his interest in human genetics. Thus, in many ways, the three of us are academic siblings with Fisher as our common parent!

The only paper I have written with Anthony Edwards is on natural selection and the sex ratio (Bodmer and Edwards 1960). Our interest in this problem was stimulated by Anthony's extensive analysis of Geissler's sex ratio data (Edwards 1958), which was a major part of his PhD thesis work, and by Fisher's suggestion in his famous 1930 book, *The Genetical Theory of Natural Selection*, that parental expenditure on their offspring was a major factor in controlling the sex ratio, at least in mammals with an X/Y sex control mechanism. This work had a significant impact on the general notion of parental care and inclusive fitness as propounded by W. D. Hamilton (1964) and others. Characteristically, with his interest in the history of genetics, Anthony, in a much later paper (Edwards 1998), pointed out that Carl Düsing, writing in 1884, had foreshadowed Fisher's idea on the control of the sex ratio, building on Darwin's earlier consideration of the problem, and had even constructed a plausible mathematical model without the benefit of being aware of Mendel's laws of genetics.

Soon after finishing his PhD in 1961, Anthony went to Pavia, where he did his seminal, pioneering work with Luca on phylogenetics. The problems posed by estimating the optimal evolutionary tree from genetic polymorphism data on human populations then led to his innovative work on likelihood as a major vehicle for statistical inference.

Evolutionary Trees From Human Blood-Group Frequency Data

The ABO blood groups, discovered in 1900, were the first polymorphism to be used for human genetic population studies in 1918. The variation in the frequencies of the A and B types in different countries, ranging from Europe through the Middle East and Africa to India, suggested that such genetic variation might be used for the definition of human populations, or ethnic groups, and their relationships. No sensible conclusions could, however, be derived from studying just one polymorphic system. It was the gradual discovery of more blood-group systems between 1940 and 1960, notably rhesus and MNS, that raised the possibility of using such data for human population relationship studies.

Fisher had an enormous influence on blood-group research in Britain through his support of the work of Race and Sanger, pioneering discoverers of new blood groups, and his influence on Mourant, who worked closely with Race and Sanger and who, with his colleagues, compiled an outstanding compendium of data on blood group and other polymorphisms (Mourant *et al.* 1954). This was the first time that worldwide data on blood-group polymorphisms were assembled together in a way that promised a much better understanding of human population relationships based on genetic data.

It was with this background that Luca and Anthony began to consider the problem of how to use the blood-group gene-frequency data to classify human population relationships objectively on a genetic basis. They argued that, rather than build classifications of similarity on arbitrary conventional taxonomical characteristics, they would build phylogenetic trees on the assumptions of a branching evolutionary model of changes in gene frequencies. However, even with the simplest assumptions of random mating and no selective effects, finding the tree that, on the basis of such an evolutionary model, best fitted an observed set of data on gene frequencies in a range of different populations, was, to say the least, very challenging.

Their first step was to define a genetic distance between any two populations based on the gene frequencies. The more similar the distribution of gene frequencies in any two populations, the less the genetic distance between them. This distance, which could be calculated between all pairs of populations to be assembled into a given tree, was additive and could be considered as a measure of the evolutionary separation between the populations. As described in Anthony's historical perspective of his work in this area ([225]), Luca suggested that one approach to estimating the best tree would be to choose one ' . . . by the method of least squares. Different topologies can be tried and the one with the least residual sum of squares adopted' (p. 269, this volume).

In the first substantive description of their method, given at the 1963 International Congress of Genetics in the Hague but only published in 1965 ([29]), Anthony introduced the new concept of choosing the tree that minimised the sum of the distances between the populations, later called the method of minimum evolution. As he put it 'Its physical counterpart would be the net formed by the minimum amount of string necessary to connect all the points in an n-dimensional space' (p. 44, this volume), where each point represents one of the populations. It was assumed that this would give a good approximation to the maximum-likelihood estimate.

In a much later paper ([146]), published in 1996, Anthony emphasised this explanation for using what he called the 'method' of minimum evolution and made it quite clear that this was not a 'parsimony principle' as he could see no reason why parsimony should be an evolutionary principle.

Even with the limited number of populations and genetic markers Cavalli-Sforza and Edwards chose for their first analysis, nothing could have been achieved without the use of electronic computers, which were only then just becoming available for the general academic user. They chose data on 15 populations with a reasonable worldwide spread and frequencies of 18 alleles from five blood-group systems. The resulting tree is shown in Figure 5 of [29] (p. 48, this volume) and has often been reproduced elsewhere. In the authors' words it showed 'three main branches. Europeans, Africans and Asiatics, the latter including Americans and splitting into a variety of subgroups' and a 'remarkable similarity between the blood group map thus obtained, and the geographical map' (pp. 48–49, this volume). It was certainly a remarkable achievement to obtain a result that was so consistent with general expectations, using relatively few genetic variants. The extent to which subsequent analyses using more sophisticated computer programs have given results largely consistent with expectations is evidence for the lack of bias in the analysis that led to this original phylogenetic tree, based only on gene-frequency data.

As a precursor to the tree analysis, Luca and Anthony did an initial principal components analysis (PCA) to provide some idea of how the populations fell into related clusters using the first two components. This was, to my knowledge, the first time that PCA was used for the analysis of gene-frequency data and it was possibly also the first time that an analogy was drawn between PCA clustering results and a tree structure (see also [30]). I wonder how many investigators now using these approaches to data analysis, especially in fields as far removed from population genetics as the genome-wide analysis of mRNA levels in a series of samples, realise the origin of these basic ideas.

An earlier paper ([27]), published in 1964, covered similar ground, but in less detail, focusing on the difference between the logic of classical taxonomic analysis and their evolutionarily based genetic approach. A series of later papers recapitulated earlier results with a gradual development of the ideas. The 1964 Cold Spring Harbor Symposium paper [28] provided evidence to support the assumption of an underlying evolutionary model based on genetic drift in the absence of any directional selection.

The book on human population genetics (Cavalli-Sforza and Bodmer 1971), which Luca and I wrote between 1963 and 1969, provided the first clear textbook descriptions of the analysis of population gene-frequency data using PCAs and of the Cavalli-Sforza–Edwards approach to the construction of phylogenetic trees.

Likelihood

In my view, the most notable of the series of phylogenetic papers that followed the initial publications from 1963 to 1965 is Anthony's presentation to the Royal Statistical Society in 1969 ([46]). In this presentation, he enlarged on the approaches to a formal solution for estimating the best tree through the use of maximum likelihood. This paper, part of which was prepared while Anthony was at Stanford University in 1964, supported by a National Institutes of Health (NIH) grant that I held jointly with Sam Karlin, reflected Anthony's careful consideration of the problems of statistical inference in relation to estimating phylogenetic trees. This led to his strong promotion of a likelihood-based approach both to estimation and, especially, to the relative evaluation of the merits of different models for explaining a body of observed data. He used the term 'support' for the natural logarithm of the ratio between the likelihoods comparing two hypotheses and argued that the way that scientists should think was to consider which hypothesis has enough support relative to others to be worth pursuing. These ideas were first clearly put forward in his 1969 paper in Nature ([42]). He drew the clear distinction between likelihood, as originally defined by Fisher, and probability, and argued strongly against a Bayesian probabilistic approach to the evaluation of a hypothesis unless there is a clear, definite rationale for the choice of the prior probability distribution. This is a view that I have held for a long time and with which

I wholeheartedly concur. These ideas on likelihood were greatly enlarged in his notable book, *Likelihood*, first published by Cambridge University Press in 1972, with subsequent reprinting and an expanded edition, in 1992, indicating the interest in his ideas.

Historical Papers

Reflecting his strong interest in the history of genetics and statistics, Anthony has written a number of historical papers, of which just a few are in this collection. Notably missing are, for example, his insightful perspective on Fisher's interpretation of the rhesus blood-group system and his perspective on the famous mathematician, G. H. Hardy, and the Hardy–Weinberg law (Edwards 2007, 2008). In his history of likelihood ([60]), Anthony starts by pointing out that Fisher first put forward the idea of maximising likelihood in a little known paper of 1912, written when Fisher was only 22, but only first used the term likelihood in 1921. Anthony emphasises the distinction between likelihood and probability, and, in particular, the method of 'maximum probability', based on Bayes' theorem.

In his path-breaking 1930 book, *The Genetical Theory of Natural Selection*, Fisher put forward what he called the 'Fundamental Theorem of Natural Selection'. This was stated as a form of maximisation of the population mean fitness, and it has often been misinterpreted as simply that, and then shown apparently to be wrong for a wide range of particular population-genetic models. In 2014, Anthony produced what is probably the simplest and most penetrating explanation of what Fisher actually meant ([238]). The key issue is that the mean-fitness increase is persistent only as a function of the effects of a gene's substitution by its alternative version and this does not include all possible sources of effect on the population mean fitness. As Anthony put it '. . . the genic rate of increase in mean fitness at any time, that is, the rate of change attributable solely to changes in gene frequency as computed from the additive genetic regression model, is equal to the genic variance in fitness at that time . . .' (p. 301, this volume).

In 1986, Anthony also contributed to what is arguably the most careful and well-documented critique of Fisher's analysis of Mendel's data ([96]), suggesting that the data were in a sense too good. This was, in simple terms, because Mendel tested for heterozygotes by self-fertilisation using only 10 plants and ignored the possibility that, with a probability of $(3/4)^{10}$, he would misclassify a heterozygote as a homozygote.

Anthony has a gift for extracting telling comments from, or relevant to, the papers he reviews. Thus, in [96], he quotes Fisher from his 1936 paper on Mendel's results as saying, 'There is no substitute for a careful, or even meticulous, examination of all original papers purporting to establish new facts' (Fisher 1936, p. 165, this volume).

Anthony's interesting perspective on the much underappreciated early Mendelian geneticist Robert Lock ([235]) shows the care and attention he gives to studying the original materials. In a copy of Lock's book, *Recent Progress in the Study of Variation*, in the Gonville and Caius college library, he found a marginal note in pencil (on p. 87), which he interpreted as being in Fisher's handwriting (p. 291, this volume). (Having seen it myself I can confirm that it is Fisher's handwriting.) From this he deduces that Fisher may have learnt much of his early genetics from Lock's book, as Morgan and his drosophilists may also have done.

Human Genetic Diversity: Lewontin's Fallacy – Comments Based on Bodmer (2015)

Traditional approaches to the characterisation of human population differences, generally influenced by racial prejudice, emphasised obvious outward features, including skin colour and facial features. This led to a view that the major differences between individual humans existed between ethnic groups rather than within them. However, Cavalli-Sforza (1966) showed that only about 15% of the overall gene-frequency variation occurred between populations compared with 85% within them (Table 12, p. 367). This result was based on a simple calculation of the variances in gene frequency (as measured by heterozygosity) for a number of polymorphisms between a set of diverse human populations compared with the variances within populations. Lewontin (1972), using a different measure of genetic heterogeneity and a larger body of data, also came to the conclusion that 85% of the genetic heterogeneity occurred within populations. He, in addition, calculated that 8.3% occurred between populations within 'races', whereas 6.3% occurred between 'races'. A similar, but completely independent calculation that I carried out in 1975, using the same approach as Cavalli-Sforza but totally different data on different populations, came to essentially the same conclusion, with an estimated between-population variance of 11% of the total. Thus, these three separate and independent assessments came to essentially the same conclusion, namely that 80–90% of the genetic variation in the human population was between individuals within populations, and only 10–15% was between populations or racial groups. Lewontin used this result to argue that there was, therefore, no meaning or value to defining human groups on the basis of genetic variation, concluding his article with the statement, 'Human racial classification is of no social value and is positively destructive of social and human relations. Since such racial classification is now seen to be of virtually no genetic or taxonomic significance either, no justification can be offered for its continuance' (Lewontin 1972, p. 398).

In 2003, Anthony produced a strong and closely argued rebuttal ([192]) of Lewontin's argument that these data negated the value of classifying human populations into groups and studying the interrelationships of these groups using gene-frequency data. Indeed, Luca, in his extensive 1966 paper where he first discussed the issue of genetic variation within and between populations, used these same data for constructing a more quantitatively supported phylogenetic tree than that originally produced by Luca and Anthony a few years earlier.

From a purely medical point of view, it can be very important to know a patient's ethnic background because that can influence whether she/he should be screened for a particular genetic disease, including, for example, Tay Sachs disease in Ashkenazi Jewish populations and sickle cell trait for those with a West African background. It is no good having a bone-marrow donor programme that does not take into account ethnic background because the frequencies of the human leukocyte antigen (HLA) types vary substantially among the various major racial groups, and not taking this into account means that certain groups would be discriminated against with respect to donor availability. Finally, most people are interested in where they come from, beyond what their genealogy can tell them, and only the genetics can eventually answer that. The mistake is not that of studying the genetic relationships between different human population groups but that of assigning a value, good or bad, to groups defined in this way.

Epilogue

Undoubtedly, Anthony Edwards's major scientific contribution was the introduction, with Luca Cavalli-Sforza, of a systematic genetically based approach to constructing phylogenetic trees of human populations using gene-frequency data. His subsequent development of likelihood-based methods laid the foundations for all future work in this and related areas. Given the obvious importance of this work, it is, perhaps, surprising that Anthony was only elected a Fellow of the Royal Society in 2015, 50 years after his first relevant major publications. Perhaps the fact that the first accounts of the phylogenetic work were in abstracts or conference proceedings has, to some extent, contributed to this delayed recognition. I remember that Fisher himself told me not to publish anything important in reviews, advice which I have, unfortunately, not always followed.

This collection of papers shows the wide and unusual scope of Anthony's interests beyond the phylogenetic work, ranging from statistical inference and the insightful analysis of Fisher's Fundamental Theorem of Natural Selection to penetrating accounts of aspects of the history of human genetics. Work not covered includes his innovative work on Venn diagrams and Pascal's triangle, and even a paper on number theory. It has been a pleasure to remind myself of such a range of Anthony Edwards's stimulating and interesting publications in the course of preparing these comments.

References

Bodmer, W. F. 2015. Genetic characterization of human populations: from ABO to a genetic map of the British people. *Genetics* 199, 267–279.

Bodmer, W. F. and Edwards, A. W. F. 1960. Natural selection and the sex ratio. *Annals of Human Genetics* 24, 239–244.

Cavalli-Sforza, L. L. 1966. Population structure and human evolution. *Proceedings of the Royal Society B* 164, 362–379.

Cavalli-Sforza, L. L. and Bodmer, W. F. 1971. *The Genetics of Human Populations*. San Francisco, CA: W. H. Freeman. (Reprinted 1999 in Dover publications)

Edwards, A. W. F. 1958. An analysis of Geissler's data on the human sex ratio. *Annals of Human Genetics* 23, 6–15.

Edwards, A. W. F. 1986. More on the too-good-to-be-true paradox and Gregor Mendel. *Journal of Heredity* 77, 138.

Edwards, A. W. F. 1998. Natural selection and the sex ratio: Fisher's sources. *American Naturalist* 151, 564–569.

Edwards, A.W. F. 2007. R. A. Fisher's 1943 unravelling of the Rhesus blood-group system. *Genetics* 175, 471–476.

Edwards, A. W. F. 2008. G. H. Hardy (1908) and Hardy–Weinberg equilibrium. *Genetics* 179, 1143–1150.

Fisher, R. A. 1930. *The Genetical Theory of Natural Selection*. Oxford: Clarendon Press.

Fisher, R. A. 1936. Has Mendel's work been rediscovered? *Annals of Science* 1, 115–137.

Hamilton, W. D. 1964. The genetical evolution of social behaviour, I. *Journal of Theoretical Biology* 7, 1–16.

Lewontin, R. C. 1972. The apportionment of human diversity. *Evolutionary Biology* 6, 381.

Mourant, A. E., Kopec, A. and Domaniewska-Sobczak, K. 1954. *The Distribution of the Human Blood Groups and Other Polymorphisms*. Oxford: Oxford University Press.

Anthony Edwards, Luca Cavalli-Sforza, and Phylogenies

JOSEPH FELSENSTEIN

Encountering Trees and Encountering Anthony

I want to concentrate here on Anthony Edwards's work with Luca Cavalli-Sforza on inferring phylogenies. I became aware of Anthony's work on this topic in 1966, when I first met him. I cannot remember seeing the 1963 abstract ([25]) or the 1964 paper ([27]) before then.

I was a graduate student in Richard Lewontin's laboratory at the University of Chicago, studying theoretical population genetics. I had become distracted from that work by being asked to help two faculty members, Jack Hubby and Lynn Throckmorton, analyze data on protein banding patterns in *Drosophila* species. They showed me Robert Sokal and Peter Sneath's important 1963 book *Principles of Numerical Taxonomy*; based on it, I had written a clustering program. I became intrigued by reconstructing trees from data. In late 1965, Robert Sokal, with Joseph Camin, published a paper in *Evolution* (Camin and Sokal 1965) on the reconstruction of phylogenies by parsimony methods from characters with discrete states. Shortly after that, Sokal visited our university, partly to see his former thesis advisor, the noted termite systematist Alfred E. Emerson. After hearing his seminar, I wrote a computer program to carry out a version of Camin and Sokal's method. About then, Harold Voris, a student of Throckmorton's, introduced me to the folks at the Division of Reptiles and Amphibians at the Field Museum of Natural History, especially Robert Inger and Hymen Marx, and I attempted an analysis of some of their data.

I was working on phylogenies, but not yet on their statistical analysis. In June 1966, I attended the Third International Congress of Human Genetics. This involved no travel or housing expenses, since it was held at the University of Chicago. At the meeting I met Anthony Edwards. He took me across the Midway to meet Luca Cavalli-Sforza, who had a room at the Center for Continuing Education. Luca wanted to know where Lewontin was. My answer bordered on the absurd: I had to tell him that Lewontin was then in Italy, on his way to a meeting. In fact, he visited Pavia, where he tried to look up Luca, only to be told that Luca was in Chicago.

From Edwards and Cavalli-Sforza's talk at the Congress, and from talking to Anthony, I became aware of their work, and of their 1964 paper ([27]).

Rereading that paper now, I am struck by how hard it works to justify interest in the phylogeny. They understood their audience, who were predominantly systematists. Overwhelmingly, systematists in the 1960s were focused on finding the correct classification. The dominant school of systematics was the "evolutionary systematics" school led by Ernst Mayr and George Gaylord Simpson. They wanted to define groups that had some consistency with the phylogeny. They allowed monophyletic and paraphyletic groups, but not polyphyletic. The classification system of amniotes then included separate classes for birds, for mammals, and for reptiles. The former two are monophyletic groups, the latter was long known to be a paraphyletic group. The ancestor of Class Reptilia was a member of that group, but was also known to be an ancestor of birds and of mammals.

Today, the phylogeny needs no justification. There is no longer a single Class Reptilia, as paraphyletic groups are not allowed by the current consensus on classification. Currently, the inference of phylogenies is central to systematics and evolutionary biology. Although present-day systematists are reluctant to admit it, the classification system above the species level is fading in importance. But, back then, the phylogeny was, at most, an aid to classification. Edwards and Cavalli-Sforza spend much of the 1964 paper arguing the relevance of the phylogeny, knowing that the audience for the volume consists of systematists.

The paper describes the reconstruction of a tree of characters undergoing a Brownian-motion process from data on the tips of the tree. The Brownian-motion process is intended to approximate changes of p gene frequencies by random genetic drift. Edwards and Cavalli-Sforza give an expression for the likelihood of the change along one branch, where the elapsed time (or branch length) is t and the distance that the Brownian-motion process moves is d, the probability density being calculated from a p-variate normal distribution with independent variates that have an equal rate of change by Brownian motion. The resulting density depends only on the elapsed time, and on the distance d between the gene-frequency coordinates at the start and finish of this change. They also discuss the calculation of a distance from gene-frequency data, using a square-root transformation to make the Brownian-motion process more accurately approximate the genetic drift of gene frequencies.

These would seem to be straightforward, but there is a hidden problem. We do not actually observe the gene-frequency coordinates produced by the Brownian-motion process at the end of each branch. Edwards and Cavalli-Sforza proposed solving this problem by estimating these gene-frequency branch coordinates at the ends of branches, considering them as parameters of the problem. In fact, they were not able to do this. This is not discussed in [27]. The tree that they present for human blood groups is one derived from a different method, a "minimum-evolution" approach – in effect by gene-frequency parsimony.

The paper and the project underlying it were pioneering. For the first time, the inference of the tree was presented as a statistical problem. Maximum likelihood was proposed as one solution, and progress made on the likelihood function. Parsimony was proposed as another solution. I am not aware of any earlier attempt to use either method for the inference of phylogenies.

A similar development is given by Cavalli-Sforza, Barrai, and Edwards in the Cold Harbor Spring Symposium paper ([28]). There (p. 30, this volume), it is acknowledged that "Estimation by maximum likelihood is still causing some difficulties, because of the very high number of parameters to be estimated and some peculiarities of the system." As in the Edwards and Cavalli-Sforza 1964 paper ([27]), gene-frequency parsimony is used for the computations instead. Less time is spent there on transformation of gene frequencies to make them be more accurately approximated by Brownian motion. There is an extensive discussion of the role of migration in villages of the Upper Parma Valley where the authors have collected blood-group data. But no method is given for combining migration with trees in a single analysis. That difficult problem is only now being tackled.

The Difficulty with Their Likelihood

Edwards and Cavalli-Sforza were finding their likelihood method ill-behaved, and they discussed this in several places ([35] and [37]). It is now easy to see the source of the problem. They computed the likelihood as a product of terms, one for each branch in the tree, assuming that they were able to infer the gene-frequency coordinates at the ends of the branches. Taking these gene-frequency coordinates at the start and end of the branch, they could compute a distance d between those points. If the length of the branch in time was t, the term in the likelihood expression for that branch would simply be the density of the normal distribution

$$\frac{1}{\sqrt{2\pi t}}\exp(-d^2/(2t)) \tag{1}$$

so that the corresponding term in the log-likelihood would be

$$-\frac{1}{2}\ln(2\pi t) - d^2/(2t) \tag{2}$$

What Edwards and Cavalli-Sforza found was that, if they made t approach zero, and the inferred coordinates at the end of the branch approached those at the start of the branch, these terms blew up in their faces. Likelihoods went off to infinity. The above expression shows why. If we took the estimates of the coordinates at the two ends of the branch to be the same, then $d = 0$.

Then, as $t \to 0$, the logarithm approaches $-\infty$ and so the whole expression rises to infinity. At first this seems strange. But recall that the term calculates, not a probability, but a probability density. Speaking informally, the probability of the change along that branch is the normal density with mean zero and variance t, multiplied by p infinitesimal quantities dx, where p is the number of dimensions in which the Brownian motion occurs. Once t reaches zero, the probability of no change becomes 1, and is not multiplied by infinitesimals. (Of course, I realize that all this is not strictly speaking mathematically kosher, but abi gezint[1]).

When the length of the branch, in time, reaches zero, the likelihood for the whole tree, which was a density function, is now a density function in p fewer dimensions, thus multiplied by p fewer infinitesimal quantities. So it is thereby infinitely higher. In effect, what we have allowed the problem to do is to merge two points that it is estimating. Since a zero branch length means that the coordinates at the two ends of that branch must be the same, the estimation is estimating one fewer internal point on the tree. In effect, the estimation problem is allowed to make its task simpler, and is infinitely strongly attracted to doing so.

A Solution

In 1967, when I was writing my PhD thesis (Felsenstein 1968), I was inspired by my contact with Edwards and Cavalli-Sforza to take a crack at the problem. Instead of trying to estimate the gene-frequency coordinates at all interior nodes of the tree, I tried to write down the density function for the tip coordinates only. This leaves out estimating the coordinates at the interior nodes; in effect, integrating over all possible interior node coordinates.

If we have a Brownian-motion process, the change in each branch is drawn from a normal distribution. The changes in each branch are independent, and the gene-frequency coordinates at a tip are then the sums of changes in the branches leading up to that tip. The set of the coordinates at the tips are drawn from a multivariate

[1] In my father's first language, Galitzian Yiddish, this means "as long as you're healthy," and is to be said with a resigned and accepting shrug.

normal distribution, whose means are equal, and whose covariances can be easily worked out from the tree topology and branch lengths. The likelihood function can be written straightforwardly as the density function of a multivariate normal distribution, which involves evaluating a matrix inverse.

There also turns out to be a simple linear transformation, which can be read off the tree by a recursive algorithm, which diagonalizes the covariance matrix. This makes the computation of the likelihood linear in the number of species on the tree and linear in the number of dimensions. All of this eliminates the infinite likelihoods and makes the inference relatively well-behaved. I later published this work (Felsenstein 1973).

Distances Too

In 1967, three papers appeared that proposed distance matrix methods for inferring phylogenies. Most widely noticed was the paper by Fitch and Margoliash (1967), which appeared in *Science* at the beginning of the year. Later in the year, two papers appeared – or perhaps it's three. The ambiguity is because Cavalli-Sforza and Edwards published the same paper ([37]) twice that year. It was submitted to *Evolution*, and appeared in their October issue. But it was also delivered at a symposium of the American Society of Human Genetics, which required publication in *American Journal of Human Genetics*. The required mutual permissions were arranged, and it appeared in their May issue, so that the same paper appeared in two journals in the same year. Finally, in the November issue of *Evolution*, a paper by Sandra Horne (1967) appeared. All three of these papers proposed choosing the tree whose predicted distances best fit a table of pairwise distances between species. Fitch and Margoliash's paper proposed a weighted least-squares fit, and the other two papers proposed unweighted least-squares methods.

At the time of the publication of [27] and [28], Cavalli-Sforza and Edwards already had their least-squares method, but they did not mention it in that paper. It was Fitch and Margoliash's paper that caught the attention of molecular evolutionists. Distance matrix methods became an important tool, particularly in the form of the neighbor-joining method of Saitou and Nei (1987), which was faster than least-squares methods, and which found trees that were close approximations to those produced by least squares.

The Number of Trees

In [37], there is also a combinatorical calculation of the number of different rooted bifurcating trees for n labeled tips. This turns out to be the product of odd integers from 1 to $2n - 3$:

$$1 \times 3 \times 5 \times 7 \times \ldots \times (2n-3) = \frac{(2n-3)!}{(n-2)!2^{n-2}} \tag{3}$$

This is a particularly straightforward expression. The quantity was earlier derived by Ernst Schröder (1870) using generating function methods, without an explicit expression for the result. The Cavalli-Sforza and Edwards result is derived much more simply. They go on to give the number of unrooted bifurcating trees and the number of tree topologies. That latter quantity had also been derived by Wedderburn (1922). Cavalli-Sforza and Edwards's 1967 paper ([37]) is the start of the modern literature on enumeration of trees with labeled tips and enumeration of tree topologies.

A More General Formulation

Anthony's exploration of the inference of phylogenies culminates in his 1970 paper ([46]) in Series B of the *Journal of the Royal Statistical Society.* Considering the Brownian-motion process, he backs up one step further, and considers the tree as arising from a pure-birth process (a Yule process) with a parameter λ for the rate of branching. This gives a prior distribution on the topologies of the tree and on its branch length. That, taken together with the likelihood functions for the data, should permit finding the posterior distribution of trees. The existence of λ makes the inference of the tree in effect an empirical Bayesian inference, since it uses a prior that has a parameter.

The result is a general expression for the empirical Bayesian inference of the tree. It abandons any attempt to estimate the coordinates at the interior nodes of the tree, instead integrating over the values at those nodes. There, he makes use of my result that the joint density function of the coordinates at the tips is multivariate normal, with a covariance matrix derived from the tree.

In the end, he retreats from this empirical Bayesian treatment, using an approximate simulation method to estimate the branching rate λ. In those years, we had no way to do integrals needed in a Bayesian treatment. Markov Chain Monte Carlo methods were years in the future. The density function of the branching times for the tree, conditional on the number of tips, would be obtained by Elizabeth Thompson in her PhD work under Anthony's supervision, and published in her Smith's Prize monograph in 1975.

A Remarkable Achievement

Taken together, the papers on phylogenies by Anthony Edwards and Luca Cavalli-Sforza, published over a period of only seven years, introduce, for phylogenies, parsimony methods, distance matrix methods, likelihood methods, and Bayesian inference. That all the major methods now in use for inferring phylogenies should spring from one collaboration is remarkable. Of course, some are easier to invent than others – parsimony apparently also occurred to Joseph Camin and Robert Sokal at nearly the same time; the development of distance methods was undoubtedly influenced by the spread of numerical clustering algorithms in the late 1950s and early 1960s. Thus, it is not surprising that Walter Fitch and Sandra Horne also published least-squares methods, and before Cavalli-Sforza and Edwards did, though Cavalli-Sforza and Edwards had the idea earlier.

In the chapter on history in my 2004 book, *Inferring Phylogenies*, I tried to tell this remarkable story. What is most remarkable about it is that it is so little known. Why? Why did it take until 2015, half a century later, for Anthony to be elected to membership in the Royal Society?

I can think of a number of reasons:

1. Anthony and Luca each moved the main focus of their work elsewhere. Luca has of course become the leader of the effort to use molecular data to illuminate human prehistory. Anthony was inspired by the problem of inferring phylogenies to consider statistical inference more generally, and he became a leading advocate of likelihood methods.
2. The data that Edwards and Cavalli-Sforza used were gene frequencies derived from blood-group loci. Starting in 1966, gene frequencies from electrophoretic variation became available, though without any way of ensuring that alleles in different species were analogous. In the 1970s, the focus shifted strongly toward variation in DNA sequences, starting with restriction-fragment and restriction-site variation. The need for Brownian-motion models largely disappeared.
3. During the 1970s work on evolution of morphological characters became focused on non-statistical parsimony approaches. All of the methods used in that field became attributed to the German entomologist Willi Hennig. Hennig's methods were not numerical or statistical, so in those controversies, Brownian-motion models were not considered.

In the long run, the centrality of their work has been acknowledged:

1. In my own papers on inferring phylogenies using the Brownian-motion model (Felsenstein 1973, 1981), and in review articles that I wrote (Felsenstein 1982, 1983), I discussed Edwards and Cavalli-Sforza's work as well as subsequent work on statistical inference with molecular sequences (Neyman 1971, Kashyap and Subas 1974; Felsenstein 1981). Interest in gene-frequency trees waned; it was the analysis of molecular-sequence data that now interested most readers.

2. In the decade of the 2000s, single nucleotide polymorphism (SNP) sites started to be used intensively for inference of trees of descent of human populations. Likelihood inference for SNP sites is difficult, though progress has been made (RoyChoudhury *et al.* 2008, Bryant *et al.* 2012). As a check on the result, RoyChoudhury *et al.* used a Brownian-motion process as a quick approximation to the full likelihood analysis. Pickrell and Pritchard (2012) also turned to Brownian-motion processes to approximate gene-frequency changes in SNPs in their more widely noticed work using trees with added loops to model admixtures between human populations.
3. Phylogenies became more important as time passed. Until the late 1970s, systematists usually assumed that they would be inferring the classification, but that they would not have good enough information to infer the phylogeny. As molecular evolutionists inferred ever more phylogenies, this information began to be used routinely. By the 1990s, even genomicists were forced to admit that they needed phylogenies and coalescents to compare multiple genomes and investigate comparative genomics. We now realize that phylogenies are not simply an odd side-issue of little interest, but the central structure that is needed to study evolution above the species level.

Gradually, the work on inferring phylogenies using Brownian-motion models is becoming known, and with it, the work of Edwards and Cavalli-Sforza in the 1960s is coming to be appreciated. This is as it should be – it is the major starting point for all work on statistical inference of phylogenies, and that problem in turn is central to evolutionary biology.

References

Bryant, D., R. Bouckaert, J. Felsenstein, N. A. Rosenberg, and A. RoyChoudhury. 2012. Inferring species trees directly from biallelic genetic markers: bypassing gene trees in a full coalescent analysis. *Molecular Biology and Evolution* 29, 1917–1932.

Camin, J. H. and R. R. Sokal. 1965. A method for deducing branching sequences in phylogeny. *Evolution* 19, 311–326.

Felsenstein, J. 1968. *Statistical Inference and the Estimation of Phylogenies*. PhD Thesis, Department of Zoology, University of Chicago.

Felsenstein, J. 1973. Maximum likelihood estimation of evolutionary trees from continuous characters. *American Journal of Human Genetics* 25, 471–492.

Felsenstein, J. 1981. Evolutionary trees from DNA sequences: a maximum likelihood approach. *Journal of Molecular Evolution* 17, 368–376.

Felsenstein, J. 1982. Numerical methods for inferring evolutionary trees. *Quarterly Review of Biology* 57, 379–404.

Felsenstein, J. 1983. Statistical inference of phylogenies. *Journal of the Royal Statistical Society. Series A (General)* 146, 246–272.

Felsenstein, J. 2004. *Inferring Phylogenies*. Sunderland, MA: Sinauer Associates.

Fitch, W. M. and E. Margoliash. 1967. Construction of phylogenetic trees. *Science* 155, 279–284.

Horne, S. L. 1967. Comparisons of primate catalase tryptic peptides and implications for the study of molecular evolution. *Evolution* 21, 771–786.

Kashyap, R. L. and S. Subas. 1974. Statistical estimation of parameters in a phylogenetic tree using a dynamic model of the substitutional process. *Journal of Theoretical Biology* 47, 75–101.

Neyman, J. 1971. Molecular studies of evolution: a source of novel statistical problems. In *Statistical Decision Theory and Related Topics*, eds. S. S. Gupta and J. Yackel, New York: Academic Press, pp. 1–27.

Pickrell, J. K. and J. K. Pritchard. 2012. Inference of population splits and mixtures from genome-wide gene frequency data. *PLoS Genetics* 8, e1002967.

RoyChoudhury, A., J. Felsenstein, and E. A. Thompson. 2008. A two-stage pruning algorithm for likelihood computation for a population tree. *Genetics* 180, 1095–1105.

Saitou, N. and M. Nei. 1987. The neighbor–joining method: A new method for reconstructing phylogenetic trees. *Molecular Biology and Evolution* 4, 406–425.

Schröder, E. 1870. Vier combinatorishe Probleme. *Zeitschrift für Mathematik und Physik* 15, 361–376.

Sokal, R. R. and P. H. A. Sneath. 1963. *Principles of Numerical Taxonomy.* San Francisco, CA: W. H. Freeman.

Thompson, E. A., 1975. *Human Evolutionary Trees*. Cambridge: Cambridge University Press.

Wedderburn, J. H. M. 1922. The functional equation $g(x^2) = 2ax + [g(x)]^2$. *Annals of Mathematics* 24, 121–140.

The Historic Split and Merger of Gene Trees and Species Trees

RASMUS NIELSEN

The early work of Edwards and Cavalli-Sforza on inferences of evolutionary trees from genetic data used models of population-level processes to infer trees. This work not only introduced the use of rigorous statistical methods for estimating trees, it also demonstrated how population-level processes could be integrated into the tree estimation procedures. Such integration of population genetics with phylogenetics is a topic that recently has received a tremendous amount of attention because it is key in the phylogenetic interpretation of large genomic datasets. However, between the time of the work of Edwards and Cavalli-Sforza in the 1960s, and recent progress on genome-wide data, phylogenetic research has only sporadically entertained the idea that population genetics might form a foundation for phylogenetic research. During much of the twentieth century, phylogenetic theory and population-genetic theory evolved primarily as separate fields. In this short commentary I explore this issue and discuss some of the possible causes and consequences of the split between phylogenetic and population-genetic theory.

Box 1: **Definitions**

This box attempts to provide short definitions for some of the vocabulary used in this paper. The reader should be aware that some of these definitions could be controversial and that proper definitions require substantially more space than provided here.

Brownian motion: a specific stochastic process that models the change in a random variable through time using Gaussian transition probabilities.

Cladistics: originally, a belief set in taxonomy and phylogenetics asserting that taxonomy should follow phylogeny based on identification of monophyletic groups using shared derived characters. In modern praxis, synonymous with the idea that the best phylogeny is the one requiring the fewest evolutionary changes (the parsimony principle in phylogenetics).

Box 1: **(Cont.)**

Coalescence theory: the study of gene trees in populations.

Coalescence tree: a graph depicting the structure of the genealogical relationship among multiple gene copies of a non-recombining marker (e.g. mitochondrial DNA or a single nucleotide position in a nuclear genome). In most cases, used synonymously with gene tree or gene genealogy.

Gaussian distribution: a generalization of the normal distribution, allowing the random variable to be vector-valued, but identical to the normal distribution for a scalar random variable.

Phylogenetics: the study and estimation of phylogenies and the use of phylogenies for evolutionary inferences.

Phylogeny: an evolutionary tree, typically intended to represent a species tree.

Polygenic traits: traits controlled by multiple genes.

Population genetics: the study of genetic variation at the population or species level, including processes of mutation, recombination, genetic drift, natural selection, migration between populations, and speciation. Models in population genetics often focus on describing changes in allele frequencies or distributions of coalescence trees.

Quantitative genetics: the use of statistical models, often based on Gaussian distributions, for understanding the effect of genes and environments on phenotypes, including segregation of alleles in pedigrees and in populations, inbreeding, mutation, recombination, and natural selection.

Species tree: the unique tree describing the hierarchical evolutionary relationship among multiple species. It is often assumed that a single true species tree exists, but introgression between lineages may, for some organisms, occasionally imply that the graph representing the evolutionary relationship between taxonomic units contains cycles (the tree is 'reticulate'). Incomplete lineage sorting, or gene-flow in a transient period after the initial split between nascent species, does not invalidate the idea of a true species.

The Split

A taxonomy of the theoretical machinery applied in evolutionary analyses might appropriately divide the field into three sub-fields: quantitative genetics, population genetics, and phylogenetics (see Box 1 for definitions). The split between population genetics and quantitative genetics can be traced back to the conflict between Mendelians and Biometricians (e.g. Provine 1971), but it has largely faded away today as the theory of individual genes in population genetics extends naturally to the analysis of polygenic traits to form a foundation for an evolutionary theory of quantitative genetics (e.g. Wright 1968, Lande 1975). Phylogenetics, however, has mostly been considered a separate field with its own theory, decoupled from population-genetic theory. This may be historically contingent on population genetics emerging from the field of genetics, while phylogenetics had its origins in taxonomic research. However, the decoupling between population genetics and phylogenetics was arguably enhanced when the cladistic (see Box 1) movement emerged, perhaps best illustrated by the claim of the so-called 'pattern cladists' that cladistic inferences can, and should, proceed without any evolutionary assumptions (e.g. Brower 2000). While many professed cladists may not subscribe to this view, a central tenet of cladism is nonetheless that phylogenetic inference should proceed only with an appeal to a particular interpretation of the parsimony principle and should not be based on explicit models of evolutionary processes (e.g. Farris 1983). Even so-called 'process-cladists' (Ereshefsky 2001), who argue in favor of the recognition of evolution as the generative process for phylogenies, have traditionally argued against the direct incorporation of these processes in the inference framework beyond just recognising the parsimony principle (e.g. Farris 1983). This dictum effectively divorced phylogenetic inference from population-genetic theory and forced the two fields to evolve more or less separately from each other for decades. Phylogenetic research in the 1970s, 1980s, and 1990s almost exclusively proceeded without taking population-level processes into account. Only as an afterthought, subsequent to the discovery of coalescence theory (Kingman 1982, Hudson 1983), would phylogenetic researchers occasionally appeal to population-genetic thinking and worry about differences between gene trees and species trees.

Coalescence Theory

The seminal papers by Kingman (1982) and Hudson (1983) revolutionised population genetics by the discovery of coalescence theory. The central idea in coalescence theory is that the genealogical relationship among copies of a gene sampled from different individuals, to a close approximation, can be described as a strictly bifurcating tree, called a 'gene tree', 'gene genealogy', or 'coalescence tree', similar to a phylogeny. This tree is random due to two factors: (1) the randomness associated with matings and offspring distribution, i.e. the randomness of the pedigree and (2) the randomness of alleles as they segregate within the pedigree. The objective of coalescence theory is to characterise the random distribution of coalescence trees under various historical and demographic processes. One of the main lessons from coalescence theory is that the shape of the gene trees may vary tremendously due to the aforementioned random processes. If we examine one locus of the genome in a population we may observe a very different gene tree than if we had examined another locus.

Most phylogenetic studies using DNA have historically proceeded by estimating a tree from a single gene or genomic region. The trees obtained in such studies are gene trees that may or may not reflect the true species phylogeny, when such a phylogeny exists (see Box 1). For closely related species, the randomness of the coalescence process may lead to discrepancies between estimated gene trees and the tree representing the species phylogeny, a phenomenon known as *incomplete lineage sorting*. Estimation of phylogeny using single genes, without accounting for the possibility of lineage sorting, may lead to inflated confidence in a wrong phylogeny (e.g. Pamilo and Nei 1988). Due to lineage sorting, strong support for a gene tree does not imply strong support for a species tree. Coalescence theory provides the natural theoretical foundation for phylogenetic inferences, because it describes the structure of individual gene trees and directly models the randomness of gene trees and the process of lineage sorting (e.g. Maddison and Knowles 2005). Uncertainty, due to lineage sorting, can therefore be incorporated into phylogenetic inference and multiple individual gene trees can, in theory, be combined using rigorous methods. Now, in the age of genomics, where phylogenetic studies usually involve many different loci, a major focus of theoretical phylogenetic research is to devise such methods for estimating the true species phylogeny based on data from multiple different loci while taking incomplete lineage sorting into account (e.g. Liu *et al.* 2009; Heled and Drummond 2010; Allman *et al.* 2010; Larget *et al.* 2010; Mirarab *et al.* 2014a, b). Coalescence theory forms the theoretical foundation for these methods. However, phylogenetic inference based on DNA data without coalescence theory is an amputated science, prone to generating false inferences.

Phylogeography

In the 1990s, the field of *phylogeography* emerged, most famously represented by Avise's book *Phylogeography: The History and Formation of Species* (Avise 2000). The aim of phylogeography was to understand population- or species-level processes by considering the geographic distribution of the sampled individuals in the context of a tree. The major tool was the gene tree, estimated as a phylogeny, and superimposed on a geographical map to help develop narratives regarding historical population-level processes. During much of the 1990s and 2000s, this approach strongly influenced many evolutionary biologists and dominated the pages of journals such as *Molecular Ecology*. Curiously, it was mostly detached from population-genetic theory. Very few efforts were made to integrate phylogeographic inferences into population-genetic theory, even though coalescence theory at this point was well-developed and familiar to most researchers. The central object of phylogeography, the phylogeny, was estimated from the data and considered fixed, and the randomness of the processes generating the gene tree was rarely taken into account. Some efforts were made to interpret phylogeographic patterns in the context of population-genetic theory (e.g. Templeton 2004); however, they typically failed to acknowledge the randomness of the coalescence process, and focused on singular estimated trees (see, e.g., Nielsen and Beaumont 2009).

Phylogeography focused attention on gene trees in populations, and should have helped the needed integration of the fields of population genetics and phylogenetics. However, it arguably had the opposite effect. It introduced phylogenetic thinking into the analysis of population-level data, but in doing so it also largely ignored one of the main results of population-genetic theory: the randomness of the gene tree. The idea that a single tree should be the object of focus in evolutionary biology, and that the estimation of this tree should proceed without reference to population-level processes, was, by the late 1990s, so ingrained in evolutionary biologists that it not only dominated the thinking of researchers in phylogenetics but it also led to a new, related, theory-free framework for interpreting population-level data. The failure of phylogeography to help span the divide between population-genetic and phylogenetic theory is a testament to the hegemonic strength of the model-free ideology that was dominating phylogenetics (e.g. Farris 1983).

The Merger

What allowed researchers to ignore population-level processes in the estimation of phylogenetic trees through the last decades of the twentieth century was that phylogenies typically were estimated based on just a single non-recombining gene, e.g. mitochondrial DNA in animals, and ribulose bisphosphate carboxylase (*rbcL*) in plants. In most studies, there was no evidence to suggest incompatibilities between different genetic markers. After the emergence of large datasets with multiple genes from each species (e.g. Mirarab 2014a), the dichotomy between phylogenetics and population genetics has slowly been vanishing, in part due to the necessity of reconciling estimates of trees obtained from different genes in the genome. Phylogenetic researchers are increasingly realising that gene trees are different from species trees; that gene trees may not reflect species trees due to the stochasticity of the coalescence process (e.g. lineage sorting); and that population genetics in the form of coalescence theory provides a theoretical foundation for estimation of species trees from DNA sequences, by providing a theory that connects gene trees with species trees. This change in the field is illustrated by the recent proliferation of new coalescence-based methods for estimating species trees from sets of gene trees (e.g. Liu *et al.* 2009; Heled and Drummond 2010; Allman *et al.* 2010; Larget *et al.* 2010; Mirarab 2014a, b). At the same time, there appears to be renewed population-genetic interest in estimating population trees as a means of illustrating genetic relationships between populations (e.g. Pickrell and Pritchard 2012). Population trees are not gene trees, but trees that reflect the diversification history of populations, similar to a species phylogeny representing the diversification process of species. The fields of phylogenetics and population genetics, therefore, appear to be merging on a common set of methods based on coalescence theory.

Edwards and Cavalli-Sforza

Today, the fields of phylogenetics and population genetics are largely merged, at least in the mind of theoreticians. It is interesting that it should take decades for this merger to happen given that early work in phylogenetics and population genetics clearly acknowledged the commonality of the two fields. The basic ideas of modern phylogenetic inference were established 40–50 years ago in papers by Edwards and Cavalli-Sforza ([25], [27], [28], [29]), Edwards ([46]), Thompson (1975), and Felsenstein (1973; 1981; 1985). In this early work, there was no dichotomy between population genetics and phylogenetics. Building on their previous paper ([27]), Cavalli-Sforza and Edwards outlined, in 1967, a theory for the estimation of population trees using allele frequencies ([37]). At this point in time there were no DNA sequence data available, but allele frequencies could be estimated for blood groups and various other genetic markers. The central idea of [37] was that the changes in allele frequencies could be used to estimate a tree representing the population history. They used Brownian motion to model the process of change in allele frequencies and developed likelihood methods for estimating the population tree based on this assumption. This is a natural framework to use because the resulting sampling distribution is Gaussian. It is an approach remarkably similar to modern, state-of-the-art approaches for analysing population-genetic data, which have been developed over the past few years. For example, the popular *Treemix* program (Pickrell and Pritchard 2012) is based on a Brownian-motion approximation similar to the one used by Cavalli-Sforza and Edwards but extended to allow admixture between populations. Related Gaussian models have, for example, also been used (see, e.g., Coop *et al.* 2010 and Bradburd *et al.* 2013) to model spatial variation. The Gaussian model has limitations; in particular, it does not model the process of allele-frequency change very well when the allele frequency is near zero or one. However, it is computationally simple, thereby providing a great advantage over many other models when analysing datasets of hundreds of thousands of variable nucleotide sites. Interestingly, the use of Gaussian distributions to model the allele frequencies when comparing multiple populations more or less disappeared from the literature for almost 40 years. Also, many of the original ideas of

[37] have yet to diffuse into the modern literature on estimating population trees. For example, [37] considered transformations of the allele frequencies that might improve on the Brownian-motion approximation to the genetic-drift process. None of these ideas seems to have entered the modern literature, and they probably deserve at least to be tested.

The work of Cavalli-Sforza and Edwards was clearly ahead of its time. It addressed questions regarding how to estimate population-level phylogenies that only otherwise have been mostly pursued the last few years, after the emergence of genome-wide data. Other ideas from Edwards's work, particularly the related ideas of using maximum likelihood in phylogenetics, were pursued by a new generation of researchers, most influentially Felsenstein (e.g. Felsenstein 1981). Felsentein was also one of the few researchers adopting Cavalli-Sforza and Edwards's idea of estimating population (or species) trees (Felsenstein 1985). However, this basic idea was otherwise mostly forgotten, or ignored, for almost 40 years.

We cannot excuse the field by claiming ignorance. The work of Cavalli-Sforza and Edwards was familiar to evolutionary biologists because of its importance for understanding human evolutionary ancestry. The efforts of Felsenstein and others, after the discovery of coalescence theory, on moving the field towards recognising the stochastic variability of trees and the lack of identity between gene trees and species trees, may have had the effect of introducing the idea of incomplete lineage sorting as a method for explaining otherwise inexplicable results in phylogenetics. However, these efforts did not, in the larger community, lead to a focus towards the estimation of species trees and towards a merger of population genetics and phylogenetics. Arguably, the ideological sway of cladistics was too strong for this to happen. We had to wait 40 years before phylogenetic and population-genetic theory finally could be merged again, driven by the emergence of genome-wide data.

Do the Best Ideas Prevail in Science?

Do the best ideas eventually prevail in science when multiple ideas are competing? Can we avoid spending our time arguing about different methodologies, as the best method will ultimately prove to be most useful and therefore win? The answer to these questions appears in this case to be 'yes'. The cladistic ideology based on process-agnostic phylogenetic inference has, at least in its most dogmatic form, largely been replaced by more pragmatic views that favor the best statistical method for estimating trees. Likewise, there is full awareness in the field of evolutionary biology today that phylogenies estimated from single gene trees may not be representative, and that methods which integrate data from multiple loci using coalescence theory should be used for phylogenetic inference. The ideas of Edwards and Cavalli-Sforza are now dominating our view of both phylogenetic and population-genetic inferences. It might be an interesting study in the sociology of science to determine why it took so long a time for this to happen.

References

Allman, E. S., Degnan, J. H., and Rhodes, J. A. 2010. Identifying the rooted species tree from the distribution of unrooted gene trees under the coalescent. *Journal of Mathematical Biology* 62, 833–862.

Avise, J. 2000. *Phylogeography: The History and Formation of Species*. Cambridge, MA: Harvard University Press.

Bradburd, G. S., Ralph, P. L., and Coop, G. M. 2013. Disentangling the effects of geographic and ecological isolation on genetic differentiation. *Evolution* 67, 3258–3273.

Brower, A. V. Z. 2000. Evolution is not a necessary assumption of cladistics. *Cladistics* 16, 143–154.

Coop, G., Witonsky, D., Rienzo, A. D., and Pritchard, J. K. 2010. Using environmental correlations to identify loci underlying local adaptation. *Genetics* 185, 1411–1423.

Edwards S. V. and Beerli, P. 2000. Perspective: Gene divergence, population divergence, and the variance in coalescence time in phylogeography studies. *Evolution* 54, 1839–1854.

Ereshefsky, M. 2001. *The Poverty of the Linnaean Hierarchy: A Philosophical Study of Biological Taxonomy.* Cambridge: Cambridge University Press.

Farris, J. S. 1983. The logical basis of phylogenetic analysis. In N. I. Platnick and V. A. Funk, *Advances in Cladistics*, New York: Columbia University Press, volume 2, 7–36.

Felsenstein, J. 1973. Maximum-likelihood estimation of evolutionary trees from continuous characters. *American Journal of Human Genetics* 25, 471–492.

Felsenstein, J. 1981. Evolutionary trees from DNA sequences: a maximum likelihood approach. *Journal of Molecular Evolution* 17, 368–376.

Felsenstein, J. 1985. Phylogenies from gene frequencies: a statistical problem. *Systematic Biology* 34, 300–311.

Heled, J. and Drummond, A. J. 2010. Bayesian inference of species trees from multilocus data. *Molecular Biology and Evolution* 27, 570–580.

Hudson, R. R. 1983. Properties of a neutral allele model with intragenic recombination. *Theoretical Population Biology* 23, 183–201.

Kingman, J. F. C. 1982. The coalescent. *Stochastic Processes and their Applications* 13, 235–248.

Lande, R. 1975. The maintenance of genetic variability of mutation in a polygenic character with linked loci. *Genetical Research* 26, 221–235.

Larget, B. R., Kotha, S. K., Dewey, C. N., and Ané, C. 2010. BUCKy: gene tree/species tree reconciliation with the Bayesian concordance analysis. *Bioinformatics* 26, 2910–2911.

Liu, L., Yiu, L., Kubatko, L., Pearl, D. K., and Edwards, S. V. 2009. Coalescent methods for estimating phylogenetic trees. *Molecular Phylogenetics and Evolution* 53, 320–328.

Maddison, W. P. and Knowles, L. L. 2005. Inferring phylogeny despite incomplete lineage sorting. *Systematic Biology* 55, 21–30.

Mirarab, S. S. Bayzid, B. Boussau, and T. Warnow. 2014a. Statistical binning enables an accurate coalescent-based estimation of the avian tree. *Science* 346, doi: 10.1126/science.1250463.

Mirarab, S., R. Reaz, Md. S. Bayzid, T. Zimmerman, M. S. Swenson, and T. Warnow. 2014b. ASTRAL: genome-scale coalescent-based species tree estimation. *Bioinformatics* 30, i541–i548.

Nielsen, R. and M. A. Beaumont. 2009. Statistical inferences in phylogeography. *Molecular Ecology* 18, 1034–1047.

Pamilo, P. and Nei, M. 1988. Relationships between gene trees and species trees. *Molecular Biology and Evolution* 5, 568–583.

Pickrell, J. K. and J. K. Pritchard. 2012. Inference of population splits and mixtures from genome-wide allele frequency data. *PLoS Genetics* 8, e1002967.

Provine, W. B. 1971. *The Origins of Theoretical Population Genetics*. Chicago, IL: University of Chicago Press.

Templeton, A. R. 2004. Statistical phylogeography: methods of evaluating and minimizing inference errors. *Molecular Ecology* 13, 789–809.

Thompson, E. A. 1975. *Human Evolutionary Trees*. Cambridge: Cambridge University Press.

Wright, S. 1968. *Evolution and the Genetics of Populations, Volume 1: Genetic and Biometric Foundations*. Chicago, IL: University of Chicago Press.

Likelihood and the Comparative Method in Evolutionary Biology

MARK PAGEL

When Paul Harvey and I sat down in 1990 to write our book *The Comparative Method in Evolutionary Biology*, we had two aims. One was to convince comparative biologists of the importance of analysing their data properly. Comparative methods are used in evolutionary biology to analyse trends among biological species. Biological evolution, as Darwin realised, is a branching process and this means that biological species normally cannot be treated as independent draws from a common underlying population – two closely related species will tend to be more similar than two distantly related ones. The challenge for comparative biology is to produce statistical methods that properly account for the non-independence in the data, and so outlining a variety of solutions became our second aim.

We wished to provide comparative statistical methods that were usable given the state of play in the field. At that time, the state of play was nothing like it is today. Phylogenies represent our best guesses as to the historical branching processes that have given rise to a group of species. It is hard to appreciate now, given the sophistication of contemporary phylogenetic inference, but there were very few good-quality phylogenies around in the late 1980s. Indeed, many comparative biologists were still using Linnean taxonomic classifications as stand-ins for the true phylogeny, or they simply ignored the problem of non-independence altogether. Those phylogenies that were available were nearly all cobbled together from parsimony methods inspired by cladistics and applied to discrete morphological characters. There was much uncertainty associated with them, they did not provide information about timings or amounts and rates of change, and the morphological traits that were used to construct the trees can often arise convergently, leading to taxa being incorrectly placed on the tree.

On the other hand, Kary Mullis's polymerase chain reaction methods, which made large-scale sequencing faster and easier, were just catching on (Mullis and Faloona 1987). This meant that *GenBank*, at that time a comparatively small repository, was poised to become a vast reserve of gene sequences, presaging, among other effects, a new era of phylogenetics. Anticipating that comparative biologists would increasingly make use of phylogenies derived from gene-sequence data, we decided that our book also needed to include a précis on phylogenetic inference. And here we benefited from the work that Anthony Edwards had, with Luca Cavalli-Sforza, done 25 years before us.

In the early 1960s the two of them were trying to make phylogenetic trees of human populations based on the frequencies of blood-group alleles. Edwards thought up a minimum-evolution method that worked directly on the gene frequencies, while Cavalli-Sforza arrived at a least-squares method that took into account population sizes and drift. Then, as Felsenstein (2004) recounts, puzzled at having arrived at two different solutions, Edwards and Cavalli-Sforza tried to reconcile their two methods by adopting a third approach, one based on likelihood. Their hope at reconciliation was, however, dashed when they discovered that the likelihood solution differed from both of their other methods.

In a 1964 paper ([27]), Edwards and Cavalli-Sforza described Edwards's minimum-evolution method along with their new likelihood approach. Their paper also included a tree derived from the minimum-evolution method that, when placed onto a map of the world, fitted well with accounts at that time of probable routes of human migration out of Africa and, eventually, around the world. The 1964 paper is noteworthy for introducing, in a single contribution, both parsimony and likelihood approaches to inferring phylogenies and shows that phylogenetic inference could be treated as a statistical problem. Harvey and I later reprinted Edwards and Cavalli-Sforza's minimum-evolution world-map tree in the chapter on phylogenetic inference in our book (Harvey and Pagel 1991, p. 76).

In retrospect, Edwards's use of likelihood is not surprising – he was, after all, along with Cavalli-Sforza, a student of Fisher's. But it was, along with Edwards's later advocacy of likelihood in his book of the same name (Edwards 1972), compelling. Likelihood approaches treat the observed data as fixed and known, rather than a random sample from some larger population. They then presume that the data arose as the outcome of a process that is described by a model of evolution whose parameter values can be estimated from the data. The maximum-likelihood solution is the set of parameter values that make the observed outcomes most probable. It is a stance that fits nicely with biological intuition: effectively, any outcome is possible but some are more likely than others given our beliefs about the way evolution proceeds.

Harvey and I adopted this perspective in our treatment of comparative methods, developing several new likelihood-based approaches of our own, and describing others that had been recently published (Felsenstein 1985) or were at that time still awaiting publication (Lynch 1991). In most cases, the methods we introduced replaced methods based on parsimony, simple counting statistics, or randomisation tests. They were designed to take account of the imperfect nature of the 'trees' to which they might be applied, but at the same time be able to accommodate a 'perfect' tree should one ever come along.

The statistical-likelihood approach was particularly useful for historical studies. The earlier parsimony and counting methods frequently adopted one or both of two approaches. First, they would reconstruct, typically using parsimony, the ancestral character states of traits on a phylogeny. Next, they would treat these reconstructions as data points, known as if without error. Not only does this latter practice grant to an inference the same epistemological status as an observation, it violates what could playfully be called a statistician's first law of thermodynamics: there cannot be more information in these inferred ancestral states than already exists in the combination of the tree, the model of evolution, and the observed data, and yet somehow additional degrees of freedom were conjured up.

The earlier methods also frequently used randomisation tests in which the observed data were reassigned at random to the tips of the phylogeny. Randomisations can, of course, be powerful and useful, but it is often lost in the context of comparative methods that the various randomisations of the data are not all equally likely under reasonable models of how these data might arise – and yet, they are treated as such in the randomisation method.

The likelihood approach avoids both of these problem areas. It is straightforward to sum the likelihood over all possible sets of ancestral character states, thereby avoiding ever reconstructing them at all, or worse, conditioning the analyses upon one set of them. And, by exploring the range of parameters – not the range of outcomes – that are consistent with the observed data, the likelihood method avoids the inconsistency implicit in the randomisation procedures.

Following our book, in later work on statistical models for discrete and continuous characters evolving on phylogenetic trees, and including analyses of comparative linguistics datasets (e.g. Pagel 1994, 1999a, b, 2000), I would adopt this same likelihood approach that Edwards had done so much to kick-start. Some of these, and later methods, I presented at workshops in the late 1990s and early 2000s at the Newton Institute in Cambridge, where, delightfully, Anthony would often be in attendance. Today, likelihood methods are more widespread and popular than ever, being well represented in both phylogenetic inference and in comparative methods.

References

Edwards, A. W. F. 1972. *Likelihood*. Cambridge: Cambridge University Press.

Felsenstein, J. 1985. Phylogenies and the comparative method. *American Naturalist* 125, 1–15.

Felsenstein, J. 2004. *Inferring Phylogenies*. Sunderland, MA: Sinauer Associates.

Harvey, P. H. and Pagel, M. D. 1991. *The Comparative Method in Evolutionary Biology*. Oxford: Oxford University Press.

Lynch, M. 1991. Methods for the analysis of comnparative data in evolutionary biology. *Evolution* 45, 1065–1080.

Mullis, K. B. and Faloona F. A. 1987. Specific synthesis of DNA in vitro via a polymerase-catalyzed chain reaction. *Methods in Enzymology* 155, 335–350.

Pagel, M. 1994. Detecting correlated evolution on phylogenies: a general method for the comparative analysis of discrete characters. *Proceedings of the Royal Society B* 255, 37–45.

Pagel, M. 1999a Inferring the historical patterns of biological evolution. *Nature* 401, 877–884.

Pagel, M. 1999b. The maximum likelihood approach to reconstructing ancestral character states of discrete characters on phylogenies. *Systematic Biology* 48, 612–622.

Pagel, M. 2000. Maximum likelihood models for glottochronology and for reconstructing linguistic phylogenies. In *Time-Depth in Historical Linguistics*, eds. C. Renfrew, A. MacMahon, L. Trask, Cambridge: The McDonald Institute of Archaeology, 189–207.

Likelihood Inference in Models for the Genetic Diversity of Populations

ELIZABETH THOMPSON

In 1970, I completed the Cambridge University Mathematics Tripos and entered the one-year graduate program, the Diploma in Mathematical Statistics. I requested "genetics" as the scientific area for my applied project diploma, and Dr. Edwards became my project supervisor. The project he assigned was to investigate how variation in rates of population increase could impact world frequencies of alleles. This not only led to my first published paper but, more importantly, taught me the significance of stochastic demographic forces in shaping population genetic diversity.

By 1971, I had no doubt that not only was statistical genetics the most ideal combination of my mathematical and scientific interests, but also that I wanted to continue to work with Dr. Edwards, and he became my PhD supervisor. Only then, did I become "Elizabeth" instead of "Miss Thompson," and at the same time he became "Anthony": this was not only a measure of his innate courtesy and sometimes intimidating formality, but also of the intellectual equality he accorded me in our meetings. Invariably generous with ideas and literature references, and always ready to listen as well as to talk, he was for me the ideal research supervisor. He was generous also with unpublished source materials; he provided me not only with Gomberg's unpublished paper on a Bayesian approach to evolutionary trees (see below) but also with C. W. Cotterman's then unpublished thesis (Cotterman 1940) on gene identity by descent and the resulting probabilities of data on related individuals. These two alone exemplify the breadth of both statistical genetics questions and statistical thinking that I was encouraged to pursue.

The year 1970 was an exciting time to enter the world of statistical population genetics. Data were becoming available and computations were becoming feasible. While R. A. Fisher and others had, over the preceding 40 years, pioneered the statistical analysis of genetic data on individuals, population genetics was still primarily the target of theoretical models rather than statistical inference. The joint work of Cavalli-Sforza and Edwards ([25], [27], [28], [29], [30], [35], [37]) developed statistical models for the analysis of population allele-frequency data, in order to make inferences about human population history. In these papers, they developed three methods for estimating the evolutionary tree of human populations: *minimum evolution, least-squares additive trees*, and *maximum likelihood*. While the first two gathered momentum in phylogenetics literature and in analyses of molecular evolution, Edwards makes clear both then ([46]) and more recently ([141], [146]) that they were to be regarded only as approximations to a likelihood-based inference. However, in these earlier papers, the problems inherent in the model-based likelihood remained unsolved.

Edwards's paper presented to the Royal Statistical Society in 1969, and published in 1970 ([46]), is a landmark in several respects. First, the earlier problem of "singularities" in the likelihood surface is resolved by the recognition that some of the "parameters" are random variables having probability distributions under the proposed stochastic model. These are the internal nodes of the phylogenetic tree, other than the root. The probability model for the node positions in the genetic space based on allele frequencies is provided by the model of Brownian motion. If a Yule process for branching is adopted, the times of split likewise are random variables. Edwards and the distinguished discussants of the paper struggled with an appropriate terminology, but we would now have no difficulty in recognising these as *latent variables*. Such unobservable (or "missing") variables are inherent in a wide variety of structured stochastic models used today in many scientific areas. It is this latent-variable formulation in [46] that led me to an iterative expectation-maximization ("EM") algorithm for maximum-likelihood estimation of evolutionary trees (Thompson 1975), solving some of the remaining open questions in [46]. "EM" is here in quotes since, as is the case for other examples in several scientific areas, the algorithm predated the formalization of EM and likelihood inference in missing-data problems (Dempster *et al.* 1977).

Several of the discussants of [46] suggested a Bayesian approach, and indeed, as noted by Cavalli-Sforza (p. 126, this volume), Gomberg (unpublished) had already adopted this approach. Edwards, however, was adamant that no probability distribution could be postulated for the root of the tree or the variance of the Brownian motion. Here, as elsewhere, he rejected a Bayesian approach.

My graduate student years with Dr. Edwards, 1970–1974, were also a time of intense development of his philosophy of statistical inference. Although, as my research supervisor, he never pushed his own ideas, it is surely his influence that has resulted in my likelihood-centered view of scientific inference. The first edition of his book, *Likelihood*, was published in 1972, and he summarizes his view in the brief paper, [56]. Likelihood is presented as the cornerstone of inductive scientific inference: and Fisher is quoted (p. 153, this volume): "[The likelihood] is a relative probability only, suitable to compare point with point, but incapable of being interpreted as a probability distribution, or of giving any estimate of absolute probability." The Neyman–Pearson frequentist approach is strongly rejected as a procedure for providing quantitative measures of credibility of explanations of scientific results.

If both Neyman–Pearson and Bayesian approaches are rejected, what remains? For a while, Edwards struggled with fiducial probability ([56]), but came back to his *likelihood axiom* that the likelihood surface alone provided measures of support for alternative hypotheses. In 1973, Edwards had a sabbatical term at the University of Aarhus, and he accorded me the opportunity to travel there with him and his family. Scientifically most memorable was the initial workshop on foundations of inference, the notes from which I still possess. Here, I was introduced to Professor Barndorff-Nielsen, and others thinking deeply and mathematically about scientific inference, and I began to apply their thinking in my own work.

On the genetics front, after completing my PhD in 1974, and again at the suggestion of Dr. Edwards, I was a postdoctoral fellow for one year in the Department of Genetics at Stanford University, under the supervision of Professor Cavalli-Sforza. Here, I learned much more of genetics, and of models that can be applied to the analysis of the genetic variation among populations. This was a time of some controversy concerning the extent of human genetic differentiation. In a later paper ([192]), published in 2003, which as much as any shows Edwards as the clear thinker, expositor, and teacher that he is, Edwards says, "This article could, and perhaps should, have been written soon after 1974" (p. 253, this volume). This brief paper makes clear the nature of the long-standing misunderstanding of the power of multidimensional data in structured stochastic systems. As genetic technologies have multiplied, and the quantities of genetic data have vastly increased, we have become familiar with the way in which data on huge numbers of allelic (Rosenberg *et al.* 2002) and, most recently, local haplotypic (Leslie *et al.* 2015) variants can reveal details of human demographic history. Some discussants of the 1970 paper ([46]) were skeptical that inferences of history could be made from a single snapshot of current population allele frequencies. Cavalli-Sforza and Edwards ([25]) recognized this potential in 1963, and indeed produced such an estimate, even with the then many times more limited genetic data of a few blood-group population allele frequencies.

As is clear from a number of his papers, e.g. [46], Edwards was an early user of Monte Carlo methods. As early as 1966, he was estimating inbreeding coefficients in a genetic isolate by simulating the descent of genes down pedigrees (a process known then and now as "gene-dropping"). This interest in simulation from the clearly defined models of genetic inheritance is closely tied to Edwards's view of inference ([141]). To paraphrase [141]: you can simulate a process if and only if you have a probability model for it, and if you have a probability model, then likelihood is the logical way to evaluate that model. In the 1990s, Markov chain Monte Carlo (MCMC) methods became ever more widely used to analyze data on complex stochastic systems. MCMC is widely viewed as a tool for Bayesian methods, sampling from the posterior distributions of parameters (Gelfand and Smith 1990). However, MCMC methods can also be used to sample from the conditional distributions of latent variables, conditional on data and on fixed parameter values. Monte Carlo estimates of likelihood surfaces based on MCMC realization of latent variables have been widely used in statistical genetics. In the interim, advances in computers and in algorithms led to ways to compute inbreeding coefficients exactly. However, 50 years later, the huge multigeneration pedigrees of some (non-human) species or breeds (for example, more than 50,000 dogs in a pedigree of 100 generations), again make the simple expedient of dropping genes down pedigrees the most effective way to estimate kinship and inbreeding coefficients.

Finally, we return again to Edwards ([146]). This paper overviews the history of his work with Cavalli-Sforza in developing methods for analysis of the genetic differences and similarities among human populations. He makes again the clear distinction between a *principle* of parsimony or minimum evolution (which he roundly rejects) and the *methods* that may minimize an objective function to provide a hoped-for approximation to the likelihood inference under a probability model. In so doing, he makes also more general and fundamental points about variation and uncertainty, and the role of likelihood inference in quantifying this uncertainty.

He first describes the rise of probabilistic inference 200 years ago. The failure of a hypothesis to account exactly for observational data was no longer seen as a fault of either the hypothesis or the data but as a measurable reflection of the random component in observations.

To the deductive reasoning of computer scientists and mathematicians, non-repeatability of inference results is often considered evidence of "error." However, Edwards, in [146], makes the point that differences are not aberrations but are *expected*. He develops the example of the positions in genetic space of two populations, for simplicity an ancestor and descendant. Conditional on these two positions, the expected position of an intermediate ancestor is on the straight line joining them, so that in this sense the most probable path of evolution will be the straight line. However, if we ask: "At a particular intermediate time, what is the most probable distance of the intermediate point from the straight line?" the answer will not be zero.

The general message of [146] is that, under any probability model, repeated realizations exhibit variation, and it is the quantifying of this variation that is the source of evidence about underlying hypotheses. This message applies not only to phylogenetic inference, but to any problem of inference under uncertainty. This is a time when "reproducible research" has become a major scientific concern.

The vast expansion of data resources, and the complexity of analyses of these data, challenge our ability to assess much in the literature. Statisticians have begun to address issues of repeatability of inductive inference results. Classical statistical measures such as p-values, Bayes factors, or indeed likelihoods, are functions of the data, and hence random variables (Boos and Stefanski 2011). Repeated realizations, even from the same probability distribution, exhibit variation. Differences are *expected*, and the assessment of differences is itself a question of inference under uncertainty. To quote the final conclusion of [146] (p. 238, this volume), "Any approach that attempts to grapple with inference under uncertainty without using ideas from the theory of probability is unlikely to command scientific respect."

I have written elsewhere (Thompson 2014) on my extreme good fortune in entering the world of statistical genetics in 1970, at a time of rapid development in genetic data, computational power, and inference for structured stochastic systems. I should also here acknowledge my equal good fortune in then coming under the supervision of Dr. Edwards, as mentor, colleague, and friend. As writing these comments have made even clearer to me, he has had a large influence on my thinking in statistical inference for complex stochastic models of fascinatingly variable genetic data.

References

Boos, D. D. and Stefanski, L. A. 2011. p-Value precision and reproducibility. *American Statistician* 65, 213–221.

Cotterman, C. W. 1940. *A Calculus for Statistico-Genetics*. PhD thesis, Ohio State University. Published in *Genetics and Social Structure*, ed. P. A. Ballonoff, New York: Academic Press, 1974.

Dempster, A. P., Laird, N. M., and Rubin, D. B. 1977. Maximum likelihood from incomplete data via the EM algorithm (with Discussion). *Journal of the Royal Statistical Society B* 39, 1–37.

Gelfand, A. E. and Smith, A. F. M. 1990. Sampling based approaches to calculating marginal densities. *Journal of the American Statistical Association* 46, 193–227.

Leslie, S., Donnelly, P., Bodmer, W., *et al.* 2015. The fine-scale genetic structure of the British population. *Nature* 519, 309–314.

Rosenberg, N. A., Pritchard, J. K., Weber, J. L., *et al.* 2002. Genetic structure of human populations. *Science* 298, 2381–2385.

Thompson, E. A. 1975. *Human Evolutionary Trees*. Cambridge: Cambridge University Press.

Thompson, E. A. 2014. A journey with statistical genetics. In *COPSS 50th Anniversary Volume: Past, Present and Future of Statistical Science*, Boca Raton, FL: CRC Press, pp. 307–316.

A. W. F. Edwards and the Origin of Bayesian Phylogenetics

ZIHENG YANG

In the early 1960s, Anthony Edwards and Luca Cavalli-Sforza made an effort to apply R. A. Fisher's maximum-likelihood method to estimate genealogical trees of human populations using gene-frequency data. They used the Yule branching process to describe the probabilities of the trees and branching times and the Brownian-motion process to model the drift of gene frequencies (after a suitable transformation) over time along the branches. They experienced considerable difficulties, including "singularities" in the likelihood surface, mainly because a distinction between parameters and random variables was not clearly made. In the process, they invented the distance (additive-tree) and parsimony (minimum-evolution) methods, both of which they viewed as heuristic approximations to maximum likelihood. The statistical nature of the inference problem was not clarified until Edwards's paper, [46], published in 1970, which pointed out that the trees should be estimated from their conditional distribution given the genetic data, rather than from the "likelihood function." In modern terminology, this is the Bayesian approach to phylogeny estimation: the Yule process specifies a prior on trees, while the conditional distribution of the trees given the data is the posterior. This article discusses the connections of the impressive paper, [46], to modern Bayesian phylogenetics, and briefly comments on some modelling decisions Edwards made then that still concern us today in modern Bayesian phylogenetics. The reader I have in mind is familiar with modern phylogenetic methods but may not have read [46], which is published in a statistics journal.

The Model and the Statistical Problem of Phylogeny Estimation

The data considered by Edwards and Cavalli-Sforza in 1964 ([27]) and 1966 ([35]) consist of gene frequencies of common blood groups from different human populations. Edwards treated different human populations while

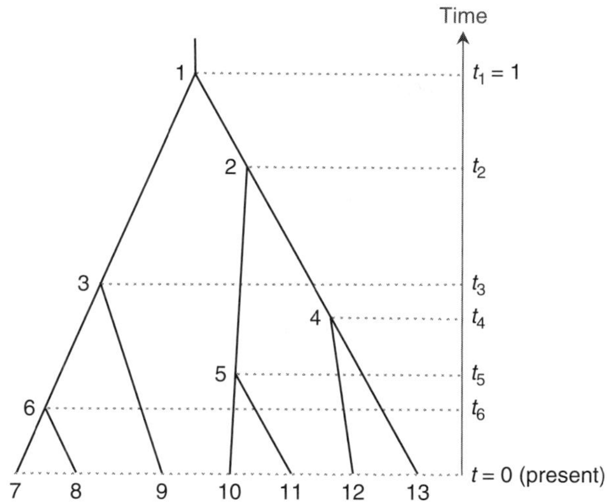

Figure 1. A phylogeny for seven ($n = 7$) species or populations used to illustrate the inference problem considered by Edwards in [46]. The tips (final particles) are numbered $n, n + 1, \ldots, 2n - 1$. The interior nodes are numbered $1, 2, \ldots, n - 1$. They represent the branching events and are ordered by time: $t_1 > t_2 > \ldots > t_{n-1}$. The time machine runs backwards, so that the present time is $t = 0$ while the age of the root (the origin at the first split) is fixed at $t_1 = 1$. The data are observed measurements in p characters from the n modern species: $\xi = \{\xi_{ik}\}$, where ξ_{ik} ($i = n, \ldots, 2n - 1$; $k = 1, \ldots, p$) is the measurement from species i in character k. Edwards considered the tree form (labelled history F), the times of non-root internal nodes, $t = \{t_1 = 1, t_2, \ldots, t_{n-1}\}$, as well as the ancestral character states $x = \{x_{ik}\}$, $i = 1, \ldots, n - 1$; $k = 1, \ldots, p$, as quantities of interest.

I focus on different species here. The data-generating model consists of two components. A Yule branching process, with a constant per-lineage rate λ of splitting, is used to describe the probability distribution, $f(F, \boldsymbol{t} \mid \lambda, n)$, of the phylogeny ($F$) and the branching times: $\boldsymbol{t} = \{t_1 = 1, t_2, \ldots, t_{n-1}\}$ (Figure 1). The Yule process assigns uniform probabilities to the labelled histories. The term labelled history, due to Edwards ([46]), refers to a rooted tree topology, with internal nodes ordered by time. For example, the rooted tree topology ((*a*, *b*), (*c*, *d*)) corresponds to two distinct labelled histories depending on whether the age of the *a*–*b* ancestor is older or younger than the age of the *c*–*d* ancestor. Note that other models of cladogenesis, such as the coalescent process (Kingman 1982) and the constant-rate birth–death process (Kendall 1948), all generate labelled histories with equal probabilities. In using the Yule process to describe the process of species formation, Edwards fixed the first branching event (the root of the tree) at time $t_1 = 1$, and conditioned on the number of species at the present time to be n. He derived the joint density for the tree form (F) and branching times ($\boldsymbol{t}$) as

$$f(F, t \mid \lambda) = \frac{2^{n-1}\lambda^{n-2}\exp\left\{-\lambda\sum_{i=2}^{n-1} t_i\right\}}{(n-1)n!(1-\mathrm{e}^{-\lambda})^{n-2}} \tag{1}$$

The second component of the model is the Brownian-motion process, used to describe the evolution of the continuous characters over time. Note that such Brownian models are now widely used in phylogenetic and phylogeographic analyses of morphological measurements from different species (Lartillotand and Poujol 2011, Solis-Lemus *et al.* 2015). The Brownian motion or random walk in one dimension gives the location of the particle time t later, given that it is at location x_0 at time 0, as a Gaussian variable, $x_t \sim N(x_0, t\sigma^2)$, where the parameter σ^2 controls how fast the particle drifts and represents the evolutionary rate for the continuous character. Gene frequencies at multiple blood-group loci are treated as a p-dimensional Brownian motion. The different dimensions (different variables) are moving independently, with the same σ^2. Given σ^2, the tree form (labelled history F) and branching times ($\boldsymbol{t}$), and the state at the root (x_{1k}), each character k observed in a modern species (a tip on the tree) is normally distributed with mean x_{1k} and variance $t_1\sigma^2$. However, the measurements for the same character observed in the modern species are correlated because they may have shared some branches. For example, $\mathrm{cov}(x_{7k}, x_{9k}) = (t_1 - t_3)\sigma^2$, where $(t_1 - t_3)$ is the time shared by the two paths from the root to the tips 7 and 9 (Figure 1). Thus, the data or the measurements of the p characters among all modern species ξ have the probability density

$$f(\xi \mid F, \boldsymbol{t}, \sigma^2, \boldsymbol{x}_1) = \frac{1}{(2\pi)^{np/2}|\boldsymbol{T}|^{p/2}}\exp\left\{\sum_{k=1}^{p} -\frac{1}{2}(\xi_k - x_{1k}\mathbf{1})'\boldsymbol{T}^{-1}(\xi_k - x_{1k}\mathbf{1})\right\}, \tag{2}$$

where $\mathbf{1}$ is a $p \times 1$ vector with all elements to be 1, and $\boldsymbol{T}$ is the variance–covariance matrix given by the tree. This is an n-variate normal density, and is nowadays known as the phylogenetic likelihood. This is equation (8) of Edwards (p. 116, this volume).

The true parameters in the model are the Yule branching rate λ, the Brownian parameter σ^2, and the initial state $\boldsymbol{x}_1$. Given those, it should be possible to simulate the process. The other unknowns, including the tree F, the branching times $\boldsymbol{t}$, and the character states at the interior nodes x_{ik}, $2 \leq i \leq n-1$, $1 \leq k \leq p$, are random variables. If one simulates the process using the true parameters, those random variables will have different realized values among simulated replicates. In [46], Edwards pointed out that the true parameters should be estimated by maximum likelihood, with the likelihood calculated by integrating over the random variables, that is, by summing over the trees (F) and integrating over the branching times ($\boldsymbol{t}$) as well as the ancestral states. The likelihood function is thus

$$L(\lambda, \sigma^2, \boldsymbol{x}_1) = f(\xi \mid \lambda, \sigma^2, \boldsymbol{x}_1) = \sum_F \int_{\mathbf{t}} f(F, \boldsymbol{t} \mid \lambda) f(\xi \mid F, \boldsymbol{t}, \sigma^2, \boldsymbol{x}_1) \mathrm{d}\boldsymbol{t} \quad (3)$$

Here, the notation is heuristic, and the sum is over all possible tree forms and the integral is over the (n – 2) branching times within each tree. The computation was deemed impossible and the model was not analyzed. Nevertheless, Edwards pointed out that random variables, F and $\boldsymbol{t}$, should be estimated from their conditional distribution given the data, with the true parameters replaced by their maximum-likelihood estimations.

The Singularity on the "Likelihood" Surface

Early attempts by Edwards and Cavalli-Sforza ([27] and [37]) treated the branching times ($\boldsymbol{t}$) and ancestral states ($\boldsymbol{x}$) as parameters. The "likelihood" function was defined as the product of the multivariate normal densities for character changes along the branches. For any branch $i \rightarrow j$, with branch length $t_i - t_j$, the density is

$$[2\pi(t_j - t_i)\sigma^2]^{-\frac{p}{2}} \times \exp\left\{-\frac{1}{2(t_j - t_i)\sigma^2} \sum_k (x_{jk} - x_{ik})^2\right\}, \quad (4)$$

given by the Brownian-motion model. As pointed out by Edwards in [46], this "likelihood" increases without bound if $x_{ik} = x_{jk}$ for all k and if $t_i \rightarrow t_j$: there are lines of singularity in the likelihood surface when there is no change along a branch for any of the p characters and when the branch disappears. Here, the mistake was to treat the ancestral states and branching times as parameters, when they are in fact random variables with statistical distributions under the model.

Some Remarks

While the Yule process of pure birth is implemented in some Bayesian programs such as BEAST (Drummond and Rambaut 2007), it is more common to use the birth–death process, which is more general by allowing species extinctions and species sampling (Rannala and Yang 1996, Stadler 2010). Often, the distribution of times is obtained by conditioning on the number of lineages at the present time and on the age of the root, as in [46], but variations exist. For example, Thompson (1975) suggests fixing $\sigma^2 = 1$, instead of fixing $t_1 = 1$, and she treats n as data (as n may be informative about the birth rate λ), rather than conditioning on n tips at the present time. Gomberg, in an unfinished report on "'Bayesian' postdiction in an evolution process" (Gomberg, unpublished work), prefers not conditioning on the present time, although the details are not so clear. The differences among those variants are not well understood.

The character states at the interior nodes of the tree are known as ancestral states. On a species phylogeny, they represent the states of the characters in the extinct common ancestors of modern species. Edwards, in [46], and also Thompson (1975, p. 119) treated the states at the root ($\boldsymbol{x}_1$) as parameters. This has the problem of introducing many parameters in the model, such that the number of parameters increases without bound when the number of characters increases. For discrete characters such as the nucleotides (T, C, A, and G) in DNA sequences, it is customary to use continuous-time Markov chains to describe transitions between character states and to assume that the process has been stationary; for exceptions, see Yang and Roberts (1995). Then the root states have a distribution given by the stationary distribution of the Markov chain. For continuous characters, the Brownian motion does not have a stationary distribution. Felsenstein (1973a) discussed an algorithm to estimate the ancestral states, eliminating $\boldsymbol{x}_1$. Statistical justifications for this procedure were discussed by Thompson (1975, p. 119).

The Yule process component of the model was dropped when Felsenstein (1973a) revisited the problem of phylogeny reconstruction using continuous characters. Thus, the tree (F) and times ($\boldsymbol{t}$) do not have statistical distributions anymore and become true parameters; these are estimated by maximizing the likelihood function, which averages over the ancestral states. Similarly, Thompson (1975, p. 60), in extending the method used by Edwards in [46], dropped the Yule process. These are the early maximum-likelihood implementations of the Brownian-motion model for continuous characters.

Origin of Bayesian Phylogenetics

Felsenstein (2004, p. 291) has included a discussion of early applications of Bayesian or Bayesian-like ideas to phylogenetics. Perhaps the most relevant is the calculation of posterior probabilities for trees by Kishino and Hasegawa (1989); see also Smouse and Li (1987), who calculated the likelihood by optimizing the branch lengths for each tree, while a fully Bayesian approach should average over the branch lengths or branching times. The modern approach to Bayesian inference of molecular phylogenies was introduced by three groups, working independently in the 1990s: Bob Mau and Michael Newton in Wisconsin (Mau and Newton 1997), a research student at Ohio State University, Shuying Li (Li *et al.* 2000), and Bruce Rannala and me in Berkeley (Rannala and Yang 1996, Yang and Rannala 1997). In all those works, the branching times are integrated out through a prior to calculate the posterior probabilities of trees. The first two groups are statisticians, applying Bayesian Markov chain Monte Carlo (MCMC) algorithms to phylogeny estimation. Note that the 1990s was the time when Bayesian MCMC algorithms were introduced into various branches of sciences, even though they were developed a few decades earlier (Metropolis *et al.* 1953, Hastings 1970).

In my case, the motivation must be owed entirely to Edwards's paper [46]. After finishing my PhD in Beijing in 1992, I went to Cambridge to work with Adrian Friday and Nick Goldman. We occasionally saw Edwards, but I believe he was working on his book on Venn diagrams, rather than phylogenetics. Adrian, Nick, and I were developing Markov models of sequence evolution for use in the maximum-likelihood method, and we had much discussion about whether the tree should be treated as a discrete parameter or a statistical model. I have provided detailed arguments elsewhere that the distinction is not a semantic one (see, e.g., Yang 1997; 2014, pp. 159–163). For example, it is not so clear how to decide whether a log-likelihood difference between two trees is due to chance. We have the rule of thumb (Edwards 1972, p. 202) that an improvement of 2 log-likelihood units (or 1.92 if we use the asymptotic χ^2 distribution) is good enough for including one additional parameter, but we lack such a "calibration" when two trees are compared. The use of bootstrap to evaluate the significance of trees has met with difficulties in interpretation (Zharkikh and Li 1992, Felsenstein and Kishino 1993, Hillis and Bull 1993). I was also concerned that the maximum-likelihood tree topology does not have the large-sample efficiency of the maximum-likelihood estimate of a conventional parameter (see, e.g., Yang 1997). Those concerns motivated my work with Bruce, when both of us were postdocs in Monty Slatkin's phylogenetics laboratory in Berkeley. We were curious to see what the alternative statistical methodology, the Bayesian, might offer, given the difficulties with the maximum likelihood method. We decided to try Edwards's prescription, but with two changes. First, we worked with DNA sequence data, using a continuous-time Markov chain model, instead of the Brownian-motion model, for continuous characters, with summation over the ancestral states achieved using Felsenstein's pruning algorithm (Felsenstein 1973b, 1981). Second, we used the birth–death process (instead of the Yule) to specify a prior on the trees and times, but this was an easy replacement. We used numerical integration to integrate over the times, so that the method is applicable to small trees only. This effort led to Rannala and Yang (1996).

This is one of the first Bayesian molecular phylogenetic analyses, and the results may be of interest even today. We applied our program to two datasets of four or five ape species (human, common chimpanzee, pygmy chimpanzee, gorilla, and orangutan). We estimated the birth rate λ, death rate μ, and the transition/transversion rate ratio κ by maximum likelihood, and used these parameters to calculate the posterior probabilities for trees, as stipulated by Edwards in [46]. The maximum a-posteriori (MAP) trees are reasonable in both datasets, grouping the humans with the chimpanzees, but the posterior probability in one dataset, at 0.9999, is uncomfortably high. This dataset, of 11 mitochondrial tRNA genes (739 base pairs), published by Horai *et al.* (1992), is fairly small, and the human–chimpanzee–gorilla relationship was a hard phylogenetic problem at the time. High posterior probabilities for trees, some decidedly spurious, continue to trouble us today, especially as datasets for phylogenetic analysis are getting increasingly large (Lewis *et al.* 2005, Yang and Rannala 2005, Yang 2007). The problem appears to have to do with the asymptotic behavior of Bayesian model selection when applied to opposing and nearly equally wrong models (Yang and Zhu 2018).

Phylogeny Estimation and Statistical Inference

I suppose my characterization of Edwards's paper [46] as the first effort to apply Bayesian statistics (rather than maximum-likelihood) to phylogeny estimation might not be looked upon by Edwards himself as reasonable. Indeed in his discussion (p. 120, this volume) he has this to say about the Bayesian approach:

> . . . and a detailed study of the present problem has, if anything, strengthened my conviction that a Bayesian approach in this instance would be a gross over-simplification. I am not prepared to give the true parameters prior probability distributions because I can see no model which would justify them. We may also note that the adoption of a Bayesian approach would not automatically resolve the dilemma of how to summarize a posterior distribution in a great many variables in terms of a few descriptive parameters, for maximizing the posterior probability would lead to the singularities . . .

A few words of clarification may thus be called for, related to the changing usage of the term "Bayesian" and the philosophy of statistical inference. The word "Bayesian" was apparently coined by R. A. Fisher in 1950 (Edwards 2004, Fienberg 2006) to refer, derogatorily, to the method of *inverse probability*, the inference method that uses Bayes theorem to derive probability distributions for parameters. The word "inverse" refers to the fact that the probability is here defined backwards, from the data to the parameter or hypothesis, or from effects to causes. Given the probability of heads for a fair coin, $\theta = \frac{1}{2}$, say, the probability of $x = 2$ heads in $n = 4$ coin tosses is $\frac{n!}{x!(n-x)!}\theta^x(1-\theta)^{n-x} = \frac{3}{8}$. This has a simple frequency interpretation: if we do many experiments, each of which involves tossing the coin four times, then in $\frac{3}{8}$ of the experiments, we will see two heads. Now suppose we have observed $x = 2$ heads in $n = 4$ tosses of a coin, what is the probability distribution of θ? To someone who is not a Bayesian statistician, this question is not meaningful. The approach of assigning a prior on θ, based on subjective beliefs or without a physical model, is the inverse method, and Edwards's formulation is not such a method.

In the tree problem, the Yule (or birth–death) process is a biologically plausible even if simple model. The true parameters are the Yule branching rate λ and the Brownian-drift parameter σ^2, which should be estimated by maximizing the likelihood function, while the tree is a random variable and its realized value should be estimated from the conditional probability, given the data. This is a standard likelihood method for estimating realized values of random variables in the model, nowadays known as empirical Bayes. Edwards would thus consider his method to be a likelihood method (although not a maximum-likelihood method) of phylogeny estimation.

Nevertheless, modern use of the term "Bayesian" does not have the derogatory tone, and the use of a physical/biological process is a common approach to specifying a prior. Of course, one may argue that the full or hierarchical Bayesian approach would assign priors on parameters λ and σ^2, rather than using their maximum-likelihood estimates, but in this commentary, I have not made an effort to distinguish between the empirical Bayes and the full (hierarchical) Bayes methods, or between a prior based on a biological model and a prior chosen for convenience.

In passing, it may be noted that the challenge of summarizing the posterior distribution of phylogenetic trees still exists today. However, the singularity in the posterior mentioned by Edwards does not exist because the posterior for a tree is calculated by integrating (rather than maximizing) over the ancestral states and branching times.

Modern Times

The early studies of Rananla and Yang (Rannala and Yang 1996, Yang and Rannal 1997), Mau and Newton (1997), and Li *et al.* (2000) assumed the molecular clock (rate constancy over time), which is often violated in comparisons of distant species. Bayesian phylogenetics really took off with the development of the program MrBayes (Huelsenbeck and Ronquist 2001, Ronquist *et al.* 2012), which adapted branch-swapping algorithms such as nearest neighbor interchange (NNI), subtree pruning and regrafting (SPR), and tree bisection and reconnection (TBR) (Swofford *et al.* 1996) into MCMC proposal algorithms to move between trees. The clock constraint was relaxed, enabling phylogenetic inference to be conducted under more realistic models. A more recent program, BEAST, infers rooted trees under the clock and relaxed-clock models (Drummond and Rambaut 2007), while PhyloBayes implements sophisticated non-stationary models to deal with substitution heterogeneity among lineages, which may be important for deep phylogenies (Lartillot *et al.* 2009). Nowadays, those Bayesian programs are standard tools in molecular phylogenetics, together with fast maximum-likelihood programs such as RAxML (Stamatakis 2006) and PhyML (Guindon and Gascuel 2003).

A brief introduction to Bayesian phylogenetics is provided in Yang (2016). More extensive reviews include Zwickl and Holder (2004) and Yang (2014, chs. 8 and 9).

An edited book by Chen *et al.* (2014) summarizes current research topics in the field. Phylogenetics may well be the largest application area of Bayesian statistics. It provides a rich testing ground for advanced Monte Carlo computational algorithms. Jerzy Neyman (1971) was certainly right to identify molecular phylogenetics as "a source of novel statistical problems."

References

Chen, M.-H., Kuo, L., and Lewis, P. 2014. *Bayesian Phylogenetics: Methods, Algorithms, and Applications*. London: Chapman & Hall/CRC.

Drummond, A. J. and Rambaut, A. 2007. BEAST: Bayesian evolutionary analysis by sampling trees. *BMC Evolutionary Biology* 7, 214.

Edwards, A. W. F. 1972. *Likelihood*. Cambridge: Cambridge University Press.

Edwards, A. W. F. 2004. Comment on Bellhouse, David R. 'The Reverend Thomas Bayes FRS: a biography to celebrate the tercentenary of his birth'. *Statistical Science* 19, 34–37.

Felsenstein, J. 1973a. Maximum-likelihood estimation of evolutionary trees from continuous characters. *American Journal of Human Genetics* 25, 471–492.

Felsenstein, J. 1973b. Maximum likelihood and minimum-steps methods for estimating evolutionary trees from data on discrete characters. *Systematic Zoology* 22, 240–249.

Felsenstein, J. 1981. Evolutionary trees from DNA sequences: a maximum likelihood approach. *Journal of Molecular Evolution* 17, 368–376.

Felsenstein, J. 2004. *Inferring Phylogenies*. Sunderland, MA. Sinauer Associates.

Felsenstein, J. and Kishino, H. 1993. Is there something wrong with the bootstrap on phylogenies? A reply to Hillis and Bull. *Systematic Biology* 42, 193–200.

Fienberg, S. E. 2006. When did Bayesian inference become "Bayesian"? *Bayesian Analysis* 1, 1–40.

Guindon, S. and Gascuel, O. 2003. A simple, fast, and accurate algorithm to estimate large phylogenies by maximum likelihood. *Systematic Biology* 52, 696–704.

Hastings, W. K. 1970. Monte Carlo sampling methods using Markov chains and their application. *Biometrika* 57, 97–109.

Hillis, D. M. and Bull, J. J. 1993. An empirical test of bootstrapping as a method for assessing confidence in phylogenetic analysis. *Systematic Biology* 42, 182–192.

Horai, S., Satta Y., Hayasaka, K., *et al.* 1992. Man's place in Hominoidea revealed by mitochondrial DNA genealogy [Erratum *J Mol Evol* 1993; 37:89]. *Journal of Molecular Evolution* 35, 32–43.

Huelsenbeck, J. P. and Ronquist, F. 2001. MrBayes: Bayesian inference of phylogenetic trees. *Bioinformatics* 17, 754–755.

Kendall, D. G. 1948. On the generalized birth-and-death process. *Annals of Mathematical Statistics* 19, 1–15.

Kingman, J. F. C. 1982. The coalescent. *Stochastic Processes and their Applications* 13, 235–248.

Kishino, H. and Hasegawa, M. 1989. Evaluation of the maximum likelihood estimate of the evolutionary tree topologies from DNA sequence data, and the branching order in Hominoidea. *Journal of Molecular Evolution* 29, 170–179.

Lartillot, N. and Poujol, R. 2011. A phylogenetic model for investigating correlated evolution of substitution rates and continuous phenotypic characters. *Molecular Biology and Evolution* 28, 729–744.

Lartillot, N., Lepage, T., and Blanquart, S. 2009. PhyloBayes 3: a Bayesian software package for phylogenetic reconstruction and molecular dating. *Bioinformatics* 25, 2286–2288.

Lewis, P. O., Holder, M. T., and Holsinger, K. E. 2005. Polytomies and Bayesian phylogenetic inference. *Systematic Biology* 54, 241–253.

Li, S., Pearl, D., and Doss, H. 2000. Phylogenetic tree reconstruction using Markov chain Monte Carlo. *Journal of the American Statistical Association* 95, 493–508.

Mau, B. and Newton, M. A. 1997. Phylogenetic inference for binary data on dendrograms using Markov chain Monte Carlo. *Journal of Computational and Graphical Statistics* 6, 122–131.

Metropolis, N., Rosenbluth, A. W., Rosenbluth, M. N., Teller, A. H., and Teller, E. 1953. Equations of state calculations by fast computing machines. *Journal of Chemical Physics* 21, 1087–1092.

Neyman, J. 1971. Molecular studies of evolution: a source of novel statistical problems. In *Statistical Decision Theory and Related Topics*, eds. S. S. Gupta and J. Yackel, New York: Academic Press, 1–27.

Rannala, B. and Yang, Z. 1996. Probability distribution of molecular evolutionary trees: a new method of phylogenetic inference. *Journal of Molecular Evolution* 43, 304–311.

Ronquist, F., Teslenko, M., van der Mark P., *et al.* 2012. MrBayes 3.2: efficient Bayesian phylogenetic inference and model choice across a large model space. *Systematic Biology* 61, 539–542.

Smouse, P. E. and Li, W.-H. 1987. Likelihood analysis of mitochondrial restriction-cleavage patterns for the human-chimpanzee-gorilla trichotomy. *Evolution* 41, 1162–1176.

Solis-Lemus, C., Knowles, L. L. and Ane, C. 2015. Bayesian species delimitation combining multiple genes and traits in a unified framework. *Evolution* 69, 492–507.

Stadler, T. 2010. Sampling-through-time in birth-death trees. *Journal of Theoretical Biology* 267, 396–404.

Stamatakis, A. 2006. RAxML-VI-HPC: maximum likelihood-based phylogenetic analyses with thousands of taxa and mixed models. *Bioinformatics* 22, 2688–2690.

Swofford, D. L., Olsen, G. J., Waddell, P. J., and Hillis, D. M. 1996. In *Molecular Systematics*, eds. D. M. Hillis, C. Moritz, and B. K. Mable, Sunderland, MA: Sinauer Associates, 407–514.

Thompson, E. A. 1975. *Human Evolutionary Trees*. Cambridge: Cambridge University Press.

Yang, Z. 1997. How often do wrong models produce better phylogenies? *Molecular Biology and Evolution* 14, 105–108.

Yang, Z. 2007. Fair-balance paradox, star-tree paradox and Bayesian phylogenetics. *Molecular Biology and Evolution* 24, 1639–1655.

Yang, Z. 2014. *Molecular Evolution: A Statistical Approach*. Oxford: Oxford University Press.

Yang, Z. 2016. In *Encyclopedia of Evolutionary Biology*, ed. R. M. Kliman, New York: Elsevier, 137–140,

Yang, Z. and Roberts, D. 1995. On the use of nucleic acid sequences to infer early branchings in the tree of life. *Molecular Biology and Evolution* 12, 451–458.

Yang, Z. and Rannala, B. 1997. Bayesian phylogenetic inference using DNA sequences: a Markov chain Monte Carlo Method. *Molecular Biology and Evolution* 14, 717–724.

Yang, Z. and Rannala, B. 2005. Branch-length prior influences Bayesian posterior probability of phylogeny. *Systematic Biology* 54, 455–470.

Yang, Z. and Zhu, T. 2018. Bayesian selection of misspecified models is overconfident and may cause spurious posterior probabilities for phylogenetic trees. *Proceedings of the National Academy of Sciences of the USA* 115, 1854–1859.

Zharkikh, A. and Li, W.-H. 1992. Statistical properties of bootstrap estimation of phylogenetic variability from nucleotide sequences. I. Four taxa with a molecular clock. *Molecular Biology and Evolution* 9, 1119–1147.

Zwickl, D. J. and Holder, M. T. 2004. Model parameterization, prior distributions, and the general time-reversible model in Bayesian phylogenetics. *Systematic Biology* 53, 877–888.

A Conversation about Fisher (1930, 1958)

WARREN J. EWENS

It is a pleasure for me to contribute to this book celebrating Edwards's work in statistics and genetics (and, most of all, their intersection). Edwards and I have had a friendly correspondence on various topics in statistical genetics for many years, usually agreeing on the various points that we have discussed, and when we disagree, doing so in a constructive manner. I shall be critical of some of Fisher's work in this paper, so I emphasize from the start that Edwards and I agree (obviously!) on the immense contributions that Fisher made to both statistics and evolutionary population genetics, essentially twin subjects in his hands. The association between these two subjects started with his famous 1918 paper on the correlation between relatives, if not earlier. The 1918 paper led to the concept, in genetics, of dividing the total variance in fitness into additive and dominance components, surely the precursor to the concept, in statistics, of the analysis of variance, which Fisher also developed.

For many years I have encouraged Edwards to write a biography of Fisher to complement that of Fisher's daughter (Box 1978), focusing on Fisher's research in population genetics. Such a biography is long overdue and Edwards is uniquely placed to write one. The rest of this commentary is intended to read almost as a conversation with Edwards, discussing points of mutual interest as they relate to the possible contents of this hoped-for biography. More specifically, I discuss various themes in Chapters I, II, and IV of Fisher's book, *The Genetical Theory of Natural Selection* (1930, 1958), which any such biography must consider. (I refer below to these books as *GTNS* (1930) and *GTNS* (1958), respectively, and *GTNS* when taken together.) It would be most desirable if my comments below were to spur Edwards on to write this biography.

In writing about *GTNS* it is clearly appropriate to start with a discussion of the background to *GTNS* (1930). In 1929, when that book was written, there was still widespread ignorance about the crucial relevance of Mendelism to the Darwinian theory of evolution by natural selection, and indeed some antipathy to the

Darwinian theory itself. The view that selection only acts negatively, by a weeding-out process, was frequently held, rather than selection acting as a positive force. Also, progressive changes in the genetic make-up of a population were frequently seen as resulting from mutation pressure rather than selection. Thus, the Darwinian theory was seen by some as being in decline. It was against this background that Fisher wrote *GTNS* (1930).

The best early criticism of the Darwinian theory was made by the polymath engineer Fleeming Jenkin. Mendel's work was of course unavailable to Darwin in 1859, and it was very bold of Darwin to write his book without having much of an idea of the hereditary mechanism. To the extent that a mechanism was currently accepted in 1859, it probably was that the value of any character in an offspring is a blending of the corresponding characters in the parents. Jenkin noted that, under the blending mechanism, the variation in a population of any character, so necessary for the Darwinian theory, would disappear over the course of a few generations. Of course variation is nevertheless observed, so that a further argument is necessary, but Darwin felt the force of Jenkin's comment. For more on this complex issue, see Winther (2000).

You are of course aware of all of the above, and I mention it only because I feel that the importance of Chapter I in *GTNS* is still not sufficiently appreciated. In this chapter, Fisher resolved the dilemma raised by Jenkin's worries by showing that the particulate nature of the Mendelian scheme implies no natural tendency to lose genetic variation once it is established. This is seen most clearly in the stability of the Hardy–Weinberg random-mating ratios p^2, $2pq$, q^2, well known today to every beginning genetics student. (Fisher, curiously, does not explicitly refer to the Hardy–Weinberg law, even in the 1958 edition: doing so and referring to the stability of the Hardy–Weinberg proportions would have made his arguments much more accessible.) The preservation of variation under a Mendelian hereditary scheme certainly needed to be pointed out in 1930, and, in Chapter I, Fisher did precisely this. In doing so, he not only reconciled Darwinism with Mendelism, but also did much more – he showed the central importance, indeed necessity, of the Mendelian system to the Darwinian theory. In this chapter Fisher also effectively demolished the view that evolution by mutation pressure is more plausible than evolution by natural selection. Please stress the importance of Chapter I, which, on its own, makes *GTNS* worthwhile.

I turn now to the "Fundamental Theorem of Natural Selection" (FTNS), the topic of Chapter II. This is a topic that you and I have discussed incessantly over the years, finally reaching agreement (and I hope understanding) on all points (e.g. [140], [238], [248]).

There are four problems (at least) with the FTNS. The first and most important is that the theorem as stated by Fisher is incorrect. A careful analysis shows that

Fisher implicitly assumed that no changes arise in the values of the average effects of the genes in the genome from one generation to the next. (Average effects are discussed below.) However these changes do occur. I later show how a correct result close to that aimed at by Fisher can be found.

The second problem is that the theorem focuses on the population mean fitness. This is a whole population concept, so that changes in population mean fitness are, to a large extent, of inter-population importance. For example, in *GTNS* (1958, pp. 45–49), Fisher discusses the changes in mean population fitness from an inter-population point of view, considering, for example, the effects of a change in mean population fitness on the growth in size of competing populations. However, so far as evolution within a population is concerned, the evolutionary focus is on changes in gene frequency *within* a population, an intra-population concept. Indeed, Fisher himself, and many others, including you and me, view these changes as the very stuff of evolution.

This leads to the third problem with the FTNS. While it unavoidably considers changes in gene frequencies, Fisher's development of the FTNS introduces them only near the end of the mathematical analysis, and does not take them as the initial quantities of interest. Apart from this, it is the relative genotype fitnesses that determine changes in gene frequencies (these changes are unchanged if all genotype fitnesses are multiplied by the same numerical constant), whereas the size of the change in mean fitness as considered by the FTNS does depend on absolute genotype fitness values. The focus should surely be on relative genotype fitnesses.

A fourth problem with Fisher's development of the FTNS is that he uses a highly unorthodox, indeed incorrect, interpretation of the fitness of any genotype, a matter discussed below.

You and I agree that perhaps the most important advance made by the FTNS is that it focuses attention on the effects of various properties of diploid populations that do not obtain for haploid populations, in particular the effects of segregation. This was an extremely important step to take. In effect, the FTNS shows that, for diploid populations, a naïve statement of the Darwinian theory that evolution by natural selection relies on variation, and that offspring resemble parents, is not appropriate. Fisher showed that a more sophisticated theory is needed, replacing the vague concept of variation by the additive genetic variance in fitness and "offspring resembling parents" by "offspring inheriting a random collection of genes from parents." This observation is arguably the most important revision of the Darwinian paradigm since Darwin's time. The negative comments made above, and below, about the FTNS do not detract from this assessment.

I now take up the question of the (viability) fitness of any genotype. You and I have noted Fisher's unorthodox view that the fitness of any genotype is a measure of the increase in frequency of that genotype from one generation to the next (or rate of increase in a continuous-time model). This unorthodox view has been criticized by several authors, most notably Charlesworth (1970) and Denniston (1978), Denniston flatly calling Fisher's interpretation "incorrect" (an adjective with which I agree). Surely, the correct interpretation is that the (viability) fitness of any genotype is a measure of the innate ability of individuals of that genotype to survive to the age of reproduction and then reproduce. Furthermore, because of segregation, this measure is not directly related to the increase in frequency of that genotype. At the whole-genome level, essentially no currently existing genotype will appear in the next generation (the uniqueness of the individual), and thus, under Fisher's definition, the fitness of essentially all currently existing genotypes would be zero. Even at the one-locus level it is easy to derive examples where a genotype that has the lowest viability of all existing genotypes nevertheless steadily increases in frequency over successive generations through the creation of individuals having this genotype from individuals of other genotypes, via Mendelian segregation.

I now outline the calculations that lead to a correct result which is as close as possible to that which Fisher was attempting to achieve in the FTNS. In doing so I also make all the same assumptions that Fisher made; for example, monoecious populations, no stochastic effects, and so on. These calculations generalize those of Denniston (1978) for the one-locus random-mating case. Mating is not assumed to be random and the fitness of any individual is allowed to depend on the entire-genome genotype of that individual. Some details omitted here are given in Ewens (2004), Section 2.8.

The analysis follows the material starting on page 63 of Ewens (2004), and the equations referred to below are the equation numbers in that book. Let Δx_u be the change in the frequency of the allele A_u from one generation to the next, given parental generation allele frequency x_u at the time of conception of that generation, and the fitness scheme as given on page 62 of Ewens (2004). Then, as given in Eq. (2.62),

$$\Delta x_u = x_u a_u / \overline{W}, \tag{1}$$

where a_u is the average excess of the alleles A_u as defined in Eq. (2.60), and $\overline{W}$, the population mean fitness, is defined in Eq. (2.56) of Ewens (2004).

It is now necessary to define the "fitness" of the allele A_u. For this, we take the value $\alpha_u + \overline{W}$, where the average effect α_u of this allele is defined through the analysis on pages 62 and 63 of Ewens (2004). No explicit formula for α_u is given in those pages, and none is necessary for the analysis below.

The changes in gene frequencies given by Equation (1) are not enough on their own. The change in the frequency of the representative allele A_u should be weighted by α_u, which can be either positive or negative. Summing these weighted changes over all alleles and multiplying by 2 to account for the diploid nature of the organism, we get, from Equation (1), $2\sum_u \Delta x_u \times \alpha_u = 2\sum_u x_u a_u \alpha_u / \overline{W}$. From Eq. (2.61) in Ewens (2004), this is identical to ${\sigma_A}^2/\overline{W}$, where ${\sigma_A}^2$ is the additive genetic variance in fitness at the locus considered, leading to

$$2\sum_u \Delta x_u \times a_u = \sigma_A^2/\overline{W}. \tag{2}$$

Despite the fact that the additive genetic variance is central both to the FTNS and the principle derived above, the latter is not a re-statement of the one-locus version of FTNS. It differs from the FTNS in not making the assumption that average effects do not change from one generation to the next, and in taking as its starting point the gene frequency changes brought about by natural selection.

The extension of Equation (2) to the whole-genome case is immediate. The analysis at the top of page 260 in Ewens (2004) shows that, in the whole-genome case,

$$2\sum_k \sum_u \Delta x_{ku} \times a_{ku} = \sigma_A^2/\overline{W}. \tag{3}$$

In this equation, the summation "$\sum_k$" is over all loci in the genome, x_{ku} is the frequency of allele A_u at locus k and α_{ku} is the average effect, or "fitness," of that allele, $\overline{W}$ is the whole-genome mean fitness, and ${\sigma_A}^2$ is the whole-genome additive genetic variance. This equation seems to be as close as one can get to what Fisher was aiming at in the FTNS.

I now move to Chapter IV of *GTNS*. Here, I think you can do much to clarify what Fisher wrote. This chapter is entitled "Variation as determined by mutation and selection," but this title is curious in that the main focus of the chapter concerns the effects of random changes in gene frequencies from one generation to the next. Perhaps "Variation as determined by stochastic factors" would have been more appropriate.

The chapter may be divided, for convenience, into two parts. The first concerns survival probabilities of favorable new mutants, and the vehicle used to assess these is the theory of branching processes. Even now, this part of the chapter can hardly be improved upon, and I am sure that you will give Fisher full credit for it. It is masterly and I make no further comment on it here.

The analysis in the second part of this chapter in *GTNS* (1958) closely follows that of the *GTNS* (1930) and concerns the stochastic changes in gene frequencies in a finite population. All references here to Fisher's work refer to *GTNS* (1958). In 1958, far more efficient approaches were available to the problems considered than those used by Fisher, and I hope that in your biography you can update and clarify Fisher's 1958 analysis by using Markov chain theory and the backward Kolmogorov equation, as discussed below.

To be specific, Fisher's first aim was to calculate the "decay of [genetic] variance" due to stochastic factors. The analysis concerns what is now called the simple Wright–Fisher Markov chain model and its diffusion-theory analog. Random mating is assumed in the analysis. Although Fisher (1958) approached this through the analysis on pages 92–94, it would have been sufficient to use Wright's well-known and elegant "heterozygosity" approach to the problem. This starts with the simple fact that if there are two alleles at the locus in question, at generation t, with frequencies $p(t)$ and $1 - p(t)$, the heterozygosity measure (under random mating) is $H(t) = 2\ \{p(t)(1 - p(t))\}$. Writing the mean of $H(t + 1)$ given $H(t)$ as $E\{H(t + 1)|H(t)\}$, Wright showed that, in the Wright–Fisher model, $E\{H(t + 1)|H(t)\} = \{1 - (2N)^{-1}\}H(t)$, where N is the population size. Thus, the "rate of decay of variance" is, exactly and immediately, $1 - (2\ N)^{-1}$. In 1958, this was already a textbook result. The analysis on pages 92–94 is not only unnecessary, it is far too long, and also gives an incorrect value for the "rate of decay."

The purpose of Fisher's "rate of decay of variation" calculation was to show that although genetical variation will eventually be lost, because of fixation of one of the two alleles in the Wright–Fisher model, where mutation is not allowed, this will tend to take a long time. In this way, the variation-preserving property of the Mendelian system, as discussed in Chapter I for the deterministic case, continues to hold. This is of course a most important aim. I think that here (and elsewhere) Fisher does not pay enough attention to Wright's "effective population size" concept. The simple Wright–Fisher model is highly idealized, and, in more complex cases (for example involving two sexes with an extreme sex-ratio), the effective population size can be quite small, so that genetic variation can be lost quickly, and gene frequencies can change far more rapidly without selection in models less idealized than the Wright–Fisher model. (For more on the effective population size concept, see Winther *et al.* (2015).)

Fisher's calculations on pages 92–94, while not necessary for the "rate of decay of variation" analysis, introduce another important matter. In effect he considers a sequence of loci, which I label here loci 1, 2, 3, . . ., each initially homogeneous for some allele "*A*". At locus j in this sequence, a single mutant allele "*a*" arises in generation j. No further mutation arises at this locus, so that, eventually, the new mutant will either be lost (probability $1 - 1/(2N)$) or fixed (probability $1/(2N)$) in the population, and variation at locus j is then lost. Fisher in effect found the mean number of loci segregating in a given stationary generation. His result is amazingly accurate (Watterson 1975). Further, he showed that, at any given time, the mean number of loci for which the mutant is represented j times is close to $2/j$. This formula leads easily to the currently widely used concept of the frequency spectrum of the process. Fisher has never, so far as I know, received credit for this, and I hope that in your book you will give it to him.

The final problem (pages 99–101) considered by Fisher concerns the stochastic behavior of the frequency of a selectively favored allele. Suppose that this frequency is denoted by p and define θ by $\cos\theta = 1 - 2p$. Fisher derived (in line 18 of page 99) the forward Kolmogorov equation satisfied by θ. (The simpler equation, given in terms of p, and first stated by Kimura [1957, equation (4.3)], is now standard and preferred.) Fisher's main aim was to find the probability of fixation of this favored allele. Although his eventual fixation probability calculation is correct, the derivation using the forward Kolmogorov equation is long and hard to follow. The approach using the backward Kolmogorov equation is simpler, more natural, and more direct, and leads cleanly to the desired result in just a few lines. It had been available, by 1958, for many years, and I wish that, in 1958, Fisher had used it. Further, the theory of Markov chains was well established in 1958, and much of Fisher's analysis could have been much more simply carried out using that theory. I hope that you can re-write this part of his book using that theory and the backward Kolmogorov equation.

Fisher's analysis shows that it is the product of the selective advantage of the favored allele and the population size that is relevant for the fixation probability, so that if the selective advantage of the favored allele is anything other than minute, the effect of selection is very significant. However, the only population size that Fisher mentions is a thousand million, surely a large one for many populations. Here, his ignoring the effective population size concept is unfortunate – a more convincing analysis would have taken this into account and led to a consideration of effective population sizes, sometimes much smaller than the actual sizes.

I have limited my comments to Chapters I, II, and IV of *GTNS*, since I am not competent to discuss the later chapters. I am aware that some of the comments made above are of the mathematical nit-picking type. However, Fisher's arguments are often mathematical, and his book will be used by future researchers for many years. I therefore think it is important for you to clarify, simplify, and, where necessary, correct, these purely mathematical arguments. Much more than that, your unrivalled knowledge of his work and his career, and above all of his status as a leading scientist of the twentieth century, make it essential that you write his biography.

References

Box, J. F. 1978. *R. A. Fisher: The Life of a Scientist*. New York: Wiley.

Charlesworth, B. 1970. Selection in populations with overlapping generations. I. The use of Malthusian parameters in population genetics. *Theoretical Population Biology* 1, 352–370.

Denniston, C. 1978. An incorrect definition of fitness revisited. *Annals of Human Genetics* 42, 77–85.

Ewens, W. J. 2004. *Mathematical Population Genetics*. New York: Springer.

Fisher, R. A. 1930. *The Genetical Theory of Natural Selection*. Oxford: Clarendon Press.

Fisher, R. A. 1958. *The Genetical Theory of Natural Selection*, 2nd edition. New York: Dover.

Kimura, M. 1957. Some problems of stochastic processes in population genetics. *Annals of Mathematical Statistics* 28, 882–901.

Watterson, G. A. 1975. On the number of segregating sites in genetic models without recombination. *Theoretical Population Biology* 7, 256–276.

Winther, R. G. 2000. Darwin on variation and heredity. *Journal of the History of Biology* 33, 425–455.

Winther, R. G., Giordano, R., Edge, M. D., and Nielsen, R. 2015. The mind, the lab and the field: three kinds of populations in scientific practice. *Studies in History and Philosophy of Science Part C: Studies in the History and Philosophy of Biological and Biomedical Sciences* 52, 12–21.

Edwards on Fisher's Fundamental Theorem of Natural Selection

SAMIR OKASHA

R. A. Fisher's "Fundamental Theorem of Natural Selection" (FTNS) was the centrepiece of his *The Genetical Theory of Natural Selection* (1930), a book whose influence on twentieth-century evolutionary biology can hardly be overstated. Yet the theorem puzzled commentators for many years, leading to seemingly endless controversy over its meaning, scope, and validity. One of Anthony Edwards's many contributions, and the one that I will mainly focus on here, is to have helped remedy this unfortunate situation by clearly expounding the logic of Fisher's argument, thus promoting a wider understanding of what the FTNS really means.

Three of Edwards's papers reprinted in this volume deal directly with the FTNS. The first, published in 1967 ([36]), is a short letter to *Nature*, taking issue with some remarks on the FTNS by the geneticist C. C. Li. The second ([140]) is a 30-page paper in *Biological Reviews* (1994), which offers a detailed account of the twists and turns of the debate over the FTNS, including a painstaking analysis of what previous commentators got right and wrong. The third ([238]), also published in *Biological Reviews*, in 2014, discusses the broader intellectual context of the theorem and contains some remarks on the post-1994 literature on the FTNS. A fourth paper in the volume ([49]) is a 1971 review of Sewall Wright's collected papers and deals with a related topic: the "adaptive landscape" concept, which Wright (wrongly) thought could be underpinned by Fisher's FTNS. Other relevant works of Edwards, not reprinted here, include two papers in *Theoretical Population Biology* (Edwards 1990, 2002) and the final chapter of the second edition of *Foundations of Mathematical Genetics* (Edwards 2000).

Fisher's original formulation of the FTNS read "the rate of increase in fitness of any organism at any time is equal to its genetic variance in fitness at that time"

This work was supported by the European Research Council Seventh Framework Program (FP7/2007–2013), ERC Grant agreement no. 295449.

(Fisher 1930, p. 37). (By "fitness of any organism" Fisher meant mean population fitness.) Since the variance in fitness, of which the "genetic variance" *sensu* Fisher is a part,[1] cannot be non-negative, the FTNS seems to imply that the mean fitness of a population will almost always increase as a result of natural selection. However, this has long been known not be true, even in models with fixed genotype fitnesses; indeed, Fisher (1941) contains a counterexample. This prompts the question: what exactly does the FTNS mean, and is it true? The answer, in short, is that Fisher was not talking about the *total* change in mean fitness at all, but rather the partial change that is "ascribable to natural selection," in Fisher's words; Fisher computes this partial change by holding fixed the average effects of all alleles (on fitness). Since these average effects will in fact change as the population evolves (unless dominance and epistasis are completely absent), the partial change in mean fitness will, in general, not equal the total (or actual) change. The FTNS, as Fisher intended it, concerns only the partial change, and is a correct theorem about it.

This "partial change" construal of the FTNS, sometimes called the "modern interpretation," is today fairly well known among evolutionists, thanks mainly to the work of Price (1971) and Ewens (1989). The route by which biologists arrived at the modern interpretation is dissected brilliantly in [140] by Edwards, who argues convincingly that it captures Fisher's intended meaning, based on extensive textual evidence from Fisher's published writings and private correspondence. In his most recent 2014 paper on the topic ([238]), Edwards offers a slightly revised historiography, noting that two papers prior to those of Price and Ewens had already understood the meaning of the FTNS – Bennett (1956) and Kimura (1958) – but had been overlooked by subsequent authors. Future historians of evolutionary genetics will owe a debt to Edwards's painstaking work on the early reception of the FTNS.

A key aspect of the FTNS for Fisher, and one which Edwards has repeatedly stressed, is the idea that the evolutionary change due to natural selection is constituted by changing allele frequencies alone; all other changes are classed as "environmental," including, crucially, changes in genotype frequencies that do not derive directly from changes in allele frequency, e.g. that are due to the mating pattern; these count as changes in the "genic environment," for Fisher. This was spelled out clearly in Fisher (1941) but left largely implicit in Fisher's original (1930) exposition of the FTNS. As Edwards explains, this rather broad sense of "environment" is crucial to Fisher's own understanding of his FTNS, for it provides the underpinning justification for his claim that the "improving tendency" of natural selection, as revealed by the FTNS, is offset by "environmental deterioration." In effect, any change in the average effects of the alleles constitutes an

[1] Fisher's "genetic variance" refers to what is today called the additive genetic variance; see [140], pp. 197–198, this volume.

environmental change, for Fisher, and is excluded from the partial change in mean fitness that the theorem is concerned with.

The difference between the partial and total change in mean fitness is at the root of the disagreement between Fisher and Wright over the latter's adaptive landscape concept, lucidly explained by Edwards (in [49], [140], and in Edwards 1990). Wright's adaptive landscape, in one of its versions, plotted population mean fitness $\overline{W}$ as a function of population-wide gene frequency; Wright argued that natural selection would lead $\overline{W}$ to increase, pushing a population up the steepest path to a peak in the landscape. Were this true, it would mean that $\overline{W}$ was a "potential function" of the evolving system, i.e. a function whose value is maximized (or minimized) at equilibrium and whose slope determines the rate of change of the system. Wright appealed to the FTNS in support of this idea, but Fisher repeatedly insisted that his theorem did not support it, and that Darwinian evolution need not be governed by a potential function at all, so the very idea of an "adaptive landscape" is suspect. In retrospect, it is clear that Wright (like many others) wrongly thought that the FTNS concerned the total change in $\overline{W}$, rather than the partial change.

Re-reading Edwards's papers on the FTNS, a number of points stand out. The first is Edwards's commitment to careful and detailed scholarship, mentioned already. The second is an admirable willingness to change his mind. Readers will note that the point of Edwards's 1967 paper ([36]) is to argue, against Li, that the FTNS is a continuous-time result and should not be used as the name for any discrete-time result. However, in 1994 ([140]), Edwards follows Ewens (1989) in presenting a version of the FTNS in a discrete-time model with non-overlapping generations. In his most recent 2014 publication ([238]), Edwards formulates the FTNS in continuous time, closely following Fisher's original formulation; Edwards tells me (personal communication) that he no longer regards the discrete-time formulations as true to Fisher's intent. This is one of a number of issues surrounding the FTNS on which disagreement persists among contemporary commentators.

A third point that emerges is Edwards's own broadly Fisherian view of evolution by natural selection. Though much of his work on the FTNS is devoted to explaining Fisher's logic and correcting misunderstandings in the literature, Edwards's own sympathies do emerge and they are generally supportive of Fisher. In particular, he endorses Fisher's critique of Wright's adaptive landscape concept, and appears sympathetic both to Fisher's conception of "genic environment" and to the specific mathematical definition of "environmental constancy" that Fisher uses. On the issue of whether the FTNS has any real biological significance (as opposed to whether it is a correct mathematical statement), Edwards plays his cards close to his chest. The main bone of contention here is whether or not the partial change in mean fitness, which the FTNS tells us is proportional to the additive genetic variance, so cannot decrease, is a quantity of any biological interest. In [140], Edwards outlines both sides of the case without endorsing either (see especially pp. 219–221, this volume). However, one suspects that his sympathies are closer to those who view the FTNS as embodying a deep truth about Darwinian evolution.

On one important interpretive issue Edwards does reveal his hand. The issue is whether the FTNS can legitimately support a view of evolution by natural selection as in some sense an "optimizing process." Grafen (2003) defends the view that the FTNS has an "optimizing interpretation"; Gardner (2009) argues likewise. These authors espouse, and attribute to Fisher, the claim that evolution by natural selection leads organisms to have phenotypes that maximize their fitness. (This is different from Wright's claim that natural selection maximizes mean population fitness $\overline{W}$.) In [238], Edwards argues persuasively that Fisher did not in fact hold the view that Grafen and Gardner attribute to him, and that Fisher was in fact critical of this view. Though he does not argue the point, one suspects that Edwards regards the view in question as an erroneous way of understanding Darwinian evolution.

Edwards is probably right that Fisher would not have endorsed the "optimizing interpretation" of his FTNS that Grafen and Gardner advocate. However, it must be admitted that Fisher's language does at times lend itself to such an interpretation, for example in his frequent talk of "improvement" and "progress" in relation to his FTNS (e.g. Fisher 1930, p. 22, p. 37). Moreover, under the heading "adaptation," Fisher talks about the "consequences to the organic world of the progressive increase of fitness of each species"; and the definition of adaptation that he gives appears to involve optimization in the sense in which Grafen uses the term (Fisher 1930, p. 38). So it is not hard to understand where Grafen's interpretation of the FTNS comes from. As Sean Rice has insightfully observed, Wright and Fisher "both seemed concerned with phrasing their results in terms of something being maximized whenever possible" (Rice 2004, p. 36).

To conclude, in his writings on the FTNS, Edwards has performed a valuable service in helping us understand the mathematics of Fisher's argument, the broader intellectual context of his theorem, and the interpretive difficulties to which it has led. Both contemporary evolutionary biologists interested in Fisher and historians of the discipline are in his debt.

References

Bennett, J. H. 1956. Population genetics and natural selection. *Genetics* 28, 297–307.
Edwards, A. W. F. 1990. Fisher, $\overline{W}$ and the fundamental theorem. *Theoretical Population Biology* 38, 276–284.

Edwards, A. W. F. 2000. *Foundations of Mathematical Genetics*, 2nd edition. Cambridge: Cambridge University Press.
Edwards, A. W. F. 2002. The Fundamental Theorem of Natural Selection. *Theoretical Population Biology* 61, 335–337.
Ewens, W. 1989. An interpretation and proof of the Fundamental Theorem of Natural Selection. *Theoretical Population Biology* 36, 167–180.
Fisher, R. A. 1930. *The Genetical Theory of Natural Selection*. Oxford: Clarendon Press.
Fisher, R. A. 1941. Average excess and average effect of a gene substitution. *Annals of Eugenics* 11, 53–63.
Gardner, A. 2009. Adaptation as organism design. *Biology Letters* 5, 861–864.
Gayon, J. 1992. *Darwin et l'après-Darwin. Une histoire de l'Hypothèse de sélection naturelle*. Paris: Kimé.
Grafen, A. 2003. Fisher the evolutionary biologist. *The Statistician* 52, 319–329.
Hamilton, W. D. 1967. Extraordinary sex ratios. *Science* 156, 477–488.
Hodge, M. J. S. 1992. Biology and philosophy (including ideology): a study of Fisher and Wright. In *The Founders of Evolutionary Genetics: A Centenary Reappraisal*, ed. Sarkar, S., Dordrecht: Kluwer Academic Press, 231–293.
Karlin, S. 1975. General two-locus selection models: some objectives, results and interpretations. *Theoretical Population Biology* 7, 364–398.
Kimura, M. 1958. On the change of population fitness by natural selection. *Heredity* 12, 145–167.
Lewontin, R. C. and Kojima, K. I. 1960. The evolutionary dynamics of complex polymorphisms. *Evolution* 14, 458–472.
Maynard Smith, J. 1985. In Haldane's footsteps. In *Studying Animal Behavior: Autobiographies of the Founders*, ed. D. Dewsbury, Chicago: University of Chicago Press, 347–354.
Mitchison, A. 2003. Jim's cool reception among the British geneticists. In *Inspiring Science: Jim Watson and the Age of DNA*, eds. J. R. Inglis, J. Sambrook, and J. A. Witkowski, New York: Cold Spring Harbor Laboratory, 49–53.
O'Donald, P. 1980. *Genetic Models of Sexual Selection*. Cambridge: Cambridge University Press.
Price, G. 1972. Fisher's 'fundamental theorem' made clear. *Annals of Human Genetics* 36, 129–140.
Rice, S. H. 2004. *Evolutionary Theory*. Sunderland, MA: Sinauer Associates.

A. W. F. Edwards, R. A. Fisher, and *The Genetical Theory of Natural Selection*

MAURIZIO ESPOSITO

In a recent article ([238]), A. W. F. Edwards traces the gene's-eye-view of evolution back to Ronald Fisher and his seminal book, *The Genetical Theory of Natural Selection* (*GTNS*). As is well known, the book presents a Darwinian theory of evolution revisited with the tools of mathematics, statistics, and genetics. The theory shows how gene frequencies change over time; uses a powerful analogy between kinetic theory and hereditary particles (genes seen as molecules of gas); and has at its core the Fundamental Theorem of Natural Selection. We can easily follow Edwards when he interprets *GTNS* as an updated version of classical Darwinism: " . . . it is illuminating to think of *The Genetical Theory* as a kind of mathematical-Mendelian appendix to the *Origins of Species*" (Edwards 2000, p. 1423). This is a fair synthesis of Fisher's important achievements, and it is also the way many other evolutionary biologists, biophilosophers, and geneticists see the book. However, this description overlooks the fact that five chapters of the monograph are dedicated to human evolution and the historical collapse of civilizations – two topics that Fisher himself considered inseparable from the previous chapters.

Drawing on Edwards and other interpreters of Fisher's *GTNS*, I argue that we need to consider these final chapters seriously if we want a deeper understanding of Fisher's evolutionary biology, especially in relation to the range and limits of genetic explanations. Indeed, Fisher's radical gene-centered view needs to be contextualized when applied to human evolution because he seriously considered the influence of extra-genetic factors on birth rates, and, therefore, evolutionary trends. In the first part of this commentary, I briefly sketch some of the principal ways *GTNS* has been read and interpreted, from Sewall Wright to A. W. F. Edwards. Then, I show how these interpretations overlooked Fisher's sophisticated attempt to connect the natural and human sciences, and I explore some of Fisher's uses of the human sciences within his overall project. Finally, I assess two crucial points of Edwards's interpretation of Fisher's biology: first, that Fisher's gene-centric view was more sophisticated than many posterior readings have led us to believe; and second, that eugenics was not necessarily the driving motive of Fisher's evolutionary biology.

Interpreting *GTNS*

Ever since its publication, *GTNS* has been read as addressing two different topics: evolutionary biology and eugenics. While many biologists tended to underscore the former and neglect the latter, many historians did the opposite. For instance, in Sewall Wright's 1930 review, Fisher's book is mainly framed around evolutionary biology, despite the fact that, as Wright readily recognized, "More than one-third of the book is devoted to discussion of the trend of evolution in man" (Wright 1930, p. 355). Reginald Crundall Punnett followed the same hermeneutical approach. He not only recognized that "A considerable part of Dr. Fisher's book is devoted to considerations on the evolution of man in his social aspects," but also that many readers " . . . will probably find this the brightest part of the book" (Punnett 1930, p. 597). Nevertheless, he dedicates one paragraph out of 11 to its analysis. J. B. S. Haldane's review certainly was more balanced (Haldane 1931). He fully recognized the relevance of Fisher's argument beyond Darwinism and Mendelism. However, he focused on Fisher's eugenic arguments, missing entirely the sophisticated discussion of the interactions between social and evolutionary forces behind the rise and fall of civilizations. For the next three decades, despite the reviews written by influential biologists, Fisher's book did not receive wide attention, whether among evolutionary biologists or social scientists. Only after the 1960s did *GTNS* know a relative revival through the influential contributions of George C. Williams, Donald Hamilton, and Maynard Smith (Edwards, 2000) – although the last part of *GTNS* was, once again, safely ignored.

From the 1970s, J. H. Bennett started publishing Fisher's correspondence, providing at the same time a detailed analysis of Fisher's achievements in *GTNS* and elsewhere (1971–1974, 1983). Bennett's scholarship opened a window on Fisher's world and contributed to the dissemination of Fisher's scientific contributions. But again, we get a picture of Fisher divided between science and ideology, or biology and eugenics. In 1990, with the centennial celebration of Fisher's birth, the geneticists James Crow and William Dove revived the same hermeneutical approach: Fisher was a mathematician and biologist, with a passing interest in social issues. "His eugenic writings," Crow declares, "have had no lasting influence on either biologists or historians. In his later life Fisher did not write about these subjects, nor did he talk about them (to me at least). I don't think he had changed his mind, but simply tired of trying to get people to take his proposals seriously" (Crow 1990, p. 209). In short, the last five chapters of *GTNS* could be considered an idiosyncratic distraction of a prolific mind. Finally, in 2000, Edwards provided a concise and competent account of *GTNS*'s history and overall contents (see Edwards 2000). He drew on Crow's interpretation of the central themes and novelties of *GTNS* and complemented the discussion with Fisher's biography and further historical information. In particular, Edwards underscored Fisher's connection with Leonard Darwin and eugenics. However, what could not be seriously accepted, Edwards added, was that " . . . Fisher's youthful enthusiasm for eugenics was the engine for his advances in statistical and evolutionary genetics" (Edwards 2000, p. 1421). Even though eugenics had to have been important to Fisher, it was neither constitutive of, nor essential to, his scientific contributions. In other words, the latter could be safely disconnected from Fisher's social interests.

If there is one fundamental lesson that we can learn from the history of the reception of *GTNS*, it is that most readers struggled to find thematic continuity between the first seven chapters and the last five. Not surprisingly, the book's structure preoccupied Fisher and the editors before its publication. The editor from Clarendon Press, Kenneth Sisam, suggested introducing each chapter with a short summary: "You see," Sisam writes to Fisher, "our whole concern is to help the reader to follow your argument easily, instead of having to double back on his track in order to pick up points he had missed at the outset" (in Bennett 1983, p. 21). To the editors, the connections between chapters were not always clear and needed to be worked out. Fisher was also worried that the chapters on humans would not fit well. As he wrote to Sisam, "I am conscious that the Chapters on Man will, from your point of view, tend to alter the character of the whole book, and I want you to feel perfectly free to change your mind about publication" (Fisher in Bennett 1983, p. 21). Sisam was neither a biologist nor a scientist, but a philologist and historian. He told Fisher that the last chapters were very significant for the book: "To me personally, your latter chapters were of very great interest, and I know no reason why biologists should not consider the ultimate ends of their science" (in Bennett 1983, p. 21).

Despite Sisam's support, Fisher's concerns were fully justified. While Fisher regarded the chapters dedicated to civilization and human evolution as inseparable from the previous chapters (Bennett 1983, p. 19), the book was not read in that way, even after the *GTNS*'s second edition appeared in 1958. For instance, historians of science have tended to see these chapters as the driving motive of Fisher's evolutionist interests. They interpreted Fisher's biological concerns of the first seven chapters as only a prelude to what really mattered to him: eugenics. Edwards, in [238], criticizes this interpretation. He laments that some historians – he has in mind Norton (1978) and Mackenzie (1981) – have seen Fisher's science as a reflection of his social concerns and ideological commitments, overlooking Fisher's genuine scientific achievement. Supporting Edwards's remarks, I think that Fisher was neither a simplistic gene-centric biologist nor a mere eugenicist using science for ideological purposes. Rather, what in my opinion has been missed in the midst of comprehensible, and often grounded, accusations of sexism, racism, biologism, and elitism, is Fisher's ambition to tie the human and natural sciences within a coherent and complex evolutionary framework (Gould 1996).

In this chapter, I use the expression "human science" to refer to the study of extra-genetic factors that Fisher deemed to be relevant to human evolution. In other words, the study of the historical, social, cultural, and economical elements that create the environments in which natural selection works. After all, Fisher explicitly referred to "sociological causation," "sociological consequences," and "sociological factors" in human evolution, and his evolutionary biology is deeply imbued with economic metaphors (Gayon 2014). Furthermore, when we go through the final chapters of *GTNS*, we realize that eugenics was part of a larger discussion. Indeed, Fisher linked eugenics with one of the most traditional themes of the late-eighteenth and nineteenth centuries: the rise and fall of past civilizations. The latter was a topic that had previously fascinated humanists like Lord Monboddo, Edward B. Tylor, Adam Ferguson, and John Lubbock, among others. However, Fisher added two further elements to such a traditional discussion: Darwinian biology and a genetical theory of natural selection. In particular, Fisher intended to show how some specific social forces created the conditions for the selection of two of the most important human characteristics: intelligence and fertility. Fisher maintained that both traits were genetically inherited. The differential distribution of those genetic traits could influence the reproductive rates in human populations and, as a consequence, human evolution itself. This led Fisher to conclude, as I will show in the next two sections, that an unbalanced distribution of these two traits could explain the repeated fall of complex civilizations. However, the fact that genetics and natural selection could in part account for the collapse of civilization did not imply, to Fisher, that human history could be plainly reduced to organic evolution. Indeed, while Fisher tried to make sense of how human evolution worked from a biological viewpoint, he was aware that biology was not enough when applied to the human beings. He argued that in genetics there were two schools: one emphasizing that genetics was an abstract science that could not be applied to humans (Fisher 1930, p. 173). The other, bigger and overambitious, attempted to reduce human evolution to the evolution of simpler species (Fisher 1930, p. 174):

> A larger, and more enterprising school, fully imbued with a sense of the universal applicability of genetic knowledge, recognize in mankind a useful field for its exploitation. They are more ready, and are perhaps better prepared, to appreciate the similarities of human inheritance to that of *Drosophila* or Maize, than they are to appreciate the special problems which the evolution of man in society presents; and will sum up the human problem in a cursory, and even superficial, chapter at the end of an elaborate, and often admirable exposition of modern inheritance discoveries.

Fisher believed that when we approach the problems of human evolution, we need to supplement genetics and Darwinism with new research tools. As he concluded: "While genetic knowledge is essential for the clarity it introduces into the subject, the causes of the evolutionary changes in progress can only be resolved by an appeal to sociological, and even historical facts" (Fisher 1930, p. 174). How can the supposed forefather of the neo-Darwinians – therefore of all the followers of the gene's-eye-view of evolution – argue that we need to appeal to sociology and history if we want to make sense of human evolution? Is the gene's-eye-view only valid for organisms except *Homo sapiens*? Is there a fundamental rupture between humans and their non-human ancestors? We know that Fisher did believe that genetics and natural selection applied to humans. *Homo sapiens*, like amoebas or butterflies, was subject to natural selection. Humans were the product of evolution, like any other organism. Yet Fisher argues that there were social mechanisms among humans that could not be explained by genetics and natural selection alone.[1] Is this view incoherent?

I think we can easily resolve this tension if we explore the possibility that Fisher, when discussing human evolution, was not the gene-centered Darwinist that many biologists and historians might expect. Perhaps we have read Fisher through the anachronistic glasses of post-1960s evolutionary biology. Perhaps, behind many neo-Darwinian analyses or postmodern reconstructions, lies a Fisher who was much more pluralistic and sophisticated, fully aware of the strengths and limits of evolutionary or genetic explanations. Even more striking is that we do not need to dig through Fisher's correspondence to realize this; we do not need any complex historical interpretation based on the analysis of Fisher's readings and relations with his contemporaries. Instead, we need only to reread Fisher's book from pages 1 to 265 as one unique and integrated argument, linking evolutionary biology with human evolution, and therefore natural science with the human sciences.

Beyond Genetics and Evolutionary Biology

Jonathan Hodge, James Moore, Michael Ruse, Stephen Jay Gould, and others have dedicated attention to Fisher's different interests (see Hodge 1992, Ruse 1996, Gould 1996, Moore 2007). In focusing on Fisher's religious preoccupations, political convictions, and philosophical beliefs, they already noted that there was

[1] Notably, Fisher did not provide examples of non-human societies except for social insects. Perhaps this is due to Fisher's assumption that, in other non-human animals, social forces, which might be able to affect the differential reproduction in a population, did not exist. In the case of eusocial insects, Fisher believed that their social behavior depended entirely on their particular mode of reproduction. In other words, the reproductive burden falls not on many different individuals (Fisher used the technical term "individualistic reproduction" for referring to human reproduction), but on a fecundated queen that establishes a colony of clones. In short, while an ant colony can be considered as a single organism; a human "colony" remains a set of genetically different individuals.

much more going on than a synthesis between Darwinism and Mendelism. Thus, while we do not need to reconsider the relation between science and ideology in *GTNS* or other works of Fisher, I think we need to reconsider the Fisherian tension and continuity between biology and human science (i.e. all the non-genetical or non-biological factors influencing particular human behaviors). For example, in Fisher's *GTNS*, is natural selection sufficient for explaining human evolution? Are sociological and historical phenomena reducible to biological phenomena? Definitely not, if we read the whole book.

Let us proceed to the evidence. One of the first surprising – and often overlooked – aspects of *GTNS* is that Fisher felt the need to discuss a good deal of sociological, anthropological, and historical literature. We find references to Edward Gibbon, Joseph Arthur de Gobineau, James Frazer, Edvard Westermarck, and other historians and social scientists.[2] But Fisher does not have the intention to show how all this literature can be reduced to a few biological principles. He discusses the advantages and disadvantages of the different proposals related to the evolution of humans and the collapse of civilizations. Then he illustrates how biology, as exemplified by genetics and evolutionary biology, can offer novel perspectives on old sociological phenomena. Human science can be enriched by, not reduced to, an understanding of natural selection and adaptation.

For instance, Fisher discusses de Gobineau's famous idea that racial mixture was the cause of aristocratic decline: "The Comte de Gobineau, to whose active mind the decline of civilizations presented itself as the greatest of human problems, believed that the disappearance of the original aristocracies was universally due to race mixture" (Fisher 1930, p. 238). While de Gobineau saw the aristocracy as a class of the best racial types of men who lost their supposed excellence through miscegenation with "inferior" races, Fisher believed that miscegenation produced genetic variance. Therefore, according to his Fundamental Theorem of Natural Selection, a mixed population would offer a higher rate of evolutionary change. We should not expect decay or degeneration from a mixed population – on the contrary. For Fisher, the problem of the disappearance of the "upper classes" could not be ascribed to genetic mixture, but to the way natural selection worked at the level of human societies. Natural selection was the central factor in evolution, in both human and non-human organisms. However, while evolution among non-human organisms could be fully explained by the rate of gene frequencies and selection, human evolution could not be fully understood in biological terms alone. The reason is simple: the human environment was more multifaceted than the natural environments to which organisms normally adapt. Humans live in a social

[2] On Gibbon, see Fisher 1930, pp. 204–205. On de Gobineau, Frazer, and Westermarck see, in order, pp. 238–239, p. 121, and p. 192.

environment that is itself the historical product of their needs, wills, and traditions. Postponement of marriage, sexual prohibitions, or any cultural practice of reproductive inhibition could have important consequences on the genetic assortment of human populations. These cultural elements are not genetically caused, but they influence the birth rate of the following generations. While fertility or infertility were conceived by Fisher as genetic characteristics, they were previously selected through what Fisher called "social selection," a phenomenon that civilization itself had set in motion.[3] For instance, considering the old classic example of peppered moth evolution, it may be observed that, just as the air pollution during the industrial revolution increased the frequency of dark colored moths, macro-cultural changes can increase or decrease the frequency of particular genetic characteristics such as fertility and intelligence. An extra-genetic cause – i.e. industrial revolution or postponement of marriage – produces the frame in which natural selection filters the "fitter" contextual options.[4] In short, Fisher believed that the social environment proposes and biology disposes so that both genetics and evolutionary biology are not enough to explain the trends of human evolution (although, of course, they may be enough to explain the evolution of non-human organisms).

It is important to specify, however, that for Fisher the relevance of extra-genetic elements affecting human evolution did not imply the acceptance of forms of non-genetic inheritance, as many recent theories of cultural evolution argue (see, for instance, Jablonka and Lamb 2005, Danchin *et al.* 2011). Indeed, in the Fisherian perspective, cultural elements produced the conditions under which the frequency of particular genes increases or decreases in a population. In brief, we do not need any other form of inheritance that is different from the genetic one (i.e. epigenetic, symbolic, or behavioral). Extra-genetic factors only shape the "cultural" or "environmental" space in which populations evolve though a change in gene frequencies.

Edwards's Warnings and the Place of the Human Sciences in Fisher's Biology

So far we have seen how complex the relation between natural selection and social environment was for Fisher. However, one could still ask; how can cultures change the frequency of inherited characteristics such as intelligence and fertility? In *GTNS*, Fisher mentions various examples showing how that might happen. For instance, Fisher examined the case of what he termed "barbarian communities." He argued that "barbarian communities" were naturally fit societies

[3] On Fisher's social selection, see Chapter XI of *GTNS*.
[4] With the proviso that while moths are affected by environmental changes they had not produced, humans are affected by the historical changes they themselves have set in motion.

because in such a social environment the genetic characteristics such as fertility and skills (the latter could be intelligence, courage, beauty, etc.) were highly valued and, therefore, selected together. Instead, in civilized environments, natural selection tended to sort out skilled and infertile people.[5] Why did that occur? Fisher argued that middle class habits, such as birth limitation through contraception or abstinence, postponement of marriage, and, therefore, the bourgeois propensity for limiting family size, had a significant statistical impact on reproductive rates. Indeed, through the analysis of censuses in Britain and the USA, Fisher observed that fertility was an inherited characteristic differentially distributed across social classes (i.e. proletarians were more prolific than middle-class people). Assuming that middle-class people were statistically more intelligent than proletarians,[6] and, conversely, that proletarians were more fertile, Fisher concluded that civilization (which, for him, was mainly exemplified by middle-class values) was creating the contextual conditions for the selection of infertile, and, at the same time, smart people. Not surprisingly, Fisher used the expression "social promotion" of infertility. The term referred to the process through which a new social environment could thwart the transmission of particular genetic traits. In other words, the social promotion of infertility produced the environmental circumstances for the genetical selection of infertility among the members of the middle class. From a more general perspective, Fisher was supposing that human social behaviors do not spring directly from human genetic endowments (as some sociobiologists might argue). Rather, they constitute the environmental conditions in which natural selection works, changing, generation after generation, the nature of the human animal. It was precisely for this reason that human sciences mattered to Fisher; they helped to identify the social forces shaping the space in which natural selection operates. Human populations, unlike other animal populations, have traditions and customs with a relevant impact on differential reproductive success.

Hence, how we read Fisher's book changes the way we interpret his evolutionary biology. The more we concentrate on Fisher's first chapters, the more we tend to consider him a radical supporter of the gene's-eye view. The more we focus on the latter five chapters, the more we tend to underscore Fisher's ideological motives. But if we read the book as one long argument, tying together natural and human sciences, we can better appreciate the boundaries and restrictions that he imposed

[5] In Fisher's words: "What is, perhaps, more important, is that a number of qualities of the moral character, such as the desire to do well, fortitude and persistence in overcoming difficulties, the just manliness of a good leader, enterprise and imagination, qualities which seem essential for the progress, and even for the stable organization of society, must, at least equally with intelligence, have led to social promotion, and have suffered, in consequence, a relative concentration in the more prosperous strata of existing populations. These classes have, however, also been selected for relative infertility" (Fisher 1930, pp. 262–263).

[6] We should keep in mind that intelligence was for Fisher an inherited trait, so that it could not be related to educational habits. As a consequence, social hierarchies were not structured according to inherited wealth, but according to inherited natural (genetic) characteristics.

on overambitious gene-centered views. I concur with Edwards's critical remark that "Fisher's reputation as having an oversimplified 'gene-centred' view of evolution is often promoted by those who, like Mayr, have not studied *The Genetical Theory* properly and have fallen into the trap of assuming that he was ploughing the same furrow as Haldane and Wright" ([227], p. 281, this volume). Accordingly, there are two principal warnings worth emphasizing in Edwards's interpretation of Fisher's work: first, that Fisher's gene-centric view was not as simple as many interpreters have maintained; and second, that Fisher's interests in eugenics cannot be taken as the driving motive of his science. Neither admonition, however, supports the conclusion that Fisher's engagement with human science was of small importance for the development of his evolutionary biology. An analysis of Fisher's overlooked attempt to tie evolutionary biology with human science could provide a richer framework for assessing his overall view of evolution – a view that is probably subtler and more multifaceted than many neo-Darwinians have imagined. As the biologist Alan Grafen pointed out about the last chapters in *GTNS* (Grafen 2003, p. 327):

> Today's biologists do not much read those chapters, I guess partly because of fear, essentially of social Darwinism and through an association with genocide, and partly because we are narrower technicians now and tend to suspect grand altruistic applications of our subjects. There is so much still to emerge from chapters 1–8 of Fisher (1930) that it would be surprising if a close study of the remaining, largely neglected, chapters did not yield important ideas.

Thus, taking seriously Edwards's warnings, it is time to reconsider Fisher's gene-centered view of evolution, but with a renewed effort to appreciate his integrated view of evolutionary biology and the human sciences.

References

Bennett, J. H. 1983. *Natural Selection, Heredity, and Eugenics*. Oxford: Clarendon Press.

Bennett, J. H. 1971–4, *Collected Papers of R. A. Fisher*. 5 volumes, Adelaide: University of Adelaide.

Crow, J. F. 1990. R. A. Fisher, a centennial view. *Genetics* 124, 207–211.

Danchin E., Charmantier A., Champagne F., *et al*. 2011. Beyond DNA: integrating inclusive inheritance into an extended theory of evolution. *Nature Reviews Genetics* 12, 475–486.

Edwards, A. W. F. 2000. *The Genetical Theory of Natural Selection*. *Genetics* 154, 1419–1426.

Esposito, M. 2016. From human science to biology. The second synthesis of Ronald Fisher. *History of the Human Science* 29, 44–62.

Fisher, A. R. 1930. *The Genetical Theory of Natural Selection*. Oxford: Clarendon Press.

Gayon, J. 2014. Population genetics, economic theory, and eugenics in R. A. Fisher. In *Evolutionary Biology: Conceptual, Ethical, and Religious Issues*, eds. P. Thompson, and D. Walsh, Cambridge: Cambridge University Press, 137–150.

Gould, S. J. 1996. The smoking gun of eugenics. In *Dinosaur in a Haystack: Reflections in Natural History*, ed. S. J. Gould, London: Cape, 296–308.

Grafen, A. 2003. Fisher the evolutionary biologist. *The Statistician* 52, 319–329.

Haldane, J. B. S. 1931. Mathematical Darwinism, a discussion of the Genetical Theory of Natural Selection. *Eugenics Review* 23, 1.

Hodge, J. S. 1992. Biology and philosophy (including ideology): a study of Fisher and Wright. in *The Founders of Evolutionary Genetics: a Reappraisal*, ed. S. Sarkar, Dordrecht: Kluwer Academic Publishers.

Jablonka, E. and Lamb, M. 2005. *Evolution in Four Dimensions: Genetic, Epigenetic, Behavioral, and Symbolic Variation in the History of Life*. Cambridge MA: MIT Press.

Mackenzie, D. A. 1981. *Statistics in Britain, 1865–1930*. Edinburgh: Edinburgh University Press.

Moore, J. 2007. R. A. Fisher: a faith fit for eugenics. *Studies in History and Philosophy of Biological and Biomedical Sciences* 38, 110–135.

Norton, B. 1978. Fisher and the Neo-Darwinian synthesis. In *Human Implications of Scientific Advance, Proceedings XV International Congress of the History of Science*, ed. E. G. Forbes, Edinburgh: Edinburgh University Press, 481–494.

Punnett, R. 1930. A review of '*The Genetical Theory of Natural Selection*'. *Nature* 126, 595–597.

Ruse, M. 1996. *Monad to Man: The Concept of Progress in Evolutionary Biology*. Cambridge, MA: HUP.

Wright, S. 1930. *The Genetical Theory of Natural Selection*: A review. *The Journal of Heredity* 21, 349–356.

Reading and Misreading Fisher: Anthony Edwards and the History of Population Genetics

JEAN-BAPTISTE GRODWOHL

> I am fond of denying that I am a historian and of claiming instead to be an informed story-teller. I sit between real historians of science (whom I greatly admire) and scientists themselves, hoping to improve their communication.
>
> *(AWFE to the author, 23 September, 2015)*

In the past two decades, Edwards's writings have compelled scientists and historians of science alike to revise some seemingly well-established points about the history of population genetics. This includes, most noticeably, their interpretation of R. A. Fisher's thinking. Indeed, the contrast between the conventional picture of Fisher, resulting from a large body of scientific and historical work, and the picture sketched by Edwards, on the basis of personal acquaintance and sustained reading of Fisher's work, can hardly be over-emphasized. In both philosophical and biological circles, the name of Fisher has become synonymous with an emphasis on natural selection to the exclusion of almost anything else, grand views about the maximizing properties of natural selection, and a neglect of the complexity of genetics for the sake of powerful generalizations (see, however, Bennett 1983 and De Winter 1997). A different Fisher appears in Edwards's writings and reminiscences. Here, we find a biologist who is anxious to anchor mathematical models in the detailed biology of study systems and is skeptical about the unbridled development of mathematical theory. This 'new' Fisher emerges as a scientist rather than as a philosopher, a problem-solver and method-builder rather than a one-theory man. Interestingly, Edwards has not only tried to clarify the thoughts of his master. He has also documented how Fisher has become the patron saint of a range of beliefs and mathematical methods that were only remotely related to his own work.

In this commentary, a number of archival documents are referenced. The author is grateful to A. W. F. Edwards for making available unpublished papers (AWFEP), the staff of the British Library for John Maynard Smith papers (JMSP) and William D. Hamilton papers (WDHP), and American Philosophical Society, Philadelphia for R. C. Lewontin papers (RCLP).

In doing so, Edwards has unveiled how the image of a leading scientist is constructed. This image is sometimes of no less historical significance than the views he actually held. Indeed, the views attributed to Fisher, as distinct from the ones he advocated, have exerted a deep influence on the history of theoretical population genetics.

Reading Fisher at Cambridge

Edwards's contributions to the historiography of population genetics are intimately linked to his involvement in Fisher's school at the University of Cambridge, in the late 1950s. It is worth reflecting upon the fact that Fisher was, briefly, in the position of leading a school of theoretical population genetics. Between 1950 and 1957, he gathered a small group of mathematically inclined students, who made models and devised experiments along lines suggested by Fisher. In this, Fisher's position was relatively unusual, especially by comparison with his colleagues and rivals, Haldane and Wright. In spite of their tremendous influence, neither Haldane nor Wright had supervised students in their own area of expertise. As observed by Provine, throughout his entire career at the University of Chicago, Wright had a number of students, but he did not supervise a single student in theoretical population genetics (Provine 1989, pp. 189–191). Primarily a physiological geneticist, Wright was a very independent researcher with respect to his work in population genetics, and unlikely to be a source of encouragement for students working on mathematical theory. At University College, London, Haldane had launched a dynamic laboratory of genetics, whose main activities were devoted to the study of the genetics of *Drosophila subobscura*. In this laboratory of experimental genetics, Haldane was *the* theoretician. An engineer having switched to zoology to work with Haldane, John Maynard Smith could have been the outstanding exception. Maynard Smith wanted to work on population-genetic theory and went to University College to fulfill this ambition. He soon gave up, feeling Haldane's presence inhibiting – Haldane was able to solve theoretical problems in a fraction of the time Maynard Smith would have needed (Maynard Smith 1985, p. 350). Hence "JMS" made himself an experimenter, only starting on population-genetic theory after Haldane's departure for India, in 1956. Maynard Smith's example points to the curious inhibiting effect a leading theorist can have on his own students, involuntarily discouraging them, by his own presence, from working in his own area of expertise.

Fisher's case was different. After settling at Cambridge, in 1943, he hired as lecturer a professional mathematician, George Owen, and tried to implement a course in genetics for undergraduates. These courses were of a very high level of mathematical sophistication, especially given the limited education in mathematics of biology students. They were, in effect, an original selection experiment, attracting students predisposed to mathematical thinking, while discouraging more biologically inclined students (on the popularity of Fisher's lectures, see Fincham and John 1995, pp. 123–124, Fisher Box 1978, p. 383). Latecomers to the Department of Genetics, Walter Bodmer and Anthony Edwards had been trained in mathematics and physics, respectively. They joined a small group of dedicated collaborators and students, which represented a distinct school. Partly because of its high mathematical standards, and partly as a result of the numerous quarrels that had isolated Fisher from his peers in the UK, this school stood relatively apart in England. An indirect effect of this isolation was the building of an *esprit de corps* among Fisher's students, who composed a close circle of devotees (see Edwards 1995, p. 808). This school started to bear fruit in the late 1950s, producing, among other things, a very technical literature on the population genetics of one-locus systems (see references in Edwards 1977).

Edwards's first independent inquiries crystallized on the problem of the evolution of parental investment. In 1957–1958, Edwards asked Fisher why, in herd animals, the sex ratio was still one half when many of the males were obviously superfluous. Preferring to make his students find the answers by themselves rather than providing them "fully baked," Fisher referred Edwards to his book. In a cryptic demonstration *ad absurdum*, Fisher had argued that the assumption of unequal parental investment between the sexes generated contradictory consequences (Fisher 1930, p. 142). This terse argument attracted little interest until the 1950s, when it became the object of two independent mathematical derivations. The first one was by an American student, Richard Shaw (see Shaw and Mohler 1953). The second one was by Edwards, who formulated explicitly the conditions underlying Fisher's argument, as offering a general theory of the evolution of parental investment (Bodmer and Edwards 1960).

In parallel with this mathematical work, Edwards decided to run an artificial selection experiment on mice, planning to change the sex ratio of his population by killing male offspring at birth. As for most experiments aiming to change the sex ratio, this small experiment failed. There would be no reason to mention it, apart from the effect it exerted on a Cambridge undergraduate assisting Edwards, William Donald Hamilton. Hamilton's initial contribution was negative. After two weeks of work, he pointed out to Edwards that the experiment was doomed to fail. According to Fisher's argument, Hamilton observed, natural selection acts on genes affecting the sex ratio at the time of conception (called the primary sex ratio). If mortality takes place *after* the period of parental investment, then it has no influence on the population sex ratio. Acting on the secondary sex ratio, by killing males, was thus unlikely to be of any effect (Edwards 2002, "W. D. Hamilton", AWFEP). Later, working at the Galton Institute and at Imperial College, London,

Hamilton mulled over various situations in which the conditions underlying Fisher's argument did not hold. Benefitting from Bodmer and Edwards's clarification, he tied together these explorations in a paper, "Extraordinary sex ratios," which effectively founded the field of sex allocation theory (Hamilton 1967; Hamilton, "Natural selection and the sex ratio," unpublished conference, WDHP, Z1X90). One of Hamilton's main points was to argue that situations in which Fisher's argument did not hold were quite widespread in nature, leading the population sex ratio to deviate from equality. Far from being an objection to Fisher's argument, Hamilton showed how this argument permitted the relation of sex-ratio deviations to specific violations of the assumptions underlying it. In doing so, he made Fisher's argument the null model of sex allocation theory (see Bull and Charnov 1988 for a discussion of the relevant literature).

In the meantime, the Fisher school had disappeared. Retiring in 1957, Fisher cohabitated for a while with his successor, John Marion Thoday, in Whittingehame Lodge. A bright, experimental geneticist, Thoday was a fervent believer in the principle that natural selection promotes the welfare and long-term survival of populations rather than the fitness of the organism (Thoday 1953). These views were finding a revival in the mid 1950s, prompting Fisher to condemn them explicitly in the second edition of *GTNS* (Fisher 1958, pp. 49–50). Cohabitation between Fisher and Thoday was thus unlikely to last. Hamilton, who had hoped to work with Fisher, wrote with anxiety: "I am afraid that he and Professor Thoday may fight like queen bees – the personal battle which the new queen always wins (Hamilton to C. Brasch, 9 Oct 1959, quoted by Segerstrale 2013, p. 54). Hamilton's fears were soon confirmed. Fisher, who had been invited by Australian geneticists for a six-month stay, decided to settle in Adelaide, and the Cambridge school dispersed. "When Fisher died, in 1962," Edwards commented, "he left behind no school of theoretical population genetics. The dispersal of the members of his department was world-wide – at about that time Bennett and Parsons were in Australia, Bodmer and Mandel in the United States, Cavalli-Sforza and Edwards in Italy, and Owen in Canada" (Edwards 1990, p. 282).

Out of Eden: Reading Fisher in a Wrightian World

Dispersal, however, does not necessarily mean extinction. These students tried to transplant Fisher's legacy to their new surroundings. Established at Stanford University, Walter Bodmer followed and encouraged the development of an influential school in mathematical population genetics, commonly referred to as the Stanford group (Bodmer 2009, p. 231). In his lectures at Brown University, Providence, RI, Peter O'Donald used *GTNS* as a basis for exercises, making computer simulations to check Fisher's mathematical estimates in the arduous chapters IV and V (P. O'Donald to WDH, Oct 19, 1970, WDHP, Z1X89; see O'Donald 1980). John Henry Bennett edited Fisher's scientific correspondence and provided one of the best overviews of Fisher's contributions to genetics and evolutionary theory (see Bennett 1983). For his part, having made himself a statistician, Anthony Edwards moved to Italy, then Stanford, before going back to the United Kingdom. He then intervened, from time to time, in population-genetic battles.

In the 1960s, the age-long battle between Fisher and Wright seemed to have been resolved in the latter's favor. In the wake of Motoo Kimura's and Richard Lewontin's landmark works, theoretical population geneticists were using and refashioning Wright's methods in exploring the mathematics of deterministic systems (e.g. Lewontin and Kojima 1960). In this Wrightian world, Edwards's incursions appeared as rear-guard reactions by one of the last remnants of the Fisher school. At regular intervals, Edwards challenged Wright's claims, spoke in defence of Fisher, and tried to set the record straight. A case in point was his review of the second volume of *Evolution and the Genetics of Populations*, in which Wright presented the results of 50 years of mathematical research (Wright 1969, [49]). Edwards's review was not the kind of respectful tribute one writes in homage to a founder of one's field: it was a detailed mathematical criticism of the basic tool used by Wright for analyzing the mathematics of genetic systems under selection, Wright's mean fitness function. A few years later, Edwards challenged Kimura's valiant attempt at putting Wright's adaptive topographies on firm mathematical ground (Kimura 1958; see Edwards 1974). In parallel with these debates, he was pondering over the meaning of Fisher's Fundamental Theorem of Natural Selection (FTNS).

Debating about Fisher

Edwards met the theorem during his Cambridge Tripos examination: "Write notes on the Fundamental Theorem of Natural Selection". He cautiously skipped the question, which became for him the cause for life-long interrogation. Edwards's problem was as follows. Fisher had stated that the rate of change in mean fitness under natural selection was equal to a certain quantity, namely, the genetic variance in fitness that existed at that time. Since a variance cannot be negative, it seemed to imply that the effect of natural selection on mean fitness could only be positive. But this seemed only to be true when most of the complications of genetics and evolution in the real world were excluded. Why had Fisher given such importance to a statement that depended on such stringent assumptions? Trying to make sense of the theorem, Edwards scribbled teaching materials and algebraic derivations, which he kept mostly to himself. In 1967, he sent a very cryptic letter to *Nature* about the FTNS ([36]), aptly summarized by George Price: "Fisher may have been correct – if only we could understand what he meant!" (Price 1972, p. 129). In the late 1960s, Price set down to work, compared meticulously the two editions of *GTNS* and established the now standard interpretation of the theorem, vindicating Edwards's conjecture: "It will be shown here," Price wrote, "that [Edwards's] view is correct. Fisher's theorem does indeed hold with the generality that he claimed for it. The mystery and the controversy [about the meaning of the FTNS] result from incomprehensibility rather than error" (Price 1972, p. 129). Price's interpretation has been abundantly commented upon: Price restricted the scope of the FTNS to describe a specific component in the change in mean fitness. The FTNS isolated the component that was caused by natural selection, in the process changing gene frequencies. It was this component, and not the full change in mean fitness, that was equal to the variance in the breeding value of fitness (or additive genetic variance) existing at that time (see Price 1972 for details, and Gayon 1992 for comments). Impressed by Price's reading, Edwards reconsidered his own algebra and added a comment: "this is mostly nonsense. See Price." However, Edwards and Price made very different uses of the theorem. After spending months on deciphering it, Price deemed it of very secondary interest. As he wrote to John Maynard Smith, "My paper on 'Fisher's "fundamental theorem" made clear' . . . is going to cause much surprise and amusement with its elucidation of the extraordinary mess Fisher made of things . . . It is not very important scientifically because Fisher's theorem is nowhere near so important as he claimed" (GRP to Maynard Smith, August 5, 1972, JMSP, Ms 86597B). Price has indicated in his paper the reasons for his disillusionment (Price 1972, pp. 139–140), which was consonant with his inordinate ability to lose interest in a topic after having worked on it for some time. In contrast to Price, Edwards considered FTNS to be important and consistently returned to it.

Initial stimulations came from repeated criticisms of the theorem, which was still circulating in its pre-Price version in the literature of theoretical population genetics. By the mid 1970s, this literature had become increasingly mathematical. Stanford mathematicians, in particular, were establishing a very productive school of their own, the overt ambition of which was to break with the tradition of vagueness and obscurity that had, in their view, dominated the field since its inception. Mathematical rigor and "exact" results were paramount. Maximizing principles then came under close scrutiny. Here, both Fisher and Wright provided targets. Because of their apparent affinities, Fisher's fundamental theorem and Wright's principle of the mean-fitness increase seemed like Siamese twins to US population geneticists: while Fisher had formulated the *law*, Wright had given the *method*. Hence, criticisms of Wright's mean-fitness method seemed to challenge also the status of Fisher's fundamental theorem. In a thorough analysis of two-locus systems, Samuel Karlin, leading force of the 'Stanford School', stated that Fisher's fundamental theorem was "in general" false (Karlin 1975, p. 372). In reaction, Edwards submitted to Karlin's journal, *Theoretical Population Biology*, a manuscript claiming that Fisher's theorem was immune from the on-going mathematical revisions. Referees were unimpressed. "I see no real reason for publishing this brief historical note, except perhaps in a book on the history of population genetics" (Anonymous review to an unpublished manuscript by AWFE, ~1976, AWFEP), the first referee commented. The second referee took things more personally: "Current-day workers are not fools nor knaves. They are valiantly trying to establish the truth in the face of very confusing and dogmatic statements by Fisher (and to a much lesser extent, Wright). They are trying, for example, to formulate precise models (a notable deficiency with Fisher) and to see what conclusions they lead to. It is certainly true that with any precise model with two or more loci, the FTNS is untrue in discrete time" (Anonymous review, ~1976, AWFEP). In reply, Edwards wrote to the editor: "perhaps I'll follow your first referee's advice and write a book on the history of population genetics . . . It could cover the period from the end of Provine's book until the arrival of Kimura on the scene, and might remove a number of my colleagues' historical misapprehensions" (AWFE to M. W. Feldman, September 27, 1976, AWFEP).

Edwards did not write this book, and the debate about Fisher repeated itself. Ten years after the reviews, he took issue with Joseph Felsenstein, this time in the columns of *Nature*. The origin of the dispute was a review of Provine's biography of Sewall Wright in which Felstenstein reminisced about his early years as a population geneticist, trained in Richard Lewontin's laboratory at the University of Chicago. "Evolutionary theory in the 1960s and 1970s was dominated by optimization; by the search for the quantity evolution would maximize . . . The picture hanging on our wall was of R. A. Fisher, who had emphasized the importance of natural selection in large populations" (Felsenstein 1986, pp. 175–176). This program of developing theoretical population genetics on maximizing principles had failed, Felsenstein explained, and Fisher's picture was taken down; extensive mathematical works had shown that, in general, selection did not lead to maxima of mean fitness. Felsenstein's reminiscences prompted Edwards to send a letter to *Nature* and object that Fisher did not share the belief in mean-fitness maximization that was so widely held among American population geneticists in the early 1960s: "indeed, his students from the late 1950s (I write as one of them) were taught *not* to think in such terms, and we developed a marked scepticism towards those many who did . . . In the matter of optimization, the picture on the wall was that of Sewall Wright" (Edwards 1987, p. 10). Replying to Edwards's letter, Felsenstein made a very interesting observation. He objected that Edwards's point "[wa]s relevant to a discussion of [Fisher's] intellectual system, but it happens to have been of no influence on subsequent researchers. My review invoked Fisher as the patron saint of this belief; this he was, even if he may not have been a believer himself" (quoted in [140], p. 220, this volume).

From this exchange, Edwards must have realized that Fisher's influence on the field did not reduce to his actual writings. Fisher's image, constructed by several generations of researchers, was certainly as influential. This image made Fisher a father of maximizing principles, arch-determinist oblivious of any complication hindering natural selection. Edwards's conclusion was original: he settled on teaching himself how Fisher had become the patron saint of maximizing principles. The result of this work was a paper bearing the appropriate title of "The Fundamental Theorem of Natural Selection" ([140]).

A Counter-History of Theoretical Population Genetics

The main conceptual aim of [140] was to disconnect Fisher's fundamental theorem from Wright's mean-fitness-increase principle. However, by relating the construction and diffusion of the avatars of Fisher's fundamental theorem in theoretical population genetics, Edwards showed in detail how the misrepresentations of Fisher's fundamental theorem influenced and shaped the development of the discipline. Edwards's paper reads as a counter-history of (American) theoretical population genetics, relating major results and discussions to failed attempts at proving or falsifying erroneous versions of Fisher's fundamental theorem. In this narrative, a world of people and problems appeared, which did not fit the clear-cut debates between selection and drift that were familiar to historians of science. Edwards's account put the spotlight on life in a modeling science. Heated debates flared up about apparently small and arcane problems, such as the reliability of mathematical methods, the value given to mathematical counter-examples, and the ambition of providing great quantitative generalizations. Surprisingly, this history was no less picturesque, no less riddled with controversies, than the classic narrative based on controversies between evolutionary theories.

Indeed, it is worth commenting on the effects this paper can exert on a scholar who has learnt the history of population genetics from the writings of William Provine, Jean Gayon, and Jonathan Hodge (Provine 1989, Gayon 1992, Hodge 1992). These pioneering works depicted the history of population genetics as a development of a grand debate between Fisher and Wright. At some point or other, every episode in the history of this field seemed to be related to it (see Skipper 2000 and Winther *et al.* 2013 for recent appraisals). Furthermore, Provine repeatedly emphasized that this debate was not of a technical nature, because Fisher and Wright had been able to solve their disagreements on quantitative questions (Provine 1992, pp. 202–203). It was a fundamental debate about evolutionary theory. To use Jean Gayon's powerful formulation, the Fisher–Wright debate may be described as a confrontation between a few fundamental "philosophemes" about the evolutionary process: attention to population structure versus overt emphasis on very large population sizes; a harmonious, integrative view on genetics versus gene reductionism; and considerations of the diversity of factors (inbreeding, epistasis, randomness) that affect the evolutionary process versus a focus on the sole deterministic effect of natural selection. It is of no surprise that the history of population genetics has been of such interest to philosophers of science: The Fisher–Wright debate was, simply, the philosophy of evolution in action. The lesson we, students in the humanities, drew (erroneously!) from these conclusions was that we could comment on the meaning of the debates without having to venture too much into the arcane mathematics. Equations did not seem that important. After all, we knew what the real issues were.

Edwards seriously challenged these comfortable conclusions. In particular, he highlighted problems. Documenting the methodological debates between Fisher and Wright, he quoted several texts in which Fisher consistently objected that Wright's mean-fitness function did *not* necessarily maximize. One of Fisher's main objections concerned the very stringent conditions for Wright's method to give reliable results. In this discussion, the positions usually attributed to Fisher and Wright seemed to be reversed: Wright was defending a general mathematical method ascribing maximizing properties to $\overline{W}$, brushing aside counter-examples and exceptions as trivia, while Fisher was claiming that this method made idealizing assumptions that would be seldom met in practice. Fisher gave the example of a violation of the random-mating assumption (self-fertilization in plants) in which Wright's method was useless (Fisher 1941). By quoting these objections, Edwards questioned the picture of Fisher as the arch-determinist, solely preoccupied with the effects of natural selection in idealized populations of infinite size. If Fisher was the selectionist he was supposed to be, only interested in the deterministic effects of selection that occurred in populations of very large sizes, why was he so hostile to Wright's deterministic equations?

Returning to the Fundamental Theorem

Edwards's account challenged the perception of Fisher, but it was not very explicit about Fisher's actual views. He was then primarily interested in the reception of Fisher, and the misrepresentation of the fundamental theorem in the American literature of theoretical population genetics. Having clarified for himself the American story, he decided to return to the roots of the fundamental theorem and to elucidate the context of its original formulation (e.g. [238], [248]).

The main results of this line of work are reproduced in this volume (e.g. [238], [248]), and I limit myself to point out his main conclusions. Edwards relates Fisher's fundamental theorem to his decade of work on the linear model. He compares the FTNS to the statistical analysis of the effect of selection on a character at a single moment in time, as in an agricultural experiment. In this light, the FTNS is an attempt to measure the quantity of progress caused by selection, as predicted by a linear-regression model of fitness on gene dosage. In formulating the theorem, Edwards argues, Fisher was still a biometrician, starting with the data and trying to isolate and measure the very variability that gave scope to selection. It was a piece of retrospective genetics. This simple observation entails several interesting consequences.

One such consequence concerns the coherence between Fisher's scientific works in statistics and in evolution. The search for what could tie together Fisher's various contributions to statistics, genetics, and evolutionary theory had been a major stumbling block for historians of science. In a thorough essay on Fisher, Jonathan Hodge declared that these contributions were unrelated. "Fisher's mathematics, that is, principally, his mathematical statistics – including what he saw as its lessons for the understanding of statistical inference and, beyond that, inductive logic – has little discernible relevance to his thinking about genetics and evolution" (Hodge 1992, p. 254). Indeed, the contrast between Fisher's alleged view of evolution and his contributions to statistics was perplexing. What struck historians of biology was Fisher's alleged emphasis on natural selection under highly idealized conditions, neglecting the complexity of genetic phenomena. By contrast, what struck researchers using Fisher's statistical methods was precisely their fertility in the study of complex phenomena, especially through the use of multiple linear regression, which permitted researchers to study several variables studied simultaneously, and to consider their interactions (see the reminiscences of an entomologist about Fisher's influence in Williams 1964, p. 305). Hence, Fisher's contributions to multivariate statistics did not seem only unrelated to his evolutionary writings, they seemed to be radically different ways of doing science.

As a matter of fact, the reader attempting to draw a coherent picture of Fisher's scattered works, even in the single field of genetics, might be disappointed by the lack of unifying theoretical agenda, especially by comparison with Wright. As shown in detail by Provine, Wright's works read as the coherent expansion of a single program, the progressive methodological consolidation of a theory of adaptive evolution whose main tenets were formulated in 1932 (Provine 1989). By comparison, Fisher's works do not present a similar unifying feature. Natural selection, which is supposed to unify his works (see Hodge 1992, p. 237), does not seem very relevant to Fisher's tedious analysis of tetraploid inheritance or to his theory of inbreeding. His book, *GTNS*, reads as a succession of insights, conjectures, and arguments, some of them based on innovative methods rather than on the exposition of a unified theory of evolution. This did not escape the attention of practitioners. In the mid 1960s, Richard Lewontin observed, in a letter to Sewall Wright: "your theory of evolution was a truly synthetic one. This, of course, is not true of Fisher's. In a sense Fisher did not really have a theory of evolution, but his contributions were of a kind of fragmented attack on a series of population genetical problems. Even *The Genetical Theory of Natural Selection* does not seem to hang together as a coherent theory" (Lewontin to Wright, December 1, 1965, RCLP).

If Edwards's interpretation of the FTNS is right, a common thread links Fisher's statistical work, the fundamental theorem, and his conjectures on the evolutionary origins of Mendelian phenomena, such as dominance and recombination. In these different areas of investigation, the emphasis is retrospective: "genetics supplies the fact as to living things as they now are, facts which, like the living things in which they occur, have an evolutionary history and may be capable of an evolutionary explanation, facts which are not immutable laws of the working of things but which might have been different had evolutionary history taken a different course" (Fisher 1932, p. 1965). In this sense, indeed, Fisher "did not really have a theory of evolution." To a certain extent, the advocacy of a particular theory of adaptive evolution may have seemed a limitation to Fisher, who did not understand why Wright felt the need to advocate an extremely specific hypothesis for adaptive evolution, dependent on very stringent assumptions about population structure (see the conditions required for the exquisite balance of factors in Wright 1932). Evolution was already there, in action, and had to be analyzed on a case-to-case basis. The aim of population genetics, as an inductive science, was to reconstruct the genetics and causes for the present phenomena, which were, after all, nothing other than "points of special intricacy" (see Fisher 1953). In this respect, Fisher was perhaps closer to Haldane than he thought. As observed by an excellent connoisseur of the small milieu of British geneticists, the immunologist Avrion Mitchison, the two grand old men of population genetics in England had much in common. Their styles of research were similar: they chose a problem, for the treatment of which they built the appropriate mathematical or statistical tools, before turning to another issue, being reluctant to speculate without an actual organism at hand (Mitchison 2003, pp. 50–51). This raises the question of the place of the FTNS in Fisher's genetics. It is an exact statement about the rate of change in mean fitness caused by selection, as formulated in a linear regression model. There is no question that Fisher considered the theorem as fundamental: the FTNS isolates a certain quantity of progress caused by selection, for any assumption about the mating system and demography of the species (see Grafen

2015). The theorem has conceptual implications, clearing away a certain number of misunderstandings about natural selection (such as that a large population is no less subject to selection than a small one, that natural selection is not limited to situations of limitation of resources), issues that were far from being clear for Fisher's contemporaries. However, it only elucidated a fundamental feature about natural selection. The rest of population genetics, for Fisher, that is, the very content of the science, was a study of details, for the understanding of which adequate tools had to be built. Hence, Fisher's *GTNS* reads, indeed, as a succession of puzzles and problems, rather than as a unified "theory" of evolution by natural selection.

References

Bennett, J. H., ed. 1983. *Natural Selection, Heredity, and Eugenics: Including Selected Correspondence of R. A. Fisher with Leonard Darwin and Others*. Oxford: Clarendon Press.

Bodmer, W. 2009. Sam Karlin: A personal appreciation. *Theoretical Population Biology* 75, 230–232.

Bodmer, W. F. and Edwards, A. W. F. 1960. Natural selection and the sex ratio. *Annals of Human Genetics* 24, 239–244.

Bull, J. J. and Charnov, E. 1988. How fundamental are Fisherian sex ratios? *Oxford Surveys in Evolutionary Biology* 5, 96–135.

De Winter, W. 1997. The beanbag genetics controversy: Towards a synthesis of opposing views of natural selection. *Biology and Philosophy* 12, 149–184.

Edwards, A. W. F. 1974. On Kimura's maximum principle in the genetical theory of natural selection. *Advances in Applied Probability* 6, 10–13.

Edwards, A. W. F. 1977. *Foundations of Mathematical Genetics*. Cambridge: Cambridge University Press.

Edwards, A. W. F. 1987. Evolution and optimization. *Nature* 326, 10.

Edwards, A. W. F. 1990. Fisher, $\overline{W}$, and the fundamental theorem. *Theoretical Population Biology* 38, 276–284.

Edwards, A. W. F. 1995. Fiducial inference and the Fundamental Theorem of Natural Selection. *Biometrics* 51, 799–809.

Felsenstein, J. 1986. Evolution in the twentieth century. *Nature* 324, 175–176.

Fincham, J. R. S. and John, B. 1995. David Guthrie Catcheside. 31 May 1907–1 June 1994. *Biographical Memoirs of Fellows of the Royal Society* 41, 119–134.

Fisher, R. A. 1930. *The Genetical Theory of Natural Selection*. Oxford: Clarendon Press.

Fisher, R. A. 1932. The evolutionary modification of genetic phenomena. In *Proceedings of the Sixth International Congress of Genetics, 1932, Ithaca, New York, Volume 1*, ed. Jones, D. F. New York: Brooklyn Botanic Garden, 165–172.

Fisher, R. A. 1941. Average excess and average effect of a gene substitution. *Annals of Eugenics* 11, 53–63.

Fisher, R. A. 1953. Croonian Lecture: Population genetics. *Proceedings of the Royal Society of London. Series B, Biological Sciences* 141, 510–523.

Fisher, R. A. 1958. *The Genetical Theory of Natural Selection*, 2nd edition. New York: Dover.

Fisher Box, J. 1978. *R. A. Fisher: The Life of a Scientist*. New York: Wiley.

Grafen, A. 2015. Biological fitness and the Fundamental Theorem of Natural Selection. *The American Naturalist* 186, 1–14.

Hamilton, W. D. 1967. Extraordinary sex ratios. *Science* 156, 477–488.

Hodge, M. J. S. 1992. Biology and philosophy (including ideology): a study of Fisher and Wright. In *The Founders of Evolutionary Genetics: A Centenary Reappraisal*, ed. Sarkar, S., Dordrecht: Kluwer Academic Press, 231–293.

Karlin, S. 1975. General two-locus selection models: some objectives, results and interpretations. *Theoretical Population Biology* 7, 364–398.

Kimura, M. 1958. On the change of population fitness by natural selection. *Heredity* 12, 145–167.

Lewontin, R. C. and Kojima, K. I. 1960. The evolutionary dynamics of complex polymorphisms. *Evolution* 14, 458–472.

Mitchison, A. 2003. Jim's cool reception among the British geneticists. In *Inspiring Science: Jim Watson and the Age of DNA*, eds. Inglis, J. R. Sambrook, J., and Witkowski, J. A., New York: Cold Spring Harbor Laboratory, 49–53.

O'Donald, P. 1980. *Genetic Models of Sexual Selection*. Cambridge: Cambridge University Press.

Price, G. R. 1972. Fisher's 'fundamental theorem' made clear. *Annals of Human Genetics* 36, 129–140.

Provine, W. B. 1989. *Sewall Wright and Evolutionary Biology*. Chicago, IL: University of Chicago Press.

Provine, W. B. 1992. The R. A. Fisher–Sewall Wright controversy. In *The Founders of Evolutionary Genetics: A Centenary Reappraisal*, Sarkar, S. ed., Dordrecht: Kluwer Academic Press, 201–229.

Segerstrale, U. 2013. *Nature's Oracle: The Life and Work of W. D. Hamilton*. Oxford: Oxford University Press.

Shaw, R. F. and Mohler, J. D. 1953. The selective significance of the sex ratio. *The American Naturalist* 87, 337–342.

Skipper, R. A. 2002. The persistence of the R.A. Fisher-Sewall Wright controversy. *Biology and Philosophy* 17, 341–367.

Thoday J. M. 1953. Components of fitness. In *Symposia of the Society for Experimental Biology*, ed. R. Brown and F. Danielli, Cambridge: Cambridge University Press, 96–113.

Williams, C. B. 1964. Some experiences of a biologist with R. A. Fisher and statistics. *Biometrics* 20, 301–306.

Winther, R. G., Wade, M. J., and Dimond, C. C. 2013. Pluralism in evolutionary controversies: styles and averaging strategies in hierarchical selection theories. *Biology & Philosophy* 28, 957–979.

Wright, S. 1932. The roles of mutation, inbreeding, crossbreeding and selection in evolution. In *Proceedings of the Sixth International Congress on Genetics, Ithaca, New York 1932*, ed. Jones, D. F., Ithaca, NY: Brooklyn Botanic Garden, vol. 1, 356–366.

Wright, S. 1969. *Evolution and the Genetics of Populations: A Treatise in Four Volumes. Volume 2: The Theory of Genetic Frequencies*. Chicago, IL: University of Chicago Press.

Variance-Partitioning and Classification in Human Population Genetics

NOAH A. ROSENBERG

In his 2003 paper on human genetic diversity ([192]), Edwards develops a simple model of allele frequencies in two populations, using it to examine the potential for classifying an individual genome into one of the two populations on the basis of its alleles. He employs the model to critique interpretations of the magnitude and significance of human genetic diversity tracing to the celebrated paper of Lewontin (1972).

Edwards draws a distinction between two questions about human genetic variation. The first is the variance-partitioning question initially posed by Lewontin (1972): in a partition of worldwide human genetic variation into components representing differences among individuals within populations, differences among populations within major geographic groups, and differences among geographic regions, what are the numerical values of the variance components? The second is the classification question: can an individual be accurately classified into one of a series of specified populations on the basis of his or her multilocus genotypes? The notable contribution of Edwards's article is the lucidity with which it uses a simple two-population model to demonstrate that the result for the former question need not imply the result for the latter; even if the within-population variance is the largest component, classification can still be possible.

As Edwards's treatment shows, the statistical distinction between the two questions lies in the way in which information is combined across multiple loci. In a variance-partitioning analysis, different loci amount to separate trials, each of which produces its own estimate of the desired variance components. The variance partition for a multilocus dataset is computed as an average across loci; additional loci merely refine the precision of the estimate and ensure that it approaches the value that would be obtained if data were available genome-wide. In a classification analysis, however, information is *accumulated* across loci, rather than simply averaged; while any particular allelic type might occur at only slightly different frequencies across groups, multilocus *combinations* of allelic types differer dramatically in their probability of originating in different source populations. Classification is possible not because any one genetic character is population-specific or even substantially different in frequency across groups, but because multilocus combinations of characters are much more likely to arise in one group than in the others. Variance-partitioning does not consider multilocus combinations; in fact, in many implementations, including Lewontin's, it is performed directly from allele frequencies – without any knowledge of individual multilocus genotypes.

Edwards writes that his article "could, and perhaps should, have been written soon after 1974" (p. 801). In a sense, it was. Shortly after the appearance of Lewontin's (1972) variance partition and the related work of Nei and Roychoudhury (1972, 1974), Spielman and Smouse introduced the classification problem through a multivariate assignment of individual members of geographically dispersed Amazonian natives by village, village cluster, or tribe (Spielman and Smouse 1976, Smouse and Spielman 1977), reporting a classification success "far in excess of random expectation" (Spielman and Smouse 1976, p. 330). Mitton (1978) then used multilocus combinations to implement a form of variance-partitioning, finding substantially greater difference among populations than in the averaging work of Lewontin (1972). This paper drew three sharp critiques, framed primarily as statistical objections (Chakraborty 1978, Lewontin 1978, Powell and Taylor 1978), to which Mitton (1978), employing principal-component analysis, responded by arguing that "multivariate analyses may provide resolution of groups that is not apparent in a sequence of single-locus analyses" (p. 1143).

A commentary by Neel (1981) characterized the distinct questions appearing in the exchange. These were first, the problem of Lewontin, Nei, and Roychoudhury (Neel 1981, p. 83): "What proportion of all the genetic variation within some large group can be apportioned to differences among subgroups and among individuals, on the average, over all known loci?" And second, the problem of Spielman, Smouse, and Mitton (Neel 1981, p. 83): "Are the levels of allelic frequency variation found between human populations sufficient to generate a useful taxonomy?" Neel wrote that "Chakraborty, Lewontin, Powell, and Taylor appear to me to be chastising Mitton for asking his question rather than the one already posed by Lewontin, Nei, and Roychoudhury." Thus, in [192], Edwards's attempt to separate a similar pair of questions occupies a parallel position in the literature to Neel's commentary (and another by Chakraborty [1982], analyzing the latest work of Smouse *et al.* [1982]), except that in Edwards's critique, it is Lewontin who is chastised.

The efforts represented in Edwards's paper, in the early literature, and in further recent attempts – including one co-authored by Lewontin – to clearly distinguish questions at issue in the study of human genetic variation (Feldman and Lewontin 2008, Barbujani and Colonna 2010, Rosenberg 2011) illustrate the ongoing challenge of providing a genetic portrayal that simultaneously represents both similarity and difference in an accurate manner. As Edwards remarks, the variance-partitioning result has been taken as the basis for statements that it does not directly imply. For example, the frequently reported summary of this result, in the form of a statement that 85% of human genetic variation is within populations, might deceptively appear to respond to a question not about variance-partitioning, but rather about the presence and absence of alleles: of the genetic variants that exist in the human genome, how many are present within a given population? This latter question has a result smaller than 85%, with quite different values occurring in different parts of the world (Rosenberg 2011).

However, a connection between formulations of the classic population-genetic statistic F_{st} in a type of variance-partitioning calculation and in an analysis that examines times of coalescence of genealogical lineages, in the same and different populations, does mean that variance components have a surprising equivalence with results on pairwise similarity and difference (Rosenberg 2011). In this equivalence, the within-population variance component for worldwide human genetic variation – computed using a heterozygosity statistic instead of the entropy statistic in Lewontin's original work – is approximately equal to the ratio of the number of pairwise genetic differences for two individuals, chosen from the same population, to the number of pairwise differences for two individuals who are chosen arbitrarily from anywhere in the world. Thus, the *Nature* statement quoted by Edwards (p. 250, this volume), "two random individuals from any one group are almost as different as any two random individuals from the entire world," if considered in terms of pairwise differences, and assuming 85% is taken to be a large number, is a legitimate – though non-obvious – consequence of Lewontin's result.

The fallacy to which Edwards refers in the title of [192] concerns the conflation of variance-partitioning and classification, but he also critiques Lewontin's view that a high value of the variance component for differences among individuals within populations argues against human racial classification. The destructive typological race theories of the past, requiring small numbers of genetic characters with dramatic biologically significant differences across groups to be typical of the genome-wide pattern, are clearly belied by the high within-population variance component. In the disagreement between Edwards and Lewontin, however, much about their senses of the meaning of "race" is left unstated in both positions. The interpretation of the magnitude of human genetic similarities and differences in relation to "race" demands more explicit context on what properties a theory of the existence or non-existence of "races" entails for human variation. It also requires an analysis of whether or not the necessary criteria are satisfied by the actual patterns in human populations. Nevertheless, Edwards's critique helps to illuminate that in assessing the applicability of any such theory of race, the answers to many separate questions about human genetic similarity and difference must be simultaneously explained.

A phenotypic perspective that extends beyond an exclusively genotypic analysis is surely relevant for such inquiries. Edge and Rosenberg (2015) extended the model in [192] to incorporate a selectively neutral quantitative phenotype, examining three additional questions related to variance-partitioning and classification. (1) How does the mean difference between two populations in the value of a phenotype change as the number of loci that influence the trait increases? (2) What proportion of variance in a phenotype is expected for the between-population variance component? (3) Does a phenotype become increasingly useful for identifying the source population of an individual as the number of loci grows? In each case, the model predicts that a single neutral quantitative trait has similar properties in terms of variance-partitioning and classification to those of a single genetic locus. This conclusion, building on Edwards's framework, solidifies the link that the variance-partitioning and classification results in the genetic studies of Lewontin and Edwards extend analogously to a class of phenotypes. Thus, [192] has contributed not only to characterizing the multiple distinct questions that are of interest in the study of human genetic similarities and differences, but also to providing a model for their further investigation.

References

Barbujani, G. and Colonna, V. 2010. Human genome diversity: frequently asked questions. *Trends in Genetics* 26, 285–295.

Chakraborty, R. 1978. Single-locus and multilocus analysis of genetic differentiation of the races of man: a critique. *The American Naturalist* 112, 1134–1138.

Chakraborty, R. 1982. Allocation versus variation: the issue of genetic differences between human racial groups. *The American Naturalist* 120, 403–404.

Edge, M. D. and Rosenberg, N. A. 2015. Implications of the apportionment of human genetic diversity for the apportionment of human phenotypic diversity. *Studies in History and Philosophy of Science, Part C: Studies in History and Philosophy of Biological and Biomedical Sciences* 52, 32–45.

Feldman, M. W. and Lewontin, R. C. 2008. Race, ancestry, and medicine. In *Revisiting Race in a Genomic Age*, eds. Koenig, B. A., Lee, S. S. J., Richardson, S.S., Piscataway, NJ: Rutgers University Press, 89–101.

Lewontin, R. C. 1972. The apportionment of human diversity. In *Evolutionary Biology*, eds. Dobzhansky, T., Hecht, M. K., Steere, W. C., New York: Appleton-Century-Crofts, vol. 6, 381–398.

Lewontin, R. C. 1978. Single- and multiple-locus measures of genetic distance between groups. *The American Naturalist* 112, 1138–1139.

Mitton, J. B. 1977. Genetic differentiation of races of man as judged by single-locus and multilocus analyses. *The American Naturalist* 111, 203–212.

Mitton, J. B. 1978. Measurement of differentiation: reply to Lewontin, Powell, and Taylor. *The American Naturalist* 112, 1142–1144.
Neel, J. V. 1981. The major ethnic groups: diversity in the midst of similarity. *The American Naturalist* 117, 83–87.
Nei, M. and Roychoudhury, A. K. 1972. Gene differences between Caucasian, Negro, and Japanese populations. *Science* 177, 434–436.
Nei, M. and Roychoudhury, A. K. 1974. Genic variation within and between the three major races of man, Caucasoids, Negroids, and Mongoloids. *American Journal of Human Genetics* 26, 421–443.
Powell, J. R. and Taylor, C. E. 1978. Are human races "substantially" different genetically? *The American Naturalist* 112, 1139–1142.
Rosenberg, N. A. 2011. A population-genetic perspective on the similarities and differences among worldwide human populations. *Human Biology* 83, 659–684.
Smouse, P. E. and Spielman, R. S. 1977. How allocation of individuals depends on genetic differences among populations. In *Human Genetics: Proceedings of the Fifth International Congress of Human Genetics, Mexico City, 10–15 October 1976*, eds. Armendares, S., Lisker, R., Amsterdam: Excerpta Medica, 255–260.
Smouse, P. E., Spielman, R. S., and Park, M. H. 1982. Multiple-locus allocation of individuals to groups as a function of the genetic variation within and differences among human populations. *The American Naturalist* 119, 445–463.
Spielman, R. S. and Smouse, P. E. 1976. Multivariate classification of human populations I. Allocation of Yanomama Indians to villages. *American Journal of Human Genetics* 28, 317–331.

Part 3

Interviews with A. W. F. Edwards

What follows is a selection from a set of interviews with A. W. F. Edwards carried out over three days by R. G. Winther in Cambridge, UK, in May 2015. Day 1 covered selection theory; Day 2, phylogenetic inference; and Day 3, Edwards and Fisher. Numbers in brackets (e.g. [28]) refer to Edwards's article numerical system used throughout this volume. Important or hard-to-find articles and books mentioned by Edwards or Winther are footnoted, but in the interest of flow, not every published work referred to is footnoted. Some of the commentators are mentioned, and the commentaries above can be consulted for further background. Amy Coffin and Virginia Peschke transcribed the interviews, and A. W. F. Edwards and R. G. Winther, and the copy-editor Zoë Lewin, checked, verified, and streamlined the transcripts. We attempted to correct for infelicities of spoken language and of the recording so as to make as faithful and readable a transcript as possible. To that end, Winther is grateful that quoting Edwards verbatim is effectively like quoting a completed and elegant typescript. However, with a view to being more concise, the following is a slightly abbreviated version of the conversation. Note that '[. . .]', embedded in a paragraph, indicates that some lines have been deleted. Where '[. . .]' is on a separate line, more substantial sections of the conversation have been removed. The full version of these interviews can be requested directly from the editor.

Day 1

Selection Theory

Edwards's Office, Cambridge, UK

20 May 2015

Selection Theory and Mathematical Population Genetics

[. . .]

RGW: What is your approach in your book *Foundations of Mathematical Genetics*?[1] [. . .]

AWFE: [. . .] What that book does is simply reflect what I knew at the time, what I thought worthwhile putting down, but more particularly, making use of trilinear coordinates which I'd learnt as an extra at school, a non-examinable extra, and because I have a fairly geometrical mind I found that approach very attractive. I could see that there was an analogy between quadratic mathematics, diploid organisms, and conic sections. So why not talk about diploid organisms in the language of conic sections?

RGW: One could say that you have added something to the literature . . . in terms of thinking about the geometry of population genetics. You have de Finetti diagrams in there. Those seem to be one of the few kinds of geometrical diagrams that other textbooks such as Hartl and Clark use.[2] But could you say a little more about whether you think that there still might be a space in population genetic theory for geometry as opposed to calculus?

AWFE: I can certainly produce an unsolved geometrical problem. [. . .] I spent a lot of time trying to find a geometrical proof of the fact that, at a single multi-allelic locus under random mating, the mean fitness is a non-decreasing function. This is Mulholland and Smith and so forth, summarised in

[1] Edwards, A. W. F. 2000. *Foundations of Mathematical Genetics*, 2nd edition. Cambridge: Cambridge University Press.

[2] Hartl, D. L. and Clark, A. G. 1989. *Principles of Population Genetics*, 2nd edition. Sunderland, MA: Sinauer Associates.

Kingman's work.[3] There must be a geometrical demonstration of that. When I was a student, I was not working on that side of things, but I shared rooms with Steve [P. H.] Mandel, and Mandel was working on precisely these problems. He told me many years later that Fisher used to come into the room, which we shared, when I wasn't there, and talk to him about this problem. Fisher would cover the blackboard with conic sections and try and explain to Steve something about the relevance of that kind of mathematics to the increasing nature of the mean fitness $\overline{W}$.[4] Now it's a great pity that it happens that Fisher used to come upstairs into that room when I wasn't there. Because if I'd been sitting there quietly listening to these two, I might have been able to remember something. But Steve Mandel, who is still alive, can tell me no more. The problem remains that somewhere, using trilinear coordinates and a bit of ingenious geometrical mathematics, there's a geometrical analogue of Kingman's proof.[5]

[. . .]

RGW: Is there anything else you would like to add about how geometry can be used in population genetics? Let me be a little more specific. We will get to this question a little later, but do you think it could at all be related to the adaptive landscape? Is the adaptive landscape geometrical and is that then just an unfortunate use of geometry?

AWFE: I'm not aware that the adaptive landscape metaphor has ever really had a geometrical backbone to it. Some writers called it the 'adaptive seascape', and that in fact is a very much better metaphor. I'd like to be able to claim that phrase for myself but I've picked it up and I don't know where it comes from.

RGW: Is it a better landscape because it's dynamic or because it can be smoother?

AWFE: It's better because it's dynamic. It's changing all the time.

RGW: But so is the landscape. It's just your geological point of view. Just change the geological scale and the landscape changes. [Both laugh]

3 See Mulholland, H. P. and Smith, C. A. B. 1959. An inequality arising in genetical theory. *The American Mathematical Monthly* 66, 673–683; Kingman, J. F. C. 1961a. A mathematical problem in population genetics. *Proceedings of the Cambridge Philosophical Society* 57, 574–582; Kingman, J. F. C. 1961b. On an inequality in partial averages. *The Quarterly Journal of Mathematics* 12, 78–80.

4 From the epilogue of the first edition of Edwards's *Foundations of Mathematical Genetics* (published by Cambridge University Press in 1977): 'I have omitted all treatment of gene-frequency change by the use of differentials of the mean viability (like path coefficients, also due to Sewall Wright) . . . This is a major departure from current text-book practice' (p. 108). Edwards's assessment remains true of the nearly identical second edition except in the added chapter, 'Fisher's Fundamental Theorem.' Interestingly, the only other substantive change in the second edition (2000) was the removal of the epilogue, which had ended as follows: 'None of the books listed goes more deeply into the basic theory than we have done; rather, they show its extension laterally and upwards. But it seemed worthwhile to try to secure the foundations' (p. 109).

5 Kingman 1961b.

AWFE: To go back to texts on population genetics. I don't write books for any other reason than that I just enjoy writing about these things. I like doing that sort of thing, so I sit down and I do it. I don't think any of my books have been written at the behest of a publisher.

[...]

RGW: If we think of the ANOVA [analysis of variance] decomposition of the phenotype into additive variance, into dominance variance, and so forth (looking at Falconer, and Falconer and MacKay, looking more at the quantitative genetics point of view[6]), is there any sense in which the way that Fisher thought about the ANOVA decomposition of phenotype is related to how he was thinking about the design of experiments and how he was thinking about ANOVAs more generally?

AWFE: There is an ambiguity in Fisher's use of the word 'statistical' in this context. In the early times of his writings, I now think that when he used the word 'statistical', he was simply referring to the statistical theory of gases. He was drawing that analogy. I've made the mistake somewhere in my writings of not understanding that, and thinking that when he used the word 'statistical' he was actually referring to his parallel work on 'statistical' inference. We needn't dwell on the problem of 'statistical' as he first meant it because that's an obvious connection and we know from his background that that was a very important part of his education, statistical mechanics. We then have to ask, what connections are there between his thinking about statistical inference and his contributions to evolutionary theory? There, I think we are on much more difficult ground. But I have some ideas from his writings in the 1930s, which might have been in the background of his mind at the time. Because he was very clear from 1930 onwards, when writing about evolutionary matters, that his approach was different from what he called the 'analytic approach', particularly as exemplified by the work of Sewall Wright and J. B. S. Haldane. He didn't criticise that; he just said that it was not the way that he thought about analysing evolution, and it's notable that he himself very rarely proceeded, apart from the odd example, in that direction. He did not do what most people would regard as normal mathematical genetics. His approach, he described in 1932, by contrast, was 'inductive'. He was interested in drawing out from what was observed what were the basic natural selection forces at work and how they could be described. And, in doing that, he was greatly influenced by his own 1918 paper. There, you have the whole 'analysis of variance' concept beginning to inform his genetics.

[6] Falconer, D. S. and Mackay, T. F. C. 1996. *Introduction to Quantitative Genetics*, 4th edition. Harlow: Longman.

The 1918 paper, contrary to what many people say about it, has nothing about natural selection in it at all. I think people get tired before they get to the end, and then a few years later they make comments about the 1918 paper containing things which, in fact, even towards the end, it doesn't actually contain. But it contains one very important statement, which is mistaken but was very important in Fisher's development of *The Genetical Theory of Natural Selection* [*GTNS*][7], on the non-stochastic side. That is the statement that if you regress the values of a character – height was a familiar example of this – on the number of genes of a particular type in a locus and you use the values, the regression values, which he called the 'average effect of the genes', then they carried the information which was passed on to the next generation. Now, we know that that is only true under certain circumstances. It's the basic regression, the simple regression model. But that informed his Fundamental Theorem of Natural Selection. That's the basis of that. So we have to be careful when thinking about what impact statistics as a whole had on Fisher's thinking in evolutionary theory, to be quite clear which bits of the statistical thinking had an influence.

[...]

RGW: [...] Just some clarification: could you give a definition or characterisation of the analytical approach, as Fisher saw Wright and Haldane doing, versus a characterisation or definition of how Fisher thought of statistical inference, in contrast. What would be very brief characterisations of the analytical approach? The statistical approach? The inductive approach?

[...]

AWFE: [...] The important paper is his paper in 1932 at the International Congress of Genetics in the presence of both Haldane and Sewall Wright.[8] And that's the congress at which Sewall Wright produced his adaptive landscapes. So what Fisher is saying there, mostly, is what he is *not* doing. He is *not* constructing a mathematical model which would enable him to predict what the position was going to be in 20 generations' time. All the students doing mathematical genetics who are my contemporaries were used to starting with that model. It seems a very sensible model to start with. But in many statistical situations, experimental situations, you don't have a model. You're doing induction. You

[7] Fisher, R. A. 1930. *The Genetical Theory of Natural Selection*. Oxford: Clarendon Press.
[8] Fisher, R. A. 1932. The evolutionary modification of genetic phenomena. *Proceedings of the Sixth International Congress of Genetics* 1, 165–172; Wright, S. 1932. The roles of mutation, inbreeding, crossbreeding and selection in evolution. *Proceedings of the Sixth International Congress of Genetics* 1, 356–66.

plant the seeds, you count the corms or whatever they are, and then you settle down to try and analyse what's going on. You look at the main effects, and you see what they do. Then you look at the two-factor interactions. Under no stage do you actually have a model of the biological process which has produced the data. One engages in inductive inference; that's what Fisher is trying to do, in my view, in the 1930s. Paramount, as his method of looking at natural selection from the point of view of the distribution of the genes, which he felt was the most important thing; paramount, in that analysis is his Fundamental Theorem.

[. . .]

RGW: Is the Fundamental Theorem an analytical approach?

AWFE: No.

RGW: Here, I get confused. I can understand Fisher's contrast, I can see why ANOVA is an inductive approach. What I have a harder time seeing is how the Fundamental Theorem, at least the way the deductions happen, once you have your assumptions, and you have the definitions of average effect and average excess . . .

AWFE: [Interrupting] Those aren't definitions. Well, they're definitions, but they're not descriptions of something which is supposed to exist. They're not analytical.

RGW: What are they?

AWFE: They are parameters, if you like. They're more like parameters.

RGW: But you set up initial . . .

AWFE: [Interrupting] He [Fisher] didn't set up anything. The point of the Fundamental Theorem is that – well not the *point* of it, but the *explanation* of the Fundamental Theorem – is that you do not have things like viabilities of genotypes. You do not have generations. You do not have mating. None of those things occur on those pages. He is analysing what has happened, not setting up a model for why it happened. Everybody is busy treating it, or even criticising it, or even misunderstanding it, as though it were a model of an analytical kind.

RGW: Warren Ewens and others have recast Fisher's description or induction of the Fundamental Theorem in terms of an analytical model. Do you think that's true?

AWFE: Yes, absolutely.

RGW: Okay, so it has been *recast* as an analytical model.

AWFE: It has, yes. And I found that very influential, because suddenly there was light thrown on the Fundamental Theorem. I, along with practically everybody else, was confused about the Fundamental Theorem. All I will claim for my approach is that I knew that it wasn't what it wasn't.

RGW: In other places in *GTNS*, he . . . does he build models in other places?

AWFE: Well, increasingly, I think, that . . . I assumed, when I was younger, that these difficult parts of *GTNS* like, for example, the bit on natural selection and the sex ratio . . . [had] some mathematics behind [them] and that the job of us research students was to dig out what that mathematics was. But nowadays, I believe that his mathematical ability was so intuitive that he didn't actually have to do the mathematics.

[. . .]

RGW: Do you think it has then been useful for you, Ewens, and almost countless others to try to work out some of the models that might be behind, or that might be interpreted from, Fisher's words?

AWFE: Yes, I think it has been useful, because even though I increasingly believe that they were not the models that were in Fisher's head, I think they've had a very important instructive effect. I think that getting out the mathematics of natural selection and the sex ratio [was useful], which incidentally was done by Carl Düsing in Charles Darwin's time, anyway, as I have pointed out in the literature, as I managed to discover.[9] That is very educative, and that's certainly true with the Fundamental Theorem, and, in particular, Warren Ewens's 1989 paper is the source of all, I was going to say *correct* comment, but all modern *helpful* comment on the Fundamental Theorem.[10] It's been very interesting to see how it was possible to get out of Fisher's ideas a model with generations and with fitnesses attached to specific genotypes, and still show the mechanics of the theorem. But then, of course, as Warren Ewens often says, 'That's not very good, is it?' Because the average effects, in fact, are always changing. That's because it's not actually the model that we need.

[. . .]

AWFE: If you want a model, you'd better go and ask Sewall Wright, and read his big books, which is an attempt to create that model, that sort of global model, but of course he has to keep making adjustments. For example, when he eventually admitted that, in the simplest cases, the population does not move up his adaptive landscape by the steepest route, he had to invent things called 'fitness functions', which were alternative views of fitness, which would themselves then fit the straitjacket which he was imposing on the system.

9 Edwards, A. W. F. 2000. Carl Düsing (1884) on *The Regulation of the Sex Ratio*. *Theoretical Population Biology* 58, 255–257.

10 Ewens, W. J. 1989. An interpretation and proof of the Fundamental Theorem of Natural Selection. *Theoretical Population Biology* 36, 167–180.

RGW: Let me ask you a very simple question, because we are going to have philosophers of science as an audience also of this book. For you, what are the functions of a model? What does a model do and what do you do with a model?

AWFE: Well, I once wrote a paper called 'Models in genetics'.[11]

RGW: Can you say more?

AWFE: When I was a research student, because I was asked at short notice to stand in for Niels Bohr, I think. So I thought, 'Well, I don't know very much about this subject, but I'll talk about models in genetics'. I think it must have been a colloquium on Models in Biology or something like that. I say this jokingly, because I don't think you'll find anything interesting there or relevant to answering the question.

[...]

RGW: Are your geometrical reconstructions, or your geometrical proofs, in *Foundations of Mathematical Genetics*, are they a model?

AWFE: Well, they're based on a model. They're based on a very simple Mendelian model. I would certainly use the word 'model' in their connection. I think there are difficulties in Anglo-American interpretations of the word 'model', as a matter of fact.

RGW: Yes, absolutely.

AWFE: I've noticed this in connection with maximum likelihood. That sometimes the American use is to talk about the 'maximum-likelihood model', but as far as I'm concerned, that's a *method*, which you might apply to a model.[12] So I have a restricted, perhaps English, notion of what a model is. Of course, physical models we can all understand, so we're really talking about mathematical models and logical models, and I think I would limit the use of the word – I think I *do* limit the use of the word in my own work – to logical models, which of course include mathematical models.

[...]

RGW: [...] I'm going to throw in another word in the mix. I think you're absolutely right. It's very interesting how you're distinguishing 'method' from 'model'. What about 'theory'? What is a theory?

[...]

AWFE: It's not an easy question, but what would I expect of a theory? Let me think aloud. Would I expect a theory necessarily to be predictive? Well,

[11] Edwards, A. W. F. 1960. Models in genetics. In *Models and Analogues in Biology (Symposia of the Society for Experimental Biology)*, Issue 14, Cambridge: Cambridge University Press, 6–12.

[12] See also Edwards's book *Likelihood*, first published by Cambridge University Press in 1972, reprinted in 1976, and reissued as a "Cambridge Science Classics" paperback in 1984 and 1987. An expanded edition of the book appeared with Johns Hopkins University Press in 1992.

I guess in the sense that anything that enables me to make predictions constitutes a theory. Are there theories which exist which don't enable me to predict anything? I don't think I've got anything useful to say on this, actually.

[. . .]

AWFE: Well, of course, I will say one thing, and that is that when we use the phrase 'mathematical theory', that's self-defining. I mean, it's a theory because it's a bit of mathematics, and the mathematics is based on axioms, and we assume it's correct. But we still call it a theory. Pythagoras's theorem, for example.

[. . .]

RGW: I'll admit one bias. I learned population genetics from Hartl and Clark. It's a generational thing. And yes, I learned it at Stanford, from Marcus Feldman, so it's in an American context. But I say that because they have an apparatus which, I guess, is fairly standard, not universal, where you have little omegas, which are the fitnesses of either a gene or a genotype, but they're individual fitnesses, and then they have capital $\overline{W}$, with a bar, which is the mean fitness. There are other kinds of fitnesses. I mean, you can talk about life stage; Charlesworth has worked on fitness [. . .]

AWFE: Yes, he's got a whole book.[13]

RGW: Yes, a whole book. Then you have to talk about components of fitness, reproduction and survival, so even in the mathematical notation, the concept is pluralistic, polysemous – some might say it's a mess. In all these mathematical symbolisms, how do you think Fisher thought about it? Do you think he had these kinds of distinctions in mind?

AWFE: No, because these are all distinctions which are important in specific models.

RGW: I see.

AWFE: He really wasn't doing analytical models like that. So insofar as he is clear about what he means by 'fitness', he introduces it in Chapter II of *GTNS*, and he's creating a theory of Malthusian fitnesses, which are sort of imaginary fitnesses that would apply once the population has reached a stable state.

RGW: And then, what is it comparative against? What is the contrast class?

AWFE: In the case of the Fundamental Theorem, it's genotypes.

RGW: Okay, at one locus? Or can it also be at multiple loci?

[13] Charlesworth, B. 1994. *Evolution in Age-Structured Populations*, 2nd edition. Cambridge: Cambridge University Press.

AWFE: He doesn't actually say (well, he probably does), but it's clear that he's thinking quite generally, you see. He doesn't have a specific model, so he doesn't really say, 'Let us start off with one locus and two genes and see what happens'. But of course, all of us who try and understand those pages get a piece of paper and we start off with one locus, and two genes, and we try to see what happens.

RGW: This is actually, I think, a deep point. It's not just that . . . Fisher's project in the 1930 book [is] not analytical modelling. Part and parcel of it not being analytical modelling also means that he doesn't have a stringent definition of fitness.

AWFE: I think that's right! So that it's more easily generalisable. Another thought on this topic occurs to me, and that is, earlier in the book, he comments most favourably on the intellectual abilities of non-mathematicians, like Charles Darwin. And he is clear that their intellectual abilities are easily as good as those of people like himself who do the mathematics. The influence of Darwin on Fisher simply cannot be exaggerated. Now, that sounds like the sort of truism that you could say about anybody, but my reading of this, which has come out in a fairly recent paper that I gave, is that you can contrast the Darwinian influence on Fisher with the Darwinian influence on Haldane and on Wright. On Fisher, it isn't just the Darwinian theory that he has absorbed so completely, it's the Darwinian mode of thought.

RGW: Playing on that idea, if we can do an intellectual 'analysis of variance' here, let's not just look at the comparison of the three great founders, but in Fisher himself, what other influences were there on him? We've already heard that there's statistical gas mechanics. There's Darwin. What else?

AWFE: And a great deal of Darwin. Well, now that's an interesting question. There are people like [Sir Arthur] Eddington, who was an influence on him, and if we look at the very first interest that he had in statistical inference, we have to go back to F. J. M. Stratton, who was the astronomer, fellow of Gonville and Caius College, who taught the young Fisher and introduced him to statistical problems. And now we can say, following my writing about Robert Heath Lock, that Robert Heath Lock's book on genetics, 1906, I now believe – and I seem to have persuaded most people that they should believe – was a very powerful influence on the young Fisher.

[. . .]

RGW: Well, now I'm going to ask you another question, which I think is fair. It could be slightly controversial, but as I told Warren Ewens, in one of my comments to him [regarding] his contributions to the volume, I said, 'Okay, Fisher's book, *GTNS*, approximately 40% of it is about the relation of evolution to society – and [humans], and reproduction, and social class, etc.' Without saying anything further right now, because obviously there's a lot of political background, what do you make of those sections in Fisher? Did they have any influence on you, and do you think in any of the same ways as Fisher does?

AWFE: I'm not sure that when I read *GTNS* for the first time, I actually read as far as that. So when I started digging into that, it was often for particular reasons – that I'd read somewhere that some concept or other, which was original to the book, was to be found in that part. Then, of course, I began to be a bit more familiar with it.

RGW: Have you since then read these latter parts? Have you ever sat down and read the entire volume?

AWFE: Of course I've read it a lot since. Its basic thesis is that fertility is inherited. Something is going to happen evolutionarily, so *what* is going to happen evolutionarily? The origin of that part is to be found in the Adelaide archives, and Fisher started writing a book in 1919. There are chunks of that typescript which appear in *GTNS*. Here is a 29-year-old, still in the very early days of Mendelian models for the genetical process, thinking, as many people were at that time: 'My goodness, this must have implications for man. What are they?' [. . .]

[. . .]

RGW: [. . .] Could one say, regardless of whether Fisher is considered a right-wing person – which certainly some commentators have argued – could one not just separate that from him trying to speak very simply to get at the truth? Him just trying to say, 'But what are the implications of biological theory on humans? Let's at least ask the questions of what are the implications.' [. . .]

AWFE: You have just said exactly what I think Fisher's position was. That's my reading of it. I think it's important as we sit here, in a Cambridge college, to realise that that is the sort of discussion that goes on in these colleges. Completely free and open discussion of questions of scientific importance which have political or social implications. When you think about the fact that there are people here, in Fisher's time, like Joseph Needham, who would be regarded as a very left-wing person, and they would be arguing these points all the time. My own sensitivities in this are that I like to think that my science informs my political views. I have a certain horror of political views infecting scientific work. Now of course I accept that they are bound to influence the *kind* of work that people do; that's inevitable. But I react quite strongly against any sort of suggestion or anything I can see in a piece of work which seems to me to have been tendentious, politically tendentious; so much so, that I am never climbing on bandwagons. I always want to go off in the opposite direction and see where the truth lies out there. I think Fisher was not politically left-wing or politically right-wing, I think he was simply politically naïve.

RGW: That's interesting. What reasons do you have for believing that?

AWFE: Because I don't think that he readily appreciated the impact of things that he might say. We're talking really about private conversations now, rather than manifestos which he produced, because he didn't produce any. He was a straightforward, God-fearing Christian who appreciated the liturgy of the Anglican Church. And here you have, in Gonville and Caius College, two men, Joseph Needham on the left, one kind of Anglican Christian, and Ronald Fisher, another kind of Anglican Christian. But very, very similar in their attitude towards religion. Fisher once gave a sermon with the title 'Faith is not credulity'. That's a very interesting sermon.

[. . .]

RGW: Did Fisher and Needham have many conversations?

AWFE: Oh, every evening. They used to spar with one another. There are Fellows in the college, still around, who will describe it. I can only describe it second-hand. They had completely different minds. Joseph Needham had this wonderfully capacious, but sort of confused, muddled mind that he had difficulty retrieving information from, but it was all there. It was just absolutely fabulous to toy with it, as we would do, because he was Master when I became Fellow. I came to Caius after Fisher had died, so my college stories of his are second-hand, but he was reputed to have this keen, analytical mind. He knew a lot, but what he really enjoyed playing with was this analytical mind of his – his ability to think very clearly about things.

RGW: An analytical mind that didn't build analytical models.

AWFE: Absolutely.

[. . .]

RGW: What was the relationship between Fisher and Haldane?

AWFE: That was quite different. That was based on, I think, considerable mutual respect. You have to remember that, in the 1930s, they worked quite closely together in London at University College. There are nice things that Haldane said about Fisher on Fisher's trips to India. Of course there were barbs, they were both capable of making barbed statements about each other. [. . .]

[. . .]

AWFE: [. . .] But I do have something to say about Fisher and Sewall Wright. Sewall Wright was a softie in some sense, I think. He was a gentle person. Fisher was a person with a capability of losing his temper, and certainly of not suffering fools gladly. I suspect that there was a simple, trans-Atlantic difference, again, in modes of scientific discourse. The custom in this country was much more like our parliamentary custom, of a sort of slanging match and then going round for a beer together afterwards.

[. . .]

AWFE: And I think this is seen quite clearly in the relations between Fisher and Harold Jeffreys on statistical inference. They clearly had the highest respect for each other's point of view. With Haldane, I think it was a bit different, but given Haldane's great contributions and Fisher's ability to appreciate other people's contributions. Fisher was a very well-informed scientist, particularly for someone who didn't read very well. He was always up-to-date with his science. We know that particularly, in this college, because he and Francis Crick were friends.

[. . .]

'Human Genetic Diversity: Lewontin's Fallacy'

[. . .]

RGW: This article is 'Human genetic diversity: Lewontin's fallacy.' Let me start most simply. Is this your title?

AWFE: Yes, with a qualification. I was worried about it when I first saw it actually set up in type, and I thought, 'Do I really want to call it "Lewontin's fallacy"?' And my editor said, 'Yes, you do'.

RGW: But you are the one who thought it first. It wasn't the editor?

AWFE: No, I think I thought it first, and then I had second thoughts about it, but I had an opportunity to withdraw it.

RGW: But the editor then said, 'You should keep it'.

AWFE: Yes – and I'll return to the editor in a moment if you want to get a story from me as to how that came to be written.

[. . .]

AWFE: Well, there are two things I would like to add to that, but they're not to the content of the paper. They are acknowledgements. First of all, I'd like to acknowledge the fact that the problem was drawn to my attention by my brother John, who was Professor of Genetics at Oxford. With particular reference to the American Anthropological Association statement, he was getting taxed by this.[14] I said, 'What is all this about?' I found out what was happening, but I didn't know where it came from. I started asking my genetical friends, 'Where does this 85%/15% split come from?' And nobody knew. So, I actually opened a file on which I inscribed my brother's description of the problem – what he called it, which was 'the death of phylogeny'. That was what my file was called, my brother's statement: the death of phylogeny. But for a few years, a couple of years, I was asking around still, 'Where does this come from?' In particular, when I saw the statements in *Nature* and in *The New Scientist*, I really did put in an effort to find out. Finally, I tracked them down to this paper of Lewontin's, and to his book, which of course was on my shelf all the time.[15] But I had no idea it came from him. So I thought about this for a while and didn't really see clearly what was going on, until one day at lunch, here in Gonville and Caius College, somehow the topic came up in conversation. It was the sort of thing that does come up in conversation at our high table, and it so happened that I had three Fellows of the Royal Society engaging me in conversation, one of whom, my namesake Sir Sam Edwards, we have just had his funeral today. Anyway, four of us scientists got talking about this, and I tried to explain what the problem was, and that, of course, as every teacher knows, is how you actually work out for yourself what the solution to a problem is. I thought I'd done reasonably well, but of course I wasn't satisfied, so I went back to my room and I typed out a better account of the explanation which I'd been giving to these three. When I'd done that, somehow the problem came up in conversation later on, a few days later, with Adam Wilkins, who was the editor of Wiley's monthly science journal *BioEssays*. He said, 'You should publish something about that'. I said, 'Well, this is not really publishable in that form', which I still have, actually. I've got a copy of it. He said, 'Well, write me an essay. Do it'. So I wrote him what appeared, with the exception of the bit on principal-component

[14] American Anthropological Association. 1998. AAA Statement on "race." *Anthropology Newsletter* 39 (September), 1.

[15] Lewontin, R. C. 1972. Apportionment of human diversity. *Evolutionary Biology* 6, 381–398; Lewontin, R. C. 1974. *The Genetic Basis of Evolutionary Change*. New York: Columbia University Press.

analysis, which he insisted on putting in, which I thought was a bit of a distraction, but there it is. So, those are the two people I should be acknowledging at the end. One was the person who drew the problem to my attention, and the other one was the person who insisted on my publishing it. So there it is.

RGW: And, how do you react, or what do you make of the fact that – I think certainly in the United States but I also know internationally, insofar as people find these topics of interest – your name has almost become a household name about these matters?

AWFE: Well of course, one doesn't mind, if one is happy with what one's written, people talking about it.

RGW: And you are satisfied with what you've written?

AWFE: I'm satisfied with what I've written, yes. I think it's the right combination of describing a statistical fallacy, or rather, describing a conclusion taken from a statistical analysis which is not strictly relevant to the conclusion which was drawn by the original author.

RGW: Correct. One question I get a fair amount when I teach – I sometimes teach this – is why call it a 'fallacy'? I mean, it could be a puzzle, it could be a statistical inference mistake, it could be a logic gap. Why 'fallacy'?

AWFE: I think it's an example of drawing a political conclusion from a set of data, which was not justified. One of the reasons I am happy with the paper is because of the quotations not just from Fisher, at the beginning, and the journal's, but also the quotations I put in it from Dick Lewontin's work.[16]

RGW: It seems that we have two different kinds of fallacies.

AWFE: Yes, absolutely. I don't think I was clear about this, as clear as I could have been. I could improve upon the paper, but I still stand by what it says.

RGW: But to be clear, there's a statistical fallacy, if you will, which is a mistaken inference of ignoring multilocus correlations, and only looking at one locus at a time

AWFE: In its essence, it's the failure to appreciate that, if you work in terms of information, that information can be accumulating. Information can be destructive, but it can be accumulating. As soon as you think, 'Goodness, across the loci, we might be accumulating information', then the statistical way of analysing this in terms of thinking about correlations and principal components becomes obviously relevant.

[16] Consult analyses and sources in Kaplan, J. M. and Winther, R. G. 2014. Realism, antirealism, and conventionalism about race. *Philosophy of Science* 81, 1039–1052; Winther, R. G. 2014. The genetic reification of "race"? A story of two mathematical methods. *Critical Philosophy of Race* 2, 204–223, reprinted in Appendix 2.

And the information way of thinking about it, which was Dick Lewontin's way of thinking about it, is potentially misleading. Now, let's give him the benefit of the doubt. I think he was misled.

RGW: But that was meant to be a measure of diversity. His Shannon entropy measure of diversity was just meant to be a measure of information at one locus.

AWFE: It was indeed, yes. The mental jump, which he must have unconsciously made – well, the mental jump which he *should* have made is that this measure of information is cumulative, is potentially cumulative above loci. And of course it turns out that it is.

RGW: And then to be clear, there's this statistical fallacy, and then there's a normative fallacy, a moral fallacy, a political fallacy.

AWFE: Yes. I think that's become clear from such subsequent discussions as I have encountered. Now I haven't gone searching the internet or anything like that, but of course I am aware of discussion having taken place. I got a lot of very, very happy geneticists as a result of that paper, because they all wrote to me and said, 'Ah, *now* we see what's going on'. Later, one of them of course wrote to me and pointed out that the basis of the fallacious interpretation had, in fact, been expounded in the 1970s by . . .

RGW: Mitton.

AWFE: Mitton, that's right. Mitton.

RGW: I was going to ask you about Mitton, and Smouse, and there were a few others. And Lewontin had written a page-and-a-half response to Mitton, back then.[17]

AWFE: Yes, really?

RGW: Where he states that Mitton is taking the wrong kind of average. He's doing a multiplicative instead of additive average.

AWFE: Well, all that was completely unknown to me, and I may say, all my colleagues, because I had been asking everywhere, 'Now where does this stuff come from?'

RGW: The person who brought that to my attention was Noah Rosenberg.

AWFE: Was it? It was brought to my attention, not by him, by somebody else.

RGW: In Noah's contribution to the volume, he obviously thinks that your statistical discussion is exactly correct.

AWFE: Yes.

RGW: He plays on your one sentence here, 'This article could, and perhaps should, have been written soon after 1974'.

[17] See references in Rosenberg's commentary.

[Both laugh]

AWFE: Yes, that's interesting of course.

RGW: And then he says, 'In a sense, it was'.

AWFE: Yes, absolutely. Quite right, quite right. So I think I can be quite relaxed about this article; well, I am quite relaxed about it anyway. I should say that one of the best commentators on it was Joe Felsenstein.

RGW: When? Where?

AWFE: Well, perhaps just in conversation. I'm not sure that he's written on it, but you see, there he is: a student of Lewontin's, very perceptive about things, a friend of mine, and entirely qualified to look at the problems behind this. I would think he would have a very good judgment on it. Maybe he will produce it.

[. . .]

AWFE: Well, I see the task of the geneticist who is not actively involved in studying any particular disease, which means people like me who think about the theory more, is to describe the properties of the huge genealogy of man as best they can using what statistical tools there are, and pointing out the consequences of Mendelian inheritance if you have a huge structure like this. [. . .]

[. . .]

AWFE: So, what is there else to say about Lewontin's fallacy? What I do think is that – insofar as I've been able to discover a reaction from him, it is essentially that he wasn't trying to . . . it is essentially that there are two problems. He really wasn't going from the variation problem to the phylogenetic problem.

RGW: Or the classification problem.

AWFE: Or the classification problem. But I've got quotations from him on that, which shows that precisely that's what he *was* doing. So there's some revisionist history going on here.

[,,,]

RGW: A few more questions here. Just noting, for the interested reader, that you have a little, can I call it a toy model?

AWFE: A little model? Oh, yes, I do indeed have a very little model.

RGW: Yes, yes, where basically you're looking at many similar loci, and you're assigning a certain probability, call it 40% . . .[18]

AWFE: . . . to make, to get the 85% to 15% . . .

[18] For an alternative, geometric interpretation, see Appendix 1.

RGW: . . . and then you're showing that as you use an increasing number of loci with that same frequency distribution, the 40/60 or whatever exactly it is, your probability of misclassification quickly diminishes until it reaches, asymptotically, zero. Already by 15 loci, if we use the Lewontin F_{ST} results, already, by 15 loci, you really don't have a significant probability of misclassification. Correct?

AWFE: I think so, yes.

RGW: Would you call this an analytical model?

AWFE: Yes. Absolutely.

RGW: This does give us insight into why there is, call it 'Lewontin's fallacy' or call it a statistical . . . it could almost be called an 'averaging fallacy'.

AWFE: Yes. It's terribly simple, isn't it?

[. . .]

RGW: [. . .] Your brother mentions, correctly, that there's this paper – I think it's from '99 but I'd have to check the exact date – this statement by the American Anthropological Association.

[. . .]

RGW: It's basically saying there aren't races, that the genetics shows that there aren't races, and then it's tracing that back to Lewontin. Now I should add that Lewontin's results have also been influential in psychology, and in philosophy, which is a much smaller field. Insofar as people think about the biology of race, less so now, but certainly in the 1990s and probably even in the 2000s, people were referring to that paper as the authoritative statement. Now these aren't necessarily similar people, but I like to call it the 'liberal consensus'. I say it sort of tongue-in-cheek. The liberal consensus says: 'Oh, race doesn't exist. Why do we want to say race doesn't exist? Well, it's because we want all to be equal, and it's because we want to have a nice socialist politics'. It seems to me that political views, and views about the reality of race, are completely independent. You could have a right-wing agenda, and also believe race didn't exist. And vice versa: you could have a left-wing agenda and believe that race exists.

AWFE: Absolutely. If I was a red-hot socialist, I would subscribe to the view that it is a dangerous mistake to premise the moral equality of human beings on biological similarity because dissimilarity, once revealed, then becomes an argument for moral inequality. That would be my good socialist position. So I would go around saying, 'Look chaps, you're going about it in the wrong way'.

[. . .]

RGW: [. . .] Do you believe races exist? That's what some people would want to ask you.

AWFE: Yes, well I think if I was asked that on a television program or something, I would say, 'Well, you tell me what you mean by this word, "race", and I will tell you yes or no.'

RGW: Okay, then let's do that. If we look at the populations, you know Lewontin has a three-level hierarchy, right? He has individuals, and then he has populations within the continents, and then he has the continents. Is there a sense, even a weak sense, in which these different [continental] populations . . . are races, or not?

AWFE: I would retreat to the position that, though I am not a naturalist, as far as I understand it, the word 'race' – well, actually, I needn't say any more on that. There is a very important, once again, trans-Atlantic difference here, that certainly when I was young, the word 'race' had none of the connotation which it has in the United States. Now, we've all learnt, of course, to be careful about using the word, but there was no sense in this country, in my education, that there was anything wrong with using the word 'race' to discuss obvious differences between groups of people that we'd encountered when we were schoolchildren. There's this huge ambiguity about the word 'race'. Somewhere, maybe in that book, probably in my little book, I point out that sometimes we talk about the *human* race.[19]

[. . .]

AWFE: I mean, it really gets down to the slightly infantile level of asking a person, in connection with the use of the word 'race', if he's prepared to answer the question at all, of showing him a lot of people from different parts of the world. Perhaps even lining them up and saying, 'Where is the point at which the person becomes a different race from your race? Where is it?' Well, of course you can't answer it.

[. . .]

AWFE: And yet, somehow, we need to be able to discuss dispassionately the structure in this great genealogy.

[. . .]

RGW: [. . .] To a first approximation, you would say that every member of *Homo sapiens* has moral equality?

AWFE: Has moral equality? Yes, I start from that position, yes.

RGW: Where do you think that moral equality comes from, in your view?

[19] Edwards, A.W. F. *Race, Reason, and Rubbish*. Unpublished manuscript.

AWFE: Oh, that's an interesting point. I know where mine comes from, I was educated to it. It was part of my culture.

RGW: But are there more general principles?

AWFE: What would the justification for it be?

RGW: Oh, well, there's a classic answer, obviously it's religion. There's God.

AWFE: Well, yes, I wasn't going to enter into that.

RGW: Yes, another answer are so-called Enlightenment Values, you know, Kant's moral theory, or Rousseau, going back a little bit further. Where, basically we . . . [emphasise shared] principles of reason, that every rational being is equal because [we all] have very generally these cognitive capacities, even if [we are] not all the same. But we're all basically, from Aristotle to Kant, [considered] 'rational creatures'. And because of that, we have moral equality.

AWFE: This is the sort of philosophy which I tend to not get involved in, but I wonder really if it's necessary to think of it in those terms at all. In the end, it's a matter of how one treats each individual. It's an extension of the treatment of individuals to producing principles which you would like to see governing actions involving any member of the species. I don't think that's very profound; I mean, it is really repeating that it's a part of my cultural education.

RGW: I hasten to add that, actually, it is a good argument, and it's what in classic philosophy is called a 'contractualist perspective'. It's what Rousseau and Kant [defend].

AWFE: Well, I haven't heard that phrase, but I understand. That's quite interesting.

RGW: Yes, contractualist. That's Rousseau, famously, and Kant read Rousseau and was very impressed by [him]. But we don't have to do more philosophy than that. Again, you don't have to answer, but I am genuinely curious. What do you think of ethnocentrism?

AWFE: What, that there are different modes of thinking in different cultures?

RGW: Something like that, yes. And that your culture is the best. It seems to be, paradoxically, semi-universal that every culture, almost, has thought that they are the best. Or the centre of the world. And obviously, that is a force against moral equality and moral universality. Is that something that worries you? Particularly in England, also, socially. You live here, in England and Great Britain, and it becomes a more . . . one term is 'multicultural society', as you have more and more people from all over the world. Both here, your students at Cambridge, and in London – you see it just walking on the street. Is that something, just morally and personally, you think about or just wonder: Can we treat everyone with moral and social equality? Do you have any views on that?

AWFE: I would be happy to say I would like to treat people with *moral* equality, but I'm not sure that I would extend that to *social* equality, whatever really that means.

RGW: Right. Or economic equality or have a system like in Sweden or Denmark.

AWFE: I think our society, whatever we do with it, it's going to be structured one way or another. If you ask me whether I think we ought to have an unstructured society, you're never going to get there. So we can discuss the structures by all means.

RGW: Good. Okay, I appreciate you talking about this because I know these aren't easy questions.

AWFE: Yes, well, I think it's a product of my age to some extent. I won't say that one gets cynical about everything, but one does tend to take a view of the problems of the world that . . . I think it's probably a well-known view amongst political scientists, who think about these things carefully, that this idea that they are soluble in principle is not a helpful way of trying to ameliorate the actual problems we see today. So, well, maybe that's enough.

[. . .]

[RGW]: Last, we'll have a little more to say about likelihood, although not terribly much, tomorrow, but I can't help but think about 'the gene's-eye view' as being one that seeks generality and maybe simplicity. Whereas Sewall Wright's views, or views about gene networks (which has become, you know, all the rage today, and I am curious what you think about gene networks), those are ones that realise that you can't really have the generality – or if you have the generality, it's at the expense of doing justice to the complexity. Do you have any views about how to deal with genetic systems, as opposed to Fisher's more generalised, simple . . .

AWFE: Oh, I think it depends on the aims of your study. If you are trying to mathematise Darwin . . .

AWFE: [. . .] If you are trying to mathematise Darwin, I think you'll go down Fisher's route. If you have particular interests in, obviously particular organisms, but you're interested in the details – interested in, shall we say to take a topic from Fisher's paper, 'The evolution of dominance'[20] – then you do need to go down that analytic route and think very carefully about it. I think Fisher was to Darwin what Maxwell was to Faraday.

[. . .]

[20] Fisher, R. A. 1931. The evolution of dominance. *Biological Review* 6, 345–368.

Day 2

Phylogenetic Inference

Edwards's Home, Cambridge, UK

21 May 2015

Statistics and Phylogenetic Inference

[. . .]

RGW: Could you help us understand the family of statistical methods of phylogenetic inference that you were so central in developing? In particular, could you speak [about] the differences of two senses of parsimony: minimum evolution and probabilistic reasoning on the one hand, and Ockham's razor and Hennigian cladistics on the other hand? Could you also speak [about] the family of statistical methods that you developed in a phylogenetic context? I'll just mention a few, and then I'd like you to give us some kind of organisation or systematised conceptual taxonomy: maximum likelihood, minimum evolution, least squares, and so forth.

AWFE: I think that's most easily done from the way that the methods developed in the minds of Luca Cavalli-Sforza and me all those years ago, because it wasn't clear at the time, as we set out to solve these problems, exactly what the problems were. We were influenced by taxonomic methods, and in particular by the work that Peter Sneath and others had done, applying the new computers to taxonomic problems. So we really start with the prospect of deriving trees from genetic data.

[Pauses for a moment] Our initial focus was the genetic data arising from the study of the blood groups by Arthur Mourant and others, which had revealed substantial differences in gene frequencies from different populations around the world.[1] Luca Cavalli had had the idea, in about 1948, which he used to reminisce about, when he was in Fisher's

[1] Mourant, A. E. 1954. *The Distribution of the Human Blood Groups*. Oxford: Blackwell; Boyd, W. C. 1950. *Genetics and the Races of Man: An Introduction to Modern Physical Anthropology*. Oxford: Blackwell.

department: rather than just drawing isogene maps, which were quite familiar from the work of W. C. Boyd and Arthur Mourant himself, maybe one could actually try and extract a tree structure from the information and see how it related to the geography of the continents. So when I went to work in Italy, this was a well-defined problem, but nothing had been published, either on the computing side or on the statistical side, which helped us. We could see rather quickly that the developments that were taking place in taxonomy, with the advent of the new computers, were not really very helpful. The taxonomists were still worrying about problems like the choice of characters and the weighting of characters, and they were simply using the computers as mechanical accountants to speed up the analyses that they were doing. But of course Cavalli-Sforza and I were geneticists, but we were also fairly well-informed statisticians. So it was very easy for us to begin applying what we knew about genetics and statistics. The first step was that Cavalli-Sforza, thinking of the structure of a tree, realised that the difference in gene frequencies between two populations must be somehow related to the connectedness on the tree. From a statistical point of view, a very simple way to consider how one might make use of that fact is to regard the populations as existing in a space so that they actually have pairwise distances between them, and to think that the branches of the tree themselves have lengths. Then, the problem can be looked at, in statistical terms, simply as a statistical estimation problem, by using least squares. So you're trying to find the best-fitting tree where the distances from one population to another along the branches of the tree are to be, as nearly as possible, chosen (in this case by least squares) to match the distances in the space of the data. This produces statistical problems itself, because you have to decide how you're going to get gene-frequency data into a metric space, preferably a nice Cartesian space, so it's easier to draw pictures and so forth.

[...]

AWFE: A straightforward space. You have to think about that. And there were ideas that Fisher had told Cavalli-Sforza about himself, which helped there. Then, you have to think about the statistical problem of using least-squares estimation on a structure when you don't know what the structure is. So you find yourself confronted with the problem of having to choose not only the branch length, but actually the topological form of the tree in the first place. All this is quite unthinkable, at least for more than five or six species, unless you've got computers. So we started computer program writing. Well, I was doing the writing of the computer programs on the basis of this way of looking at the data. Now, I have to think a little bit about the next stage because we're going to move from that stage to the concept of minimum evolution. Least squares is a sort of minimisation procedure of one kind, but I wasn't very happy, the more I thought about it, with what we were doing with Luca's least-squares procedure. Unless we fixed the joint points, the nodes of the tree, into the space by some constraint (which would have been a very difficult thing to do, I think), the tree that we were ending up with did not actually exist. It was not representable in the space of the data that we'd started with. Yet the idea was that the gene frequencies had been changing as the populations had evolved, but all the time the gene frequencies from the past would have been representable in that space. If you do a Luca least-squares tree, that's not true.

RGW: Can you say a little more about why that's not true?

AWFE: It's not true for the good reason that there's no reason why it should be true. [Laughs] There's no reason why, if you start to work out – in estimating a least-squares tree like that – you don't, come to think of it, end up with coordinates for the nodes. So they might be anything. Just a minute, I'm not saying that clearly. I think my first answer is the best one, that there's no reason why the nodes should be in that space. Well, when, 50 years ago, we stopped and thought about it, I wanted them to be in the space, because I envisaged that space as being where all the evolution had taken place. What we were looking at was the plane, or the space, of the present-day data.

RGW: The space of present-day data is a summary of all the loci and the gene frequencies in all the loci.

AWFE: Yes, all the genetic data somehow ... see the earlier part of the conversation. If you had managed to get it into Euclidean space ... it's convenient to think of it as just two-dimensional because then we can actually imagine what's going on. But it can have any dimensionality, so long as it's a nice, workable Euclidean space.

RGW: And you have a nice figure of that in one of your early papers ([28], p. 29, this volume).

AWFE: Yes, I'm talking [about] that picture. That picture has the time dimension in it, too. So it has a space of two dimensions for the data, and a time dimension. Well, when you look at the tree just in the two-dimensional space, which contains the data, and if you believe that could you see the evolving tree, without thinking about the time dimension, you would see it in that space. Then it's very easy, and it suddenly happened to me in the middle of one night, to think there must be a shortest tree, which joins all those points together. Now, I was ready-programmed for that idea because I was interested in recreational mathematics, and I knew about Steiner trees – that's the minimum-road problem between three cities. So I knew that this was a situation in which, in two dimensions, you had branches of roads intersecting at 120°. All that was second nature to me. I thought, now what's difficult about many dimensions? You must be able to have these Steiner trees in many dimensions. That sounds very complex, but, in fact, it's no more complex than two dimensions. So I immediately sat down to compute Steiner trees in an arbitrary number of Euclidean dimensions. I wrote out a piece of paper the morning after I thought of this. I still possess it. It's headed 'Antonio's theorem', and the first thing it says is '1: In what sense it [the taxonomic tree] is 'best' has yet to be determined'.[2] That was a very good thing to have written.

[. . .]

RGW: [. . .] Now why did you call it 'Antonio's theorem'?

AWFE: Well, because I was Anthony living in Italy. [Laughs] Simple as that.

[. . .]

AWFE: [. . .] Let's go to the distinction between algorithmic methods and statistical methods.

RGW: Good.

AWFE: Algorithmic methods are those that are not based on any particular model for the evolutionary process, they just come into the sort of 'what if'/'bright idea' category. Like minimum evolution, people talk about Ockham's razor. That was known to me, but not part of the thinking. The whole of physics, and a lot of biology, is suffused with the idea that some things are being maximised or some things are being minimised. You really don't need to engage in sort of William-of-Ockhamry to justify this. It is part of our upbringing – without, I think, going back to any sort of philosophical principle – that you should use the minimum number of entities in an argument, or something like that. It's very natural. If you have these algorithmic methods, they're wonderful for computers, but they don't actually have a real justification to them unless you think a bit more deeply about it. I think it's fair to say that I wasn't very happy with Luca's half-statistical method because he didn't have a model for the generation of the trees – he merely had a well-known statistical method for doing estimation on those trees, which is a rather different thing.

[. . .]

2 'Antonio's theorem' is reprinted in Appendix 4.

RGW: Would another distinction also be useful, between making a model or constructing a model, on the one hand, and confirming or testing a model?

AWFE: Yes, I think that will automatically come out in the conversation. I should think it's always very uppermost in my mind. So I wasn't too happy with that because of the question of what was happening to these nodes. I suppose that led me to begin thinking in terms of the process that had constructed the tree. I'm not being specific at all, now. I think, subliminally, my mind would then have naturally been starting to think that this tree must have been developed according to some process. Let's leave that side there for the moment.

[Pauses for a moment] Luca ... didn't really like my minimum-evolution trees, which I was computing quite rapidly because it's a very easy computer program to write. I should say here that the style of tree problem in an arbitrary number of dimensions is one of those NP [non-deterministic polynomial time] problems (or whatever they're called) that doesn't have a direct solution, but a great deal is known about it, far more than is necessary for playing with it in a biological playground. So that problem is essentially solved. But he didn't like it, I suspect for the same reason that he began to think, 'Well, what's behind this? What's the justification for this?' Right about that time, we were attending lectures by Motoo Kimura, who was staying in Pavia for a month, and he was going through his differential equations, describing the ways gene frequencies change and so forth. There were Luca and I, beginning to wonder about some sort of model for the process. So I can remember precisely the day, but not the date, when we both came to the same conclusion. When I say independently, it's clear that our thoughts were converging on the basis of all the discussions that we'd had. What actually happened was that we had to go down to Naples for a meeting of the committee that was running the Laboratorio Internazionale di Genetica e Biofisica, of which Luca's group in Pavia was a subgroup. Luca was always in a hurry, so he was going to fly down. I didn't like the flying, so I was going to go down by the overnight train. And when we met at Naples station, we opened our mouths simultaneously and said, 'Well, obviously, what we've got to do is to use that system that Motoo Kimura's been teaching us to have the simplest biological analogue of what nature is doing in allowing all these gene frequencies to change. Let's just suppose we have random genetic drift on gene frequencies in the gene-frequency space, and we have a process of random splitting of populations, then we've got a model.' I don't suppose we will have walked out of the front of Naples station before we will have said, 'and of course we will then do estimation by maximum likelihood'.

[...]

RGW: Can you say, to help the reader, what a likelihood function tells you in this case and what it means?

AWFE: Well, a likelihood function is simply the basis of the method of maximum likelihood. You're trying to find the maximum of the likelihood, the top of the hill of the likelihood function. What is the likelihood function? The likelihood function is simply the function of the probability of what you have actually observed, given the particular values of the parameters which you've chosen. As those choosable parameters have a lot of different possibilities, this describes a space. For every point in that space (that is, for every collection of parameters), there is a likelihood which you compute. If the parameters are in a continuous space, you end up with a likelihood function which is continuous. With a bit of luck, it's a nice function that has a maximum.

RGW: What determines the shape of the surface?

AWFE: The algorithm of the problem. The probability model, which is the basis of your assumptions of the way that the data were produced in the first place. In the case of, shall we say, looking at the sex ratio in families, you have a model of a binomial distribution which has one parameter, so your likelihood function is very easy to plot. If you've still got the dust jacket of your copy of *Likelihood* from all those years ago, you will find it plotted. So likelihood functions are very simple things in principle.

[...]

AWFE: When I got interested in sex-ratio data from reading Fisher's *Statistical Methods for Research Workers*,[3] I came across his example of the binomial distribution, which has too great a variance. That is, a greater-than-binomial variance. The binomial variance is npq, and the data showed a greater variance than that, which was not explicable by the inclusion of twin births, who are often, of course, of the same sex. I thought, well, surely, if that parameter p is actually varying a little bit, then one could find the resultant distribution, provided one had a specific model for the variability of that parameter p. So why not choose a nice little beta distribution, give that to the parameter p, do some integration, and find out what the resulting distribution is? I found that out. Of course, I was still an undergraduate. I didn't then know it had been used quite a lot before. So, that looked very agreeable. And then,

[3] Fisher, R. A. 1925. *Statistical Methods for Research Workers*. Edinburgh: Oliver and Boyd.

because the foundation beta distribution has two parameters, between them they are allowing for more variability than you get in a binomial distribution. You then . . . if you apply maximum likelihood . . . find that the problem is not simply analytically soluble. Well, it might be, but it didn't seem to me to be. I got stuck, so I thought I'd better dream up another method of estimation.

[Pauses for a moment] As a young research worker, supposed to be doing experimental work with mice, teaching myself statistics, I thought I knew about maximum likelihood, because I knew about the binomial distribution. But the equations didn't seem to work out. I got two equations with two parameters. Then, I thought, if I equate the mean of the theoretical distribution to the mean of the observed distribution, that will take care of one of the degrees of freedom. That's what happens when you do that with a single-parameter binomial distribution. You go through all the maximum likelihoods, and you find it just tells you to equate the observed mean with the theoretical mean. So, that's great. Then I thought, I've got another parameter or another degree of freedom with my two parameters. Why don't I just equate the observed variance with the variance of the theoretical distribution? Well, of course, I didn't know that this was Karl Pearson's method of moments. [Laughs] One day I was chatting to other research students, and I said I was using this method to do the estimation. I don't remember which one it was, but he said, 'You better not tell Fisher that. That's Karl Pearson's method of moments.' A few days later, I was using the one calculator which the department had, which was kept in the library. A manual calculator, using Karl Pearson's method of moments, on the data from Fisher's book, when he came in and asked me what I was doing. I don't remember the conversation exactly, but I do remember, and have put on paper somewhere in my reminiscences, that he said to me, . . . he didn't see exactly what I was doing. I must have told him I was looking at the data which I got from his book . . . 'You know, you really ought to be studying for your examinations now, not doing things like this.' But he said it in such a half-hearted way [Laughs], that I'm sure now, looking back, he was very pleased to see a student working on his own, [that] something that he just thought up might be interesting. So that's a little bit about maximum likelihood.

[. . .]

AWFE: Okay, so now with these very elementary computers I am trying to compute a likelihood function. You can think of it in two or three dimensions. Well, think of it in two dimensions. Think of it like one of those pictures of Sewall Wright's adaptive landscapes.

RGW: Well, I was about to make this analogy, but I almost feared your reaction.

AWFE: Any sort of seaside picture of sand dunes will do. Let's go to Denmark and have some nice sand dunes. Or take the yacht out and have some unbroken waves, and what we're trying to do is to find the maximum. If the surface turns out to be quadratic, then you know that you've got an easily analytically-soluble problem, and similarly for various other simple functions. But my function wasn't like that, so I had to write a program using the standard methods of hill ascent, meaning that a point that I started from would set off on a journey, and it would crawl its way, necessarily increasing the likelihood, up the steepest ascent of the hill until it got to the top, which it would do if there was only one maximum. [. . .] I have these little programs which solve the problem for just four populations. Then I'd done my three years in Pavia, and I left for a year in Stanford. At Stanford, I had much better computing facilities. I had this little prototype program. I had been to London from Italy, and been to see Arthur Mourant, the great tabulator of world blood-group gene frequencies, the top man in London, top man in the world at that time. Luca had told me to ask Arthur Mourant to choose 15 populations, spread around the world, for which he had access to reasonable amounts of data. Not all his, personally, but in his tabulations. So he and I chose 15 populations, three from each continent, that sort of thing. After all, we were only trying to get some data to see if the programs could be made to work. We weren't, at that time, thinking that this was a sort of publishable activity.

AWFE: [Pauses for a moment] I, at least, as the technologist, wasn't too bothered with what these populations were. There is a difference here between Luca Cavalli and me, because *he* really wanted to know the answers: How did the human population spread round the world? *I* really wanted to know the answers for philosophical and statistical problems which were worrying me. Well, this was great because it made us a wonderful working pair. We both wanted very much to solve the same problem, but the actual motivation for the two of us was rather different. Yet, we were both geneticists. I understood his motivation perfectly well. He and I knew about the same amount of statistics without actually being formally trained in it. He understood my interests perfectly well. So that was really a great partnership. Anyway, off I went to Stanford, where there was a nice big IBM computer, and it was very easy to expand my program from four populations to 15 populations.

I started running it on the 15 populations whose data I'd got from Mourant and had reduced to a space of 14 dimensions, probably. So there I was trying to find the maximum of a likelihood function in a space of 14 dimensions. I forget how many parameters that leads to, but anyway, a lot. I could not get the program to converge. It simply never occurred to me that this was anything other than a program error. These were the very early days of programs, early days of FORTRAN. Indeed, the first programs that I'd written for four populations before machine languages, before higher-level languages like FORTRAN and ALGOL had been produced. I was actually programming in machine language right from the beginning. It never occurred to me that the program was doing precisely what I wanted it to do, but arriving at a result which was so surprising to me that I thought it must be something wrong with the computer – that used to happen too, of course – or wrong with my programming. So I was held up quite a long time.

[...]

RGW: Would you say that maximum likelihood is the method that you perfected and perhaps did the most work on?

AWFE: Well, I think it was Joe Felsenstein who really perfected it. [...] I developed my 15-population program and coped with the problem arising from these two sorts of parameters; then, a program called EVOTREE, which I finally worked on in Cambridge after having arrived from Aberdeen in October 1968. The Royal Statistical Society paper ([46]) does not contain any practical calculations, even as examples, but I do possess those, the background to the paper. It's a very technical paper. I shall never – well, I certainly won't now – but I've never written anything as technical as that.

[...]

RGW: [...] As I understand you, maximum likelihood is a paradigm example of a statistical method.

AWFE: Yes, because it is based on a stochastic model.

RGW: It's based on a model, and it has estimation.

AWFE: Let's call it stochastic–statistical, because you're both requiring a model which is probabilistic or stochastic, and then you're using the king of estimation methods, namely the method of maximum likelihood.

RGW: Right. Going back to this distinction between model construction and model testing: model construction, what gives you the model? Well, it's the branching diffusion process. What gives you the estimation? Well, it's maximum likelihood. But what allows us to test the model in the end? Is that a separate step? Conceptually, even?

AWFE: It's a sep[arate] . . . well no, it's not really a separate step, is it? Because, from the point of view of likelihood, you would be interested in how good your solution is, as measured by the curvature of likelihood functions. But then, you get onto the next stage of thinking of ways in which to extract more information about whether the rates have been the same in different branches, and so on. You get into modern evolutionary-tree theory.

RGW: Right, but couldn't you then say that that's the next iteration? That would be a new cycle?

AWFE: Yes, that's the next generation, really. When you've solved the simple problem, or what we thought was going to be a simple problem, then you begin to think, 'Oh well, you know, it won't be quite like that in this space. There will be differences.' And actually, for me, that goes back to my 1963 paper to the Genetical Society, when I put forward a principle of minimum evolution ([25]). In the typescript, which, alas, I threw away in Italy, so it's not one of my ancient typescripts, I can remember part of my presentation at the Genetical Society, which will have been about 20 minutes. You know, one of these annual society presentations. I wouldn't have just got up and said [Laughs], 'Let's do minimum evolution'. I will have already been thinking . . . I was already thinking about the next stage. I imagined that the space in which we were constructing our evolutionary tree was no longer isotropic. That is, that there were bits of rapid evolution, or less viscous space, and there were barriers which would be difficult to overcome.

[. . .]

AWFE: And you might suppose that in some parts of the space, selection was progressing more slowly than in other parts of the space. So, you have no longer a simple isotropic model. I was, I remember, using the analogy of the road builders' problem in Switzerland.

RGW: Can you say more about that? What's that analogy?

AWFE: Well, suppose you want to build the minimum road network in Switzerland. It wouldn't be much good taking a map without contours on to plan your network because in reality, some bits are much more difficult to drive a road across than other bits. So you'd better put the contours in it. Those are the areas of slower building – not to mention other difficulties. I remember using that metaphor.

[. . .]

AWFE: [. . .] When I was in Aberdeen, I had good support from the Medical Research Council, running a little unit called the Human Genetics Computer Unit.

[...]

AWFE: I had a nice little in-house computer provided by the Medical Research Council, which I ran, with the assistance of an excellent Danish programmer ... [t]rained by IBM in Copenhagen, so she did FORTRAN. We were getting on very well there, but I made this academic move to Cambridge. When I established myself in Cambridge, my application for money from the Medical Research Council was turned down.

RGW: I see.

AWFE: At that point, I should perhaps have changed research council, because after all what I was doing was not yet of any great medical importance, but they had supported me very well previously. So, I found myself without the support that I needed, and incidentally, without an IBM computer to use anyway, because Cambridge didn't have one. It was still in the home-build.

RGW: So you had to scratch that research program?

AWFE: So, well, I found it difficult, not only once, but continually difficult to do that sort of work. I quite deliberately said to myself, 'Other people are getting on very well with it. They have assets that I don't have. They can get money which I can't get, apparently.' Though, I think I was probably rather poor at trying to get money. I don't like going begging for things which [Laughs] I think should be provided anyway.

[Pauses for a moment] So, I just sat down one day, and I thought, 'Well, here are all my friends here. They seem to spend all their time writing grant applications, and then they produce lots of students that get unemployed. They have to worry about their next jobs and so on and so forth.' So I thought, 'Why don't I look at the assets that I've got here. I have a stipend. I have fabulous colleagues, in that they know absolutely everything, around in Cambridge to go and talk to. I have superb library facilities. I have now got, as things developed in Cambridge, adequate computing facilities. Why don't I just go down the thinking, book-writing, academic path, and leave the technology to the computer wizards who can write all these programs and so forth?' So, academic pressures intervened at that stage.

RGW: Okay, yes, because if you look at your CV, then the articles on phylogenetics tail off.

[...]

RGW: So, one might say that it started with the data and it ended with the computer. [Laughs]

AWFE: Yes, in a way that's true. It is true of various fields which statisticians were themselves expert in, that the computer scientists came along, and partly, they spent time reinventing what the statisticians knew already, but partly, they used their ever-increasingly powerful machines to start, not so much doing inference, but – winnowing might be the right word – big data.

RGW: Yes.

AWFE: It's very interesting talking to human geneticists, of course. I find now that the modern human geneticist, with the vast amount of data that he has, is quite fascinated to learn that, all those years ago, our problem was that we had such a small amount of data that the very best methods of statistical inference were honed in the field of human genetics, and genetics generally.[4] That was the lead field, largely because of Fisher. The method of maximum likelihood was first used by him in 1922, in a paper I imagine deliberately put in the *American Naturalist* to advertise the method in connection with linkage in *Drosophila*.[5] Indeed, anybody who looks at this development carefully will come to the obvious point – which I didn't mention, but it's so obvious I could do it extremely quickly – that I was trained in the statistical estimation of genetic linkage. I used to say to Luca Cavalli when he put this problem to me, 'There isn't going to be enough data to solve your problem, because I know how difficult linkage is. In linkage, you have a linear tree. The first thing you have to do is to work out what the order of the loci is, before you can do ordinary maximum-likelihood estimation. You're asking me to do the same thing with a chromosome in 15 dimensions, with branches all over the place. I don't think you'll ever have enough data to do that.' He wore me down – he was my employer, after all – and said, 'You go and try.' It turns out, there's plenty of data. Now, of course, we're into really big data.

RGW: What's your perspective on big data?

AWFE: Well, to be able to choose a particular chromo[some] – a particular gene, a particular allele I should say, and follow it across the world. And that's absolutely stunning. So you really don't need maximum-likelihood trees anymore, I suspect. What has happened now, of course, is that taxonomy is beginning to look, really, a little bit backward in its philosophy. Taxonomy always was an essentially algorithmic activity. Here come

4 Consult, for instance: Cavalli-Sforza, L., Menozzi, P, and Piazza, A. 1994. *The History and Geography of Human Genes*. Princeton, NJ: Princeton University Press.

5 Fisher, R. A. 1922. The systematic location of genes by means of crossover observations. *The American Naturalist* 56, 406–411.

the geneticists, and say, 'No, we're not going to do this algorithmically. We're going to have a stochastic process for what's happening, because we know that before very long we're going to be able to sequence amino acids. In the course of time, surely we're going to be able to get out DNA sequences. Then we're going to need these statistical methods for drawing evolutionary trees from molecular data of various kinds, but particularly DNA sequences.' I remember saying to Luca one day, and this will have been about 1962, I suppose, when we were beginning to think of these things. There we were, nine years after the discovery of the structure of DNA, reflecting on the fact that, sooner or later, our biochemical friends, our molecular friends, are going to be able to sequence this stuff.

[. . .]

AWFE: I remember saying to Luca, 'Okay, so you say we're going to have these DNA sequences in due course. What's going to happen,' I jokingly said, 'when we find that the sequence for the whale is the same as the sequence for *Drosophila*?'

[. . .]

Human Genetics and Phylogenetic Inference

RGW: [. . .] The theory that you developed for statistical theory and phylogenetic inference, do you think you could have started with any kind of data? It needn't have been human data. It could have been mouse data. It could have been *Drosophila* data. It could have been pea data. Or, do you think somehow there was something special about the structure of human genetic variation?

AWFE: It did, in fact, start with other data.

RGW: It *did* start . . .

AWFE: [Interrupting] It started with some biochemical data provided by a friend of Luca Cavalli's in Milan called Silvestri, I think. I presented that work at the 1963 International Biometrics Conference, which happened to be here in Cambridge. I still have the preprints from that. So, that was just presence or absence data. You know, you have twenty characters. You have 15 different types of bacteria, and each bacterium either has or has not each of these characters. So you've got a row of noughts [zeros] and ones, noughts and ones; what do you do with it? So, that, in a way, was an initial test of how one should think about these things. From there, very quickly – because the original incentive for working on this was the human blood-group data – from there we moved very quickly to the continuous case of gene frequencies. In a sense, that was unfortunate from a technical point of view, because if you have discrete data, like the noughts/ones, you don't get any singularities in the likelihood function. So you're not confronted with that problem, which held me up for two or three years. [. . .]

[. . .]

AWFE: To go back to an earlier comment this morning, my primary interest is in the methodology. My interest in the results is more that they should be reliable, and that their reliability should be measurable by statistical methods. I don't much care what the answer is. Now maybe this is just the background of a Welshman with no known ancestry, coming out of the valleys with a common name, and not bothering too much about that sort of thing.

RGW: [Laughs] Yes, that's good.

AWFE: The one thing I've often wondered about, speculated about though – When did the first Eskimo, going east-about in Greenland, meet the first European, going west-about? I mean, that's a specific moment, when they came round the corner of an iceberg and they saw each other. When and where was that? That would interest me.

RGW: It was probably some Scandinavian – right? – who saw the Eskimo. Yes. [Laughs] More to ask there, but . . . So you haven't compared the sort of genealogy of the Cavalli-Sforza trees from your first tree to later in 1988, 1994?

AWFE: Well, I've read what they were doing all the time. Yes, and how it has slowly changed from . . . as the magnitude of the dataset gradually increased.

[. . .]

RGW: Of course, because, you know, I like maps, I get to ask you one question. Which is . . . you mentioned the Swiss road-builders that need contours. I like that example. This map . . .

AWFE: [Laughs] Oh, this.

RGW: Yes.

AWFE: Now, what have you got this from?

RGW: I think it's the 1964 paper ([27]).[6]

AWFE: This is the actual original?

RGW: Yes, that's the original.

AWFE: It gets distorted . . . by people who don't quite realise how to copy it, and I think in one case because Luca wanted to change it a little bit. [Laughs]

[6] See also the back cover of this volume.

What actually happened was that I was sitting in my room in Pavia one day, and I got hold of a sort of Mercator-type – it obviously isn't Mercator, but anyway . . .

[. . .]

AWFE: I don't know where I got it from. It's a sensible sort of map, maybe I even had to glue two bits together to make sure that the international dateline wasn't at the edge of the paper. That's the sort of thing I would have done, you see. I sat there, and I had this problem as to where the origin of this species might be. We had a method of determining where to put an origin, simply by, I think, dividing the total branch lengths on each side. We couldn't draw the thing unless we had some sort of origin. We had that. I remember sitting there, drawing this, when Luca came into the room. He said, 'What's that?' I said, 'Well, you know, I thought a nice tree might, from a topographic point of view, just go on a world map.' He liked it very much, so of course we published it here. Later on, of course, it's become so obvious that we could have saved the embarrassment of all those Maoris coming from Alaska.

[. . .]

RGW: [. . .] To your best knowledge, isn't it the first time that genetically inferred population structure in humans is superimposed on a map?

AWFE: I think that's probably true, yes. I certainly didn't get the idea from anybody else.

RGW: You *didn't* get the idea from anybody else?

AWFE: No, no. I just sat there and somehow thought, 'Oh, well that would be fun, I wonder if I can find a map?' I mean, that's very typical of the sort of way my mind works. I'd like a picture, I'd like a bit of geometry.

RGW: But you're also saying you didn't think of it *too* much, because like you say, you could have put the branching somewhere in Russia . . .

AWFE: Yes.

RGW: . . . In Western Russia, right?

AWFE: Yes, absolutely you could. [Laughs] I think that was the very first drawing. I mean, it wasn't that Luca came in, and said, 'Oh, well look, why don't we . . . Yes, that's a very good idea, that principle, we could do this, that, and the other.' He just looked at it and said, 'That's it! Use that in your lecture!'

RGW: Well, I said I would only ask you one question about maps, but let me ask you a question about space. Of metric space. Did you see the analogy because there's something special about metric space here that operates the same way in physical linkage, with chromosomes, as it might with a tree, or a branching tree . . .

AWFE: No, there is a difference there, isn't there? Because behind the metric map that you're trying to create by methods like this, for genetic linkage, there lies a real chromosome. But behind our evolutionary tree, what is there? There is something we're trying to estimate, but we can't see it.

RGW: No, you can't see it. But doesn't Felsenstein, and others, don't they speak about the true tree?

AWFE: Well, yes, but what does the phrase 'true tree' mean? It's unknowable.

RGW: You don't have to convince me.

AWFE: You might be able to do it if bones were your data, but gene frequencies, no.

RGW: So there's a physical chromosome behind linkage.

AWFE: Yes. And now we know a great deal about the position of these genes. Often, when I talk to modern geneticists, I quiz them about the continuing importance (or otherwise) of the metric map. I mean, does it matter anymore? It doesn't seem to matter very much. So I always point out, well, if you want to make predictions about number of offspring where the alleles at two particular loci will be manifest in the same individual, don't you need it for that? But, of course, they sort of shake their heads, and say, 'Well, well, not . . . No, not really.'

RGW: That's fascinating, but hang on, then. You said earlier that the data, that data on bacteria and human genetic data are what got you motivated to think about the problem. But now you're also telling me that you were also motivated by an analogy with linkage analysis.

AWFE: Yes. Yes, I am. But it's a methodological analogy.

RGW: So the data [are] what screamed to you 'Make sense of me!'

AWFE: Yes.

RGW: And then the analogy with linkage analysis is what got you thinking, 'Well, maybe this is how we could start thinking about it.'

AWFE: It's fair to say that I must have been using the linkage analogy in the stage of my conversations with Luca when I was still very sceptical about his project for the reasons I've explained. It might have accelerated our work quite a bit if I'd continued the use of the analogy by pointing out that we should do the same sort of mathematics as one does with linkage. Now that was clearly in my mind, but I don't think it was ever expressed quite so directly. We might have cut out a lot of thinking about minimum evolution and so forth, and gone straight to the Kimura gene-frequency model. You know, these things take time. Also, I think, when it comes to minimum evolution, that really took off in quite a big way amongst people trying to construct trees. It was easy to understand, it was fairly easy to program. We had the Fitch discrete data case to use.

A great deal was made of it. It was only when people started to write that it was all based on Ockham's razor that I began to think, well, no, actually it wasn't. It was just a sort of guess as to what a proper solution would be.

RGW: So, it could have quickened the work if you had taken the analogy further. I want to get to Ockham's razor in one minute, but still, talking about the analogy: Why didn't you take it further? Did you just . . . did Luca distract you or something?

AWFE: The analogy of the linkage?

RGW: Yes.

AWFE: Well, in a way, it only came back to me, in some strength, as an explanation of what we were doing quite a few years later. I knew it was important in the original discussions, because I remember that specifically, but then we went the way that we did, which I've described. I think it's only later when I've found myself trying to describe, to geneticists and others who might have some familiarity with the linkage problem, that actually we're doing the same as people do, or did – I don't think they do it anymore, with the linkage problem. When I say I don't think they do it anymore, I don't think there's much pressure to create those linkage maps, at least in *Homo sapiens*. There may still be some pressure in bacterial genetics or something like that. I mean, here I sit in Cambridge with people doing clever things all around me, and nobody has ever come to me and said, 'Look, I've got a bit stuck with a linkage problem. I believe you know about this.'

RGW: No one, not even in the 1970s, 1980s?

AWFE: No. No. Well maybe they didn't get stuck, or maybe they didn't know that I knew about it.

RGW: Returning to Ockham's razor: I guess I missed asking you an obvious question earlier. I mentioned it, but then it got lost. Hennig's methods, and cladistics, and parsimony – I assume that's what you're referring to when you say you're using Ockham's razor.

AWFE: No, I wasn't connecting Hennig with it at all. When it comes to Hennig, I think I have to retreat into a position that I looked in some detail at Hennig's work, and I read the book,[7] and I tried to see whether the straightforward principle of minimum evolution was actually contained in that work. Some people were saying that it was. Everything I know about that, or knew about that, is in the paper that I wrote about the

[7] Hennig, W. 1966. *Phylogenetic Systematics*. Urbana, IL: University of Illinois Press.

origin of the method of minimum evolution ([146]). I don't have anything more to say about it. If anyone else has anything to say about it, I should be very interested to hear it. Evidentially, treading on Hennig's toes was a dangerous thing to do.

[...]

RGW: Weren't there a set of methods also using Bayesian methods?

AWFE: Ah, now there is a story. You will find statements in the literature that it was I who introduced Bayesian methods into evolutionary trees.

RGW: [Laughs] Don't let Ronald [Fisher] hear that.

AWFE: [Laughs] Now this I deny.

RGW: Yes, I'm sure you do.

AWFE: And it's a misunderstanding of the word 'Bayesian'. Now, the word 'Bayesian' was introduced into statistics, that very word, by R. A. Fisher in 1950.

RGW: He used to call it 'inverse probability', right?

AWFE: Yes, even then there's some slight difficulty as to how 'Bayesian' inverse probability was, because if you go right back to James [Jacob] Bernoulli, inverse probability can be used to describe any attempt to use probability arguments to arrive at an estimate of something. So the whole sort of significance testing and narrowing of confidence intervals and so forth, is, in a sense, inverse probability.

AWFE: The word 'Bayesian' itself, actually, it's in the book. The big book, published by John Wiley, of Fisher's statistical papers in 1950,[8] where Fisher was invited to contribute just a paragraph or two, introducing each paper. There, he introduces the word 'Bayesian'. What he means by it is a method of arguing about a probabilistic model. Where the unknown parameters, or some of the unknown parameters of the model, are assumed to have probability distributions, even though there is no other justification than a great need to have a probability distribution because, without them, you can't make the whole system of producing final probabilities work. That's what he meant by it. It's a point well taken by geneticists, because we have a Mendelian mechanism which produces the coin-tossing, which justifies certain assumptions. But that's hardly being Bayesian. The 'Bayesian' word is used for assuming such probability distributions, or isolated probabilities, for which there is no justification at all, merely for the sake of operational convenience. The difficulty then arose, that people took this word 'Bayesian', and they thought it referred to Bayes's theorem. It *doesn't* refer to Bayes's

[8] Fisher, R. A. 1950. *Contributions to Mathematical Statistics*. New York: Wiley.

theorem; it refers to Bayes's *postulate*. In Bayes's paper, at the end of the discussion, there is a scholium which contains the postulate that it might be acceptable to postulate a uniform distribution. This is separated out from the main paper. Bayes wasn't certain about it. So people use this word 'Bayesian' just to mean 'using Bayes' theorem'. This is particularly true in the medical statistical literature. So there's great excitement about the introduction of Bayesian methods. All they're using is conditional probability. They call it Bayesian, in opposition to significance testing and so forth. So, there was a problem about what the word 'Bayesian' means. So, if you now look at my 1970 paper ([46]), which has a complex mathematical model, and then proceeds to try and estimate the parameters of the model by maximum likelihood, people call that Bayesian, simply because I've got a decent mathematical model. From their point of view, it's Bayesian, because you could look at the operation of that model from a conditional probability point of view, and you could look at the conditional probability point of view as using Bayes's theorem, but there's no Bayes's theorem in it.

RGW: We don't have to go down this path, but I mean, do you think then that the discussion between so-called frequentists and Bayesians, is that a red herring?

AWFE: As far as evolutionary trees are concerned, yes. Most of the time it is, because it's based on a misunderstanding of both points of view by the other.

RGW: That's a very interesting take. I see. So, just returning to Fisher for a moment, isn't he well-known for taking issue with what he called 'inverse probability'?

AWFE: Yes, and that comes about partly because of his understanding of the Mendelian mechanism as a physical process for producing probabilities of a half. So, he is not prepared, in other contexts, to make an assumption about a parameter about which nothing is known, except that it's either nought or one, or shall we say 10 or 20. He is *not* prepared to assume a probability of a half for each of those without any ulterior justification of the kind that he's used to in Mendelian genetics. From my point of view, it's an unnecessary thing to discuss anyway. It must be the case that the information in a sample is contained in the likelihood function for that sample, in the absence of Bayesian prior probabilities. That must be the case, because we know that if you have prior probabilities, because of Bayes's theorem, we know that the information in the posterior probabilities is entirely governed by the likelihood function. We have the prior probabilities, we multiply by the likelihood function, and we get the posterior probabilities. Now why, philosophically, it should be thought that it's necessary to have some prior probabilities to interpret this likelihood function, which you've already persuaded yourself from a Bayesian point of view carries all the information from the sample, I can't understand. That argument seems to me to be completely washed out. So we'd better go down to Heffer's [Booksellers], and buy a book called *Likelihood*, and see what we can make of it. When we find that we get stuck because there are things that we don't know, which if we did know, would enable us to cope with bivariate likelihood functions and so forth. Instead of retreating into an untenable position, to make the system work, why don't we sit down and ask ourselves, 'Well, why do we get stuck now? What's going on here?'

RGW: Good.

Connections between Phylogenetic Inference and Population Genetics

RGW: [. . .] There are relatively few people who have connected, or simultaneously worked, in the fields of phylogenetic inference and phylogenetics, on the one hand, and evolutionary genetics or population genetics or quantitative genetics or call it what you will, on the other. Few people have worked in both [families of] fields as much as you have. From what we've talked about yesterday and today, it sounds you were led to phylogenetics, in part, by the linkage analogy, but also by Luca knocking on your door. Population genetics, you were led to by thinking about the Fundamental Theorem, and being influenced quite heavily by Fisher. [. . .] Do you see any relation between these two fields? One last thing, Felsenstein, in some quote somewhere, I think it's in *Inferring Phylogenies*[9] – and certainly Rasmus Nielsen, in his commentary – says that there has tended to be a tension in evolutionary theory [between] people working on phylogeny and people working on population genetics. They don't seem to understand one another, and the future, roughly, is people who can work on both. But we have a person sitting here in front of us, who *has* worked on both.

AWFE: Yes, I don't see any great connection, other than the absolutely fundamental and original one. That in our case, of using gene frequencies, we're adopting a model which comes straight out of the theory of population genetics. So, I mean, that is a very, very close and

[9] Felsenstein, J. 2004. *Inferring Phylogenies*. Sunderland, MA: Sinauer associates.

fundamental connection. When it comes to the analysis of data on the basis of that model, then we have quite a separation. If you're given the model, if you accept that the model is the real link with population genetics, and ask yourself, 'Well, what about the rest of the development?' The rest of the development has the real link with statistical theory, statistical estimation theory. Population genetics, whether on the evolutionary biology data-processing side, or on the theoretical side, doesn't really have such a strong link with statistical inference theory.

[. . .]

RGW: Let me give you some suggestions. Let me suggest some connections. One is just this idea that there are maximisation and minimisation principles.

AWFE: Well, I'm quite clear on saying that those are not very important.

RGW: Yes, you are.

AWFE: So that the connection there is somewhat negative. I have a chapter in some book on maximisation principles in biology, in which I argue that the well-known and long-accepted principles of maximisation in the physical sciences, and the principles of minimisation of least squares and other such things in statistical theory, do not necessarily mean that when you start thinking in biological terms you can leap from a Darwinian theory of evolution and the survival of the fittest to formal maximisation and minimisation principles in biology. Or at least, in evolutionary biology. There, of course, what is coming out is my thinking about the Fundamental Theorem, and the extent to which the Fundamental Theorem has been misconstrued as a maximisation theorem.

RGW: Yes, and in your 1996 piece in *Systematic Biology*, you cite an earlier [from 1983] comment of yours – I'm not going to read the entire passage – but it's basically a critique of parsimony. I'll read just two sentences ([146], p. 234, this volume): 'It has repeatedly been argued that such methods are justified because evolution proceeds parsimoniously. Beneath this assertion . . .'

AWFE: Ah, yes.

RGW: Yes, you remember, '. . . lurks a philosophy that has tainted much of [modern] evolutionary biology, namely the false analogy with physics or with game theory according to which biological systems are striving to achieve certain ends, to win certain contests, to minimize this or to maximize that.' And then, this is a wonderful sentence: 'It is the fascination of the potential function, the lure of the directed force, the mirage of the goal, that misleads.'[10] You would agree with that?

AWFE: Yes, that's me getting vocal on the subject.

[...]

AWFE: [...] You see, there are two threads of thought there. One is simply taking from Fisher the statement that he was not trying to develop what I've been describing as analytic methods. He was trained in the physical sciences and was well aware of the possibility of proceeding by a parallel route in the biological sciences, but he rejected it. And the second, of course, is my own experience in trying to understand the Fundamental Theorem and immediately reacting against those people who misinterpreted it.

RGW: Good, but now I am going to put a mirror to you. Surely, when you're building these statistical models with Cavalli-Sforza, especially the maximum-likelihood estimation, and using this branching-diffusion process as the model construction, and using maximum likelihood as estimation and testing, then, there you *are* doing analytic modelling.

AWFE: Absolutely.

RGW: Then, you say, in 2009 in *Genetics* in 'Statistical methods for evolutionary trees' ([225], p. 271, this volume) ... 'In retrospect, the real contribution from Pavia in 1962–1964 was the embedding of modern phylogenetic analysis in the Fisherian tradition of stochastic model building and efficient statistical estimation by maximum likelihood and the use of genetical data.'

AWFE: I'm talking about linkage, aren't I?

RGW: Yes. You're talking about linkage, but the Fisherian tradition of stochastic model building ...

AWFE: Yes, I see. People might misinterpret that because the word 'Fisherian' has got so attached to significance testing, but that's not how we geneticists saw it.

RGW: No, but you have said that Fisher didn't have this ... okay, to put it most bluntly and confusedly, I would say, given what you said yesterday, it almost sounded like there *was* no Fisherian tradition of stochastic model building.

AWFE: This is a misunderstanding. Which bit of Fisher are we talking about?

RGW: Okay, good. *GTNS*.

AWFE: Ah, yes. But that's a particular aspect of his work that was deductive.

[10] This quote is found verbatim in Edwards's intervention on p. 263 in the Discussion section of Felsenstein, J. 1983. Statistical inference of phylogenies. *Journal of the Royal Statistical Society. Series A (General)* 146, 246–272.

RGW: Inductive?

AWFE: Inductive, I beg your pardon, that's *inductive*. But if you look at other aspects of his work, and in particular, the training that we youngsters had from him, what Fisher was very good at, and what we learnt sort of automatically from him ... and this, particularly in linkage, but all through the work on ... the mouse work, and the theory of inbreeding and so forth, he was very good at stochastic modelling. Allied to that excellence in stochastic modelling was a great facility with transformations. He knew, much better than anybody else, what kind of analytical transformation to use, you know, the *z* transformation or something like that, to get what was in a rather unpalatable form into a palatable form, mathematically. It was one of his, I was going to say, tricks. One of his skills was to think up transformations.

RGW: Would you say – even then, even in the Fisherian stochastic model – I wonder, whether there might be an important distinction. Might there be an important distinction between two kind[s] of analytic modellers? One is someone like Fisher, who not only is doing stochastic modelling, and [is] good at transformation, but he's very problem-oriented. He can look at a whole range of different problems, and, to use modern parlance you may not like, he was good at 'hacking'. He was a good hacker. He was good at just seeing many different questions and problems, and then attacking them each with whatever appropriate statistical tools and analytic models were necessary. As opposed to another kind of analytic modeller, that wasn't so driven by questions or problems, but was driven more by mathematical elegance.

AWFE: I think that's right. I do think that's right. Let us see if we can think of anything that Fisher did, purely for the delight of extending an analytical model in a perhaps useful direction. You really have to take down the volumes and thumb through them, asking yourself the question as you go along. There's a quotation in Joan Box's biography,[11] I think it must've been 1929, and Fisher is writing to his old schoolmaster at Harrow school, replying to a letter of congratulations on his election to the Royal Society. He says in this letter that all his work has been driven by his biological interests.

[...]

RGW: Again, I guess, another map question does emerge. What was your interest in map projections? Was it just intellectual curiosity?

[11] Box, J. F. 1985. *R. A. Fisher: The Life of a Scientist*. New York: Wiley.

AWFE: It was spherical trigonometry. I just thought spherical trigonometry was lovely. This idea, after you've been trained in plane trigonometry for 15 years or something, 10 years, of discovering triangles with three right angles in the corners, that's all very nice.

RGW: Are there common themes between population genetics ... between drawing trees and thinking about evolutionary processes that can come via geometric thinking?

AWFE: I think there is an element in common, from a geometrical point of view. Some people are at ease thinking about things geometrically, and some people not so much. Famously, Fisher didn't mind how many dimensions he was thinking in at all. I just enjoyed things like conic sections at school. I just enjoyed that sort of analytical geometry, so it's a very easy part of my thinking.

RGW: Thank you. One last question which came up earlier, and I didn't have a chance to ask you then, but it is in the context of geometric thinking. When you were talking about maximum likelihood and such, when we were talking about that in the earlier session, and you started talking about a topography and you did, jokingly, refer to Wright and the seascape, just to be clear. Here's a hypothesis, tell me if I'm right or wrong: It's not that Edwards is opposed to using geometrical visualisations, or to using a landscape or seascape, it's just that he thinks that Wright's picture is incoherent for a variety of reasons.

AWFE: Yes.

RGW: Okay. But that the seascape or landscape is perfectly reasonable when you're thinking about maximum likelihood.

AWFE: Oh, I'm sure that if I'd been around, when a contemporary of Wright in 1930, when I was asked to review Fisher's book, I would probably have said exactly the same thing that Wright did.

[Both laugh]

AWFE: And probably live to regret it.

[...]

Day 3

Edwards and Fisher

Edwards's Office, Cambridge, UK

22 May 2015

Edwards's Biography

RGW: [. . .] Today . . . we're doing the last session of interviews with Professor Edwards. Today, it's a mix of questions. There are some biographical questions, and then some final questions I have about Fisher. I'll start by asking you about your formal education. What was your formal education like? In particular, I'd like to ask you three aspects. How much science did you have, and what kinds of science? How much mathematics did you have, and what kinds of mathematics? And, how much philosophy, or history and philosophy, of science did you have? How did each of these areas influence you?

AWFE: [. . .] As far as the formal education is concerned, I was fortunate to go to one of these British public schools. I had the same kind of education, went to the same kind of school, as Fisher and Haldane and the others of that generation. These schools provided absolutely first-rate scientific education [. . .]

RGW: What about mathematics?

[. . .]

AWFE: Geometry and trigonometry I always enjoyed. Right the way back to the school before I went to my public school, I became very good at trigonometric identities. Most people don't even know what those are, now, I suppose. Because I was always the best boy in the class, when the class was set problems to do, I would always finish them before the others. I was given these pages of trigonometrical identities to prove, and I just found that a very interesting, pleasant way of passing the time – a sort of crossword. I became very good at them, and I also started to learn about some spherical trigonometry then. I found that very amusing, so there is a side of me which is really quite geometric.

I think perhaps the problem was that I wasn't very good at rigorous proofs and arguments, and so forth. One of my schoolmasters wrote in a report, 'Edwards is a somewhat speculative mathematician', which is a sort of two-edged compliment really, because it's that speculative side which has led me to wonder how to make Venn diagrams with lots of sets, and, you know, that kind of thing.[1]

RGW: Yes.

AWFE: I'm a mathematician at heart, but not by profession.

[. . .]

AWFE: My formal education in history and philosophy of science is limited, going back to my first two years in Cambridge. [. . .] I got on very well with that because I was very interested in the history. I was already interested in the philosophy because I was interested in the philosophy of induction, through having got interested in statistics. When I found that the lectures on induction didn't recognise the inductive arguments of statistics, I stopped going to them.

RGW: Is that when you met Ian Hacking?

AWFE: That's a very interesting question. I think the answer must be no. My contact with Ian Hacking was when his book *Logic of Statistical Inference* was published.[2] That must have been 1965. I bought it on a visit to Cambridge, but I didn't really study it until I got to Cambridge in 1968. I think I will have gone immediately to find Ian Hacking, and we've known each other since then.

RGW: Do you think that is useful work? Is that useful for a statis[tician] . . .

AWFE: [Interrupting] His book?

RGW: Yes.

AWFE: I think it's an absolute foundation book in studying the problem of induction, as viewed by statisticians. It is so fundamental because he was the first person, truly, to understand the relationship between Fisher's fiducial probability and Harold Jeffreys' work.[3] By isolating what he called the principle – no, just a minute, I mustn't call it the 'principle of indifference' – Ian Hacking has a related principle whose name doesn't come to me at the moment, which is part of a wonderful explanation of fiducial probability.

RGW: This isn't the pre-trial versus post-trial?

[1] References numbered 114, 115, 116, 128, 130, 137, 163, 180, 196, 211, 212, 219, 220, 224, 229, and 240 in the list of publications of A. W. F. Edwards in Part 4. For purposes of consistency and also due to length constraints, A. W. F. Edwards's work on Venn diagrams was not included in this volume.

[2] Hacking, I. 1965. *The Logic of Statistical Inference*. Cambridge: Cambridge University Press.

[3] Jeffreys, H. 1939. *Theory of Probability*. Oxford: Clarendon Press.

AWFE: That's tremendous from that book, too. I like that very much, and I often quote, that distinction that Ian makes there. No, it's 'principle of irrelevance'. He has a 'principle of irrelevance' in that book. Only if this principle of irrelevance holds can you make the fiducial argument work. So he really isolates the problem with the fiducial argument.

[. . .]

RGW: [. . .] Have you ever disagreed with an important point Fisher has made?

AWFE: Yes, that's a very risky thing to do, because you usually regret it a couple of years later. But there *is* a point where I say he's hoist with his own petard over randomisation tests, where it seemed to me (and I haven't reflected on it since, but I should say it still seems to me) that he's introducing randomisation at the wrong point. His introduction of randomisation, for which he's so famous in experimental design, is part of the way that he justifies the significance tests which come out of that. But there's a separate point at which he invokes randomisation, so I thought that was wrong.

[. . .]

AWFE: [. . .] [M]y PhD thesis is called 'The Human Sex Ratio.' That was a little bit of initiative where I departed from what I was supposed to be doing, and invented this task for myself of writing about the human sex ratio.

[. . .]

AWFE: [. . .] I couldn't contribute to an understanding of the Fundamental Theorem until, in the 1970s, I got interested in it. I think that's because I was given Sewall Wright's gene-frequency volume . . .

RGW: [Interrupting] Volume 2. Of his collections . . .[4]

AWFE: . . . Volume 2, to review ([49]). I thought, 'I'd like to take that very seriously', so I took it as the only book on the travel, by train, across Europe to the Black Sea to a scientific meeting. I took that very seriously, with the result that you know. I then gave a paper in 1971 to the Biometric Society on the difference between Fisher's Fundamental Theorem and Wright's adaptive landscapes. But I wasn't yet ready to write about that because I still didn't have a clear idea of what Fisher's Fundamental Theorem was. If you look in my *Foundations of Mathematical Genetics*, there's a little apologia at the end, saying why I left it out. I left it out because I didn't yet understand it. That was as late as 1977. Then, you might ask, 'Well, why didn't you understand it from Price's paper in 1972?' The answer is that I wasn't working in that line

[4] Wright, S. 1969. *Evolution and the Genetics of Populations. Volume 2: The Theory of Gene Frequencies*. Chicago, IL: University of Chicago Press.

of things at that time. Although I noticed Price's paper, I didn't spend time reading it. I was extremely interested in trying to understand fiducial probability at the time. My career was still in statistics and, particularly, statistical inference. Then I spent three months as a visiting professor in Aarhus University, again concentrating wholly on problems of statistical inference, and, incidentally, at that point, understanding fiducial probability for the first time. So that's the first part of my career: it's statistical inference, application of statistics, applied statistics to evolutionary trees – but it's basically statistical. Then, I did begin to get interested in evolution again.

[. . .]

RGW: [. . .] I do want to ask you, because I personally consider it a very important book, a very influential book, and I think many of us do, even if it's a controversial book, what you think about *The Selfish Gene*, the book by Dawkins from 1976.[5] What do you think about its metaphor of 'selfish' gene?

AWFE: I think that's a fair piece of journalism.

RGW: [Laughs]

AWFE: I think the message is a good one and an interesting one, and it's, as Richard Dawkins himself knows, a very Fisherian one.

RGW: Yes. How is it Fisherian, in your analysis?

AWFE: It's Fisherian in the sense that the message that Fisher is conveying in *GTNS* is that the important thing to measure is how gene frequencies change. There are places in that, and certainly in his correspondence, when he's saying, 'I just do not consider anything to have changed unless the gene frequencies change. That, to me, is what constitutes natural selection.' So, to create the phrase, the 'selfish gene', is really quite a clever piece of journalism. As to the actual content of the book, it came out just at the time that I was beginning to be interested in that sort of thing, so I'm sure it made a considerable impression on me. I don't actually remember.

RGW: What would you say to someone who might say that you could also do these equations in terms of genotypes?[6] That you could look in terms of genotypic frequencies and assign fitnesses to genotypes?

AWFE: Well, of course that was my training.

RGW: Right, exactly.

AWFE: But then you scratch your head, and you think, 'Okay, well, but it's the genes [not genotypes] that go from one generation to the next, so how

[5] Dawkins, R. 1976. *The Selfish Gene*. Oxford: Oxford University Press.

[6] Sober, E. and Lewontin, R. C. 1982. Artifact, cause and genic selection. *Philosophy of Science* 49, 157–180.

are we going to handle this?' The answer is the Fundamental Theorem of Natural Selection.

RGW: Then what might you say to someone who would maybe accept, okay, 'selfish gene', but what about the 'cooperative gene'? What about the 'collaborative gene'?

AWFE: They're selfish in a different way. I mean, I think you must allow the metaphor to have a very broad interpretation. Personally, I've always found arguments about the levels of selection as completely vacuous.

RGW: How so?

AWFE: You can look at selection at any level you like. To argue that any particular level somehow has priority seems to me to be vacuous. Except, of course, when you take into account the bit that is transmitted from one generation to the next. That, clearly, is a special level, but it doesn't stop you talking about selection at any other level that you think of.

[. . .]

AWFE: [. . .] You ask about students. I was extremely fortunate in the students I had, though they were very small in number – again, because of the slightly odd position I was in. In Cambridge, I was in the medical faculty, for example. The first student I had who has made a good contribution was Christopher Cannings, who was a student of mine in Aberdeen and has had a distinguished career since then, retiring as a professor in Sheffield. The second student I had was here in Cambridge, Elizabeth Thompson, E. A. Thompson, who's had an extremely distinguished career, and has an enormous number of students herself. It's sort of compensated me for not having had very many students. Chris Cannings worked on fairly straightforward population genetic models, but he had a much greater degree of insight into these models, and into questions of convergence and stability, and doing them really properly in a way that it hadn't been done before. So his was a very competent PhD. Elizabeth's was essentially two whole PhDs stuck together. It could have been two completely different people writing two PhDs. She started off on the evolutionary-tree work, and that ended up with a little book, actually, called *Human Evolutionary Trees*, because it saw the applications were with human data.[7] That was basically a very deep statistical analysis of the inference problem with evolutionary trees. She was looking at things from a theoretical, statistical point of view, where Joe Felsenstein in the States tended to look at

[7] Thompson, E. A. 1975. *Human Evolutionary Trees*. Cambridge: Cambridge University Press.

things more from a practical, computing point of view, and so he was successful in finding the tricks which would enable the solutions to come forward. Elizabeth's book is at a deeper level than that.

[. . .]

Fisher Revisited

[. . .]

RGW: [. . .] Was Fisher happy, or was there some kind of bitterness, about him leaving the UK and going to Australia?

[. . .]

AWFE: You retire at the age of 67. So he retired from Cambridge, but he could have stayed on. I mean, I'm retired from Cambridge, but here I am, sitting in Caius College, talking to you. That's fine. It's not really a place you have to retire from, but Fisher did retire. That gave him the freedom to travel, which he liked. Now, there is a negative side. I said there was no negative side to his departure. There is a negative side, which I know from the time that I got to know him. He was not as well-recognised in statistics and in genetics as he should have been, because the two are subjects that moved on quite rapidly. Although he was very good at keeping up with the latest work – that was very much a characteristic of him – people had sort of thought that those subjects had moved on. They didn't appreciate, and certainly on the evolutionary biology side, they didn't appreciate what a great man he had been in his working life. He must have sensed that he was a little bit isolated academically. That comes out in the Fisher memorial lecture I gave, which is called 'Fiducial inference and the Fundamental Theorem of Natural Selection'.[8] These are both very close to his heart, as ideas that he had, which were so much in advance of the way that other people were thinking that they came in for a great deal of criticism. Well, ignorance or criticism. Indeed, in the end, they turn out to be such difficult concepts that one has to qualify them or present them in ways which are more limited than I think perhaps he had hoped. That's certainly true with statistics. There's no doubt that, in his prime, he thought that there was going to be a complete, global theory of statistical induction.

[. . .]

[8] Fiducial inference and the Fundamental Theorem of Natural Selection. *Biometrics* 51, 799–809 (reference 145 in the list of publications of A. W. F. Edwards in Part 4.). Consult also Edwards's 'My Fisher' in Appendix 3 of this volume.

RGW: Last question, as I understand it and as I am told. Fisher worked for the smoking industry, is that correct?

AWFE: Well, he didn't actually work for the smoking industry, no. That's not right. Now, I have spoken of this to the Biometric Society, but I haven't thought about it for quite a long time. He was engaged as an adviser to the Tobacco Research Council. The Tobacco Research Council was, at least in part, funded by the tobacco industry. He got tangled up in it publicly because he was a bit naïve about it, and he was using the argument put forward by Doll and Hill as an example of a bad statistical argument.[9] That is, he criticised it on statistical grounds. As far as his relation with the tobacco industry was concerned, he was known to have been furious when it was suggested, I think by somebody in the industry, that his role was other than as a dispassionate observer, giving them scientific advice. That's what he felt he was doing, but it didn't work out like that because the advice that he was giving was essentially that the statistical work was faulty in some respect, and that not sufficient consideration was given to the difference between correlation and causation.

RGW: As I understand it, there is a [narrative] that Fisher was claiming that there isn't evidence for causation in smoking. That there is no evidence of smoking causing cancer.

AWFE: Well I think, in brief, that is probably true. He felt that the statistical evidence was not satisfactory. Not being a doctor, he didn't realise that a lot of doctors, round about that time, must surely have had that idea. It's just incredible that it required a statistical analysis to do it. You have to go back to that time, when what we see as absolutely so obvious that it must have been seen generations before then, had not been seen, had not been accepted. That is why Doll and Hill was so valuable. In the atmosphere of that time, to query it was not unreasonable.

RGW: If I'm understanding correctly, you're saying that, based on the data, Fisher's argument is intellectually valid, from a statistical point of view, but that doesn't mean anything about what the mechanistic causes might be.

AWFE: No, so he was free to speculate about the mechanistic causes. He was pointing out that different genotypes might have different liabilities to lung cancer, that they might have different susceptibilities to damage caused by smoking, and so on. You can see the general argument. He

9 Doll, R. and Hill, A. B. 1950. Smoking and carcinoma of the lung: preliminary report. *British Medical Journal* 2, 739–748.

was clearly, certainly politically, naïve about that. Also, in a sense, scientifically naïve, in that he got involved in matters in which he was not professionally competent, in the way that a couple of epidemiologists who were actually doctors would be. They must have had good reason to be investigating smoking and lung cancer. The rest of the world didn't know about that at the time.

[...]

RGW: Is there any final thing you want to say?

AWFE: Just thank you for taking an interest. I will actually add a rider to that. I am very impressed with philosophers of science, how carefully they read the literature. So far as I observe it, you even read newspapers as carefully as that. This is the great thing about philosophers of science, they actually read stuff which most people don't read. They read it very carefully.

RGW: [Laughs] Yes.

AWFE: That's a pretty good compliment to a scientist in the end. It's very agreeable to have something that I did 50 years ago to have been noticed – to actually revisit that time and talk to people about it, and try and help them understand what was going on at the time. It's very good.

RGW: Thank you.

Part 4

A. W. F. Edwards

Curriculum Vitae

ANTHONY WILLIAM FAIRBANK EDWARDS

Born London, 4 October 1935

Educated at Uppingham School and Trinity Hall, University of Cambridge (Scholar 1957). Natural Sciences Tripos: Part I – Class II.2, Part II (Genetics) – Class I. BA (1957), MA, PhD (Genetics, 1961).

Posts

Eugenics Society Leonard Darwin Research Fellow, Department of Genetics, University of Cambridge, 1960.

Research Associate, International Laboratory of Genetics and Biophysics (LIGB, Naples), Institute of Genetics, University of Pavia, Italy, 1961.

Acting Assistant Professor, Departments of Genetics and Mathematics, Stanford University, California, USA, 1964.

Senior Lecturer, Department of Statistics, University of Aberdeen, 1965.

Berkeley Bye-fellow in Medicine, Gonville and Caius College, Cambridge, 1968.

Assistant Director of Research, Department of Human Ecology (later Department of Community Medicine, later Department of Public Health), University of Cambridge, 1970; promoted Reader in Mathematical Biology (later Biometry), 1978, Professor of Biometry, 2000. (Retired 30 September 2003.)

Fellow of Gonville and Caius College, Cambridge 1970 to present.

Visiting Professor of Mathematics, Department of Theoretical Statistics, University of Aarhus, Denmark, 1973.

Visiting Professor of Mathematics, Department of Mathematics, Massey University, New Zealand, 1996.

Offices, etc.

Treasurer of the Sir Ronald Fisher Memorial Committee, 1969–2009.

Senior Proctor, University of Cambridge, 1978–79.

President of the British Region of the International Biometric Society, 1992–94.

Chairman of Cambridge University Library Syndicate, 1993–98.

Chairman of the Christiaan Huygens Committee for the History of Statistics of the International Statistical Institute, 1999–2003.

Sometime member of the Council and of the General Board, University of Cambridge; of the Council of the International Biometric Society; of the Cambridge District Health Authority.

Named Lectures

Buehler Lecturer, University of Minnesota, Minneapolis, USA, 1992.

II Fisher Memorial Lecturer, University of Adelaide, Australia, 1992.

XVIII Fisher Memorial Lecturer, UK, 1994.

Snedecor Lecturer, Iowa State University, Iowa, USA, 1996.

Galton Lecturer, Galton Institute, UK, 1997.

VII Adriano Buzzati-Traverso Lecturer, University of Pavia, Italy, 1999.

Zyskind Lecturer, Iowa State University, Iowa, 1999.

Fellowships, etc.

Fellow of the Royal Statistical Society, 1962.

Member of the International Statistical Institute, 1976.

Fellow of the Institute of Mathematics and its Applications, and CMath, 1991.

Fellow of the Linnean Society, 1994.

Honorary Professor of the University of Pavia, Italy, 1999.

Honorary Member of the Genetics Society, 2002.

Fellow of the Royal Society, 2015.

Higher Degrees and Awards

ScD Cambridge, Genetics, 1972.

Medaglia Teresiana, University of Pavia, Italy, 1999.

LittD Cambridge, History and Philosophy of Science, 2005.

Cedric Carter Medal of the Galton Institute, 2010.

Gold Medal of the Telesio-Galilei Academy of Science, 2011.

Scientific Publications

A. W. F. EDWARDS

(Including only major book reviews and abstracts)

* *Refers to papers included in this volume*

1957

1 The proportion of umbra in large sunspots, 1878–1954. *The Observatory* 77, 69–70.

1958

2 Number of mating types required for balance in multipoint linkage programmes *Nature* 181, 503–504.

3 An analysis of Geissler's data on the human sex ratio. *Annals of Human Genetics* 23, 6–15.

4 The sex distribution in the offspring of 5,477 Swedish ministers of religion, 1585–1920. *Hereditas* 44, 447–450. (With M. Fraccaro)

1959

5 Some comments on Schützenberger's analysis of data on the human sex ratio. *Annals of Human Genetics* 23, 233–238.

1960

6 The meaning of binomial distribution. *Nature* 186, 1074.

7 Natural selection and the sex ratio. *Annals of Human Genetics* 24, 239–244. (With W. F. Bodmer) Reprinted in *Group Selection*, ed. G. C. Williams, Chicago: Aldine-Atherton, 139–149 (1971) and in *Demographic Genetics*, eds. K. M. Weiss and P. A. Ballonoff, Stroundsburg, PA: Dowden, Hutchinson and Ross, 88–93 (1975).

8 Distribution and sequences of sexes in a selected sample of Swedish families. *Annals of Human Genetics* 24, 245–252. (With M. Fraccaro)

9 Models in genetics. In *Models and Analogues in Biology (Symposia of the Society for Experimental Biology)*, Issue 14, Cambridge: Cambridge University Press, 6–12.

10 Natural selection and the sex ratio. *Nature* 188, 960–961. Reprinted in: *Group Selection*, ed. G. C. Williams, Chicago: Aldine-Atherton, 158–160 (1971).

11 On the size of families containing twins. *Annals of Human Genetics* 24, 309–311.

12 On a method of estimating frequencies using the negative binomial distribution. *Annals of Human Genetics* 24, 313–318.

1961

13 A factorial analysis of sex ratio data. *Annals of Human Genetics* 25, 117–121.

14 The population genetics of sex-ratio in *Drosophila pseudoobscura*. *Heredity* 16, 291–304.

1962

15 A factorial analysis of sex-ratio data: a correction to the article in vol. 25, 117. *Annals of Human Genetics* 25, 343–346.

16 The number to breed to be reasonably certain of being able to reject one of two hypotheses. *Heredity* 17, 277–278.

17 Genetics and the human sex ratio. *Advances in Genetics* 11, 239–272.

1963

18 Migrational selection. *Heredity* 18, 101–106.

19 The measure of association in a 2 × 2 table. *Journal of the Royal Statistical Society A* 126, 109–114.

20 Estimation of the parameters in short Markov sequences. *Journal of the Royal Statistical Society B* 25, 206–208.

21 Human sex ratio and maternal immunity to male antigen. *Nature* 198, 1106–1107.

22 The limitations of population models. *Proceedings of the 2nd International Congress of Human Genetics, Rome, 1961*, 222–223.

23 A linkage for drawing the Normal distribution. *Applied Statistics* 12, 44–45.

24 Natural selection and the sex ratio: the approach to equilibrium. *American Naturalist* 97, 397–400.

*25 The reconstruction of evolution. (Abstract) *Heredity* 18, 553, and *Annals of Human Genetics* 27, 104–105. (With L. L. Cavalli-Sforza)

1964

26 Infinite coprime sequences. *Mathematical Gazette* 48, 416–422.

*27 Reconstruction of evolutionary trees. In *Phenetic and Phylogenetic Classification*, eds. V. H. Heywood and J. McNeill, The Systematics Association Publication No. 6, London: The Systematics Association, 67–76. (With L. L. Cavalli-Sforza) Reprinted in *Cladistic Theory and Methodology*, eds. T. Duncan and T. F. Stuessy, New York: Van Nostrand Reinhold (1985).

*28 Analysis of human evolution under random genetic drift. *Cold Spring Harbor Symposia on Quantitative Biology* 29, 9–20. (With L. L. Cavalli-Sforza and I. Barrai)

1965

*29 Analysis of human evolution. Proceedings of the 11th International Congress of Genetics, The Hague 1963. *Genetics Today* 3, 923–933. (With L. L. Cavalli-Sforza)

*30 A method for cluster analysis. *Biometrics* 21, 362–375. (With L. L. Cavalli-Sforza)

1966

31 Sex-ratio data analysed independently of family limitation. *Annals of Human Genetics* 29, 337–347.

*32 Studying human evolution by computer. *New Scientist* (19th May) 30, 438–440. Reprinted in *Evolution*, ed. M. Ridley, Oxford: Oxford University Press (1997).

33 Mathematical genetics. In *Teaching Genetics*, eds. C. D. Darlington and A D. Bradshaw, Edinburgh: Oliver and Boyd, 121–124. (Reprinted with Appendix by A. W. F. Edwards)

34 Linkage on a circular map. *Genetics* 54, 1185–1187.

*35 Estimation procedures for evolutionary branching processes. *Bulletin of the International Statistical Institute* (Proceedings of the 35th session) 41, 803–808. (With L. L. Cavalli-Sforza)

1967

*36 Fundamental Theorem of Natural Selection. *Nature* 215, 537–538.

*37 Phylogenetic analysis: models and estimation procedures. *Evolution* 21, 550–570. (With L. L. Cavalli-Sforza) Also in *American Journal of Human Genetics*, Supplement 19, 233–257.

38 Automatic construction of genealogies from phenotypic information (AUTOKIN). (Abstract) *Bulletin of the European Society of Human Genetics* 1, 42–43.

1968

39 Natural selection and the de Finetti diagram. *Annals of Human Genetics* 31, 421–428. (With C. Cannings)

39a Simulation studies of genealogies. (Abstract) *Heredity* 23, 628.

1969

40 A genetical application of homogeneous coordinates. *Eureka* 32, 5–9.

41 Expected genotypic frequencies in a small sample: deviation from Hardy–Weinberg equilibrium. *American Journal of Human Genetics* 21, 245–247. (With C. Cannings)

*42 Statistical methods in scientific inference. *Nature* 222, 1233–1237.

43 Genetic taxonomy. In *Computer Applications in Genetics*, ed. N. E. Morton, Honolulu: University of Hawaii Press, 140–142.

44 Tristan da Cunha – the human biology of isolation. *The Listener* (16th January) 81, 65–68. (With H. Lewis, D. Roberts, G. Keir, M. Richardson, J. Fisher, and D. Tyrrell)

1970

45 The search for genetic variability of the sex ratio. *Journal of Biosocial Science*, Supplement, 2, 55–60.

*46 Estimation of the branch points of a branching diffusion process. *Journal of the Royal Statistical Society B* 32, 155–174.

47 Likelihood. *Nature* 227, 92.

1971

48 Estimation of the inbreeding coefficient from ABO blood-group phenotype frequencies. *American Journal of Human Genetics* 23, 97–98.

*49 Review of *Evolution and the Genetics of Populations*, Volume 2, by S. Wright. *Heredity* 26, 332–338.

50 Distances between populations on the basis of gene frequencies. *Biometrics* 27, 873–881.

51 Science, statistics and society. *Nature* 233, 17–19.

52 Mathematical approaches to the study of human evolution. In *Mathematics in the Archaeological and Historical Sciences*, eds. F. R. Hodson, D. G. Kendall, and P. Tautu, Edinburgh: Edinburgh University Press, 347–355.

1972

*53 Affinity as revealed by differences in gene frequencies. In *The Assessment of Population Affinities in Man*, eds. J. S. Weiner and J. Huizinga, Oxford: Clarendon Press, 37–47. (With L. L. Cavalli-Sforza)

54 *Likelihood*. Cambridge: Cambridge University Press.

55 Biological problems, and opportunities, of isolation among the islanders of Tristan da Cunha. In *Population and Social Change*, eds. D. Glass and P. Revelle, London: Arnold, 383–417. (With H. E. Lewis and D. F. Roberts)

*56 Likelihood. *Bulletin of the Institute of Mathematics and its Applications* 8, 329–331.

1973

57 The likelihood treatment of linear regression. *Biometrika* 60, 73–77. Reprinted in *Likelihood* (expanded edition). Baltimore, MD: Johns Hopkins University Press.

57a Paperback polymorphism. *Nature* 246, 438.

1974

58 A stochastic cross-country or festina lente. In *Stochastic Geometry*, eds. E. F. Harding and D. G. Kendall, London: Wiley, 287–290. (Originally published in *Sailplane and Gliding* 14, 12–14, 1963)

59 On Kimura's maximum principle in the genetical theory of natural selection. *Advances in Applied Probability* 6, 10–13.

*60 The history of likelihood. *International Statistical Review* 42, 9–15. Reprinted in *Likelihood* (expanded edition). Baltimore, MD: Johns Hopkins University Press (1992).

61 A problem in the doctrine of chances. *Proceedings of Conference on Foundational Questions in Statistical Inference*, 1973, eds. O. Barndorff-Nielsen, P. Blaesild, and G. Schou, 43–60. Aarhus: University of Aarhus. Reprinted in *Likelihood* (expanded edition). Baltimore, MD: Johns Hopkins University Press.
62 Population estimates from recapture studies. *Nature* 252, 509–510.
63 Interval estimation for the two-parameter double exponential distribution. *Technometrics* 16, 641–642.
64 Distance measures for phylogenetic trees. In *Genetic Distance*, eds. J. F. Crow and C. Denniston, New York: Plenum Press, 41–43.

1976

65 *Likelihood* (revised edition). Cambridge: Cambridge University Press.
66 Fiducial probability. *The Statistician* 25, 15–35.
67 Confidence limits. *British Medical Journal* 2, 472.

1977

68 *Foundations of Mathematical Genetics*. Cambridge: Cambridge University Press.
69 *Foundations of Mathematical Genetics* (Special printing for submission for the Adams Prize, with a new preface). Cambridge: Cambridge University Press.
70 Statistical inference – Bayesian and non-Bayesian. *American Journal of Human Genetics* 29, 107–108.
71 A geometric formulation of the stability condition at a triallelic locus. In *Measuring Selection in Natural Populations* (Lecture Notes in Biomathematics 19), eds. F. B. Christiansen and T. M. Fenchel, Berlin: Springer-Verlag, 291–300.
72 Selection at a multiallelic locus: Feller's transformation. *Annals of Human Genetics* 41, 219–224.

1978

73 Commentary on the arguments of Thomas Bayes. *Scandinavian Journal of Statistics* 5, 116–118.
74 Review of *Stochastic Models in Population Genetics*, ed. W.-H. Li. *Heredity* 41, 239–247.
75 R. A. Fisher's work on statistical inference. In *I fondamenti dell'inferenza statistica*. Florence: Parente, 117–124. Reprinted in *Likelihood* (expanded edition), Baltimore, MD: Johns Hopkins University Press.

1979

76 Parsing a genealogy. *Advances in Applied Probability* 11, 2–3.

1980

77 Bayes's theorem and the evaluation of tests. *British Medical Journal* 281, 1285.

78 Parsing a genealogy. In *Population Structure and Genetic Disorders*, ed. A.W. Eriksson. London: Academic Press, 273–283.

1981

79 Hereditary genius. Review of *Statistics in Britain, 1865–1930: The Social Construction of Scientific Knowledge* by D. MacKenzie. *London Review of Books* 3, 12–13.

1982

80 Sums of powers of integers: a little of the history. *Mathematical Gazette* 66, 22–28.

81 Pascal and the Problem of Points. *International Statistical Review* 50, 259–266. Reprinted in *Pascal's Arithmetical Triangle*. London: Griffin.

82 Grouping Darwin. *Nature* 300, 572. (Correction in 301, 192).

82a Review of Biometry, 2nd ed. by R. R. Sokal and F. J. Rohlf; Statistical Tables, 2nd ed. by F. J. Rohlf and R. R Sokal. *Systematic Zoology* 31, 229–230.

1983

82b When Gould meets Galton. Review of *The Mismeasure of Man* by S. J. Gould. *London Review of Books* 4, 13.

83 Pascal's problem: the 'gambler's ruin'. *International Statistical Review* 51, 73–79. Reprinted in *Pascal's Arithmetical Triangle*. London: Griffin.

84 Pierre de Fermat. In *Encyclopedia of Statistical Sciences*. New York: Wiley, volume III, 62–63.

85 Fiducial distributions. In *Encyclopedia of Statistical Sciences*. New York: Wiley, volume III, 70–76. Reprinted in *Likelihood* (expanded edition). Baltimore, MD: Johns Hopkins University Press.

86 How to run elections. *Nature* 303, 568.

1984

87 The nonequivalence of likelihood and conditional-likelihood solutions in multinomial sampling. *American Journal of Human Genetics* 36, 229–232. (With E. A. Thompson)

88 Approximating the tetrachoric correlation coefficient. *Biometrics* 40, 563 (With J. H. Edwards)

89 *Likelihood* (paperback edition – Cambridge Science Classics) Cambridge: Cambridge University Press.

1985

90 Pascal's triangle. In *Encyclopedia of Statistical Sciences*. New York: Wiley, volume VI, 628–630.

90a Where did Pascal learn of the arithmetical triangle? *45th Session of the International Statistical Institute* 1, 159–160.

1986

91 Is the reference in Hartley (1745) to Bayesian inference? *The American Statistician* 40, 109–110.

92 A quick route to sums of powers. *American Mathematical Monthly* 93, 451–455.

93 More on the too-good-to-be-true paradox and Gregor Mendel. *Journal of Heredity* 77, 138.

94 Evolutionary relationships of human populations. *Nature* 323, 744.

95 The problem of points. In *Encyclopedia of Statistical Sciences*. New York: Wiley, volume VII, 253–255.

*96 Are Mendel's results really too close? *Biological Reviews* 61, 295–312.

97 How many reactor accidents? *Nature* 324, 417–8.

1987

98 *Pascal's Arithmetical Triangle*. London: Griffin.

99 Evolution and optimization. *Nature* 326, 10.

100 Defending Mendel merely perpetuating a myth. *Nature* 326, 449.

101 Loopy catenaries. *Mathematical Spectrum* 19, 65–71.

102 How many nuclear reactor accidents? *Nature* 327, 663.

103 *Likelihood* (paperback edition reprinted – Cambridge Science Classics). Cambridge: Cambridge University Press.

104 What Fisher meant. *Nature* 329, 10.
105 Bernoulli, James. In *The New Palgrave: A Dictionary of Economics*, eds. J. Eatwell, M. Milgate, and P. Newman, London: Macmillan, volume I, 232.
106 De Moivre, Abraham. In *The New Palgrave: A Dictionary of Economics*, eds. J. Eatwell, M. Milgate, and P. Newman, London: Macmillan, volume I, 802.
107 Fisher, Ronald Aylmer. In *The New Palgrave: A Dictionary of Economics*, eds. J. Eatwell, M. Milgate, and P. Newman. London: Macmillan, volume II, 376–377. Reprinted in *Time Series and Statistics*, eds. J. Eatwell, M. Milgate, and P. Newman. London: Macmillan, 95–7 (1990).
108 Likelihood. *The New Palgrave: A Dictionary of Economics*, eds. J. Eatwell, M. Milgate, and P. Newman, London: Macmillan, volume III, 184–185. Reprinted in *Time Series and Statistics*, eds. J. Eatwell, M. Milgate, and P. Newman, London: Macmillan, 126–9 (1990).
109 Pascal, Blaise. *The New Palgrave: A Dictionary of Economics*, eds. J. Eatwell, M. Milgate, and P. Newman, London: Macmillan, volume III, 814–815.

1988

110 Pascal's triangle – and Bernoulli's and Vieta's. *Mathematical Spectrum* 20, 33–37.
111 Lod score or log-likelihood? *Nature* 333, 308.
112 Computers and genealogies. *Biology and Society* 5, 73–81.

1989

113 The distribution of navigational errors. *Journal of Navigation* 42, 151.
114 Venn diagrams for many sets. *New Scientist* 121, 51–56.
115 New 3-set Venn diagram. *Nature* 339, 263. (With C. A. B. Smith)
116 Venn diagrams for many sets. Bulletin of the International Statistical Institute, 47th Session, Paris, 29 August–6 September, Contributed papers, Book 1, 311–312.
*117 Probability and likelihood in genetic counselling. *Clinical Genetics* 36, 209–216.

1990

118 Random thoughts. *Nature* 346, 419–20.
119 R. A. Fisher, 1890 – 1962. Proceedings of Invited Papers, International Biometric Conference XV, Budapest, 2–6 July, 3–6.

120 Fisher, $\overline{W}$, and the Fundamental Theorem. *Theoretical Population Biology* 38, 276–84.
121 R. A. Fisher: Twice professor of genetics: London and Cambridge *or* "A fairly well-known geneticist". *Biometrics* 46, 897–904. (*See also* 193)

1991

122 Patterns and primes in Bernoulli's triangle. *Mathematical Spectrum* 23, 105–109.
123 A note on Waring's rule. *Historia Mathematica* 18, 177.
124 Means and trees. Note 75.18. *Mathematical Gazette* 75, 201–202.
125 Bayesian reasoning in science. *Nature* 352, 386–387.
126 Summing powers of integers. *Mathematical Spectrum* 24, 25.
127 How to iron a hypercube. *Mathematical Gazette* 75, 433–436.

1992

128 Rotatable Venn diagrams. *Mathematics Review* 2, 19–21.
129 Celebration of the centenary of the birth of Sir Ronald Aylmer Fisher, 1990. *Historia Mathematica* 19, 81–82.
130 Metrical Venn diagrams. *Annals of Human Genetics* 56, 71–75. (With J. H. Edwards) Reprinted in *Cogwheels of the Mind: The Story of Venn Diagrams*. Baltimore, MD: The Johns Hopkins University Press, 2004 (*see also* 196).
131 The structure of the Polar Eskimo genealogy. *Human Heredity* 42, 242–252.
132 *Likelihood* (expanded edition). Baltimore, MD: Johns Hopkins University Press.

1993

133 Bayes, Thomas. In *The Dictionary of National Biography; Missing Persons*, ed. C. S. Nicholls. Oxford: Oxford University Press, 50–51. (*See also* 199)
134 Bromhead, Sir Edward Thomas Ffrench. *The Dictionary of National Biography: Missing Persons*, ed. C. S. Nicholls. Oxford: Oxford University Press, 89. (*See also* 200)
135 Galton, Karl Pearson and modern statistical theory. In *Sir Francis Galton F.R.S.– The Legacy of his Ideas*, ed. M. Keynes, London: Macmillan, 91–107.
136 Mendel, Galton, Fisher. Second Sir Ronald Fisher Lecture, University of Adelaide, 12th November 1992. *Australian Journal of Statistics* 35, 129–140.
137 John Venn and R. A. Fisher. *The Caian*, November 1993, 64–66.

1994

138 Probability and statistics in genetics. In *Companion Encyclopedia of the History and Philosophy of the Mathematical Sciences*, ed. I. Grattan-Guinness, London: Routledge, volume 2, 1357–1362.

139 R. A. Fisher on Karl Pearson. *Notes and Records of the Royal Society of London* 48, 97–106.

*140 The Fundamental Theorem of Natural Selection. *Biological Reviews* 69, 443–474.

1995

*141 Assessing molecular phylogenies. *Science* 267, 253.

142 Biometry from Galton to Fisher. In *Applied Statistics – Recent Developments*, eds. J. Frohn, U. Gather, W. Stute, and H. Thöni, Göttingen: Vandenhoeck and Ruprecht, 85–94.

143 The foundation of the European Society of Human Genetics. *Journal of the European Society of Human Genetics* 3, 63–64. (With J. H. Renwick)

144 Forced evolution. *Nature* 376, 11.

145 Fiducial inference and the Fundamental Theorem of Natural Selection. *Biometrics* 51, 799–809.

1996

*146 The origin and early development of the method of minimum evolution for the reconstruction of phylogenetic trees. *Systematic Biology* 45, 79–91.

147 The early history of the statistical estimation of linkage. *Annals of Human Genetics* 60, 237–249.

148 Is the Pope an alien? *Nature* 382, 202.

149 Is the frontispiece of *Gulliver's Travels* a likeness of Newton? *Notes and Records of the Royal Society of London* 50, 191–194.

150 W. D. Hamilton's Darwinian precursors. *Times Literary Supplement*, 6th December, No. 4888, 17.

1997

151 Three early papers on efficient parametric estimation. *Statistical Science* 12, 35–47.

152 The early history of the statistical estimation of linkage. In *Genetic Mapping of Disease Genes*, eds. I.-H. Pawlowitzki, J. H. Edwards, and E. A. Thompson, San Diego, CA: Harcourt Brace, 9–14.
153 What did Fisher mean by "inverse probability" in 1912–22? *Statistical Science* 12, 177–184.
154 Fisher Memorial Lectures. *Biometrics* 53, 1557–1558.
155 Natural selection and the sex ratio. *Nature* 389, 780.

1998

156 The Eugenics Society and the development of biometry. The 1997 Galton Lecture, London, 19th September. *Essays in the History of Eugenics*, ed. R. A. Peel, London: The Galton Institute, 156–172.
157 Bernoulli family. In *Encyclopedia of Biostatistics*, eds. P. Armitage and T. Colton, Chichester: Wiley, volume 1, 297–298.
158 Fisher, Ronald Aylmer. In *Encyclopedia of Biostatistics*, eds. P. Armitage and T. Colton, Chichester: Wiley, volume 2, 1523–1529. (With J. Fisher Box)
160 Galton, Francis. *Encyclopedia of Biostatistics*, eds. P. Armitage and T. Colton, Chichester: Wiley, volume 2, 1599–1601.
161 Pascal, Blaise. *Encyclopedia of Biostatistics*, eds. P. Armitage and T. Colton, Chichester: Wiley, volume 4, 3275–3276.
162 Yule, George Udny. *Encyclopedia of Biostatistics*, eds. P. Armitage and T. Colton, Chichester: Wiley, volume 6, 4783–4784.
163 Seven-set Venn diagrams with rotational and polar symmetry. *Combinatorics, Probability and Computing* 7, 149–152.
164 Natural selection and the sex ratio: Fisher's sources. *American Naturalist* 151, 564–569.
165 Statistical inference. In *Handbook of Defeasible Reasoning and Uncertainty Management Systems, Volume 1: Quantified Representation of Uncertainty and Imprecision*, ed. P. Smets, Dordrecht: Kluwer, 357–366.

2000

166 *Foundations of Mathematical Genetics*, 2nd edition. Cambridge: Cambridge University Press.
167 Sewall Wright's equation $\Delta q = (q(1-q)\partial w/\partial q)/2w$. *Theoretical Population Biology* 57, 67–70.
168 *The Genetical Theory of Natural Selection*. *Genetics* 154, 1419–1426.
169 Gilbert's sine distribution. *Teaching Statistics* 22, 70–71.

170 Carl Düsing (1884) on *The Regulation of the Sex Ratio. Theoretical Population Biology* 58, 255–257.

2001

171 Valinnan voima. *Tieteessä tapahtuu* **2**, 28–34.
172 *Annotated Readings in the History of Statistics*. New York: Springer. (With H. A. David)
*173 Darwin and Mendel united: the contributions of Fisher, Haldane and Wright up to 1932. In *Encyclopedia of Genetics*, ed. E. C. R. Reeve, London: Fitzroy Dearborn, 77–83.
174 Fisher, R. A. In *Encyclopedia of Genetics*, eds. S. Brenner and J. H. Miller, San Diego: Academic Press. Second edition: *Brenner's Online Encyclopedia of Genetics*, Elsevier (2013).
175 Blaise Pascal. In *Statisticians of the Centuries*, eds. C. C. Heyde and E. Seneta. New York: Springer, 17–22.
176 G. U. Yule. In *Statisticians of the Centuries*, eds. C. C Heyde and E. Seneta. New York: Springer, 292–294.
177 Likelihood in statistics. In *International Encyclopedia of the Social and Behavioural Sciences*, eds. N. J. Smelser and P. B. Baltes. Oxford: Pergamon, 8854–8858.
178 Fisher, Ronald A. (1890–1962). In *International Encyclopedia of the Social and Behavioural Sciences*, eds. N. J. Smelser and P. B. Baltes. Oxford: Pergamon, 5698–5701. (*See also* 242)
179 Occam's bonus. In *Simplicity, Inference and Modelling*, eds. A. Zellner, H. A. Keuzenkamp, and M. McAleer, Cambridge: Cambridge University Press, 128–132.

2002

180 Dr Venn and his diagrams. *Cambridge – The Magazine of the Cambridge Society*, No. 49, 16–18.
181 La ricostruzione degli alberi evolutivi. In *Le radici prime dell'Europa*, eds. G. Bocchi and M. Ceruti, Milan: Mondadori, 23–30.
182 Hamilton built on work by Haldane and Fisher. *Nature* 416, 581.
183 Fisher, Ronald Aylmer. In *Encyclopedia of Evolution*, ed. M. Pagel. New York: Oxford University Press, 351–352.
184 Fundamental Theorem of Natural Selection. In *Encyclopedia of Evolution*, ed. M. Pagel. New York: Oxford University Press, 371–372.

185 Commemorative windows in the Hall of Gonville and Caius College, Cambridge, England. *The Mathematical Intelligencer* 24, 43–44.

186 Fisher information and the Fundamental Theorem of Natural Selection. *Rendiconti Scienze B (2000), Istituto Lombardo Accademia di Scienze e Lettere, Milano* 134, 1–15.

187 *Pascal's Arithmetical Triangle* (paperback edition). Baltimore, MD: Johns Hopkins University Press.

188 Fundamental Theorem of Natural Selection. *Theoretical Population Biology* 61, 335–7.

189 Professor C. A. B. Smith, 1917–2002. *The Statistician (Journal of the Royal Statistical Society D)* 51, 404–405.

190 *Pavage anallagmatique isochromatique*. In *Géométrie des tissus. Mosaïques. Échiquiers* by A-M. Décaillot. *Revue d'histoire des mathématiques* **8**, 145–206 (p. 206).

2003

191 Pascal's work on probability. In *The Cambridge Companion to Pascal*, ed. N. Hammond, Cambridge: Cambridge University Press, 40–52.

*192 Human genetic diversity: Lewontin's fallacy. *BioEssays* 25, 798–801.

193 R. A. Fisher – Twice Professor of Genetics: London and Cambridge *or* "A fairly well-known geneticist". *The Statistician (Journal of the Royal Statistical Society D)* 52, 311–318. (See also 121; reprinted in *Genetics Society News* Issue 55, July 2006, 33–39)

2004

194 *A Century of Mendelism in Human Genetics*. Boca Raton, FL: CRC Press. (Edited with M. Keynes and R. Peel).

195 Mendelism and Man 1918–1939. In *A Century of Mendelism in Human Genetics*, eds. M. Keynes, A. W. F. Edwards, and R. Peel, Boca Raton, FL: CRC Press.

196 *Cogwheels of the Mind: The Story of Venn Diagrams*. Baltimore, MD: The Johns Hopkins University Press.

*197 Parsimony and computers. In *Milestones in Systematics*, eds. D. M. Williams and P. L. Forey, The Systematics Association Special Volume Series 67, Boca Raton, FL: CRC Press, 181–190.

198 Comment on 'The Reverend Thomas Bayes, FRS: A biography to celebrate the tercentenary of his birth' by D. R. Bellhouse. *Statistical Science* 19, 34–37.

199 Bayes, Thomas (1701–1761), mathematician and Presbyterian minister. In *Oxford Dictionary of National Biography*, Oxford: Oxford University Press, volume 4, 440–1. (*See also* 133)

200 Bromhead, Sir Edward Thomas Ffrench, second baronet (1789–1855), mathematician and landowner. In *Oxford Dictionary of National Biography* Oxford: Oxford University Press, volume 7, 818. (*See also* 134)

201 Darwin, Leonard (1850–1943), scientist and eugenicist. In *Oxford Dictionary of National Biography*, Oxford: Oxford University Press, volume 15, 213–214.

2005

202 R. A. Fisher, *Statistical Methods for Research Workers*, First Edition (1925). In *Landmark Writings in Western Mathematics, 1640–1940*, ed. I. Grattan-Guinness, Amsterdam: Elsevier, 856–870.

203 Defying symmetry. *Significance* 2, 87–88.

204 Variance – What the $n - 1$? *Mathematics Today* 41, 99.

205 System to rank scientists was pedalled by Jeffreys. *Nature* 437, 951.

206 Catastrophe theory applied to Darwinian evolution. *The Linnean* 21, 12–14.

207 Linkage methods in human genetics before the computer. *Human Genetics* 118, 515–530.

208 Sexes and statistics. *Significance* 2, 185–186.

2006

209 The estimation of linkage in the absence of parental phase information. *Annals of Human Genetics* 70, 971–973.

210 Fisher, Demetrius and Wright: contending models. *BioEssays* 28, 440.

211 An eleventh-century Venn diagram. *Bulletin of the British Society for the History of Mathematics* 21, 119–121.

212 Euler diagrams and Venn diagrams. *Mathematical Intelligencer* 28, 3.

213 Statisticians in stained glass houses. *Significance* 3, 182–183.

214 Bernoulli's solution. *Mathematics Today* 42, 222.

2007

215 Maximisation principles in evolutionary biology. In *Philosophy of Biology*, ed. M. Matthen and C. Stephens, Amsterdam: North-Holland, 335–347.

216 Review of The Art of Conjecturing *together with* Letter to a Friend on Sets in Court Tennis by Jacob Bernoulli translated with an introduction and notes by Edith Dudley Sylla. *The Mathematical Intelligencer* 29, 70–72.
217 R. A. Fisher's 1943 unravelling of the Rhesus blood-group system. *Genetics* 175, 471–476.
218 A cautionary tale. *Significance* 4, 47–48.
219 Cover art: genetic code Edwards–Venn diagram. *Nature Genetics*, cover, September 2007. All bases covered, *Nature Genetics* **39**, 1061.
220 *Cogwheels of the Mind: The Story of Venn Diagrams* (second printing with corrections). Baltimore, MD: The Johns Hopkins University Press.

2008

221 Comments on Zabell's paper 'On Student's 1908 paper "The probable error of the mean"'. *Journal of the American Statistical Association* 103, 11–13.
222 *Ending the Mendel–Fisher Controversy*. Pittsburgh: University of Pittsburgh Press. (with A. Franklin, D. J. Fairbanks, D. L. Hartl, and T. Seidenfeld)
223 G. H. Hardy (1908) and Hardy–Weinberg equilibrium. *Genetics* 179, 1143–1150.

2009

224 Erroneous Euler and Venn diagrams. *The Mathematical Intelligencer* 31, 2–4.
*225 Statistical methods for evolutionary trees. *Genetics* 183, 5–12.

2010

226 Christmas list. *Mathematics Today* 46, 99.

2011

*227 Mathematizing Darwin. *Behavioral Ecology and Sociobiology* 65, 421–430.
228 One more class: Martin Gardner and logic diagrams. In *A Bouquet for the Gardener: Martin Gardner Remembered*, ed. M. Burstein, New York: The Lewis Carroll Society of North America, 160–174. (With A. Moktefi)
229 Cover Art: Mendelian ratio Venn diagram. *Nature Genetics*, cover, September 2011.
230 Letter. *Significance* 8, 188.

2012

231 Punnett's square. *Studies in History and Philosophy of Biological and Biomedical Science* 43, 219–224.

231a Punnett's square. *Genetics Society News* Issue 67, 29–31.

232 Reginald Crundall Punnett: First Arthur Balfour Professor of Genetics, Cambridge, 1912. *Genetics* 192, 3–13.

233 Letter. *Significance* 9, 44.

2013

234 The arithmetical triangle. In *Combinatorics: Ancient and Modern*, eds. R. Wilson and J. J. Watkins, Oxford: Oxford University Press, 167–180.

*235 Robert Heath Lock and his textbook of genetics, 1906. *Genetics* 194, 529–537.

236 *Ars conjectandi* three hundred years on. *Significance* 10, 39–41.

237 Bayes and binomials. In *A Book for Dennis: To Dennis Lindley in celebration of his 90th birthday, 25th July, 2013.* Privately printed, Blurb.co.uk, 44–46.

2014

*238 R. A. Fisher's gene-centred view of evolution and the Fundamental Theorem of Natural Selection. *Biological Reviews* 89, 135–147.

239 Historical notes: mathematical stained glass. *Mathematics Today* 50, 318.

240 *Cogwheels of the Mind: The Story of Venn Diagrams* (Chinese edition). Shanghai: Scientific & Technological Education Publishing House.

2015

241 Statistics and genetics between Italy and Britain in the 20th century: an interview by A. W. F. Edwards (by Luigi Palla). *Metron* 73, 45–55.

242 Fisher, Ronald A (1890–1962). In *International Encyclopedia of the Social and Behavioural Sciences*, 2nd edition, ed. J. D. Wright, Oxford: Elsevier, 222–225.

243 Robert Heath Lock. In *Oxford Dictionary of National Biography.* (Online)

244 Reginald Crundall Punnett. In *Oxford Dictionary of National Biography.* (Online)

245 Letter. *Significance* 12, 46.

246 Lieutenant Lawrence Bragg. Letter, *Mathematics Today* 51, 145.

247 Tennis. Letter, *Mathematics Today* 51, 313.

2016

*248 Analysing nature's experiment: Fisher's inductive theorem of natural selection. *Theoretical Population Biology* 109, 1–5.

249 Punnett's square: A postcript. *Studies in History and Philosophy of Biological and Biomedical Science* 57, 1–2.

Appendix 1

A Geometric Interpretation of [192]

AMIR NAJMI

In his 2003 contribution in *BioEssays* ([192]), Edwards identified a seeming paradox: how can we reliably classify and robustly cluster samples even when, on average, any one sample measurement or feature may be rather uninformative, with most (say, 85%) of the feature variance being within clusters? Edwards's answer is that classification and clustering improves, the more features we take into account. But why is this the case? An algebraic model motivates [192].[1]

Here, a simple geometrical demonstration illustrates the point. Like so many simple things, the resolution to the seeming paradox is hard to explain to someone who is not already convinced. I therefore make a few simplifying assumptions whereby the proof lies in the Pythagorean theorem.

Let us consider two normal distributions with variance 1, the first with mean $-d$ and the other with $+d$. These represent the distribution of any property or feature from one of two classes which we will call C– and C+. These classes are presumed to have equal prevalence. Also, the features are presumed to be *independent*, within each class (the more correlated, the less this argument applies).

Let us suppose we observe Feature x_1 to have value v and we wish to classify the sample. The best we can do is to pick the class for which Feature x_1 is more likely to take on value v. Let us say that samples from Class C– are twice as likely to have $x_1 = v$ as Class C+. Our guess of Class C– would be right ⅔ of the time and wrong ⅓. The "majority class" is defined at each value of Feature x_1 to be the more likely class (this may vary at each value v). If we were to pick the majority class for every value of Feature x_1, we would be wrong to the extent of the prevalence of the "minority class" at each value of x_1.[2] The total error is the integral of the

[1] Others have extended and supplemented Edwards's model, developing their own analytical models, including: Tal 2012; Edge and Rosenberg 2015. On the basic logic of Edwards's model, and its contrast to Lewontin's variance-partitioning methodology, see Winther 2014 (reprinted in Appendix 2), Rosenberg commentary.

[2] Picking the class whose distribution assigns a higher likelihood to Feature $x_1 = v$ is a form of maximum-likelihood classification. It works well in this case because the classes have equal prevalence. In general, we would weight the likelihood of each class by its prevalence, which gives the Bayes classifier.

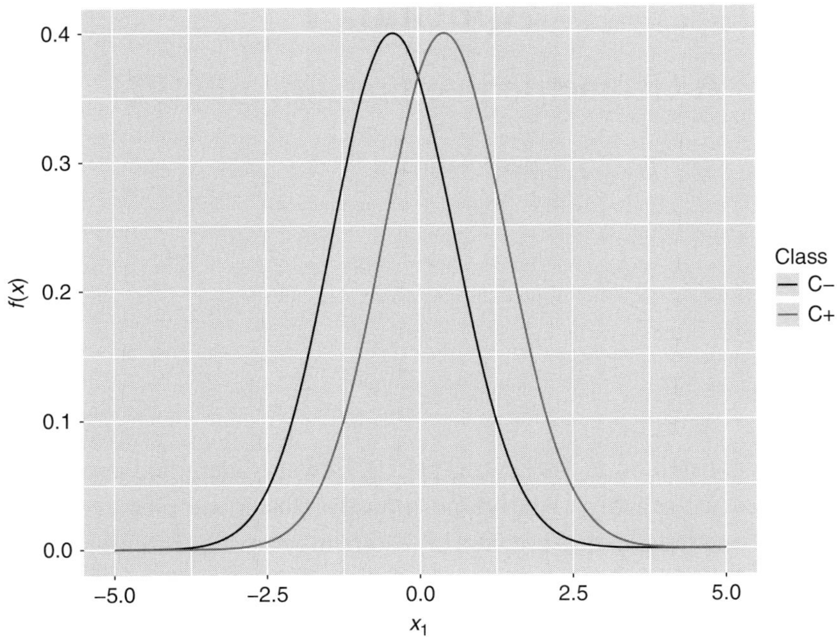

Figure 1. We can see, by symmetry, that for any value of $x_1 < 0$, C– is more likely than C+.

distribution of the minority class, which is the overlap between the distributions. In case of total overlap of the distributions of the two classes, there is no majority and hence either class can be picked. This results in the overlap being 50%.

Given only Feature x_1, our best decision rule to guess the class is:

$$x_1 < 0 => C-$$
$$x_1 > 0 => C+$$

(We can ignore $x_1 = 0$, since the probability of this happening is 0.) Why is the rule above the right one? Consult Figure 1.

For the case of interest in this volume, d = 0.42. This is what gives the approximate and *average* 85% within and 15% between group differences, per the general human genetic diversity data. More generally, this is because total variance = within-group variance + between-group variance. In particular, within-group variance = 1 (by stipulation) and between-group variance = d^2 (by definition of variance, and by symmetry: $\frac{1}{2}[(0-(-d))^2+(0-d)^2]$). Thus: within/total = 0.85 = $1/(d^2+1)$. Therefore, $d = \sqrt{(1/0.85 - 1)} = 0.42$. Let us assume also that *all* features (e.g. different allele frequencies across populations, for distinct genetic loci) are distributed this way, with $d = 0.42$. Some characteristics have a much higher d, but then others have a much lower d. The demonstration here is for an average, typical

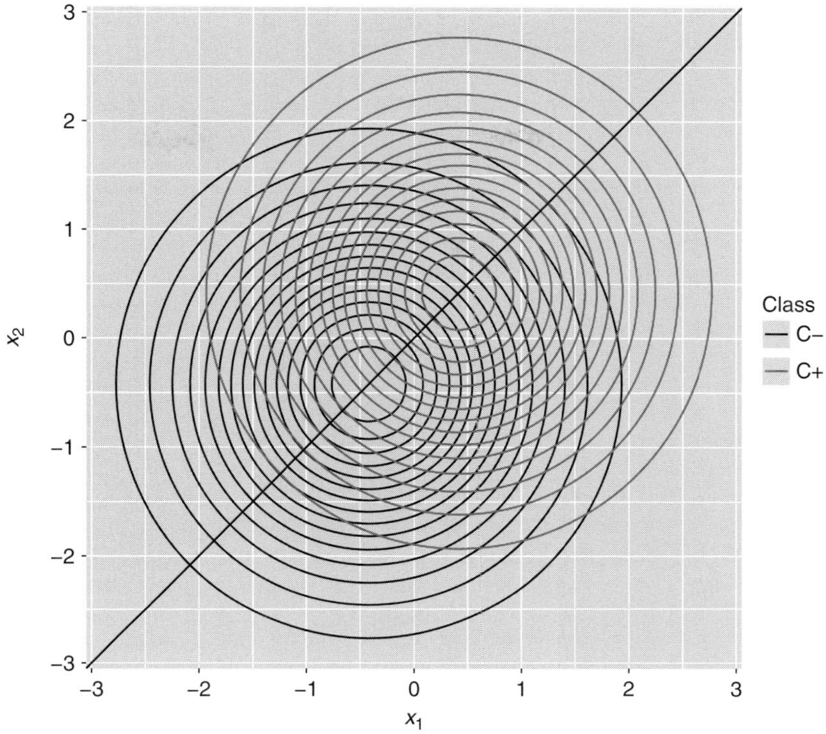

Figure 2. Probability density functions of two features of two sample classes. Since the features x_1 and x_2 are independent, the probability density at any point is just the product for that class.

case. Regardless of differences in *d* across features, the behavior of increased classification reliability with increasing dimensionality is *always* true, and is intended as the main take-home lesson.[3]

Now imagine we use two features, x_1 and x_2. What is the best decision rule if we were to combine both?

Figure 2 shows contour plots of the density of each of the two classes. If we know what the normal looks like in one dimension, it is not too hard to discern the general pattern of the contours (or think of a topographic map). What we see is that there is circular symmetry about the center of each distribution, i.e. around (–*d*, –*d*) and (+*d*, +*d*). The probability monotonically decreases as we move further away from the origin.

[3] A caveat might be necessary. This is true if the average informativeness of the features stays constant as features are added or if the features are weighted appropriately. The unweighted version of Edwards's classifier can be worse with more markers than with few, if the allele frequencies vary site-to-site, as Tal 2012 shows. For example, an unweighted sum from five very noisy markers and one near-perfect one is not as good a classifier as the near-perfect one alone. The problem can be fixed by upweighting the feature that is already a near-perfect classifier.

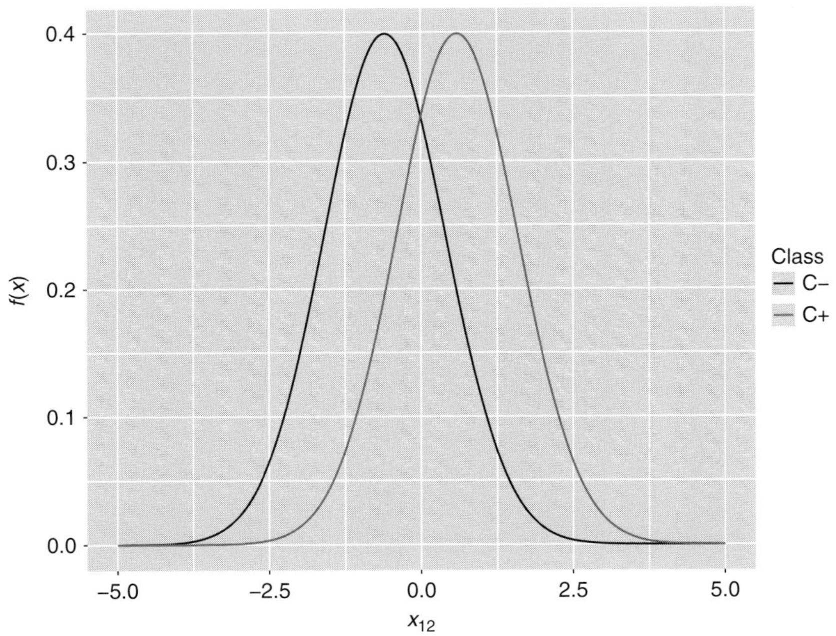

Figure 3. By the Pythagorean theorem, the two distributions are now $2d\sqrt{2} = 2.8d$ apart. In contrast, before using only x_1, the means were just $2d$ apart.

As before, given (x_1, x_2), we want our classification rule to pick the more likely class. We can turn this two-dimensional problem into a one-dimensional problem as before, if we simply project onto the plane $x_1 = x_2$. Let us not argue if this is the best we can do but simply note that we can do this. The projection of each two-dimensional normal onto this plane will yield exactly the unit normal (i.e. variance 1). This makes sense because projecting onto either x_1 or x_2 would have given us a unit normal (that is how we started). We can still make the decision rule of the new coordinate x_{12} (Figure 3):

$$x_{12} < 0 \Rightarrow C-$$
$$x_{12} > 0 \Rightarrow C+$$

On scrutinizing Figures 1 and 3, it can be noted now how far the means of the projected one-dimensional distribution are from each other. Consulting Figure 2, confirm that the hypotenuse ($2.8d$ in this case) is *always* longer than either side ($2d$ in either orthogonal projection or marginal, in this case). The Pythagorean theorem generalizes to higher dimensions. If we inspect the pictures carefully we can see that there is increased separation. This is the effect of having more information. We have essentially combined two pieces of knowledge and improved classification. Overlap of the two distributions went from 67.4% to 55.3%.

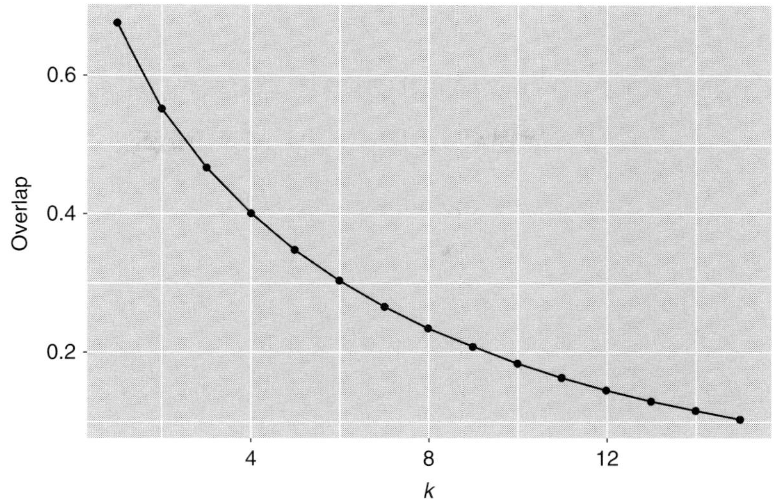

Figure 4. Interestingly, Figure 1 on p. 799 of [192] is similar but subtly different. That one is set up as a number of binary observations, whereas the example here employs normal distribution. The values are not quite the same. The normal-distribution example lends itself better to geometric arguments.

In general, combining k predictors will push cluster means of the projected one-dimensional distribution to $2d\sqrt{k}$. This can be seen in Figure 4. Combining 15 independent features gives an overlap of just 10%.

To summarize this appendix, if you start with two normal distributions, each with variance 1 and with means 0.84 units apart, they will overlap considerably, about 67.4%. Thus, based solely on this feature, our estimate would be wrong about 67.4%/2 of the time, which is not much better than 100%/2 of guessing. But now imagine we have two such features. We end up with two two-dimensional normal distributions. Because of the Pythagorean theorem they overlap less. And so on. In higher dimensions, overlap keeps decreasing and hence misclassification rates decrease. These will never be 0%, but they can be arbitrarily small.

References

Edge, M. D. and Rosenberg, N. A. 2015. A general model of the relationship between the apportionment of human genetic diversity and the apportionment of human phenotypic diversity. *Human Biology* 87, 313–337.

Tal, O. 2012. The cumulative effect of genetic markers on classification performance: insights from simple models. *Journal of Theoretical Biology* 293, 206–218.

Winther, R. G. 2014. The genetic reification of "race"? A story of two mathematical methods. *Critical Philosophy of Race* 2, 204–223.

Appendix 2

The Genetic Reification of "Race": A Story of Two Mathematical Methods

RASMUS GRØNFELDT WINTHER

Reprinted from *Critical Philosophy of Race* 2, 204–223, The Pennsylvania State University Press, 2014.

THE GENETIC REIFICATION OF "RACE"? A STORY OF TWO MATHEMATICAL METHODS

Rasmus Grønfeldt Winther, University of California, Santa Cruz and University of Copenhagen

Abstract

Two families of mathematical methods lie at the heart of investigating the hierarchical structure of genetic variation in *Homo sapiens: diversity partitioning*, which assesses genetic variation within and among predetermined groups, and *clustering analysis*, which simultaneously produces clusters and assigns individuals to these "unsupervised" cluster classifications. While mathematically consistent, these two methodologies are understood by many to ground diametrically opposed claims about the reality of human races. Moreover, modeling results are sensitive to assumptions such as preexisting theoretical commitments to certain linguistic, anthropological, and geographic human groups. Thus, models can be perniciously reified. That is, they can be conflated and confused with the world. This fact belies standard realist and antirealist interpretations of "race," and supports a pluralist conventionalist interpretation.

Keywords: mathematical methods; genetic classification; diversity partitioning; clustering analysis; reification

1. Introduction

Two mathematical methods lie at the heart of genetic classifications of human groups: *diversity partitioning* and *clustering analysis*. They are two sides of the same mathematics coin. Both are legitimate and consistent methodologies. As probability theory and statistics modeling practices, they cannot be questioned. This is the constructive part of the article.

Importantly, neither of these methodologies necessarily implies anything about the *reality* of human groups. This is meant in two ways. First, claims about the robustness of groups involve conventional choices based on assumptions about, for instance, how much genetic distance is required on average between groups for groups to be considered genuinely different and be granted ontic status. Different modeling assumptions yield different outcomes. Second, the very data that are fed into the methods are neither unbiased nor theory-neutral. For instance, linguistic, archaeological and anthropological data are used to (pre-)define human groups; moreover, there may be visual or unconscious bias regarding which phenotypes of a particular group are sampled for genotyping.[1] Running models based on data taken from different sets of individuals of each population would yield distinct group diversity partitionings and cluster assignment of particular individuals, at least in principle if not in practice. Thus, model assumptions and interpretations, as well as data input, are subject to theory-ladenness and bias. This is the critique.

The basic argument? The modeling machinery may be well-oiled. The variables, functions, and derivations of diversity partitioning and clustering analysis are appropriately articulated and consistent, as we can expect from mathematical, formal systems. Yet model results are sensitive to (1) input, (2) particular assumptions made, and (3) output or interpretation. Reification of our expectations, biases, and preexisting theoretical maps (e.g., linguistic or geographic human groups) can occur in many places. In other words, we can make concrete things out of our abstract representations (i.e., reify) in various ways. That is, we might see or infer groups, natural kinds, and subdivisions when these do not exist. More broadly, arguments about the realism vs. reification of race are philosophical insofar as philosophy is critique (e.g., John Dewey, Michel Foucault). What is being critiqued in this article are the data and models used to answer empirical questions about whether our species is subdivided into groups (alternatively: natural kinds). As Bas van Fraassen recently pointed out to me (May 2014), these arguments are not philosophical in the broader sense of illuminating the aims (e.g., truth vs. empirical adequacy) and structure (e.g., syntactic vs. semantic) of scientific theory.

The goal? This article motivates the logic and procedure of these two methods and identifies places where reification can occur. The conclusion lists a set of open research questions that could inspire students of these topics.

2. A Story of Two Methods

What is the logic of these two families of methods? Very briefly, diversity partitioning assesses the amount of genetic differentiation present among distinct human groups. How much more genetically similar are two randomly chosen individuals from the same group on average as compared to a randomly chosen individual from that group and an individual from another group (either from the same or a different continental region)? If there is no genetic differentiation among groups, then two individuals from the same group will, on average, be as different as two individuals from different groups. If there is some variation among groups, then two individuals from the same group will, on average, be more alike than two individuals from differing groups. The robust empirical result repeatedly found over the last forty years, using different molecular techniques on genes and proteins, and extensive global sampling, is that (approximately) 85 percent of all genetic variance is found within human subpopulations (e.g., Han Chinese or Sami), 10 percent across subpopulations within a continental region, and only 5 percent of the genetic variance is found across continents (i.e., "Negroid," "Caucasoid," and "Mongoloid"—Lewontin 1972 terms) (e.g., Lewontin 1972; Nei 1973; Barbujani, et al. 1997). No one questions these results, nor should they.[2]

In contrast, clustering analysis assigns particular individuals to groups (clusters) or to a specific weighting of more than one group. It also determines the gene allele frequencies of the clusters, under the assumption of a particular number of clusters. An individual is assigned probabilistically to the cluster or set of clusters that most closely matches what we call the "presence profile" of alleles in the individual. Intuitively, if an individual is *AA*, we would (ceteris paribus) prefer to assign it to a group with an *A* frequency of, say 90 percent rather than one with 50 percent *A*. But how do we assess the allele frequencies of groups? One way is to use Bayesian modeling. That is, we start with prior assumptions about gene frequencies in different populations (e.g., assume extremely high frequency of *A*, *B*, and *C* in some clusters, and extremely low frequency in others), and readjust the priors upon sampling of individual genotypes (e.g., Pritchard, Stephens, and Donnelly 2000). As long as a sufficiently high number of genetic loci are used (roughly 20 to 50), individuals can be assigned to clusters with extremely low probability of misclassification. The reliability of correct group assignation is very high. The computational power and programs for this task have been developed over the last ten years (e.g., Pritchard, Stephens, and Donnelly 2000; Edwards 2003; Rosenberg et al. 2002).

2.1. Diversity Partitioning

Diversity partitioning assesses how much of the total genetic diversity or variation in a species exists among individuals within predefined groups at various hierarchical levels. That is, the amounts of genetic diversity found among individuals at the following three levels of groups are compared: (1) within single, local groups, (2) within different intra-continental groups, and (3) within different inter-continental groups. In order to assess the hierarchical structure of genetic variation, we need to develop measures thereof. Two broad types of properties are used to ground measures of diversity or variation: *group* and *individual* properties. Classic population genetic theory (e.g., Sewall Wright 1965, 1969; Cockerham 1969, 1973; and Lewontin 1972, 1974a) focuses on group properties, especially heterozygosity (see below). Other recent theoretical and empirical work uses detailed genetic sequence information to compare the genomes of individuals, within and across groups. Excoffier, Smouse, and Quattro (1992) developed a technique (Analysis of Molecular Variance, or AMOVA) that uses standard analysis of variance (ANOVA) theory to partition genetic variation at three hierarchical levels: individual, groups, and across groups of groups (e.g., continental regions) within a species (e.g., Barbujani, et al. 1997 used this method; on ANOVA, see Lewontin 1974b, Winther 2014a).

A common measure of variation used in these statistical methodologies is Wright's *F*-statistics. The basic method is to assess the levels of heterozygosity (e.g., the frequency of 2*Aa* rather than *AA* or *aa*, for a single biallelic locus) at different loci, in distinct groups. The more similar heterozygosity levels are to each other across groups, the more genetically similar the groups are.[3] *F*-statistics are a group property. One of Sewall Wright's *F*-statistics is as follows:

$$F_{ST} = \frac{H_T - \overline{H}_S}{H_T}$$

Where H_T is the total heterozygosity of the entire population, and $\overline{H}_S$ is the heterozygosity within each subpopulation (group), averaged across all subpopulations. The fundamental idea here is to compare total heterozygosity to average subpopulation heterozygosity. Total heterozygosity is calculated by first averaging all the allelic frequencies, for different loci, across all groups, and then using those allele frequencies to determine the *expected* heterozygosity of the total population. Average subpopulation heterozygosity is calculated by taking the *actual* heterozygosity of each group and averaging those across all groups. In the extreme case where all groups are fixed for either one or the other allele of biallelic loci, the F_{ST} measure will be at its maximum of 1, since the average group heterozygosity will be 0. Groups will be maximally different genetically. Conversely, when groups have exactly the same levels of heterozygosity, for one or more biallelic loci, the averaged actual heterozygosity will be the same as the expected total heterozygosity (H_T) and F_{ST} will be 0. Groups would be genetically identical. The Shannon information theoretic measure of variation used in Lewontin (1972)"strong[ly] resembl[es]" the heterozygosity measure, as Lewontin recognizes (388). For our purposes here, we can consider Lewontin to have employed a kind of F_{ST} measure.

An example might help. Recall the mythical high school biology story of eye color genetics, in which two blue-eyed parents (each *bb*) always have a blue-eyed child, but two heterozygote brown-eyed parents (*Bb*), can have a child with either eye color. Setting aside the fact that eye color actually involves many genes—it is *polygenic*—this simple example helps motivate the meaning of heterozygosity, as well as the Hardy-Weinberg principle (HWP). Imagine a single population with thousands of only heterozygote brown-eyed parents (*Bb*). After a single generation of mating, their children will have genotype frequencies of very close to 25 percent *BB*, 50 percent *Bb*, and 25 percent *bb*; the corresponding allelic frequencies will be 50 percent *B* and 50 percent *b*. Barring the action of mutation and random genetic drift, or other evolutionary forces, genotypic and allelic frequencies will remain the same throughout future generations of mating. Now consider a separate second group with thousands of only blue-eyed parents. You already know what the allele frequency distribution will be for their offspring: 100 percent *aa*. In general, HWP states that percentages do not change after the initial bout of mating. Now, in this hypothetical case $\overline{H_S}$ is 0.25 because it is simply the average heterozygosity across the two groups, calculated from genotypes: (0 + 0.5)/2. In contrast, H_T is the expected heterozygosity in a single large population with allele frequencies equal to the mean allele frequencies in the two subpopulations. In the first population, with both brown-eyed and blue-eyed individuals, these allelic frequencies are 0.5 and 0.5, while in the second they are 0 and 1, giving *pooled* allele frequencies of 0.25 and 0.75. H_T thus turns out to be 2(0.25)(0.75) = 0.375. Thus, F_{ST} is (0.375 - 0.25)/0.375 = 1/3. The example chosen is among the simplest possible, and is intended to make the basic logic of F_{ST} explicit. Whenever allelic frequencies differ across two or more populations, the pooled allelic frequency average will be different than the average of the allelic frequencies of each individual group, and F_{ST} will be non-zero. There is inter-group variation. Consider two final observations. If the first group had consisted of *BB* parents, there would be no heterozygosity within *either* group (i.e., $\overline{H}_S = 0$), but the pooled heterozygosity would be 0.5. In this case, F_{ST} would be 1, its maximum. Second, whenever F_{ST} is non-zero, there will always be excess homozygosity (alternatively: deficient heterozygosity), per the Wahlund Effect.

Let us turn to the way *F*-statistics relate to diversity partitioning. F_{ST} and the two other hierarchical inbreeding coefficients related to it—i.e., individual-to-total population, F_{IT}, and individual-to-group F_{IS}—are used to calculate hierarchical variance partitioning. In fact, with the three group levels considered—in diversity partitioning of *Homo sapiens*—i.e., intra-group, intra-continental groups, and inter-continental groups—F_{ST} must be calculated hierarchically, twice. Setting these complications aside, there are clean, relatively simple, and well-documented relations among inbreeding coefficients and variance components.[4] Indeed, F_{ST} is equal to the between-group variance, at a given level. The general result using these typical methods of diversity partitioning in humans is the (approximately) 85 percent/10 percent/5 percent variance measures mentioned above.

A simple table adapted from actual data in Lewontin (1972, 1974a)[5] helps further motivate intuitions (see table 1).

TABLE 1 ALLELE FREQUENCIES OF TWO DISTINCT GENES AS USED IN LEWONTIN (1972) AND (1974A), ACROSS "RACES"

Gene	Alleles	Caucasoid	Negroid	Mongoloid
Duffy	Fy	0.03	0.94	0.1
	Fy*a*	0.42	0.06	0.9
	Fy*b*	0.56	0	0
Auberger	Au*a*	0.62	0.64	
	Au	0.38	0.36	

Note: Frequencies are rounded from four to two significant figures. Empty cells indicate lack of data. See esp. Lewontin 1974a, 153.

I choose the Duffy and Auberger genes of *Homo sapiens* from among the seventeen genes Lewontin used because they are instructive contrast cases. Duffy shows very high variation among the standard "races" (terms from Lewontin 1972), with one allele, Fy, being practically absent in Europeans but almost omnipresent in Africans. In contrast, Auberger exhibits very little interracial variation. Lewontin and others since him have discovered that most of our genes are like Auberger. This is another way of saying that approximately 95 percent of genetic variation is found within races. On the other hand, if most of our genes had been like Duffy, F_{ST} would indeed be significant, potentially approaching 1. In that case, most genetic variation would have been between races. Emphatically, genes similar to Duffy in frequency distribution are relatively rare in the human population.[6]

Which ontology is inferred from the (approximately) 85 percent/10 percent/5 percent result? Many interlocutors have argued that these results show that there is not very much genetic differentiation between groups defined on geographic, anthropological, or linguistic criteria. They reject these categories as "real," or at least as "biologically relevant." In particular, racial categories are *testable empirical hypotheses* that the data ultimately rejects. (Or so the interlocutors argue.) Diversity partitioning methodologies indicate that the abstraction of "race" is neither grounded in, nor justified by, genetic data.

2.2. Clustering Analysis

Given the assumption of particular individuals belonging to either a single group (cluster) or to a specific weighted combination of more than one group (when multiple population ancestry— i.e., admixture—is suspected), and supposing that there is a certain specific number *K* of groups, how can individuals be assigned to their appropriate groups or weighted fractions of groups? Most basically, the presence profile of alleles in the individual is matched to the group or mix of groups that most closely matches it. Genetic structure across loci is used as information to infer cluster membership. "Structure" here does *not* mean (but it *can* mean, particularly for phenotypic "racial" characters; see section 5 below) that if an individual has allele *A*, she will also tend to have alleles *B* and *C* (alternatively: if she has a certain facial morphology, she will also tend to have a particular skin color and a specific hair type). After all, such correlation assumes significant linkage disequilibrium (i.e., statistical non-independence across loci) among the three loci, for which there is no guarantee, and which empirically is often not the case, certainly not for the neutral microsatellites, RFLPs, and SNPs that are used in many of these studies. Rather, structure here means that if the alleles of a sufficient number of loci in a given individual are identified, then we can classify that individual as belonging to a particular cluster with high probability.

A brief thought experiment might help motivate intuitions about the logic behind clustering analysis. Consider two groups Δ_1 and Δ_2, with systematically different gene frequencies. For three biallelic loci, A, B, and C, respective frequencies of the dominant allele {*A, B, C*} are {0.9, 0.4, 0.49} for Δ_1 and {.05, 0.7, and 0.5} for Δ_2. We now actually have cross-loci information about the likelihood that an individual belongs to a certain cluster. Think about it: if I told you that an individual has haplotype *Abc*, what would you *bet* is her cluster membership? The answer is Δ_1. After all, *A* is practically absent in Δ_2 and *b* is significantly more likely in Δ_1 than in Δ_2; admittedly, whether the individual has *C* or *c* provides very little information. More generally, model-based statistical analysis (through either maximum likelihood or Bayesian statistical methods, e.g., Pritchard, Stephens, and Donnelly 2000; Rosenberg et al. 2002, 2005) tells us with which (high) probability an individual belongs to Δ_1.[7] I leave it as an exercise to the informed reader to evaluate how much money you would bet on in this case. The point is that with sufficient cross-loci information about the haplotype of individuals, we can safely identify the clusters to which an individual belongs.

The point is actually a bit trickier because the population allele frequencies are often not actually known but are themselves estimated from the data. This may seem viciously circular or overly cumbersome. But it is neither. As long as we decide a priori on a fixed number of clusters, *K*, the mutual fitting of cluster allele frequencies and individual genotypes can be calculated in a straightforward manner. Pritchard, Stephens, and Donnelly (2000) do this through iterative sampling. This is their simplest algorithm, also built into the computer program *STRUCTURE* (947):

Step 1. *Sample* $P^{(m)}$ *from* Pr $(P|X, Z^{(m-1)})$.

Step 2. *Sample* $Z^{(m)}$ *from* Pr $(Z|X, P^{(m)})$.

Where *m* indicates the step, *P* is the random vector of population allele frequencies, *Z* is the random vector of populations of origin, and *X* is the random vector for individual genotypes. Pritchard, Stephens, and Donnelly write: "Informally, step 1 corresponds to estimating the allele frequencies for each population assuming that the population of origin of each individual is known; step 2 corresponds to estimating the population of origin of each individual, assuming that the population of allele frequencies are known" (2000, 947). In other words, we start with prior assumptions about genotype frequencies (e.g., for three biallelic loci, assume extremely high frequencies of *A, B,* and *C* in some clusters, and low frequencies in others—i.e., high frequencies of *a, b,* and *c*), place individuals in the clusters that best match their genotype frequencies, recalculate genotype frequencies for the clusters and adjust priors accordingly, and repeat this entire computational process until we have clusters in which Hardy-Weinberg expectations hold (per locus), and linkage equilibrium (across loci) exists.

Which ontology is inferred from the fact that individuals can be reliably assigned to robust clusters? This fact could be taken as evidence for the strong reality of human groups, though few interlocutors have made this exact sort of statement, perhaps because of the potentially reactionary political repercussions such utterances may have or may imply (but see the recent book, Wade 2014). In discussing "Lewontin's Fallacy," Edwards (2003) claimed that the argument that the "division of *Homo sapiens* into these [racial] groups is not justified by the data" is fallacious because it "ignores the fact that most of the information that distinguishes populations is hidden in the correlation structure of the data and not simply in the variation of the individual factors" (798). He does not make ontologically strong pronouncements in this article, but his arguments are not inconsistent with a position stating that continental region classifications are real. The background argument here seems to be that significant inferential reliability of assigning individuals to clusters supports the reality of human group classifications. The results of diversity partitioning and clustering analysis pull in opposite ontological directions. Even so, their mathematics is mutually consistent, as we will now see.

3. Internal Methodological Consistency?

The two methods reviewed in section 2 are distinct ways of *characterizing the hierarchical structure of genetic variation*. The former assesses the hierarchical composition of genetic variance by exploring how similar groups are to one another, at a given level of the hierarchy. The latter assigns individuals to clusters, or finds groups, through the use of Bayesian modeling strategies or other methodologies (e.g., Principal Components Analysis). The two methods are used with the same overall aim of assessing the hierarchical genetic structure of (human) populations, but they answer different questions and use distinct methods to do so.

They are mutually consistent. Indeed, *differentiation among groups (as measured by* F*-statistics) is the logical outcome of similarity of individuals within groups (as found with the program* STRUCTURE*), and vice versa*. Thus, neither method is "wrong" or invalid, though each may be used inappropriately if employed to answer a question it was not designed to answer.[8] Diversity partitioning correctly indicates that there is very little genetic differentiation among races, and just a little more among populations within a race. (N.b., there is extremely little variation among human beings in general. We are all basically identical across most of our genome.[9] Only the sequences that vary are here considered.) Even though the vast majority of genetic variation exists among individuals rather than across groups, clustering into groups can *still* be done, if we assume that a certain number of clusters exist. All you need is a little variation (read: non-identity) among the allelic frequencies in different populations, and among the distinct continental regions. Indeed, to take the argument to its extreme, if two groups or clusters of individuals had identical frequencies at 9,999 loci, but differed in frequency at just one locus, they would be different groups. As observed in the last paragraph of Rosenberg et al. (2002): "The challenge of genetic studies of human history is to use the small amount of genetic differentiation among populations to infer the history of human migrations" (2384). One could add: "and to infer the group memberships of any particular individual (under the assumptions of the statistical model)."

While our two mathematical methods are consistent, their aims, questions of interest, and basic assumptions are distinct. Diversity partitioning is particularly useful for evolutionary analyses of the opportunity for selection and random genetic drift in hierarchical populations. Clustering analysis can be used for making medical predictions (e.g., Burchard et al. 2003, Kumar et al. 2010). (But see the first item in the numbered list in the concluding section 6.) In fact, as Helen Longino put it to me recently (May 2014), discussion about "the metaphysics of race" is only interesting and important in so far as it connects with work in other areas such as biomedicine, forensics, and physical and cultural anthropology. Whether race is taken to exist or not—and in which sense (e.g., biological and/or social)—impacts the practices and commitments of these fields. Naturally, highly charged ethical, social, and political questions enter. Moreover, it is interesting that either method can be used for inferring migration patterns, because structured populations (with sufficiently high F_{ST}) and clusters (inferred populations in Hardy-Weinberg and linkage equilibrium) correspond, to an extent at least, to historical lineages. In summary, the two methods are consistent but rely on distinct assumptions and have different purposes.

4. Loci of Reification

The mathematical methods described above are just naked mathematics. Their logic is watertight, but they are highly dependent on the assumptions used in their construction (see also Gelman 2008; Winther 2014a and references therein). In exploring loci or places where pernicious reification can occur, let us start with the data input stage of modeling. Which presuppositions are made about the homogeneity of sample sizes across groups, and how representative are the samples of the group (e.g., consider the difference between sampling Han Chinese with a population size of approximately 1.2 billion, and the roughly 150,000 Samis of northern Europe)? Are the sample data points independent of one another and homoscedastic (i.e., error variance in data samples is constant across loci and across clusters)? Is phenotypic appearance used as a conscious or unconscious sieve for which blood and tissue to sample? (If so, insofar as there is any correlation between genotype and phenotype, the data points are not independent.) With which cluster/group definitions do or should we start, either in collecting data and defining the groups to be tested in diversity partitioning, or in collecting data and setting the Bayesian priors in our cluster analysis? Do geography, archaeology, anthropology, and linguistics provide a priori information for genetical studies? These are all questions about the reliability of the *data input* to our models: who is sampled and which groups are presupposed? Defining the reification of race as conflating our theoretical expectations stemming from other fields such as anthropology and linguistics—as well as our phenotypic biases—with the (genetic) world, then reification can easily occur in this stage of modeling.

In each of these two mathematical methods, an irreducible theoretical element exists vis-à-vis defining populations. Indeed, groups are presupposed—from linguistic and anthropological data—or at least highly predetermined. In diversity partitioning, the starting point is the set of groups already identified by phenotypic, geographic, or cultural (e.g., linguistic) characteristics. In other words, there must be a set of properties, invariably correlated to culture, that *gives* the classification against which genetic variation is compared. In clustering analysis using the computer program *STRUCTURE*, assumptions stemming from linguistics and anthropology may be used to help set the strong Bayesian priors. The parameter space of multilocus information is too large, and clustering possibilities too massive, to *not* narrow down—guide—the possible clusters using the abstract maps of culturally defined groups. Indeed: "[The Bayesian approach] also eases the incorporation of various sorts of prior information that may be available, such as information about the geographic [or linguistic, anthropological, etc.?] sampling location of individuals" (Pritchard 2000, 947). Might there, though, be purely acultural priors in *STRUCTURE*? Perhaps, but I suggest that unconscious (which individuals are actually sampled) and even practical (e.g., which genes are most easily sequenced) biases could be an unavoidable part of the modeling effort. A representative, independent, and random sample of each and every human population (whatever these may be, given the *clinal, gradating relation* of human variation) is an extremely challenging undertaking. Of course, these claims regarding reification and the conventionality of group definitions in diversity partitioning, and in the Bayesian priors of clustering analysis, require substantiation through examination of particular case studies. As a thought experiment, consider what modelers might do if they found a clustering that cut across culture. Would they accept it or might they tweak it by putting in more cultural or geographic priors? A number of sociology of science and philosophy of science research projects await.[10] The pernicious reification of our preexisting theoretical maps about groups and populations is difficult to avoid.[11]

Let us now turn to model *output*. Once we have our F_{ST} measures and variance decompositions, how much difference is enough? Is the human F_{ST} of roughly 0.15 sufficient for calling intra-continental and inter-continental groups distinct and attributing ontic status to them, or must we meet the typical boundary F_{ST} of between 0.25 and 0.30 (see Templeton 1998)? Second, given that "the problem of inferring the number of clusters, *K*, present in a data set is notoriously difficult" and because the "posterior distribution can be peculiarly dependent on the modeling assumptions made" (Pritchard, Stephens, and Donnelly 2000, 949), it is unclear exactly how to interpret the reliability of the clustering for any particular *K*. Indeed, note that a *K* of 3 or 5 (see Bamshad et al. 2003 and Rosenberg et al. 2002, respectively) is sometimes perceived to be a true and natural cut of human genetic variation, reflecting continental regions. But this may itself be a reification. After all, *STRUCTURE* identifies "multiple ways to divide the sampled individuals into *K* clusters when $K > 6$ (Rosenberg et al. 2002). For example, in 10 replicates, *STRUCTURE* found 9 different ways to divide the sampled individuals into 14 clusters . . . (N. Rosenberg, pers. comm.)" (Bolnick 2008, 76). Thus, the apparent naturalness of $K = 3$ or $K = 5$ is actually a conventional choice about how to interpret the robustness of modeling results, rather than a mirror of nature.[12] The problem is worsened by the fact that for high *K*, there is not even a robust clustering. And, we *need* to assume a particular *K* to do the iterative sampling mentioned in section 2.2 above. How to interpret the model output depends on conventional judgments: choose the appropriate cut-off point of F_{ST} and choose the appropriate *K*.

Reification of our biases and our theoretical interpretations can occur in many places in this modeling process, especially in the input and output phases. The argument here is not that the two modeling methods lack merit or that the mathematics is wrong. But in certain places of the modeling stream, it is hard to know how to interpret input and output, or how to apply the modeling machinery. An *assumption archaeology* is necessary (Winther, under contract). Suppositions of various sorts (methodological, ontological, data-analytical, etc.) need to be stated clearly and self-reflectively. Questions and aims must be explicitly articulated and understood. Both for epistemic and ethical reasons, critical care is required.

5. Can There Be Phenotypic "Races" without Genetic "Races"?

In all of this, the question remains whether phenotypic "race" could exist even if genetic "race" is a reification. Regardless of the genetic facts, are there phenotypic races? Consider this passage from Feldman and Lewontin (2008):

> Using skin color, facial shape, and hair form, all obviously largely genetically determined . . . no one has any difficulty in differentiating between a random person taken from West Africa, from China, from Norway, or from the tropical rainforest of the Orinoco basin. With only a little more subtlety one can differentiate Amharic-speaking natives of Ethiopia from Zulus, Chinese from Japanese, and villagers of Andhra Pradesh from Afghanis by external morphology. (90–91)

As we saw above, Lewontin is hardly a realist about genetic race, yet he seems to be endorsing phenotypic race here. The measurement and metaphysical status of groups may have to be assessed separately at the genetic and phenotypic levels. That is, even if genetic race turns out to be a reification, phenotypic race could be a "human kind" (Hacking 1995), something about which I remain completely neutral. Moreover, mapping the causal links between these two levels is also a potential research project (e.g., phenotypic race could impact genetic—i.e., heritable—population structure via sexual selection, as Charles Darwin argued). These empirical, technical, and philosophical questions about phenotypic race are worth further consideration.

6. Whither Two Mathematical Methods and Genetic Reification?

The overarching research project of which this article is a small piece is the search for methods for identifying promises and dangers of scientific abstractions (Winther, under contract). I wish to find criteria and norms that will allow us to differentiate between the generative and productive use of our scientific abstract maps (such as cultural or linguistic groups), and the dangerous application—i.e., pernicious reification—of such maps. In other words, are particular groups of certain sorts *grounded* in patterns of genetic variation, or is an ontological interpretation of a clustering a biased and "viciously abstract"—to use William James's locution[13]—imposition on genetic data? This article's case study of abstraction and reification in science allows us to think about whether (1) biologists are actually measuring something significant in their clusters, (2) they are justified in their *kind-making* (see Hacking 1995, 2007), and (3) there is any knowledge/power (empirically and normatively) to their use of clusters in making predictions and formulating explanations of human evolution, capacity ascriptions (e.g., intelligence and athleticism), and disease proclivities.

Why should we care? It seems obvious that the political and social stakes are high. Our very understanding of what it means to be human is under question, as are (our understandings of) human freedom, potential, and dignity. The main purpose of this text is to help clarify the two mathematical methods before turning to deeper questions of the normativity and metaphysical nature of potential natural kinds of *Homo sapiens*.

Here are a few research projects that could be developed going forward:

1. *Classification and Function.* Do the genes we use in classifying human groups have any functional or mechanistic relevance? Our material and theoretical technologies do not yet allow us to assess the *causal* relevance of whatever genes do differ in systematic ways across at least some clusters. Presumably *some* of these genes are involved in the making of phenotypes that seem to differ across human groups. (This is a controversial claim.) But until we have found ways of describing actual genetic causal networks, the functional relevance of cluster-distinct genes (i.e., "private" alleles or genes that have extremely different frequencies in the two populations) remains unclear. Which sorts of empirical studies must be made to identify genetic causation? Can our classificatory investigations shed any light on mechanistic questions?

2. *The SMEO-P Account of Modeling.* In previous work (Winther 2006a, b), I provided a simplified and linear account of the modeling process. Modeling consists in setting up, mathematically manipulating, explaining, objectifying, and pluralizing. The SMEO-P account emphasizes the importance of ontological and methodological assumptions in modeling. I have only just begun to excavate the assumptions at play in the two mathematical methods of diversity partitioning and clustering analysis.
3. *The Role of Biologists in the "Race Debates."* Regardless of empirically robust modeling outcomes, and regardless of whichever definition or criteria of group reality biologists may have, social and political considerations may force scientists to keep a low profile about which sort of realism/constructivism/eliminativism interpretation to give (Haslanger 2008 discusses these interpretations). Put differently, is the history of racial discourse and of the biologization, objectification, and reification of racial categories so violent that everyone must engage in an enlightened dialogue that states not only the biological facts and methodologies, but also the historical, political, and social context? Should we even keep the categories of "race," ethnicity," and "population" in the face of historical baggage? Do our mathematical methodologies suggest that we can ascribe group reality at least sometimes, and with some justification? What can and should biologists add to the discussion of the reality/reification of "race"?

NOTES

This article is an edited version of an earlier book chapter, which appeared in Spanish as "¿La cosificación genética de la 'raza'? Un análisis crítico," in *Genes (&) Mestizos. Genómica y raza en la biomedicina Mexicana*, ed. C. López Beltrán (Mexico City: UNAM, 2011), http://philpapers.org/rec/WINLCG. The chapter was mostly written in Spanish, and Dr. Fabrizzio McManus Guerrero, my former PhD student, helped with final editing. Given that Spanish is not as much a lingua franca as English, that this book chapter has garnered significant interest, and, finally, that Paul Taylor, editor of *Critical Philosophy of Race*, kindly solicited a contribution from me on "analytic race theory," the piece here appears reprinted in English, with some changes. Prof. Taylor asked me to introduce the article.

While still a professor at UNAM (Universidad Nacional Autónoma de México), I was part of a "critical genomics" reading group led by Dr. Carlos López Beltrán. Discussions about the discourses and histories surrounding concepts of "race" and "mestizo," both within and without science, were lively, detailed, and insightful.

Upon moving to UC Santa Cruz in 2007, I had the pleasure of having Ian Hacking as a colleague for a few quarters. I am grateful to him for pointing me to Edwards's now-classic 2003 article, and for discussion about what might be called "the Lewontin-Edwards conundrum." (I have since discussed these matters both with Professors Richard C. Lewontin and Anthony W. F. Edwards.) Finally, a number of lengthy and energetic exchanges with Google statistician Dr. Amir Najmi, an old friend, added immeasurably to my understanding of the statistical and probability theory issues at stake. Thus, when Prof. Carlos López Beltrán generously invited me to contribute to his 2011 volumes, I had been amply primed to write on the topic by his reading group and by discussions with many, but especially Prof. Hacking and Dr. Najmi. Research support was provided by UC Santa Cruz and by both the Biocomplexity Center at the Niels Bohr Institute and the Center for Philosophy of Nature and Science Studies at Copenhagen University, during a research stay in Denmark. Prof. López Beltrán's constant encouragement and critique are and will always be appreciated. I am grateful to John Dupré, Carlos Galindo, Peter Godfrey-Smith, Eduardo García Ramirez, Cathrine Winther Jørgensen, Ian Hacking, Fabrizzio McManus, Amir Najmi, Christina Okai Mejborn, Francisco Vergara Silva, and Michael J. Wade for discussion. Alex Dor provided research assistance.

The book chapter seeded further work via three routes. First, upon sharing the book chapter, I received invitations to give lectures at Cambridge University, UC Berkeley, the University of Cape Town, and the University of Copenhagen. Second, shortly before the book chapter went to press, I sent it to Dr. Jonathan M. Kaplan. We exchanged several emails and he subsequently presented some of the book chapter's ideas, together with his own, at the Konrad Lorenz Institute in the summer of 2011. We decided to work together and I have greatly enjoyed our collaboration, which has resulted in three publications to date, in *Biological Theory*, *Philosophy of Science*, and *Theoria: A Journal of Social and Political Theory* (South Africa). Each of these articles has been written fully jointly and each was peer-reviewed. We look forward to further collaboration, perhaps including a book. Finally, the Lewontin-Edwards conundrum stimulated the writing of research grants, and I am now Principal Investigator for a trans-university research cluster (UC Berkeley, UC Davis, Stanford University, and UC Santa Cruz) on "Philosophy in a Multicultural Context," focusing during the 2013–14 academic year on "Genomics and Philosophy of Race," and institutionally rooted via the Institute for Humanities Research at UC Santa Cruz, http://ihr.ucsc.edu/portfolio/philosophy-in-a-multicultural-context/?id=15003.

Two workshops (Stanford University and UC Davis) and a big public conference (UC Santa Cruz) will result in at least a collection of articles. I look forward to further collaborations and many more learning experiences on the topics I first had the pleasure to address in this article.

In editing this translation, various stylistic and grammatical infelicities have been addressed. Examples, conceptual clarifications, and a few footnotes and bibliographic items have been added. Insightful comments by Doc Edge, Helen Longino, Mette

Smølz Skau, Bas van Fraassen, and an anonymous reviewer were addressed to the extent possible in this translation for *Critical Philosophy of Race*.

1. Since writing this piece, work surrounding implicit bias and the "implicit association test," IAT (e.g., Greenwald, McGhee, and Schwartz 1998), was pointed out to me. The relation to my argument is not that biologists acquiring data are "racist," associating different moral weights to different groups. Rather, the link to the IAT is that there may be unconscious mechanisms essentializing and singling out particular phenotypes from each human population as the appropriate exemplar of individual phenotype from which to draw blood samples for genotyping.
2. Importantly, neither Edwards (2003) nor Smouse, Spielman, and Park (1982) dispute these results. Indeed, the latter article's central argument is that "if one utilizes a multiple-locus approach, one will discover that human subspecific taxonomy is quite efficacious, even with the sort of marker loci alluded to above [i.e., the results of Lewontin 1972]" (Smouse, Spielman, and Park 1982, 445).
3. Lewontin (1972) provides four conditions that correctly describe the characteristics of *any* diversity measure: "(1) It should be a minimum (conveniently, 0) when there is only a single allele present so that the locus in question shows no variation. (2) For a fixed number of alleles, it should be maximum when all are equal in frequency—this corresponds to our intuitive notion that the diversity is much less, for a given number of alternative kinds, when one of the kinds is very rare. (3) The diversity ought to increase somehow as the number of different alleles in the population increases. Specifically, if all alleles are equally frequent, then a population with ten alleles is obviously more diverse in any ordinary sense than a population with two alleles. (4) The diversity measure ought to be a *convex function* of frequencies of alleles; that is, a collection of individuals made by pooling two populations ought always to be more diverse than the average of their separate diversities [the *Wahlund Effect*], unless the two populations are identical in *composition*" (388).
4. These were pioneered by Cockerham (1973); Holsinger and Weir (2009) is a recent instructive review of these relations.
5. Lewontin (1974a) attributes the data to Cavalli-Sforza and Bodmer (1971).
6. See Rosenberg (2011) for a synthetic review of state-of-the-art knowledge on patterns of human genetic variation.
7. Pritchard, Stephens, and Donnelly (2000) put it well: "Our main modeling assumptions are Hardy-Weinberg equilibrium within populations and complete linkage equilibrium between loci within populations. . . . Loosely speaking, the idea here is that the model accounts for the presence of Hardy-Weinberg or linkage disequilibrium by introducing population structure and attempts to find population groupings that (as far as possible) are not in disequilibrium. . . . *Under these assumptions each allele at each locus in each genotype is an independent draw from the appropriate frequency distribution, and this completely specifies the probability distribution* Pr $(X|Z, P)$. . . . [Where] X denote the genotypes of the sampled individuals, Z denote the (unknown) population of origin of the individuals, and P denote the (unknown) allele frequencies in all populations" (946, emphasis mine; sentence order slightly rearranged, as indicated with ellipses). Note that upon estimating the allele frequencies in actual populations, and sequencing individual genotypes,

we can use Pritchard et al.'s (2000) Bayesian clustering approach, embodied in the computer program *STRUCTURE*, to infer the population(s) *Z* of origin of the individuals. N.b., this article has been cited 11,753 times according to Google Scholar, May 14, 2014; on the same day, Rosenberg et al. (2002) had been cited 1768 times.

8. See Feldman and Lewontin (2008), 89–90, for another way to make this general point.
9. Subsequently to this piece, this point was made in note 3 of Kaplan and Winther (2013), 404 and note 14 of Winther and Kaplan (2013), 75. See also Barbujani, Ghirotto, and Tassi (2013).
10. Having recently learned about actual genetic protocols used in, e.g., Noah Rosenberg's lab, it seems that some of my worries here were unwarranted. See also section 3 of Kaplan and Winther (forthcoming).
11. Weiss and Fullerton (2005), and Kaplan (2011) provide brief, useful discussions of these points.
12. Kalinowski (2010) points to another related problem with *STRUCTURE*, which can also lead to reifications of clusters: "STRUCTURE is also frequently used to identify the main genetic clusters within species. In this second type of analysis, individuals are assigned to clusters . . . but *K* is deliberately set to be smaller than the actual number of populations. . . . The mathematical model used by STRUCTURE was designed for clustering individuals into Hardy-Weinberg/linkage equilibrium populations. It was not designed for clustering individuals into groups of populations, and may not work as its users intend when this is done" (1–2). Kalinowski's simulations show that when too few clusters are chosen, *STRUCTURE* pools individuals who would be pooled with other individuals if a higher *K* were chosen. Again, this is not a problem with the probabilistic mathematics, but is a problem with the interpretation of nature that we impose from our modeling result.
13. See Winther (2014b).

WORKS CITED

Bamshad, M. J., S. Wooding, W. S. Watkins, C. T. Ostler, M. A. Batzer, and L. B. Jorde. 2003. "Human Population Genetic Structure and Inference of Group Membership." *American Journal of Human Genetics* 72: 578–89.

Barbujani, G., A. Magagni, E. Minch, and L. L. Cavalli-Sforza. 1997. "An Apportionment of Human DNA Diversity." *Proceedings of the National Academy of Sciences* 94: 4516–19.

Barbujani, G., S. Ghirotto, and F. Tassi. 2013. "Nine Things to Remember about Human Genome Diversity." *Tissue Antigens* 82: 155–64.

Bolnick, D. A. 2008. "Individual Ancestry Inference and the Reification of Race as a Biological Phenomenon." In *Revisiting Race in a Genomic Age*, ed. B. A. Koenig, S. S.-J. Lee, and S. S. Richardson, 70–85. New Brunswick, NJ: Rutgers University Press.

Burchard, E. G., et al. 2003. "The Importance of Race and Ethnic Background in Biomedical Research and Clinical Practice." *New England Journal of Medicine* 348: 1170–75.

Cavalli-Sforza, L. L., and W. F. Bodmer. 1971. *The Genetics of Human Populations*. San Francisco: Freeman.

Cavalli-Sforza, L. L., P. Menozzi, and A. Piazza. 1996. *The History and Geography of Human Genes* (abridged). Princeton, NJ: Princeton University Press.

Cockerham, C. C. 1969. "Variance of Gene Frequencies." *Evolution* 23: 72–84.

———. 1973. "Analyses of Gene Frequencies." *Genetics* 74: 679–700.

Edwards, A. W. F. 2003. "Human Genetic Diversity: Lewontin's Fallacy." *BioEssays* 25: 798–801.

Excoffier, L., P. E. Smouse, and J. M. Quattro. 1992. "Analysis of Molecular Variance Inferred from Metric Distances among DNA Haplotypes: Application to Human Mitochondrial DNA Restriction Data." *Genetics* 131: 479–91.

Feldman, M. W., and R. C. Lewontin. 2008. "Race, Ancestry, and Medicine." In *Revisiting Race in a Genomic Age*, ed. B. A. Koenig, S. S.-J. Lee, and S. S. Richardson, 89–101. New Brunswick, NJ: Rutgers University Press.

Gelman, A. 2008. "Variance, analysis of." In *The New Palgrave Dictionary of Economics*, ed. S. N. Durlauf and L. E. Blume. Basingstoke, Hampshire: Palgrave Macmillan. http://www.dictionaryofeconomics.com/article?id=pde2008_A000098.

Greenwald, A. G., D. E. McGhee, and J. L. K. Schwartz. 1988. "Measuring Individual Differences in Implicit Cognition: The Implicit Association Test." *Journal of Personality and Social Psychology* 74 (6): 1464–80.

Hacking, I. 1995. "The Looping Effect of Human Kinds." In *Causal Cognition: An Interdisciplinary Approach*, ed. D. Sperber et al., 351–83. Oxford: Oxford University Press.

———. 2007. "Natural Kinds: Rosy Dawn, Scholastic Twilight." *Royal Institute of Philosophy Supplement* 82, no. 61: 203–39.

Haslanger, S. 2008. "A Social Constructionist Analysis of Race." In *Revisiting Race in a Genomic Age*, ed. B. A. Koenig, S. S.-J. Lee, and S. S. Richardson, 56–69. New Brunswick, NJ: Rutgers University Press.

Holsinger, K. E., and B. S.Weir. 2009. "Genetics in Geographically Structured Populations: Defining, Estimating and Interpreting *FST*." *Nature Rev. Genetics*: 639–50.

Kalinowski, S. T. 2010. "The Computer Program STRUCTURE Does not Reliably Identify the Main Genetic Clusters within Species: Simulations and Implications for Human Population Structure. *Heredity* 1, no.8. http://www.ncbi.nlm.nih.gov/pmc/articles/PMC3183908/.

Kaplan, J. M. 2011. "'Race': What Biology Can Tell Us about a Social Construct." *Encyclopedia of the Life Sciences* (ELS). Chichester: John Wiley & Sons.

Kaplan, J. M., and R. G. Winther. 2013. "Prisoners of Abstraction? The Theory and Measure of Genetic Variation, and the Very Concept of 'Race.'" *Biological Theory* 7: 401–12.

———. Forthcoming. "Realism, Antirealism, and Conventionalism about Race." *Philosophy of Science*.

Kumar, R., et al. 2010. "Genetic Ancestry in Lung-Function Predictions." *New England Journal of Medicine* 363: 321–30.

Lewontin, R. C. 1972. Apportionment of Human Diversity. *Evolutionary Biology* 6: 381–98.

———. 1974a. *The Genetic Basis of Evolutionary Change.* New York: Columbia University Press.

———. 1974b. "Annotation: The Analysis of Variance and the Analysis of Causes." *American Journal of Human Genetics* 26: 400–11.

Nei, M. 1973. "Analysis of Gene Diversity in Subdivided Populations." *PNAS* 70: 3321–23.

Pritchard, J. K., M. Stephens, and P. Donnelly. 2000. "Inference of Population Structure using Multilocus Genotype Data." *Genetics* 155: 945–59.

Rosenberg, N. A. 2011. "A Population-Genetic Perspective on the Similarities and Differences among Worldwide Human Populations." *Human Biology* 83: 659–84.

Rosenberg, N. A., S. Mahajan, S. Ramachandran, C. Zhao, J. K. Pritchard, and M. W. Feldman. 2005. "Clines, Clusters, and the Effect of Study Design on the Inference of Human Population Structure." *PLoS Genetics* 1(6)e70: 660–71.

Rosenberg N. A., J. K. Pritchard, J. L. Weber, H. M. Cann, K. K. Kidd, L. A. Zhivotovsky, and M. A. Feldman. 2002. "Genetic Structure of Human Populations." *Science* 298: 2381–85.

Smouse, P. E., R. S. Spielman, and M. H. Park. 1982. "Multiple-Locus Allocation of Individuals to Groups as a Function of the Genetic Variation within and Differences among Human Populations." *The American Naturalist* 119, no. 4: 445–63.

Templeton, A. 1998. "Human Races: A Genetic and Evolutionary Perspective." *American Anthropologist* 100: 632–50.

Wade, N. 2014. *A Troublesome Inheritance. Genes, Race and Human History.* New York: Penguin.

Weiss, K. M., and S. M. Fullerton. 2005. "Racing Around, Getting Nowhere." *Evolutionary Anthropology* 14: 165–69.

Winther, R. G. 2006a. "Fisherian and Wrightian Perspectives in Evolutionary Genetics and Model-Mediated Imposition of Theoretical Assumptions." *Journal of Theoretical Biology* 240: 218–32.

———. 2006b. "On the Dangers of Making Scientific Models Ontologically Independent: Taking Richard Levins' Warnings Seriously." *Biology and Philosophy* 21: 703–24.

———. 2014a. "Determinism and Total Explanation in the Biological and Behavioral Sciences." *Encyclopedia of the Life Sciences.* http://philpapers.org/rec/WINDAT-4.

———. 2014b. "James and Dewey on Abstraction." *The Pluralist* 9, no. 2 (Summer 2014): 1–28.

———. Under Contract. *When Maps Become the World: Abstraction and Analogy in Philosophy of Science.* Chicago: University of Chicago Press.

Winther, R. G., and J. M. Kaplan. 2013. "Ontologies and Politics of Bio-Genomic 'Race.'" *Theoria: A Journal of Social and Political Theory* (South Africa) 60, no. 136: 54–80.

Wright, S. 1965. "The Interpretation of Population Structure by *F*-Statistics with Special Regards to Systems of Mating." *Evolution* 19: 395–420.

———. 1969. *Evolution and the Genetics of Populations. The Theory of Gene Frequencies, Vol 2.* Chicago: University of Chicago Press.

Appendix 3

'My Fisher'

A. W. F. EDWARDS

Introduction

To have studied the works of R. A. Fisher intensively is to know him better than to have passed brief periods in his company. Moreover it has never been easier to do so, since his *Collected Papers*[1] and important books are available on the internet as well as much of his correspondence. Substantial commentaries help the scholar to follow his dense and intricate arguments, and later developments emphasise the breadth and originality of his work. Personal reminiscences can add little to the understanding of so great a scientist, but to those of us lucky enough to possess them there is something to be said for recording them for the benefit, or at least the pleasure, of our friends and descendants, while historians of science urge us to do so just in case they illuminate the workings of a mind of such incomparable brilliance.

Indeed my own introduction to Fisher was through one of his books, *Statistical Methods for Research Workers*. Eighteen years old, I had stayed on an extra term at school to take the scholarship exam of a Cambridge college, Trinity Hall. This had left me the first nine months of 1954 free before going up to Cambridge to read natural sciences, and, as an enthusiastic schoolboy-astronomer I wrote to the Astronomer Royal, Sir Harold Spencer Jones, to ask if I might spend six months working voluntarily at the Royal Greenwich Observatory at Herstmonceux in Sussex. I expect I mentioned that I was interested in the analysis of data, for in agreeing to my proposal he assigned me to the Solar Department, where I was to assist in the reduction of sunspot data from the daily photographic images of the sun, mostly taken in India.

This was done by assistants using algorithms enshrined in specially-designed forms, together with the use of tables of logarithms. I was the computer and the

Talk given by A. W. J. Edwards at the Royal Statistical Society meeting on 7 June 2012 – *'A 50th anniversary commemoration of R. A. Fisher (17 February 1890 – 29 July 1962)'*.

[1] Bennett, J. H. 1971–4, *Collected Papers of R. A. Fisher*. 5 volumes, Adelaide: University of Adelaide.

forms were the program. Part of the output was the ratio of the umbral to penumbral area of each regular sunspot, and in the evenings, after work, I started to plot this ratio year by year and soon discovered that it varied according to the 11-year solar cycle. This was disputed by the Head of Department on the grounds that, if it were true, someone would have discovered it before, but he was near retirement and his appointed successor, Patrick Wayman, was already in post as his deputy. Patrick was horrified, and introduced me to the notion of a significance test by lending me *Statistical Methods for Research Workers*. With his support I wrote my first scientific paper. But it had to wait until the retirement of the long-serving Head of Department in 1957 before it could be published, when one of Patrick's first acts was to write and ask me whether I still possessed the manuscript (Edwards 1957).

Thus by the time I was an undergraduate in Trinity Hall I knew about *Statistical Methods* by R. A. Fisher. I did not know that he was still alive as Professor of Genetics and a Fellow of the neighbouring college, Gonville and Caius. To describe my subsequent interaction with him will inevitably be in part autobiographical, but I shall try to limit myself to such background as is necessary.

Choosing to Read Genetics, 1956

The Cambridge natural sciences Tripos has a two-year Part I with a large list of subjects from which one must choose a prescribed number, followed by a one-year single-subject Part II. My Part I was physics with a clutch of allied subjects, but I found the mathematics too difficult and the rest rather boring. As a result, in mathematics, I was down-graded to the 'half-subject' and had to find another subject in lieu. I opted for history and philosophy of science, but soon gave up the philosophy lectures because, though they were on induction, they seemed ignorant of statistical inference: I had read *Statistical Methods* and thought I knew better.

I had only been placed in the third class at the end of my first year, and my plan had been not to take a scientific Part II but do Part I of the mathematical Tripos, which I thought I could cope with, and which would get me an honours degree. Then I would be well-qualified for a teaching career. However, someone, I forget whom but I now think it was probably my brother John, told me that to do Part I natural sciences but not a Part II would be to waste Cambridge and that I should look at the Part II options. I bought a copy of the complete lecture list (no longer possible, incidentally, because the university has stopped printing it) and studied the possibilities. One was genetics. I glanced at the titles of the lectures and the names of the lecturers. To my astonishment one of the latter was Sir Ronald Fisher. I put two and two together: so he was still alive! And (as I soon discovered) not Professor of Statistics but Arthur Balfour Professor of Genetics!

There was a note which read:

Students who may wish to read Part II genetics in the following academic year should see Dr Owen or Dr Wallace sometime before the division of the term.

The first half of the Easter term was not yet over, so I jumped on my bicycle and set off for the department, housed in what had originally been the Arthur Balfour Professor's designated residence at 44 Storey's Way. The front door was open, and so was the door of the room immediately across the hall, where sat a small bearded man with thick glasses. He asked me whom I was looking for, and, clutching my lecture list, I replied 'Dr Owen or Dr Wallace'. On hearing my reason for asking he said that he was Professor Fisher and would be glad to discuss the possibility of my taking Part II genetics himself. He invited me out into the garden, and for a couple of hours we walked up and down the grass paths between the experimental plots of primroses and purple loosestrife, of garden peas grown to demonstrate Mendelian segregation, and strawberry beds. I said very little, but had to be alert as to which way Fisher was going to turn at the end of each path. What he said was totally fascinating, though I have no recollection of it. He ended by saying that I should come back the following afternoon 'to continue the conversation'.

I eagerly returned as instructed, and after some further discussion he said he was prepared to take me as a student for the next academic year, in October, and that, as I did not have a biological background I should come up to the department for a week during the long vacation term 'to look at the flowers'. There are several remarkable aspects to this decision. One is that the only examination to my name was the 'third' in my first year, so I was hardly promising material. Another is that I had not consulted my tutor or director of studies or anyone else but had acted on my own initiative, just as when I had written to the Astronomer Royal. Had I tried the orthodox route I would have found that genetics was regarded as a very dubious subject considered to be trespassing on the preserves of the departments of Botany and Zoology, headed by a rather eccentric man called Fisher with a sharp temper, who would in any case no doubt be winding down in his retirement year. That I was the single student Fisher admitted to his Part II course during this last year can only be attributed to the fact that no-one responsible for advising students in any college of the university had recommended it, and I had simply by-passed the regular system.

When my admirable Trinity Hall tutor, the historian Charles Crawley, heard of my plan I fear he may, after consulting his scientific colleagues, have felt it his duty to query Fisher's invitation to me, for I later heard that Fisher had called to see him in his room and had berated him for whatever he had done. Of this I knew nothing at the time. My father, Harold Edwards, was also concerned about my choice, and wrote to Fisher, whose reply, dated 26 June 1956, is a model of a how a university teacher should reply to an anxious parent. In due course, the two men were to meet, father entertaining Fisher in London.

I should like to add that the reason I now think it was my elder brother John who advised me to do a Part II is that during his own Cambridge career as a medical student he spent three years over Part I and never progressed beyond a third in his examinations. He regretted having missed the opportunity to study a Part II subject. Even so, he was to become a Fellow of the Royal Society and Professor of Genetics at Oxford, the codiscoverer of Edwards's syndrome, trisomy 18.

When the summer came I duly turned up for my week in the department. 'Who are you?' asked Fisher. I replied that he had told me to come and look at the flowers. He pulled a large buff offprint from his desk drawer and said that I should study it and, when I had understood it, come back for something more. It was called 'The theory of linkage in polysomic inheritance' by R. A. Fisher, from the *Philosophical Transactions of the Royal Society* (Fisher 1947). In retrospect, I am sure he knew perfectly well what he was doing.

Part II Genetics, 1956–1957

Lectures started in October. I have described elsewhere (Edwards 1990a, b) some features of the department and its members. The formal lectures, as listed in the university lecture list, took place in the centre of Cambridge in the Agriculture Building (later to house, as it happens, the Department of Genetics). The convention was that the whole department attended, even sometimes the other lecturers and Fisher himself. This often led to amusing situations. Dr R. C. Campbell was invited to give a course entitled 'Animal improvement'. He used as his text Mather's book *Biometrical Genetics* (Mather 1949) from which he often appeared to be reading. Mather, who had worked with Fisher before the war, had learnt his biometrical genetics from Fisher, adding some developments of his own. The circle was completed by Fisher sitting in the front row taking notes.

Fisher gave introductory lectures entitled 'Bases of genetics' at 5 in the afternoon of the Michaelmas term (Monday, Wednesday, and Friday), presumably intended for those thinking of reading Part II the following year and thus very suitable for me. After the lecture he and the few students often repaired to the nearby pub called 'The Bun Shop', already famous as the site of his expounding of his theory of the genetics of the rhesus blood-group locus to R. R. Race in 1943 and 1944 (see Edwards 2007).

In the Part II course itself Fisher lectured on linkage ('Theory of recombination'), and after the first couple of lectures or so I went up to him and admitted I knew little statistics. What should I do? 'Well', he said, stroking his beard, 'I have written one or two books on statistics that you could try'. I went immediately to

Heffer's [Booksellers] and bought *Statistical Methods for Research Workers, The Design of Experiments*, and *Statistical Methods and Scientific Inference*, all very reasonably priced. When I graduated I asked him to sign them, which he gladly did, together with my copy of his *Statistical Tables*. Many years later, when I had got to know Frank Yates, his co-author for the latter, I asked for his signature too, and I surely possess the only copy of *Statistical Tables* signed by both authors.

In the Lent term, his lectures were on 'Polysomic inheritance' and involved an immediate dive into the mathematical theory of partitions, as I had already discovered from my exposure to his *Philosophical Transactions* paper.

I took Fisher's lectures very seriously, making careful notes which I later transcribed into a fair copy with the convention that nothing went into it which I had not understood. I pestered the research students for explanations of difficult points. As a result I own a very presentable record of Fisher's lecture course for the benefit of historians. He started each lecture with a determined attempt to be comprehensible to us students, but before long he would cease to be aware of our existence and would follow his own thoughts unfettered. When adding up a column of figures on the blackboard he would sometimes proceed from left to right, somehow organising the carries in his head. As he finished his very last lecture I thought 'This is a historic moment. He has been a professor at University College London and here for 35 years, and it is the end of his formal lecturing.' I don't think anyone else present shared the thought (except perhaps Fisher himself).

Meanwhile, I started again with *Statistical Methods* and became very interested in Fisher's comments on Geissler's data on the human sex ratio, which he uses as an example of the binomial distribution. I soon invented the two-parameter beta-binomial distribution for myself as a means of accommodating variation in the binomial parameter between families, and thought I could fit it by equating its mean and variance to the sample mean and variance. Anxious research students warned me that this was called Pearson's method of moments and was not to be mentioned in Fisher's presence as it was inefficient. But it seemed sensible to me, so I carried on using it, working with the mechanical calculating machine in the library. One afternoon Fisher came in and asked me what I was doing. I showed him nervously. He told me off so half-heartedly for not concentrating instead on my Tripos studies that I am sure in reality he was glad to see a student doing something on his own initiative. He also told me to look at Corrado Gini's book *Il sesso dal punto di vista statistico* (Gini 1908) in the library. I did so immediately and found it signed by him and dated 'August 4th, 1925'. Probably Gini had presented it to him. I subsequently learnt that it was still available and bought my own copy (dated 'April 9th, 1959', when I received it). Sometimes, when he came past I would ask for an explanation of a point in the lectures, and I have kept the little notes he made in reply. One amusing consequence of an interest in Geissler's data was that I noticed that

Fisher's reference to it in *Statistical Methods* omitted Geissler's initial 'A' (for Arthur), so I advised him of it and the correction stands as my contribution to the thirteenth and last edition.

As the academic year went by I took to my new subject like a duck to water. The biology was not too difficult, and the mathematics within my compass. Dr Owen lectured on the more mathematical side, and Dr Wallace involved us in the practical work with mice for which she was responsible. Every weekday we students would meet in Fisher's study – the room where I had first met him – for tea, and the conversation was always as lively as it was serious.

Of course nothing could put off the approaching Tripos examination. I practised hard with past papers and did not hesitate to ask Fisher directly about points that still puzzled me. I soon realised that when he passed my question on to one of the research students it probably meant it was not uppermost in his mind and therefore not likely to be a topic in the examination. I also knew by then of his very poor eyesight, so I bought a fountain pen with the widest of nibs together with a bottle of jet-black ink, and practised painting my mathematical answers with it. Came the examination and I had little difficulty. David Jones (who had been admitted the previous year and was taking two years over Part II) and I were both placed in the first class. Fisher was delighted. But George Owen was puzzled and would say no more than 'You weren't supposed to get a first, Anthony'.

It turned out that at the very time Fisher was one of my examiners in Cambridge, my father Harold Edwards was one of the examiners in surgery at Trinity College Dublin where Rose, one of Fisher's daughters, was taking her exams, but fortunately this coincidence was not discovered until later. She passed.

Fisher's actual retirement was marked by a departmental dinner for him, held in Trinity College. I was running a little hand printing-press at the time and printed the menus, which were circulated for signing at the end of the evening.

1957–1959

With a 'first' beneath my belt, the opportunity of doing research for a PhD was opened up, and I was advised to apply for a Medical Research Council Scholarship for three years in the department, which I obtained. I have kept a letter from Fisher, dated 6 July 1957, agreeing to act as a referee. Although he formally retired at the end of September, no successor was appointed for a while, and he stayed in Cambridge for a further eighteen months, until March 1959, to our great benefit. Tea in his room continued. One day I asked him why, in herd animals, evolution had not ensured that there were more females than males, and he told me to look for the explanation in his 1930 book *The Genetical Theory of Natural Selection*, which I did. At about this time the 'Dover' paperback second edition was published, and

I happened to be standing in the departmental office when the secretary opened the parcel of author's copies in Fisher's presence. When he saw the garish cover he remarked 'Looks like a book by [Lancelot] Hogben'. I bought a copy, and regret not having asked him to sign it. It was many years before I was able to obtain a first edition.

There was a general understanding that we students could help ourselves to offprints from the boxes in Fisher's room, though where this idea came from I do not know. One day, when I was on hands and knees doing just that, he came in, but rather than reprimanding me, as I feared he would, he said something friendly. I think he was wise enough to know that if you want your work to be remembered your best agents are young enquiring students.

One of the other research students was Walter Bodmer, to whom I sold a 1927 Austin 7 which I had bought for £7.10.0 [7 pounds and 10 shillings in old British money] for spares for my own 1929 Austin 7. I have to admit that I sold it to him for £15, but in self-defence I can add that when it broke down I had it mended and we remain friends. We used to vie with each other for the privilege of giving Fisher a lift from Storey's Way to Caius. One day when I won he was returning some heavy journal volumes to the centre of town and after he had squeezed into the small seat I loaded these on top of him so he could hardly see out. I think they might have been volumes of *Biometrika*, Pearson's journal to which he was unwilling the department should subscribe, but which was taken by Agriculture. I later found a marginal note in his hand in one of these volumes. My Austin was further distinguished by having once ferried J. B. S. Haldane. I had bought it (for £20) from my brother John, who had given Haldane a lift in it in London. Haldane had been smoking a cigar and John preserved the ash in the ashtray as a memento just as Haldane had left it.

My PhD work was supposed to be on the experimental modification of dominance in the mouse. I was to help continue the experiment started by Fisher and S. B. Holt at University College London many years previously (Fisher and Holt 1944). But I was much more interested in the statistical analysis of human sex-ratio data arising from reading *Statistical Methods* and from Fisher's encouragement. It had been a characteristic of research in his department that students were expected to investigate a number of different topics for their PhDs as being educationally more valuable than concentrating on a single project, so I had no difficulty in changing tack. The first academic talk I gave was on the binomial distribution and its generalisations, with Fisher sitting at the back. Years later I was to give a seminar in Madison in the presence of Sewall Wright, and a joint paper with Luca Cavalli-Sforza ([29]) with Haldane in the audience (who submitted a question on paper, which I have preserved). So I can claim to have had Fisher, Wright, and Haldane in my audiences.

Contact with Fisher continued on a daily basis, mostly at tea. He was asked to review A. E. Mourant and colleagues' *The ABO Blood Groups: Comprehensive Tables and Maps of World Distribution* and he passed it on to me to do (Mourant *et al.* 1958, Edwards 1959). He would often walk from Storey's Way to Caius at lunch time, and occasionally I would accompany him. One day when the weather was particularly suitable for flying gliders, which I was very keen on, I remarked that it was difficult to decide whether to spend the afternoon working or at the gliding club, and he immediately advised the gliding club. He was very welcoming to visitors to the department, and those of us with new babies to show off were sure of a warm reception. My first daughter Ann was born in May 1959 and was soon introduced to receive the professorial blessing, Fisher cradling her in his arms (whilst no doubt thinking back to his own daughters). Many years later, after Ann had qualified as a librarian, I asked her to invigilate a display of Fisher's books in the Caius Library which I had prepared for a visiting society but was unable to attend myself. When she doubted whether she knew enough about Fisher and his work to be able to introduce the display I pointed out that all she had to say was that Fisher had once held her in his arms.

I was only once or twice in Fisher's rooms in Caius. His set was above the Senior Combination Room, complete with the cupola over the centre of the living room. When I myself became a Fellow of Caius in 1968, I picked up a lot of stories about him, but of course they were not my own reminiscences and therefore not for recalling now. One of the occasions he invited me was to introduce me to Frank Yates, who was visiting him.

At Christmas, 1958, I discovered that Fisher was staying in college over the holiday, and thought that he might appreciate being offered an evening at our home. Catharina and I had invited my father to spend Christmas with us at our cottage in Barton, just three miles from Caius. Mother had had to be admitted to hospital with a chest complaint, and we thought the two men might enjoy each other's company. Father had arrived with a brace of pheasant given him by a grateful patient, but when Catharina came to pluck them it was clear that one of them was ripe for immediate burial. I collected Fisher from Caius, and we settled down to a rather thin meal, not helped by us having kept the news from Father who innocently offered Fisher a second helping which he accepted. Disaster was averted by a preliminary issue of Swedish akvavit and absorbing conversation which included a discourse by Fisher on the Hanging Gardens of Babylon. Before dinner he helped me light the sitting-room fire, which he insisted could only be done with precisely five pages of *The Times*. Before I returned him to Caius I asked him to conform to the Swedish habit of signing our guest book, so we have a record of the exact date: 27 December 1958.

When Fisher finally left the department there were a number of books of indeterminate ownership looking for homes. It must be remembered that the departmental library, and especially its fine offprint collection, had evolved out of the personal library of R. C. Punnett, the first Arthur Balfour Professor, augmented by Fisher's own material (such as the book by Gini already mentioned), and although the library might now be considered the property of the university, the same was not necessarily true of books elsewhere in the department which were originally, and probably still, Fisher's property. One of these was his *Contributions to Mathematical Statistics* published by Wiley in 1950. There was already a copy in the library, and this one was inscribed 'To be kept in Dr Owen's room'. Realising that the future direction of the department's interests was not likely to embrace advanced statistical theory, I took it to Fisher and asked him to sign it. He asked me where I had found it, and when I told him, he thought for a moment, autographed it, and I have used it ever since. After he had left for Australia I later took into safe-keeping his interleaved copy of the 1953 reprinted sixth edition of *The Design of Experiments* which I found lying around, containing amendments and insertions in his hand. Years later, on consulting his correspondence files in Adelaide, I was mortified to find a letter from Fisher's former secretary in the Cambridge department saying no, she could not find the interleaved copy for which he had asked. Happily, the corrections all seem to have found their way into the seventh edition anyway, but I am placing this copy with the other material from the Fisher Memorial Trust to be housed in the archives of Gonville and Caius College.

Fisher was, I think, quite lonely and academically isolated at the time, though he lived in Caius College and enjoyed the companionship of the other Fellows. I have already mentioned that his department was viewed as something of an upstart by Botany and Zoology, but this is not the occasion to elaborate the general theme other than by recalling an anecdote of my own which I have already published (Edwards, 1990a):

> We were sitting on the lawn of Whittingehame Lodge, the home of the Cambridge University Department of Genetics, in the early summer of 1957, and Fisher was telling half a dozen of us students about his relations with the University. 'I am a well-known statistician', he said, 'and a fairly well-known geneticist, but I have never been consulted on any appointment to a University post in statistics or genetics outside my own Department.'

1959–1962

My three-year MRC Scholarship ended in September 1960 and I applied for a Darwin Research Fellowship of the Eugenics Society. Not until many years later did I discover that it was Charles Darwin's fourth son Leonard who was being commemorated, and not Charles himself, and that Fisher had been the instigator of these Fellowships before the war. I do not recall whether it was at his suggestion I applied, but I am sure he must have been delighted that I succeeded, and may indeed have played his part in ensuring that I did.

After Fisher went to live in Adelaide in 1959 I saw him on his visits to Cambridge, for the last time in 1961. We were also together at the Second International Conference of Human Genetics in Rome in September that year and met frequently. One day Luca Cavalli-Sforza and his wife Alba invited us both to lunch at a hotel. As we entered it I remarked about the word ALBERGO over the door, that I thought it had an 'H' in it ('ALBERGHO'). Fisher said 'Perhaps they drop their aitches in Rome'. (I was learning Italian and was confused by the plural 'alberghi'.)

At the conference Fisher was much in evidence, no more so than when he walked out of the session on blood groups as soon as A. S. Weiner started to speak (Edwards 2007). On one occasion, he and I were chatting with Peter Parsons, who had been a Cambridge PhD student with Fisher, and Peter noticed Haldane standing all by himself a little distance away. He suggested to Fisher that we might go across and talk to him, but there was no response. After one session, Fisher told us he was going to the Vatican to visit the Pontifical Academy of Sciences, to which he had just been elected, and off he went, walking-stick in hand. It is my last memory of him. Quite possibly he had not warned them of his visit, and I do not know if he gained access.

In Cambridge Fisher spent some of the summer staying in Caius, and the time is remembered as one in which the dining in college, normally thin at that time of the year, gradually expanded as Fellows realised that their prized conversationalist was once more among them. I was, of course, not yet of their company, but I did visit Fisher in college so as to discuss a typescript he had written on the St Petersburg Paradox in probability, sitting with him in the Senior Combination Room. This has not, so far as I know, been published, but I possess a copy and I consulted an expert on the mathematics of the Paradox about it, to be advised that it was of interest but not quite correct. My last letter from Fisher, dated 14 March 1962, was in reply to one of mine about the Paradox.

'My Edwards'

These recollections, as accurate as selective and fallible memory can make them, must stand on their own feet as a record of my personal interaction with Fisher. I cannot judge what impression they make on others, but writing them has refreshed my memory of a man to whom I owe a great debt. I never had a cross word from him, and from our first meeting he always seemed interested in my future, probably contributing to it behind my back in ways I shall never know. Having studied mathematical physics himself, perhaps he saw I needed rescuing from it and that he was in a position to do so. When he saw his rescue bear fruit, and I saw how much I owed him, we became quite attached to each other, in spite of the age difference, as had he and Leonard Darwin.

I shall close these reminiscences with an anecdote that reflects this. In a letter to R. R. Race on 27 September 1960 Fisher wrote:

> The Edwards who messed up Cleghorn's data, and is formally thanked for it in his letter to *Nature* seems not to be my Edwards from Cambridge. It was the thought that it was he that annoyed me, for the estimates published in *Nature* were manifestly incompetent, and I feared that one of my own pupils was running amok and adding unnecessarily to darkness and confusion. However, I understand he is only one of Hogben's, so all is explained.

'Hogben's Edwards' was my brother John, who had been influenced by Lancelot Hogben [author of *Mathematics for the Million*] when their careers overlapped at the University of Birmingham. He allowed me to use this quotation (Edwards 1998). John had written one of the first general computer programs for maximum-likelihood estimation and had helped T. E. Cleghorn by using it to estimate the frequencies of the four chromosomes for the blood group system *MNSs*, namely *MS, Ms, NS* and *Ns*, on Cleghorn's data from English blood donors. Unfortunately neither he nor Cleghorn had noticed that when all four antisera are used all nine genotypes *MM, MN, NN* times *SS, Ss, ss* are identifiable, so that the numbers of the genes can actually be counted. Consequently, the maximum-likelihood estimates of the four chromosome frequencies are necessarily constrained to match the observed gene proportions. Only the double-heterozygotes *MNSs* are of uncertain phase, *MS/Ns* or *Ms/NS*, which is why maximum-likelihood estimation is needed at all. Fisher had immediately spotted that John's estimates for the chromosomes did not correspond to the gene counts. Naturally, with sufficient iterations, his program, if written correctly, should have converged to them, and I can only suppose that it did not do enough iterations – or perhaps did not iterate at all. But his was a pioneering application of computers to estimating gene-frequencies, and after 50 years the error should only be remembered as having generated a fine example of Fisher's barbed style in private correspondence to a close friend. I was writing my first computer programs in genetics at the time (Edwards 1961) and might easily have fallen into the same trap. But I escaped unscathed, and years later, when John showed me the letter, I was happy to find Fisher had claimed me as 'My Edwards'.

References

Edwards, A. W. F. 1957. The proportion of umbra in large sunspots, 1878–1954. *The Observatory* 77, 69–70.

Edwards, A. W. F. 1959. Review (unsigned) of *The ABO Blood Groups: Comprehensive Tables and Maps of World Distribution* by A. E. Mourant, A. C. Kopeć and K. Domaniewska-Sobczak. *Journal of Pathology and Bacteriology* 78, 609–610.

Edwards, A. W. F. 1961. The population genetics of sex-ratio in *Drosophila pseudoobscura*. *Heredity* 16, 291–304.

Edwards, A. W. F. 1990a. R. A. Fisher: Twice professor of genetics: London and Cambridge *or* "A fairly well-known geneticist". *Biometrics* 46, 897–904. (Also *The Statistician* (*Journal of the Royal Statistical Society D*) 52, 311–318, 2003, and *Genetics Society News* Issue 55, July 2006, 33–39).

Edwards, A. W. F. 1990b. Fisher, $\overline{W}$, and the Fundamental Theorem. *Theoretical Population Biology* 38, 276–284.

Edwards, A. W. F. 1998. The Eugenics Society and the development of biometry. The 1997 Galton Lecture, London, 19th September. *Essays in the History of Eugenics*, ed. R. A. Peel, London: The Galton Institute, 156–172.

Edwards, A. W. F. 2007. R. A. Fisher's 1943 unravelling of the rhesus blood-group system. *Genetics* 175, 471–476.

Fisher, R. A. 1947. The theory of linkage in polysomic inheritance. *Philosophical Transactions of the Royal Society B* 233, 55–87.

Fisher, R. A. and Holt, S. B. 1944. The experimental modification of dominance in Danforth's short-tailed mutant mice. *Annals of Eugenics* 12, 102–120.

Gini, C. 1908. *Il sesso dal punto di vista statistico*. Milan: Sandron.

Mather, K. 1949. *Biometrical Genetics*. London: Methuen.

Mourant, A. E., Kopeć, A. C, and Domaniewska-Sobczak, K. 1958. *The ABO Blood Groups: Comprehensive Tables and Maps of World Distribution*. Oxford: Blackwell.

Appendix 4

1963 Antonio's Theorem

(1)

The data for n species may be exactly represented as points in a space of $(n-1)$ dimensions.

<u>Antonio's theorem</u>

The best taxonomic tree is formed by that string network which connects all the points using the least total length of string. *

<u>Notes</u>. (1) In what sense it is 'best' has yet to be determined.

(2) The network is obviously a tree.

(3) The network is basically a hexagonal structure, the three lines at a point always being coplanar and at 120° to each other.

(4) Since it could turn out that all the points lay in two dimensions, we must learn how to construct the net for an array of points in two dimensions. This is the road-surveyors problem: he has n towns to be joined by the minimum total length of road. The solution is hexagonal; all road junctions should be of the 120° type.

Index

Paper titles are in 'single quotes'; book and journal titles *are in italics*; FTNS is 'The Fundamental Theorem of Natural Selection'; *GTNS* is *The Genetical Theory of Natural Selection*; AWFE is A.W.F. Edwards; AWFE citations are in the form 'year [paper number]'; single asterisk * denotes joint papers with Cavalli-Sforza; double asterisk ** for other multi-authored work; all other citations are in the form Name (year). Page numbers in **bold** locate the Selected Papers.